The Earth In Context
A Guide to the Solar System

Springer
*London
Berlin
Heidelberg
New York
Barcelona
Hong Kong
Milan
Paris
Santa Clara
Singapore
Tokyo*

David M. Harland

The Earth In Context

A Guide to the Solar System

 Springer

Published in association with
Praxis Publishing
Chichester, UK

David M. Harland
Space Historian
Kelvinbridge
Glasgow
UK

SPRINGER–PRAXIS BOOKS IN ASTRONOMY AND SPACE SCIENCES
SUBJECT *ADVISORY EDITOR*: John Mason B.Sc., Ph.D.

ISBN 1-85233-375-8 Springer-Verlag Berlin Heidelberg New York

British Library Cataloguing-in-Publication Data
Harland, David M.
   The earth in context: a guide to the solar system. –
   (Springer-Praxis books in astronomy and space sciences)
   1. Planetology   2. Solar system – Origin
   I. Title
   532.2

ISBN 1-85233-375-8

Library of Congress Cataloging-in-Publication Data
Harland, David M.
   The earth in context: a guide to the solar system/David M. Harland.
      p. cm. – (Springer-Praxis books in astronomy and space sciences)
   Includes bibliographical references and index.
   ISBN 1-85233-375-8 (alk. paper)
   1. Solar system.   I. Title.   II. Series.

QB501.H363 2001
523.2–dc21   20001049398

Apart from any fair dealing for the purposes of research or private study, or criticism or review, as permitted under the Copyright, Designs and Patents Act 1988, this publication may only be reproduced, stored or transmitted, in any form or by any means, with the prior permission in writing of the publishers, or in the case of reprographic reproduction in accordance with the terms of licences issued by the Copyright Licensing Agency. Enquiries concerning reproduction outside those terms should be sent to the publishers.

© Copyright, 2001

Printed by MPG Books Ltd, Bodmin, Cornwall, UK

The use of general descriptive names, registered names, trademarks, etc. in this publication does not imply, even in the absence of a specific statement, that such names are exempt from the relevant protective laws and regulations and therefore free for general use.

Project Copy Editor: Alex Whyte
Cover design: Jim Wilkie
Typesetting: BookEns Ltd, Royston, Herts., UK

In memory of

Carl E. Sagan

who died on 20 December 1996, after successfully engaging the public in the excitement of space exploration since the dawn of the 'Space Age'.

"The world's science community, and the field of planetary exploration in particular, have lost one of its most gifted minds and eloquent voices."

Dr Edward C. Stone
Director of the Jet Propulsion Laboratory, Pasadena, California

"Of all the people I have met in the course of my scientific career, no one was more gracious, understanding, respectful and encouraging towards me than Carl. From my very first professional presentation at the age of 21, to my current position on Cassini, Carl was always there with a kind, gentle word of support. I believe that he genuinely cared for people in that special way that distinguishes great humanitarian leaders. And I believe that underlying his life's work was a bedrock faith in the fragility, dignity and goodness of all humankind. His passing is a heartbreaking loss ... for his family, for the community of scientists to which he belonged, and for the world.
We who remain on Earth have lost our guardian.
He is part of the Cosmos now."

Professor Carolyn C. Porco
Department of Planetary Sciences
Lunar and Planetary Laboratory
University of Arizona

Contents

List of figures xi
List of tables xv
Foreword xvii
Author's preface xxi
Acknowledgements xxv

1 Origin of the Solar System 1
   Formation 1
   Out there 8
   Notes 8

2 Discovering the Earth 11
   Geological time 11
   Drifting continents 14
   The sea floor 18
   Sea floor spreading 23
   Supercontinent cycles 33
   Notes 38

3 The lure of the Moon 43
   The astronomers 43
   The photogeologists 45
   The field geologists 47
   Crustal formation 53
   Whence came the moon? 57
   The whole story 62
   What next? 62
   Notes 64

4 The Red Planet ... 69
- Early studies ... 70
- The first close look ... 73
- Follow up ... 75
- A global view ... 77
- Volcanoes ... 79
- Line of dichotomy ... 79
- Valley networks ... 82
- Craters ... 83
- Basins ... 84
- Cratering rates ... 85
- Highland volcanism ... 86
- Volcanic plains ... 89
- Tharsis ... 91
- Olympus ... 95
- Elysium ... 98
- Canyons ... 100
- Outflow channels ... 101
- On Mars ... 103
- Pieces of Mars ... 106
- Pathfinder ... 107
- The core ... 110
- Plate tectonics ... 111
- The search for water ... 113
- Oceanus Borealis? ... 117
- Recent activity? ... 123
- An open book ... 126
- Notes ... 126

5 Mysterious Venus ... 143
- Early observations ... 143
- Radar ... 148
- Close looks ... 149
- A strange atmosphere ... 160
- Ancient oceans? ... 162
- An active surface ... 162
- Notes ... 175

6 Ancient Mercury ... 183
- Early observations ... 183
- Mariner 10 ... 184
- Massive core ... 185
- Physiography ... 188
- Magnetic field ... 196
- Exosphere ... 197

	Since Mariner 10	197
	Notes	197
7	**Precambrian Earth**	203
	The basement	203
	Canadian shield	204
	Barberton Mountainland	207
	Greenland	211
	Superior province	213
	Western Australia	214
	Archean processes	217
	The Hadean	226
	The onset of plate tectonics	227
	Notes	228
8	**Giant Jupiter**	235
	Mr Galileo	235
	Gas giant	236
	Jupiter's moons	237
	Asteroids	240
	Pioneering deep space	242
	Voyagers at Jupiter	245
	Io	246
	Europa	262
	Ganymede	271
	Callisto	275
	The 'free lunch'	278
	Notes	279
9	**Saturn's retinue**	285
	Ringed planet	285
	A system of moons	289
	Mimas	292
	Enceladus	293
	Tethys	294
	Dione	295
	Rhea	297
	Titan	297
	Hyperion	299
	Iapetus	300
	Phoebe	301
	Moonlets	302
	And then there was one	304
	Notes	305

Contents

10 Planets beyond 309
    'George's star' 309
    Rings 312
    Fly-bys 312
    The green giant 313
    Oberon 316
    Titania 317
    Umbriel and Ariel 318
    Miranda 319
    Ring erosion 323
    The tilt 324
    The discovery of Neptune 325
    Neptune's moons 328
    Neptune's rings 328
    Final fly-by 329
    Rings revealed 333
    Triton 334
    Mission accomplished 338
    Distant Pluto 338
    Ten? 342
    Notes 343

11 Life and death 349
    Dim Sun 349
    Biogenesis 349
    Extinctions 354
    Cycles? 359
    Violent planet 362
    Dangerous environment 371
    Notes 377

Appendix 1: Planetary data 383
Appendix 2: Space missions 411
Glossary 419
Index 445

List of figures

Cover
Frontispiece. xxiii
Celestial sphere 2
Tycho's observatory 3
A list of extrasolar planetary systems 6
Structure of extrasolar planetary systems. 7
Earth's hypsographic curve 16
Gondwanaland 17
Tracks of oceanographic research ships. 20
The Earth exposed 22
Magnetic stripes on the sea floor off California. 25
How the San Andreas fault evolved 28
The San Andreas fault in the Los Angeles area. 29
Floor of the North Pacific Ocean 30
World orogeny 34
The Aqaba fault line 35
Before the North Atlantic opened 36
Break-up of the Pangean supercontinent 37
Galileo's sketches of the Moon 44
G.K. Gilbert's drawing of Imbrian 'sculpture'. 46
Apollo 15's Hadley-Apennine landing site seen from orbit. 49
Hadley Rille at ground level 50
A geological cross-section of Hadley-Apennine 51
Orientale basin 54
A crustal cross-section of the Orientale basin 55
How the lunar crust formed 56
The Moon as viewed by Clementine 58
A lunar field geologist. 63
Schiaparelli's 1888 map of Mars 71
Mars through a telescope 72
Mariner 6/7 imaging coverage 76

xii **List of figures**

Crustal thickness across the line of dichotomy................................. 81
Tyrrhena Patera 87
Isidis basin 88
Volcanic cones on Amazonis Planitia 90
Tharsis Province 92
Stress patterns in the Tharsis Province 94
The aureole of Olympus Mons 97
Elysium volcanoes 99
Outflow channels debouching onto Chryse Planitia 102
When 'Glacial Lake Missoula' flooded 103
Mars Pathfinder and Sojourner 109
MOLA's topographic map of Mars 112
Nanedi Valles 114
Sedimentary deposits on Mars 115
Layered deposits on the floor of West Candor Chasma 116
'Oceanus Borealis' 119, 120
The 'Oceanus Borealis' shoreline on Lycus Sulci 121
The 'Oceanus Borealis' shoreline on Cydonia Mensae 122
Evidence of recent seepage in crater walls 124
Stratification in the wall of Coprates Chasma 125
An early radar view of Venus's surface 148
Venus as seen by Mariner 10 152
Venera 9's landing site 153
Venera 10's landing site 153
Pioneer Venus Orbiter's radar map 155
Trajectories of the Pioneer Venus atmospheric probes 155
Venera 13's landing site 156
Venera 14's landing site 157
Geographic distribution of landing sites on Venus 157
Standard model for Venus's atmosphere 161
Lakshmi Planum 164
Sacajawea Patera 165
Venus's north polar region 166
Is western Aphrodite Terra a spreading ridge? 168
Distribution of troughs in Aphrodite Terra 169
Devana Chasma 170
A meandering lava channel 173
Mercury as revealed during Mariner 10's approach 186
The Caloris basin in context 187
Low-lying smooth plains northeast of the Caloris basin 190
A radial across the rim and ejecta blanket of the Caloris basin 191
Low-lying smooth plains southeast of the Caloris basin 192
A generalised map showing the Canadian shield's structure 205
The Yellowknife fault 206
Cratons of southern Africa 208

The Barberton Mountainland's Archean stratigraphic sequence	209
Magma chemistry	210
Greenland's Archean stratigraphic sequence	211
Cratons of western Australia	215
Pilbara's Archean stratigraphic sequence	216
Geographic distribution of cratons	218
A volcanic model for the greenstone-granite structures	220
A tectonic model the greenstone-granite structures	221
Grades of metamorphism	222
Accreting cratons form prototcontinents	223
A topographic map of the United States	225
E.J. Reese's map of Ganymede	238
Jupiter through the telescope	239
Pioneer's Galilean moons	243
The 'Grand Tour' gets underway	244
Voyager 1 reveals Jupiter's dynamic atmosphere	246
Io's sub-Jovian hemisphere	248
Prometheus on Colchis Regio	255
Prometheus plume	257
Eruptions in Tvashtar Catena	258
B.F. Lyot's map of Europa	262
Europa revealed by Voyager	263
Iceberg 'rafts' at Conamara	266
Conamara in close-up	267
Thrace and Thera	270
Ganymede's fractured surface	272
Saturn through the telescope	287
Voyager Saturn encounter trajectories	288
Close view of Saturn	289
Mimas	292
Enceladus	294
Tethys	295
Dione	296
Titan	298
Iapetus	300
Saturn's ring system revealed	303
A telescopic view of Uranus and its retinue	311
Voyager 2's Uranus encounter trajectory	313
The best pre-Voyager image of Uranus	314
Uranus's outer moons	318
Miranda overview	320
Miranda's cliff	322
Uranus's rings	323
The orbits of Uranus and Neptune	326
Voyager 2's Neptune encounter trajectory	330

xiv **List of figures**

Neptune in close-up 332
Triton's southern hemisphere 335
Triton's basins. 337
Pluto and Charon 339
Albedo variations on Pluto's surface 341
A plot of atmospheric carbon dioxide concentration 352
The Yucatan impact 355
The stacked up lava flows of the Deccan Plateau 357
Lakagigar, Iceland. 358
A plot of 'mass extinctions'. 359
The 'great dying' at the end of the Permian. 360
Impacts at the end of the Permian. 361
A volcanic eruption viewed from orbit 363
How Mount St Helens erupted 364
Mount St Helens afterwards 365
The scale of the Yellowstone eruptions 366
The Yellowstone caldera 367
How Yellowstone erupted 368
Topography of the US Pacific Northwest 369
The Snake River Volcanic Plain of southern Idaho 370
Columbia River 'flood basalts' 371
Meteor Crater, Arizona. 373
The distribution of impact craters around the globe 374
Manicouagan Crater, Canada 375
An artist's impression of an asteroid striking the Earth 376
Types of igneous rock 431

List of tables

4.1	Viking surface chemistry	105
5.1	Landing sites on Venus	159
8.1	The Titius–Bode 'law'	240
8.2	The first ten asteroids	241
9.1	Saturn's moons	291
10.1	Giant planet energy budgets	331
11.1	Volcano Explosivity Index	369

Foreword

Like most science journalists, I have been fascinated by planets from a very early age. As a child I poured over books showing brightly-coloured and often fanciful 'artists' impressions' of the surfaces of Mars and Venus, the swirling clouds of Jupiter and the rings of Saturn, but these images were quickly supplanted by the real thing by courtesy of the amazing space probes that, starting in the 1960s, were sent out to explore the Solar System. At last these artists' impressions can be consigned to history as the Solar System begins to take shape. Objects which in the best telescopes were little more than fuzzy discs have been resolved into complex and often enigmatic worlds, replete with volcanoes and geysers, sulphur lakes and deep oceans.

This revolution has allowed the development of a hitherto impossible exercise: the comparison of the geology, history and processes operating today and in the past on the Earth with those of the other planets and their moons. In this timely synthesis, David Harland has brought together the latest results from space probes, including Galileo and Mars Global Surveyor, and with these data, combined with findings from earlier probes, telescopic observations and our knowledge of the Earth, we can now begin to outline some general principles of what planets are and how they evolve.

For a few hundred years, the only planet under close investigation was, of course, the Earth, and even then it took a long, long time to put the basic picture in focus. Plate tectonics is the geological equivalent of the (still-awaited) theory of everything (physics) and the understanding of the mechanism of genetic inheritance (biology). Accepted only recently, plate tectonics underpins our understanding of the evolution of the Earth's crust.

But what about the other planets? Why does the Moon look so different to the Earth? Why are there no equivalents of the Rockies or the Andes on Mars? Do the 'rules' that apply to the Earth also work on Venus and Mercury, Titan and Io? Is there, in short, a grand unified theory that will explain the evolution of all terrestrial planets in our Solar System, and perhaps in others too?

The findings from the Pioneers, the Voyagers, Galileo and the various Martian and Venusian probes show that the Solar System is a strange place perhaps much stranger than was once imagined – and that the geological laws that apply here do not necessarily apply elsewhere. Nevertheless, intriguing analogues to terrestrial processes are found in the farthest reaches of the Solar System.

We now know that there are worlds with bigger and more active volcanoes than any seen on Earth. Worlds with crushing atmospheres and temperatures high enough to melt lead. Worlds whose crusts resemble that of the very early Earth, and worlds where water, sulphur and nitrogen replace liquid rock as the dominant volcanic material. Take, for example, the most earthlike planet, Mars. Does the Red Planet have plate tectonics? It seems not ... at least not today. Losing heat more quickly than the Earth, the Martian crust has 'frozen' into what is effectively one giant plate. That doesn't mean that nothing ever happened on Mars; far from it, this world is home to the Solar System's biggest volcano, Olympus Mons, a huge gently-sloping cone of solidified runny lava that dwarfs its Hawaiian analogue.

Volcanoes may indeed still be active today. Thanks to Mars Global Surveyor and its Orbital Camera, which has mapped parts of the Martian surface in unprecedented detail, we know that some areas are very young indeed – implying that active remodelling of the surface is still going on. Most excitingly, pictures from the Mars Global Surveyor appear to show recent water-carved gullies in crater walls. If liquid water flows on Mars today, the fourth planet's suitability as an abode for life will no doubt be reassessed. Together with the announcement of 'life' in the Martian meteorite ALH84001, the happy (for journalists) discovery of water channels on Mars in 2000 managed to push the Red Planet onto the front pages once more, after decades of being dismissed as a dead planet.

What about Venus? Despite its superficial similarity to Earth, being almost exactly the same size and density, Venus is, it seems, more like Terra's ugly sister than its twin. Thanks to a stagnant 90-bar atmosphere and temperatures of 480 °C, Venus has no standing water and little aeolian erosion. Venus has correctly been describes as a hellish world, although nothing written by Dante does justice to the extreme nastiness of this place. The lack of water could be a contributing factor in Venus's apparent lack of plate tectonics. Instead, lacking a terrestrial-style regime of moving plates and linear volcanic fissures to release heat from the interior, it seems that Venus's surface undergoes regular half-billion-year resurfacings – events of almost unimaginable ferocity. Venus is still poorly understood, however, and it is possible that scientists are misreading the radar maps, which are all we have of this cloud-shrouded planet.

If you want real drama, then travel with David Harland to the outer Solar System. The Pioneer probes, with their primitive imaging systems, gave a tantalising glimpse of what was to come with the two Voyagers, whose spectacular fly-bys through the Jovian system began the revolution in our understanding of this icy realm. Thanks to the Voyagers, we learned that at least one other body in the Solar System – Jupiter's moon Io – is home to active volcanism. The bright plumes being ejected from Io's pockmarked surface were some of the most extraordinary photographs of these missions.

Then came Galileo. Io has now been revealed to be a world of almost unimaginable ferocity. Thanks to tidal heating, Io's surface is being repeatedly resculpted by literally hundreds of volcanoes, some of which span hundreds of kilometres. We have seen, on Io, lava lakes at temperatures of nearly 2,000 °C; fire fountains a kilometre high; and strange sulphurous geysers sending plumes of material hundreds of kilometres into space. What processes could be creating this extraordinary drama?

David Harland shows how the concerted effort to make sense of the Galileo data, together with new telescopic observations has shed light on this restless little moon. Surprisingly, Io's volcanism resembles that of the very early Earth, with ultra-high temperature silicate lavas. So much heat is being liberated from Io's interior that its crust has not had a chance to 'settle down' and develop a stable regime of moving plates and continental masses. Instead, Io's landscape is marked by jagged blocky mountains twice the height of Everest or more, which result from the seething turbulence beneath its surface.

If Io is dramatic, then Europa, Jupiter's second large moon, is enigmatic. Europa is now suspected of harbouring the Solar System's largest ocean, a colossal under-ice reservoir of brine perhaps 80 kilometres deep, kept warm by the same gravitational tidal forces that generate so much fire and splendour on Io. If Europa has an ocean – and recent data from Galileo suggests that Jovian moon Callisto may also have hidden depths – then this ocean could harbour life. A lot of ifs and maybes perhaps, but forty years ago who would have even thought of looking for life around Jupiter?

The spectacular photographs from Europa taken by Galileo – and the speculations regarding possible lifeforms – have ensured that this small moon has become one of the most well-known places in the Solar System. Pictures of Europa and stories about its ocean have appeared in most newspapers, and, for science journalists, Europa is a staple source of stories.

In fact it seems that everywhere you go in the Solar System there are surprises. Saturn's moon Titan turns out to have a nitrogen atmosphere thicker than the Earth's – the only moon to have a substantial gaseous envelope. Titan's surface remains something of a mystery, although we will hopefully know much more in late 2004 and early 2005 when the Cassini spacecraft arrives at the ringed planet, dropping a small probe, Huygens, through the petrochemical smog that swathes Titan's surface.

Moving out through the Solar System, to the dim twilight worlds of Uranus, Neptune and Pluto, there is still much to astonish. For example, Uranus's moon Miranda looks as if it has been shattered and re-assembled, with a surface that is a jumble of seemingly unrelated terrains.

Nor does the excitement stop at Uranus. Triton, Neptune's big moon, is, like Io, undergoing active volcanism. When Voyager 2 swept past, it took extraordinary pictures of geysers erupting from Triton's surface, geysers driven not by boiling water or sulphur compounds, but by venting nitrogen. This nitrogen-based volcanism was something completely unsuspected before the advent of deep space probes.

This book paints an exciting picture of our Solar System. From fiery, pockmarked Mercury, to hellish Venus, our own warm and wet home, tantalising Mars and the strange landscapes of the Outer Planets, the third millennium will, we must hope, be an era when humankind finally starts to explore these strange new worlds in person.

So David Harland has not only put Earth into context with its neighbouring worlds, he has painted a picture of what future generations of explorers, human and machine, can expect to find.

Michael Hanlon
Science Writer, Daily Mail
August 2001

Author's preface

The 1960s was a remarkable decade for science. As it began, geologists were beginning to appreciate the dynamic form of the ocean floor. By the decade's end, they had developed a global model for the processes which control the Earth's surface. This unifying insight drew in other disciplines to create 'the Earth sciences'. At that same time, we were gearing up to mount the first human expeditions to the lunar surface and were starting to dispatch robotic probes to study our neighbouring planets. In the decades that followed, the synergy between these two parallel strands gave rise to the discipline of 'planetology'.

The theme of this book is comparative planetology. As such, it provides a contextual 'big picture' for my earlier *Exploring The Moon: The Apollo Expeditions* and *Jupiter Odyssey: The Story of NASA's Galileo Mission*. In addition to a historical perspective, it provides a timely round-up of the state of knowledge in this rapidly advancing field. However, with such a broad canvas, it has been necessary to focus only on specific objects, namely the rocky planets and icy moons. By heavily footnoting the text, I have endeavoured to make the book serve as a gateway to the literature.

We may never know all the answers, but, thanks to our robotic probes, when finally we venture forth in person, we will hopefully be sufficiently aware to ask the right questions; so let us look forward to the time when we can discuss the crustal forms of other Solar System objects in the same detail that we adopt when analysing and discussing the composition of the Earth.

Anyone who believes that human explorers are redundant, and that robotic probes can explore as effectively, fails to appreciate how the Apollo astronauts on the lunar surface functioned as field geologists. The Soviet Union sent remotely controlled Lunokhod rovers to the Moon in the 1970s. Each travelled a few dozen kilometres, pausing from time to time to take a panoramic picture, to test the strength of the surface and to determine the chemistry of the regolith. In fact, while the first was trundling over the western plains of Mare Imbrium in July 1971, Apollo 15's Dave Scott and Jim Irwin landed at the foot of Mount Hadley in the Apennine Range on the eastern rim of the Imbrium basin. Over a three-day period, they travelled further and examined more rocks than Lunokhod did over several months. As they drove around on their own rover, they not only reported upon their environment from a field geologist's point of view, but they also brought their specifically selected samples home for analysis. The robots had been outperformed by their human counterparts. However, a critic of human planetary exploration will doubtless argue that the Lunokhods were 1960s technology, and we can do much better today — as indeed we can. The main problem in operating a rover on a distant planet is the time lag imposed by the speed of light. It cannot be controlled directly, in real time, as the Lunokhods were on the Moon. It must be semi-autonomous, accept general instructions on its objectives and then undertake the tasks by itself. The electronics and software for this degree of onboard intelligence are now becoming available, and future rovers will undoubtedly become very sophisticated. But will robots ever render their masters redundant? Unless we follow our rovers, we will never appreciate how limited their viewpoints were.[1]

David M Harland
Kelvinbridge, Glasgow
May 2001

---

[1] 'An argument for human exploration of the Moon and Mars', P.D. Spudis, *American Scientist*, p. 269, May June 1992.

Sending robots is merely the first step towards exploration.

Acknowledgements

I would like to express my thanks to Alex Blackwell of the University of Hawaii; Jim Head of Brown University, Rhode Island; Jeffrey Kargel of the Lunar and Planetary Institute of the University of Arizona; Marc Rayman of the Jet Propulsion Laboratory in Pasadena, California; Paul Spudis of the Lunar and Planetary Institute in Houston, Texas; Geoffrey Marcy of the University of California at Berkeley; and Nick Hoffman of La Trobe University in Australia.

Thanks are also due to Jason Perry for assisting with tabular data; Ken Glover for reading draft chapters; Lynn Charron of the Centre for Topographic Information in Canada, Andrew Chaikin, Kelly Beatty, David Woods and Laszlo Keszthelyi for assistance with illustrations. Roger Launius of the NASA History Office is to be thanked for providing some very interesting books. Mike Hanlon of the Daily Mail is to be thanked for providing the Foreword.

Last, but by no means least, I must thank Clive Horwood of Praxis for his boundless optimism over the years.

1

Origin of the Solar System

FORMATION

For thousands of years, it seemed obvious that the Sun, Moon and planets circled the Earth, and were viewed against the background of fixed stars in the 'firmament'. It seemed that the Universe comprised a sophisticated system of 'celestial spheres' centred upon the Earth. However, as measurements of the motions of the planets became more accurate, the 'harmony' of the spheres became rather strained.

Nicolaus Copernicus, a Polish astronomer, realised that only the Moon moved around the Earth – all the planets, including the Earth, moved around the Sun. This 'heliocentric' theory was outlined in a book that was published upon his death in 1543.[1] Although Copernicus had dedicated the book to Pope Paul III, the Church frowned upon the idea as being a diminution of the status of the Earth in God's Universe. Although insightful, Copernicus retained the Ptolemaic idea of a 'perfect' circular motion. In 1609, in extending the work of Tycho Brahe, Johann Kepler realised that the planets actually orbit the Sun in an elliptical manner[2] and formulated his results as three 'laws': (1) the planets move in elliptical orbits with the Sun at one focus; (2) a line drawn between a planet and the Sun 'sweeps out' equal areas in equal times; (3) the square of the orbital period of a planet is proportional to the cube of its average distance from the Sun.

Deducing the nature of the Solar System by methodical measurements of planetary motions was a remarkable achievement for naked-eye astronomy. When Isaac Newton developed his theory of Universal Gravity in 1687, this explained the motions of the planets around the Sun in exquisite detail.[3]

In analysing the 'proper' motions of stars, William Herschel realised in 1805 that the Sun is moving in the general direction of the constellation of Hercules. When his research revealed that the Sun is simply one of a multitude of stars in a vast disk that we observe in the sky as the Milky Way, he introduced a revolution as profound as that of Copernicus. The planetary system whose details had preoccupied so many generations of astronomers was spectacularly insignificant in the great scheme of things. But *how* had the Sun acquired its system of planets?

In 1745, the French naturalist G.L.L de Buffon had suggested that the planets

2 Origin of the Solar System

This early 16th-century woodcut depicts the 'celestial sphere' as a fantasy in which a man is thrusting his head through the starry firmament into the infinity beyond.

condensed from the material that remained after a comet collided with the Sun.[4] He did not know that comets, while voluminous, are mostly tenuous gas and hence insignificant, so the prospect of one hitting the Sun had appeared to him to constitute a titanic event.

Immanuel Kant, the German philosopher, suggested in 1755 that the Sun formed from a cloud of interstellar gas – a nebula – which gravitationally imploded and became hot enough to glow as a result. Simon Laplace independently proposed a similar hypothesis in 1796, but in greater detail.[5] He was prompted by what he saw as remarkable facts: (1) all the planets and moons (as far as he knew) orbited in the same direction; (2) the planets rotated in the same direction; and (3) with minor departures, the planets all moved in the same plane. He reasoned that if a gas cloud was initially rotating, then its rate of rotation would have increased as it contracted in order to conserve its angular momentum. As it speeded up, he reasoned, it would have thrown off an equatorial ring of material, relieving it of its 'excess' angular momentum. However, as the collapse continued, the rotation would have increased again, and by the time it reached a stable state, the cloud would have shed a number of concentric rings. These, he argued, would have individually coalesced to form a series of planets in near-circular orbits. Furthermore, if this process was repeated on a smaller scale, then the planets would also develop systems of moons.

The 'Andromeda Nebula' (so-called because it lies in the constellation of Andromeda) can be seen by the naked eye on a clear night. It was first observed telescopically by the German astronomer Simon Marius, in 1612. In 1781 the French comet-hunter Charles Messier compiled a list of nebulous objects in order to prevent

This engraving, dated 1587, depicts Tycho Brahe's observatory at Urania on the Danish island of Hven. Tycho's finger points to a slit in the wall through which an assistant (upper right) is sighting along the large quadrant. Immediately above Tycho's raised hand is a triquetrum (one of the instruments developed by Nicolaus Copernicus). In the background are three storeys of the observatory, with assistants engaged in astronomical observations.

astronomers from mistaking them for comets. The Andromeda Nebula, therefore, is often referred to as M31, but what is it? The Orion Nebula, discovered in 1656 by Christiaan Huygens, was clearly a cloud of glowing gas. William Herschel was able to resolve individual stars in some nebulae that seemed to be 'globular clusters'.[6,7] Other nebulae, however, were only seen as fuzzy patches. If they, too, were composed of stars, then they had to be at some *incredible* distance. In fact, in 1755, Kant had speculated that the 'fuzzy patches' might constitute 'island universes'. To investigate them further, William Parsons (the 3rd Earl of Rosse) constructed a 72-inch telescope – at that time the largest telescope in the world – at Birr Castle in Ireland, and in 1845 saw a spiral structure in some of these 'mysterious' nebulae.[8] Because stars could not be resolved, the spiral form was interpreted as a disk of gas swirling around a forming star, and thus was considered to be proof of Laplace's nebular hypothesis.[9]

Saturn's ring system had initially been considered to support the theory, but a mathematical study by J.C. Maxwell in 1859 revealed that gravitational stresses would prevent the ring material accreting to form a moon. Furthermore, his mathematics also ruled against the planets having coalesced from material shed by the forming Sun. There were also other dynamical problems with the theory. It was difficult to explain why the planets managed to obtain most of the system's angular momentum. The planets account for no more than 0.1 per cent of the mass of the system, but possess 98 per cent of its angular momentum. In fact, Jupiter, which is larger than all the other planets combined, possesses 60 per cent of the Solar System's angular momentum. If the formation of moons occurred by the same process, why was most of the momentum in the Jovian system retained by the planet, rather than its moons? Such problems undermined the theory, which did not actually explain *how* the angular momentum was transferred as the rings of material were released. The nebular hypothesis had received a boost in 1854, when the German physicist H.L.F. von Helmholtz suggested that the Sun derived its energy from continuing collapse, but by 1938, when H.A. Bethe realised that the Sun is actually powered by energy released by the fusion of hydrogen into helium, the theory had long-since been abandoned.

In 1901, in America, T.C. Chamberlain began to explore the idea that the planets were the result of the Sun having had a close encounter with another star. In working out the details in collaboration with F.R. Moulton, Chamberlain decided that it must have been a near collision.[10,11] At the closest point of approach, they said, the gravitational attraction of one star on the other would have drawn out a cigar-shaped streamer of gas. This became detached and isolated in space as the stars moved apart. As the streamer cooled, the material would have condensed to create a large number of 'planetesimals'. The largest planets would have accreted in the centre of the stream, with the smaller ones at either end. In part, the impetus for this line of argument may have been a spectacular nova in Perseus in 1901. At that time, novae were believed to be stellar collisions. The relic was subsequently found to have become surrounded by a shell of glowing gas, which appeared to prove that collisions prompted the ejection of debris.

In England in 1917, James Jeans and Harold Jeffreys addressed the thorny issue of how most of the angular momentum was transferred to the planetary system by suggesting that the presence of the other star would have drawn the gaseous material considerably farther into space than would have been the case if it had been spun off by the Sun in isolation.[12,13,14] There were problems with the dynamics of the way in which the planets would have formed however, and in 1929, Jeffreys was forced to the conclusion that the Sun must have had a *grazing* encounter.[15] In 1939, Lyman Spitzer demonstrated that any material drawn from the Sun — by whatever scenario — would be so hot that it would rapidly expand, in the process becoming more tenuous; it would not cool and condense.[16] Later, in 1964, in an attempt to work around this obstacle, M.M. Woolfson proposed that the Sun encountered a 'protostar'.[17,18] Being less substantial, the protostar would have been more severely influenced by the Sun's gravity and most of the material in the streamer would have been derived from the protostar which, being cooler, would have more readily condensed. In 1946, Fred Hoyle offered a novel solution[19] by proposing that the Sun had once been the junior member of a binary star system in which its companion had evolved more rapidly and exploded in a supernova, leaving behind a cloud rich in 'metals' which had then condensed to form the planets.[20]

But these protostars and evolved companions were *ad hoc* models, and were rather contrived. In fact, by the 1940s the tide was turning back towards the nebular hypothesis. Carl von Weizsacker realised in 1944 that the condensing cloud would have been fragmented by turbulence, with each fragment developing into a smaller separate vortex.[21] Hannes Alfvén considered what would have happened to any magnetic field that the cloud might have possessed, and discovered that the field would actually have *promoted* the transfer of angular momentum to the disk and in so doing would have served to slow the central mass's rotation. He thereby avoided the 'flaw' in the original hypothesis. As the temperature of the contracting gas cloud increased and the atoms became ionised to form an electrically conducting plasma, this would have 'locked in' the magnetic field, and the intensity of the field would have increased as the contraction continued. As rings of material were shed, the plasma would have carried the magnetic field with it. In effect, therefore, the Sun was magnetically 'coupled' to the disk from which the planets later condensed.[22] Ironically, the basis for the 'revival' of the nebular hypothesis had become available several years after it had been abandoned as unworkable. After Pieter Zeeman's discovery in the 1890s that a magnetic field induced 'splitting' in Fraunhofer spectral lines, G.E. Hale of the Mount Wilson Solar Observatory realised that sunspots are associated with intense magnetic fields. This prompted him to speculate that the Sun might have a general magnetic field, so when the sunspots were at their 'minimum' in 1939 he confirmed that the Sun does indeed possess a dipole field.[23] Furthermore, the field's polarity reverses periodically. On average, the sunspot cycle peaks every 11 years, but the magnetic cycle is twice as long, at 22 years.

Over the last half-century, the nebular hypothesis has been considerably refined. It is now believed that planet formation involves a two-stage process – known as the 'nucleation and collapse' hypothesis – in which the accretion of icy and rocky planetesimals first forms a solid core, and only if this acquires sufficient mass does it

Masses and Orbital Characteristics of Extrasolar Planets
Last Updated 9 January 2001

using stellar masses derived from Hipparcos, metalicity, and stellar evolution

	Star Name	Msini (Mjup)	Period (d)	Semimajor Axis (AU)	Eccentricity	K (m/s)
1	HD83443	0.35	2.986	0.038	0.08	56.0
2	HD46375	0.25	3.024	0.041	0.02	35.2
3	HD187123	0.54	3.097	0.042	0.01	72.0
4	HD179949	0.86	3.092	0.043	0.0	112
5	TauBoo	4.14	3.313	0.047	0.02	474.0
6	BD-103166	0.48	3.487	0.046	0.05	60.6
7	HD75289	0.46	3.508	0.048	0.00	54.0
8	HD209458	0.63	3.524	0.046	0.02	82.0
9	51Peg	0.46	4.231	0.052	0.01	55.2
10	UpsAndb	0.68	4.617	0.059	0.02	70.2
11	HD168746	0.24	6.400	0.066	0.00	28.0
12	HD217107	1.29	7.130	0.072	0.14	139.7
13	HD162020	13.73	8.420	0.072	0.28	1813.0
14	HD130322	1.15	10.72	0.092	0.05	115.0
15	HD108147	0.35	10.88	0.098	0.56	37.0
16	HD38529	0.77	14.31	0.129	0.27	53.6
17	55Cnc	0.93	14.66	0.118	0.03	75.8
18	GJ86	4.23	15.80	0.117	0.04	379.0
19	HD195019	3.55	18.20	0.136	0.01	271.0
20	HD 6434	0.48	22.0	0.15	0.3	37.0
21	HD192263	0.81	24.35	0.152	0.22	68.2
22	HD 83443c	0.16	29.83	0.17	0.42	14.0
23	GJ876*c*	0.56	30.12	0.13	0.27	81
24	RhoCrB	0.99	39.81	0.224	0.07	61.3
25	HD168443b	7.73	58.10	0.29	0.53	473
26	GJ876b	1.9	61.02	0.21	0.1	210
27	HD 121504	0.89	64	0.32	0.13	45.0
28	HD16141	0.22	75.80	0.351	0.28	10.8
29	HD114762	10.96	84.03	0.351	0.33	615.0
30	70Vir	7.42	116.7	0.482	0.40	316.2
31	HD52265	1.14	119.0	0.493	0.29	45.4
32	HD1237	3.45	133.8	0.505	0.51	164.0
33	HD37124	1.13	154.8	0.547	0.31	48.0
34	HD169830	2.95	230.4	0.823	0.34	83.0
35	UpsAndc	2.05	241.3	0.828	0.24	58.0
36	HD12661	2.83	250.2	0.799	0.20	89.3
37	HD89744	7.17	256.0	0.883	0.70	257.0
38	HD202206	14.68	258.9	0.768	0.42	554.0
39	HD134987	1.58	260.0	0.810	0.24	50.2
40	IotaHor	2.98	320.0	0.970	0.16	80.0
41	HD92788	3.86	337.7	0.97	0.27	113.1
42	HD177830	1.24	391.0	1.10	0.40	34.0
43	HD 27442	1.13	426	1.15	0.02	34
44	HD210277	1.29	436.6	1.12	0.45	39.1
45	HD82943	2.3	442.6	1.2	0.6	73
46	HD222582	5.18	576.0	1.35	0.71	179.6
47	HD 160691	1.87	743	1.6	0.62	54.
48	16CygB	1.68	796.7	1.69	0.68	50.0
49	HD10697	6.08	1074.0	2.12	0.11	114.0
50	47UMa	2.60	1084.0	2.09	0.13	50.9
51	HD 190228	5.0	1127	2.3	0.43	91.0
52	UpsAndd	4.29	1308.5	2.56	0.31	70.4
53	14 Her	5.55	2360.0	3.5	0.45	98.5
54	Epsilon Eridani	0.8	2518.	3.4	0.6	19.0
55	HD168443c	17.1	1770	2.87	0.20	289

A listing of extrasolar systems as of early 2001 known to have planets. (Courtesy of Geoffrey W. Marcy of the University of California at Berkeley.)

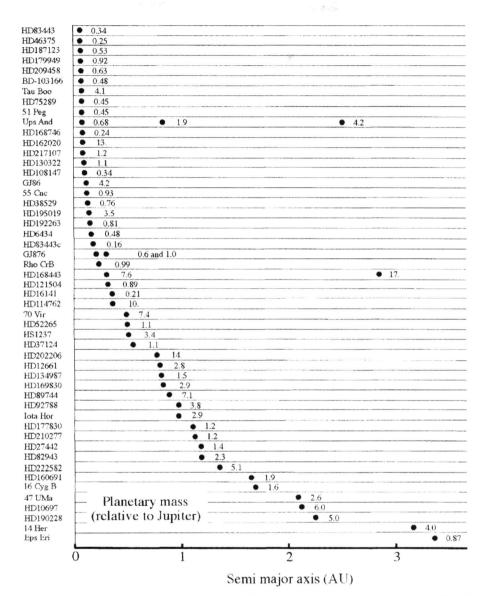

A plot of extrasolar systems as of early 2001 showing the semi-major axes of their planets. (Courtesy of Geoffrey W. Marcy of the University of California at Berkeley.)

draw in gas to form a gaseous envelope. The gas remaining in the nebula when the Sun 'switched on' was blasted back into interstellar space.

OUT THERE

By invoking a chance event, the collision hypotheses implied that our Solar System might be unique, but the nebular hypothesis raised the prospect that planetary systems could well be commonplace.

In fact, the T Tauri type of variable star, which are believed to be newly formed stars in the act of clearing out the cobwebs, provided indirect evidence that young stars do indeed produce disks. These stars have masses similar to the Sun, but are just a few million years old. In the 1980s, it was found that many of them display an 'infrared excess' from dust that is absorbing short-wave radiation from the central star and reradiating it at longer wavelengths. It was apparent that the dust was not distributed symmetrically as this would have blocked out the visible view of the star; in fact, it is a flattened disk inclined with respect to the line of sight. Recently, the Hubble Space Telescope spotted several protoplanetary disks (a term that has been contracted to 'proplyds') in the visible range, some of which were seen in silhouette against a background of hot gas in a 'stellar nursery' in the Orion Nebula. Verifying that young stars produced disks of gas and dust did not prove, however, that planetary systems would inevitably result.

After decades of fruitless surveys and frustrating false alarms, dozens of planets have been discovered over recent years, many of them by Geoffrey Marcy of the University California at Berkeley and Paul Butler of the Anglo-Australian Observatory. These planetary systems are a strange assortment. It was evident that the detection method would favour gas giants that orbit close to their stars, but the proximity of the massive planets to their parents came as a surprise. In fact, their unusual characteristics are severely testing the accepted theory of how a planetary system forms. When only one planetary system was known, it was impossible to separate the peculiarities of our own system's formation from the generic processes. As the list grows, it is becoming possible to accumulate statistics, so our insight is increasing rapidly.[24,25,26,27,28] So far, the most populous system is Upsilon Andromedae, with three giants,[29] and the smallest planet detected is Saturn-sized.[30] The race is now on to find terrestrial-type planets.[31]

NOTES

1 *De revolutionibus orbium coelestium*, N. Copernicus. 1543.
2 *The lord of Uraniborg: a biography of Tycho Brahe*, V.E. Thoren. Cambridge University Press, 1991.
3 *Principia*, I. Newton. England, 1687.
4 *De la formation des planetes*, G.L.L. de Buffon. Paris, 1745.
5 *Exposition du system du monde*, S.P. Laplace. 1796.

6. *William Herschel*, J.B. Sidgwick. Faber & Faber, 1953.
7. *William Herschel and the construction of the heavens*, M.A. Hoskin. Oldbourne, 1963.
8. *The astronomy of Birr Castle*, P. Moore. Mitchell Beazley, 1971.
9. That this was not so did not become apparent until 1924, when Edwin Hubble turned the Mount Wilson Observatory's newly commissioned 100-inch Hooker telescope on the Andromeda Nebula and resolved stars in its periphery.
10. F.R. Moulton. *Ap. J.*, vol. 22, p. 165, 1905.
11. *The origin of the Earth*, T.C. Chamberlain. University of Chicago Press, 1916.
12. J.H. Jeans. *Mon. Not. Roy. Astron. Soc.*, vol. 62, p. 1, 1917.
13. H. Jeffreys. *Mon. Not. Roy. Astron. Soc.*, vol. 78, p. 424, 1918.
14. *The universe around us*, J.H. Jeans. Cambridge University Press, 1930.
15. H. Jeffreys. *Mon. Not. Roy. Astron. Soc.*, vol. 89, p. 636, 1929.
16. L. Spitzer. *Astrophys. J.*, vol. 90, p. 675, 1939.
17. M.M. Woolfson. *Proc. Roy. Soc. Series A*, vol. 282, p. 485, 1964.
18. *The origin of the Solar System: the capture theory*, J.R. Dormand and M.M. Woolfson. Ellis Horwood Ltd, 1989.
19. 'Origin of the Earth and the planets'. In *The nature of the universe: a series of broadcast lectures*, F. Hoyle. Blackwell, p. 69, 1950.
20. To an astronomer involved in spectroscopy, any element farther up the periodic table than lithium (of atomic number 3) is considered to be a 'metal'.
21. C.F. von Weizsacker. *Z. Astrophys.*, vol. 22, p. 391, 1944.
22. H. Alfvén. In *The origin of the Solar System*, S.F. Dermott (Ed.). Wiley, p. 41, 1978.
23. *Guide to the Sun*, K.J.H. Phillips. Cambridge University Press, p. 32, 1992.
24. For a lively review of the lean years of searching for extra-solar planets, the detection techniques, and the first few successes, see *Planet quest: the epic discovery of alien solar systems*, K. Croswell. Oxford University Press, 1997.
25. 'The diversity of planetary systems', G.W. Marcy and R.P. Butler. *Sky & Telescope*, p. 30, March 1998.
26. 'Nurseries', R. Jayawardhana. *Astronomy*, p. 62, November 1998.
27. 'Forging the planets', J.A. Wood. *Sky & Telescope*, p. 36, January 1999.
28. To keep up to date, regularly check the website 'http://exoplanets.org/index.html'.
29. In 1996 Geoff Marcy and Paul Butler reported discovering a near-Jupiter-sized planet orbiting the star Upsilon Andromedae. In April 1999, after analysing 11 years worth of observations, they identified the presence of two other giant planets further from the star.
30. In February 2000 Geoff Marcy, Paul Butler and Steve Vogt reported identifying two 'Saturn-sized' planets – one orbiting HD46375 in the constellation Monoceros and the other orbiting 79 Ceti (also listed as HD16141) in Cetus. They are very close to their stars and so have short orbits, circling their primaries in 3 and 75 days respectively. Being so close in, they are very hot and so are not considered to be abodes for life as we know it.
31. *Looking for Earths: the race to find new solar systems*, A. Boss. Wiley, 2000.

2

Discovering the Earth

GEOLOGICAL TIME

In 1658 the Irish Archbishop James Ussher calculated from the Bible's Old Testament that the Earth was created at 8 pm on Saturday, 22 October 4004 BC. When the King James version of the Bible published in 1701 made reference to this calculation it became dogma.[1,2,3,4] Even so, the French naturalist G.L.L de Buffon felt obliged to suggest in 1778 that the Earth was rather older, and he argued that during seven 'epochs' spanning 100,000 years it cooled from incandescence into a molten state and then developed a solid crust and, ultimately, life.[5]

In reading *A Theory of the Earth* to the Royal Society in Edinburgh in 1785 the Scottish naturalist James Hutton cited evidence to show that the hills and mountains of the present day were sculpted by the slow processes of erosion.[6] His motto for geologists was "the present is the key to the past". Sedimentary rock bears the hallmarks of having accumulated in precisely the same way as alluvial sediment accumulates on the sea floor today.

The prevailing doctrine of 'catastrophism' held that the Earth had been created in much the same state that it now exists, but was then shaped by a series of supernatural cataclysms, of which Noah's Flood (as related in the Bible) was just the most recent. Because it was based upon the uniformity of nature and the continuity of its physical processes, Hutton's doctrine was called 'uniformitarianism'.

Hutton's insight into the power of erosion had been prompted by his study at Siccar Point on the coast near Edinburgh of a formation of a type that he later called an 'unconformity'. He had found a gently tilted sandstone sequence resting on top of a nearly vertical stack of shales and siltstones. All of the rocks were of a sedimentary nature, which led him to make a series of deductions. Firstly, the vertical strata must have been horizontal when they were deposited in water. After being folded, the tilted strata had been heavily eroded. Erosion meant that the sea floor had been transformed into dry land either by a vertical displacement of the land or by the withdrawal of the sea. After being inundated again, the sandstone had been laid on top of the exposed cross-section of the folded strata. It had re-emerged as dry land and erosion was now wearing down the sandstone. In relation to the rates at which the processes of sedimentation and erosion occur today, this activity had clearly taken more than a few thousand years.

Hutton's ability to 'read the rocks' was remarkable. When he showed his unconformity to John Playfair, a friend, Playfair was awestruck. "The mind seemed to grow giddy by looking so far into the abyss of time," he reflected.[7]

A.G. Werner, a mining engineer in Saxony in the late 18th century, argued that most rocks were of sedimentary origin. Extrapolating from salt precipitation, he thought that crystals could be created only by precipitation from solution. Having observed crystalline rock such as basalt in horizontally bedded stacks, he had concluded that it was a sedimentary rock. He argued that volcanic eruptions marked where burning coal seams broached the surface, and although he acknowledged that the intensity of this 'subterranean fire' could melt rock, he would not accept that a lava extrusion was the same material as the basalt he had observed in rock strata.

Werner had noticed a sequence in the rocks through which he mined. The basement of the sequence was crystalline granite, schist and gneiss. The evident ubiquity of sedimentary rock had led him to conclude that, early in its history, the Earth's 'nucleus' was contained in a global ocean. He therefore rationalised that granite was first to precipitate, and hence was the earliest type of rock. Visiting Glen Tilt in Scotland in 1788, Hutton found granite intruding into both limestone and shale, which clearly showed that the granite had *not* formed first. Indeed, there were blocks of limestone caught up as inclusions within the granite. Furthermore, because it had baked the rocks with which it came into contact, the granite had been hot and had certainly not precipitated from water. This led him to the conclusion that crystalline rocks resulted from igneous processes within the Earth, so he introduced the term 'plutonic' (as a reference to their formation in the underworld) to describe them.

Although the debate between the 'neptunists' and 'plutonists' raged for half a century, the publication of British geologist Charles Lyell's massive tome in the 1830s finally resolved the issue in Hutton's favour.[8] Three main types of rock were identified: igneous, sedimentary and metamorphic. The igneous rocks crystallised from molten rock. Sedimentary rocks comprised fragments of shattered and otherwise eroded rocks and were deposited in layers by eolian or aqueous processes. Metamorphic rocks were rocks that had been sufficiently treated by the heat and/or pressure within the Earth to induce chemical alteration, but not remelting of their minerals.

Once it was acknowledged that the Earth was considerably older than Ussher's calculation the question became: How old? If the processes of erosion and sedimentation were continuous and proceeded at a fixed rate, American palaeontologist Charles Walcott concluded in 1893 that some sequences would have taken tens of millions of years to build up. If the salt in the ocean is solely due to fluvial erosion, Irish geologist John Joly calculated in 1899 that it would have taken a hundred million years to accumulate the observed concentration.[9] By human standards this was an almost unimaginably long time.

There was a complication. What made the Sun shine and how long had it been active? The interior of the Earth was popularly believed to be burning coal. If the Sun was burning coal, then it could be no more than a few thousand years old. In 1854 the German physicist H.L.F. von Helmholtz pointed out that if the Sun was in a state of slow but progressive collapse, it might be powered by the conversion of gravitational energy. However, physicist William Thomson (later Lord Kelvin) calculated that if this was true, then the Sun could only have shone for 50 million years if its initial diameter had been as large as the Earth's orbit, but this implied that Venus and Mercury started out by orbiting *inside* the Sun.[10] By 1897, Kelvin's considered opinion was that the Earth was 20 million years old.[11,12,13]

Just as there seemed to be an impasse, an unexpected discovery offered the prospect of a solution. In 1896, while studying the X-rays which had been discovered by W.K. Roentgen the previous year, A.H. Becquerel in Paris realised that they were emitted only by certain elements. In 1898 Marie Curie named this phenomenon radioactivity. For some time, therefore, it appeared that the Sun was powered by the energy emitted by the decay of heavy elements. However, in 1938 American physicist H.A. Bethe realised that the Sun's energy derives from the fusion of hydrogen into helium,[14] and since the Sun is composed mainly of hydrogen, it may well have been active for *billions* of years.

Did the Earth form at the same time as the Sun? There were competing theories, one of which suggested that the planets condensed from the nebula from which the Sun was formed; and if this was true, then they were the same age. Another theory proposed that the planets condensed from a 'streamer' of material that was drawn out as another star made a close pass to the mature Sun. If this was true then the Earth could be considerably younger than the Sun.

In 1902, Ernest Rutherford observed that the nuclear fission process which is responsible for radioactive decay involved a 'half-life' law. In 1904 he realised that this provided a natural 'clock' with which to measure the time since a rock crystallised. His measurements of the rate at which uranium decayed into lead in some rocks conclusively established that the Earth is at least a billion years old. In announcing this to the Royal Society (with Kelvin in the audience) Rutherford tactfully observed that radioactivity invalidated Kelvin's assumption that the Earth was simply cooling from an initial state; in fact, it was continuously generating heat, and this was why it was older than Kelvin had calculated. Several years later, the Cambridge physicist R.J. Strutt (later Lord Rayleigh) established that a rock from Ceylon was 2.4 billion years old. At Rayleigh's suggestion, in 1911 the English geologist Arthur Holmes launched a long-term project to radiometrically date the entire geological sequence.[15] In 1927 he concluded the Earth is at least 3.6 billion years old, and, using a similar technique, Claire Patterson established in 1955 that ordinary chondrite meteorites are 4.55 billion years old, which finally established the age of Solar System.[16]

DRIFTING CONTINENTS

In his *Novum Organum* in 1620 the English philosopher Francis Bacon commented on the general similarities in shape between the Old World and the New, but he drew no specific inference. In 1666 Frenchman François Placet rationalised that the continents were split at the time of Noah's Flood, citing as evidence the fact that the Bible says that prior to this the Earth was "one and undivided". In 1749 G.L.L de Buffon suggested that the Atlantic Ocean formed as the legendary 'lost continent' of Atlantis was flooded and sank. In 1756, German theologian Theodore Lilienthal pointed out that the coastlines of the Atlantic dove-tailed, and he cited "in Peleg's days, the Earth was divided" in the Book of Genesis as a reference to the continental break-up in the aftermath of Noah's Flood.

Various non-biblical suggestions were also offered. In 1801 Alexander von Humbolt, the German explorer, announced that the Atlantic was "nothing other than a valley scooped out by sea water". Rocks in outcrop in mountainsides were realised to be sediments which were formed horizontally and then intensely folded and eroded. But what force compressed the crust sufficiently to uplift the Alps? In 1829 Elie de Beaumont said that this meant that the Earth was shrinking and in response to the compressional force its crust was undergoing vertical displacements. It was presumed that crustal movements were catastrophic; it was also assumed that when some areas were uplifted others were depressed. In this view, the ocean was simply water that had pooled in low-lying areas. Apart from having being inundated, the ocean floor rocks were expected to be the same as those ashore. In 1844 Evan Hopkins, a British mining engineer, suggested that the Earth's magnetic field was drawing the continents towards the north pole.[17] In 1857 Richard Owen, a professor of geology in Tennessee, suggested that the Earth was initially tetrahedral in shape, and had expanded in a cataclysm that fragmented the crust into the continental masses that we see today – and ejected the Moon in the process.[18] Antonio Snider, an American theologian living in France, argued in 1858 that the Earth was expanding and that the singular continent was shattered cataclysmically when the internal pressure opened rifts.[19] In 1882 Osmond Fisher, a geophysicist, suggested that the Pacific marked where the Moon had been torn from the Earth and that this had split the continental landmass whose various fragments are now in the process of drifting 'downhill' into the cavity from which the Moon was torn.[20]

The term 'geology' had been coined in 1778 by J.A. Duluc, but it was 1810 before it was granted an entry in the *Encyclopaedia Britannica*. Even so, until the late 19th century, geology was dominated by pontification. With so many theories on offer it was apparent that *physical evidence* would be required to sift the wheat from the chaff.

In the 1890s, the Austrian geologist Edward Suess, citing similarities in local rocks as his evidence, proposed that South America, Africa, Australia and India had once been united in a 'supercontinent' in the southern hemisphere which he called Gondwanaland.[21] Then in 1910 Alfred Wegener, a German meteorologist, went further, arguing not only that North America and Eurasia had been united, but also that at one time they had been connected to the southern continent, forming an integrated supercontinent which he called Pangea ('all land'). Several years later, having considered all the evidence (matching coastlines, similarities in large-scale geological structures, rock types, and fossils) he presented his conclusions in a book entitled *The Origin of Continents and Oceans*.[22]

Although the similarity in the coastlines of Africa and South America was striking, the line of the coast is not a very good indicator of the edge of a continent because if the sea level was to rise or fall by 100 metres the waterline would follow the topography, and the outline of the coast would change irregularly.[23] Nevertheless, as offshore soundings were charted on a large scale it was observed that the continents do indeed have sharp edges. In some cases, these are close to the coast, but they can be several hundred kilometres offshore. There is a fairly steep plunge from the 'shelves' at a depth of no more than a few hundred metres down to the ocean floor, typically some 6 kilometres below sea level. With the continental 'margins' mapped, the Tasmanian geologist Warren Carey made a visual reconstruction of South America and Africa in 1958 using the 2,000-metre contour which was an even better 'fit' than that based on their shorelines. In 1965 Edward Bullard at the University of Cambridge in England made a series of fits using a computer and found that the optimal fit was at the 1,000-metre contour.[24] The results indicated that if Africa's position had been fixed, South America had been rotated in a clockwise manner by some 57 degrees around a point near the Azores. There were, however, several gaps and overlaps. One of the overlaps was a massive build up of sediment from the River Niger, but this could only have been accumulated after the ocean formed. The Walvis submarine ridge to the south marked another overlap, but as this sits on the ocean floor it was created once the ocean had opened. Overall, therefore, the evidence for the two continents having once been a single landmass was compelling. Subsequently, Bullard demonstrated that Greenland fitted against Canada and the continental margins of Norway and northern Scotland were in relationships that simply could not be flukes.

An early study of similar structures on the east coast of South America and the west coast of Africa by Alexander du Toit had identified mountain belts which seemed once to have been joined.[25] Du Toit could only date such structures stratigraphically, but when radiometric dating techniques were developed Patrick Hurley of the Massachusetts Institute of Technology was able to confirm that these orogenic events did indeed occur simultaneously.[26]

One remarkable piece of evidence derived from glaciation patterns. The ancient mountain belts which ran across the divide from Africa into Brazil had been heavily glaciated. In itself, this meant that the continents moved, because for equatorial mountains to have been glaciated they would of necessity have been much further south than they are today. Glaciers scour the bedrock as they flow, and thus the direction of travel can be inferred from such marks. It was noted that whereas the glaciers in the ancient mountains of Brazil had travelled west, there was no source; the source was in Africa. Furthermore, glaciers pick up boulders, carry them along, and deposit them when the ice melts. In fact, the discovery that the Earth's climate varies and had suffered at least one 'ice age' had been inferred from the fact that Europe was littered with rocks – dubbed 'erratics' – that had been carried south by

16 Discovering the Earth

```
Mountain belts
      ↓
      ─────── 8,000 m         Average land elevation is 840 m
                              Average ocean depth is 3,800 m


- - - - - - 840 m                                    Sea level
          ↑      ↑                                   ──── 200 m
Continental platform
     Continental shelf                               ──── 3,800 m
       Continental slope ──→
                                                     ──── 6,000 m
                     Abyssal plain ↗
                    Ocean trench ──→      ╲╱ ── 11,000 m
```

This plot of the distribution of the Earth's surface in terms of elevation shows that most of the surface is either abyssal plain on the ocean floor 6 kilometres below sea level, or continental platform a few hundred metres above sea level. The extremes of elevation are insignificant in terms of areal extent. The tallest mountains rise 8 kilometres and the deepest oceanic trenches plunge 11 kilometres. In terms of means, dry land is 840 metres above sea level and the floor of the ocean is 3,800 metres below sea level. The summit of the mid-ocean ridge system (not shown) is typically at a depth of 2 to 3 kilometres.

ice. The erratics in Brazil included distinctively coloured quartzite, dolomite and chert for which there were no sources to the east on that continent; the sources were in southwest Africa.[27] Glaciers also deposit piles of rock fragments which (upon subsequent compaction into masses called tillites) indicate where glaciers once were. The distributions on either side of the ocean made a compelling case for South America and Africa having split apart, but when did this occur? A study of the offshore sediment from the River Niger revealed that it had been accumulating for 50 million years, so the continents must have split apart prior to this. A large formation of 200 million years old sedimentary rock spanned the divide, so the split must have occurred after this. In fact, a study of marine deposits determined that the split began around 120 million years ago.

The wider picture
Wegener and du Toit reinforced Suess's Gondwanaland theory with a mass of evidence of ancient orogeny running from South America across Antarctica and on to southeastern Australia, and dolerites running from southeast Africa across Antarctica and on to Tasmania. In general terms, Australia is older in the west and

there are progressively younger north–south trending belts to the east. A similar pattern in Antarctica can be rotated to 'fit' the two continents together. The Antarctic peninsula fits between the southern extremities of Africa and South America. There are also massive seams of coal in Antarctica which indicate that it was not always at the south pole. Coal derives from luxuriant tropical vegetation on land that is slowly sinking. The detritus is drowned before it can rot. This makes peat, and to convert this into a coal seam the detritus has to be overlain by several kilometres of rock so that the pressure can force the water and gas out of the peat to form a carbonaceous rock. Clearly, therefore, Antarctica must have once had a tropical climate. There was even the possibility of an interplay between the movement of continents and climate.

In 'reconstructing' Gondwanaland, India presented a problem. If it had split off of Antarctica far to the south, then to have travelled so far it must have moved at an astonishing rate. It was decided that it most probably originated from near Africa, alongside what is now the island of Madagascar. In part, this was inferred from the distribution of ancient anorthosites that form an arcuate band from South America across Africa, Madagascar, India and into Australia. The fact that the formations of western Australia are truncated by the coast indicates that a large section of ancient land is missing, but this may be present as the plateau that rises from the Indian Ocean floor and currently lies only a few hundred metres below sea level.

While it was reasonable to assume that buoyant granitic continents might 'float' on a global basaltic 'basement', as Suess had proposed, how could they 'sail' though

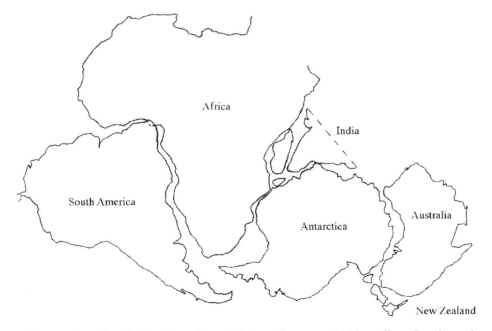

The southern hemisphere's continents fit together remarkably well as Gondwanaland. For ease of identification, this reconstruction shows the best fit (as determined by the continental slopes) in terms of the shorelines extant today.

it like ships at sea as Wegener had advocated? Reconciling the patent impracticability of the continents moving, with the vast amount of evidence that they had once been united and were now dispersing, became a serious issue for geologists in the first half of the 20th century.

In 1600, William Gilbert, a physician at the court of Queen Elizabeth, had discovered that the Earth possesses a magnetic field. The axis of the field is actually inclined 11.5 degrees from the planet's rotational axis, with the magnetic north pole offset towards Canada. Then in 1895 Pierre Curie discovered that magnets lose their magnetism when heated above a specific temperature, called the 'Curie point'. Furthermore, upon cooling below this temperature, the crystals of a solidifying iron-bearing mineral will align themselves along the ambient magnetic field. This is called 'remanent' magnetism.[28] When it was found that the fields in sediments of various ages in England diverged systematically from today's field,[29] Keith Runcorn said that this meant that while the Earth's rotational axis remained fixed with respect to its orbit around the Sun, the planet had 'rolled' erratically with respect to this axis.[30] When data from America became available, Runcorn opined that although the evidence was in rough agreement with his thesis, the case was not conclusive.[31] However, when the 'polar wandering' track from India became available it was totally inconsistent.[32] The only way that the data could be reconciled was if the continents had moved *independently* on the Earth's surface.

By 1960, it was clear from palaeomagnetic studies that the various continents had travelled considerable distances, at differing rates.[33] There had to be something strange about the ocean floor, and the best way to understand how the Atlantic Ocean had opened was to survey its floor.

THE SEA FLOOR

In the 1830s, English biologist Edward Forbes began dredging the North Sea to discover the kinds of life that inhabited the perpetually dark sea floor. The first clue to the nature of the deep ocean floor was dredged up in 1852 by a ship laying a telegraph cable across the North Atlantic.[34] This revealed that the ocean floor fell away dramatically at some distance offshore, but rose somewhat in the middle. In the 1860s George Wallich dredged living starfish from water as deep as 2 kilometres, far beyond the depth at which it had been thought that any life could survive. In the mid-1870s, during HMS Challenger's pioneering research expedition, Charles Thompson surveyed its voyage across the North Atlantic by periodically lowering a 'sounding' line.

In 1922, in the preface to the third edition of *The Origin of Continents and Oceans*, Alfred Wegener reflected[35] that "whilst the continental blocks are wrinkled in all directions by ancient and recent folded mountain chains, we do not know, in spite of all the soundings that have been made, a single feature from the enormous area of deep sea which we could claim with any certainty as a chain of mountains". But in that same year, a German oceanographic research ship started an 'echo sounding' survey of the Atlantic. Echo sounding was much more productive than making measurements using a cable. The 'tracks' that accumulated during the next few years revealed a rugged terrain. Surprisingly, there was a continuous range of mountains tracing a swath in the middle of the ocean. Ascension Island and the Azores were simply peaks where this mid-Atlantic ridge broached the surface.

Supporting continental drift, Arthur Holmes, the renowned British geologist, suggested in 1930 that the mantle might be convective and this circulation might provide the traction which makes the continents move. Geophysicist Harold Jeffreys vehemently countered that this was impossible. In 1933 Holmes argued that the median ridge existed because the ocean floor had been stretched so thin in the 'wake' of the withdrawing continents that the hot mantle was able to erupt from fissures.[36]

When C.S. Piggot developed an apparatus in 1934 to retrieve a 3-metre long core sample, and a survey was conducted from Newfoundland to Ireland, surprisingly little sediment was found in the deep ocean. If the oceans had existed in their present locations for billions of years they should have accumulated up to 5 kilometres of sediment.

The way in which seismic waves propagated during an earthquake in the Balkans in 1909 led Croatian geologist Andrija Mohorovicic to the discovery that there is a distinct transition to dense rock at a depth of about 35 kilometres. He interpreted this boundary (which carries his name as the Mohorovicic Discontinuity) to be the interface between the buoyant crust and the dense mantle. As mountain belts are formed by lateral compression, they project not only tall peaks but also deep 'roots', and the discontinuity can plunge to a depth of 65 kilometres as it traces their outline.

A lack of seismic stations on islands made studies of the way in which shocks from earthquakes were propagated through the ocean floor imprecise at best, but as Arthur Holmes expressed the state of knowledge in 1944 in the first edition of *Principles of Physical Geology*, the ocean floor "seems to be like the deeper parts of the basaltic layers beneath the continents". In effect, "it appears that the granitic layer is missing" and "the materials have the same properties beneath oceans and continents" at greater depths. Clearly, therefore, the ocean floor is not simply inundated continental crust.[37] The fact that the granitic continents 'floated' on such rock reinforced the dilemma of how the landmasses move, and the ocean floors became the focus of ever more intensive research.

A survey in the late 1940s revealed that the floor of the Pacific is littered with flat-topped mountains. By the time 10,000 such 'seamounts' had been charted a correlation was apparent – the depth of water above their summits increased to the northwest. The floor of the Pacific had been expected to be a vast abyssal plain, but it was found to have hosted a great deal of volcanic activity. Another survey revealed that Hawaii's Big Island is actually the summit of a mountain that is taller than Mount Everest in the Himalayas. In fact, Hawaii is a vast volcanic 'shield', as are the other members of this northwest-trending island chain.

Vast trenches were discovered running for thousands of kilometres paralleling the western coast of the Americas, and in conjunction with the mid-Pacific archipelagos. The deepest (12 kilometres deep) was charted in 1951 by the Glomar Challenger. Nevertheless, trenches can span 100 kilometres, so they have fairly subtle profiles. Except along the island arc of the West Indies, trenches were found to be absent in

20 Discovering the Earth

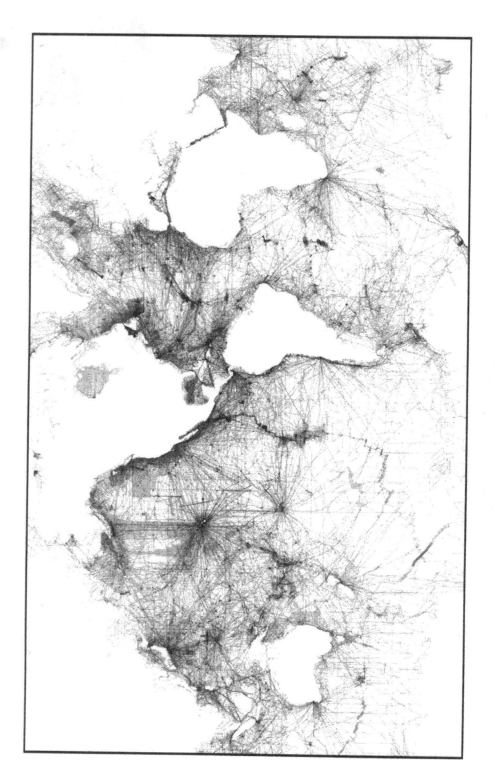

This map shows the individual tracks sailed by ships involved in the massive task of sounding the ocean floor. (Copyright W.H.F. Smith and D.T. Sandwell of the University of California at San Diego 1997.)

the Atlantic, however. The onset of the Cold War in the 1950s prompted the US Navy to inititate a programme to map the ocean floor as a preliminary to operating its fleet of nuclear submarines. By extending Mohorovicic's pioneering analysis of how the continental crust refracts seismic waves from earthquakes, the deep structure of the ocean floor was probed using shock waves generated by explosives. The results revealed that the discontinuity is typically only 12 kilometres below sea level, so the crust that forms the ocean floor is only 6 or 7 kilometres thick.

In 1953, Bruce Heezen and Marie Tharp of the Lamont-Doherty Observatory, New York, discovered a canyon running the length of the mid-Atlantic ridge. By 1956, these researchers had integrated the flood of results from sounding surveys and produced a comprehensive map of the ocean floor worldwide. In fact, the mid-ocean ridge is not restricted to the Atlantic – it curves around the tip of Africa, across the Indian Ocean, runs between Australia and Antarctica and on to the south of New Zealand before swinging northeast (as the East Pacific Rise) to intersect the Americas in the Gulf of California. Although spurs branch off from the main chain (with one running north across the Indian Ocean into the Arabian Sea) the entire structure forms a single 60,000-kilometre-long feature. It is typically 1,000 kilometres across and its crest stands a few kilometres above the flanking abyssal plains. It displaces so much water that it has a significant effect on global sea level. Currently, the oceans account for about 71 per cent of the Earth's surface. However, in terms of the continental margins the ocean floor forms only 60 per cent of the crust, the rest is inundated continent, as was once thought to be the case for the ocean floor itself. If it were not for the ridge system displacing water, large areas of continental shelf would be exposed.

The fact that there is a continuous mountain range winding its way around the world beneath the ocean's waves is awe inspiring. When seismic studies showed that the structure of the adjacent ocean floor is undeformed, it became clear that the mid-ocean ridge is not a compressional structure. It is structurally different to a mountain range on land. Although the rugged ridge is bare rock, the active seismic soundings showed that this topography is buried at the base of the continental slope by up to a kilometre of sediment, which is why the abyssal plains are so flat. As sounding surveys improved, it was found in 1958 that although a deep rift runs the entire length of the ridge's crest, this is not a continuous feature; it consists of many short linear segments which are offset from one another by perpendicular faults. On one of the US Navy surveys off the west coast of North America H.W. Menard found several east–west oriented step-like scarps in the ocean floor. One scarp was traced for 5,000 kilometres, and its height progressively diminished westward from a maximum of about 2,000 metres. As the ocean floor appeared to have been sliced into narrow strips and displaced, Menard named these features 'fracture zones'. It was also found that whereas the heat flow through the ocean floor is generally comparable to that through a continent, it is several times greater over the crest of the ridge,[38] and where the rift valley is hottest it is volcanically active. The heat flow is lowest over the oceanic trenches. Such observations clearly told a story; the trick was to achieve the perspective required to see the entire puzzle.

This map of the ocean floor was produced by Geosat, which was built by the US Navy and launched in 1985 into a polar orbit. However, the data remained classified for ten years. The satellite measured sea level variations and charted the gravitational variations, from which the seabed's topography could be inferred. It shows the 60,000-kilometre-long mid-ocean ridge system. (Courtesy of W.H.F. Smith of the National Ocean and Atmospheric Administration and D.T. Sandwell of the Scripps Institute of Oceanography, 'Global sea floor topography from satellite altimetry and ship depth soundings', W.H.F. Smith and D.T. Sandwell. *Science* vol. 277, p. 1956, 1997).

SEA FLOOR SPREADING

In 1960, Princeton University geologist H.H. Hess realised that the ocean floor is a very dynamic structure in which the material to either side of the mid-ocean rift is *diverging*, and the hot mantle is welling up to form a succession of dykes inside the rift. For the continents to move apart there must be a spreading ridge on the ocean floor between them. Although Hess's paper[39] was not formally published until 1962, it was widely circulated in preprint form and he presented the theory on a lecture tour. Robert Dietz of the Scripps Oceanographic Institute had reached a similar conclusion and (after discussing the matter with Hess and giving appropriate credit) published his own paper which, as it happened, appeared before Hess's, and it was in Dietz's paper that the phrase 'sea floor spreading' was coined.[40]

Hess's brilliance was to realise that the ocean floor is the surface expression of the mantle. At first, Hess believed that the ocean floor was simply hydrated mantle in the form of a mafic (that is, magnesium- and iron-rich) basalt called serpentinite and he thought that basalt was limited to the volcanic edifices constructed above the ocean floor, but as more and more samples of basalt were dredged up it became clear that the uppermost layer of the lavas that form the ocean floor are basalt. This is the buoyant product of thermal differentiation in the upper mantle. The denser mafic minerals are in the base of the oceanic crust. Most significantly, there is a correlation between the age of the ocean floor and its distance from the ridge. It is the buoyancy of the hot rock that prompts the formation of the ridge, and the ocean floor falls off to either side because the cooling rock is progressively less buoyant. The observations that the depth of water above the Pacific seamount summits increased away from the East Pacific Rise reflected the fact that as the ocean floor cooled and densified, it settled isostatically. The extent to which the ocean floor has settled is evident from the fact that some seamounts are now under several thousand metres of water. Furthermore, their summits are flat because they broached the surface when they were young and were subjected to wave erosion. At that time, they would have resembled the atolls that we see today atop more recent volcanic edifices. A spreading rate of a few centimetres per year was sufficient to have made all of the currently extant ocean floor in 200 million years. Considering that the Earth is a few billion years old, this was an amazing discovery because it meant that most of the planet's crust was *new*.

In proposing sea floor spreading, Hess was faced with two choices. Perhaps the Earth was smaller in the past and the continental landmass had originally covered the globe and had been fragmented when the planet began to expand. In this view, the ocean floors were made as the mantle welled up to fill the gaps. However, there was no independent evidence that the Earth was expanding at all, let alone that it had undergone a five-fold increase in volume during the last 200 million years. Alternatively, if the Earth's size had remained constant, it was evident that ocean floor could not be produced continuously without being consumed elsewhere. In another insight, Hess realised that when the old densified ocean floor abuts against a more buoyant crustal block it is forced down into the mantle; it is this process of 'subduction' that occurs at an ocean trench.

In 1906 French physicist Bernard Brunhes discovered some rocks that were magnetised in the opposite polarity to the Earth's field. As one of the few early findings of palaeomagnetic research, this fact remained an oddity for half a century. By then it was evident that this was a global phenomenon. The obvious inference that the planetary dipole field reversed its polarity from time to time was resisted because there was no obvious reason for it to do so.

In 1936 it was realised from the way seismic waves propagate through the Earth's centre that there is a solid 'inner' core within the fluid 'outer' core. The energy released by the phase change associated with solidification would have induced convection in the surrounding fluid. The circulating metallic fluid constituted an electrical current. This in turn induced a magnetic field. Initially, the phase change would have been induced by the immense pressure but as the inner core irrevocably cools the process is self-sustaining. Nevertheless, the strength of the magnetic field today indicates that the convection in the fluid part of the core is still vigorous.[41]

A computer simulation of the dipole developed in 1995 by Gary Glatzmaier of Los Alamos National Laboratory in New Mexico and Paul Roberts of the University of California at Los Angeles predicted not only that the field would reverse its polarity, but also that the solid inner core ought to be rotating faster than the rest of the planet. Paul Richards and Xiao-dong Song of the Lamont-Doherty Observatory then analysed the archive of data from seismic waves that passed through the Earth's centre and confirmed that the inner core is a true dynamo, rotating almost one degree per year more rapidly than the outer core. It is the interaction between the two that generates the dipole field.[42]

When Arthur Raff trailed a magnetometer behind a ship during a US Navy survey off the west coast of the United States in 1955, his plan was just to extend a continental survey onto the ocean floor. The 'magnetic anomalies' in the Murray fracture zone off California displayed sharply defined magnetic stripes. Nothing like it had been seen on land. It was as if a magnetic pattern was imprinted on the ocean floor. Intriguingly, the stripes were offset by the fractures in the ocean floor.[43] Another survey found that there was a strengthening of the magnetic field directly over the rift in the mid-Atlantic ridge.[44]

In 1963, at Cambridge in England, Fred Vine and Drummond Matthews realised that such magnetic anomalies strongly supported Hess's theory of sea floor spreading.[45] Being rich in iron, mafic basalts take on a particularly strong remanent signature. If the field is periodically reversed, the progressively older rocks further from the mid-ocean ridge would have literally 'recorded' the magnetic cycles.

However, the case would not be conclusive until it could be proved that the polarity of the Earth's magnetic field undergoes frequent reversals. By the early 1960s, improvements in a method for dating rocks in terms of their relative proportions of potassium and argon made it possible to date basalt flows thought to have been extruded on land in the last few million years. Allan Cox of the US Geologic Survey charted a number of such flows.[46,47] In each case the orientation of the sample was noted before it was taken. The remanent magnetic field gave the polarity of the Earth's field when the rock crystallised and the isotopic mix indicated when this had been. It was soon evident that there was a global pattern with rocks of

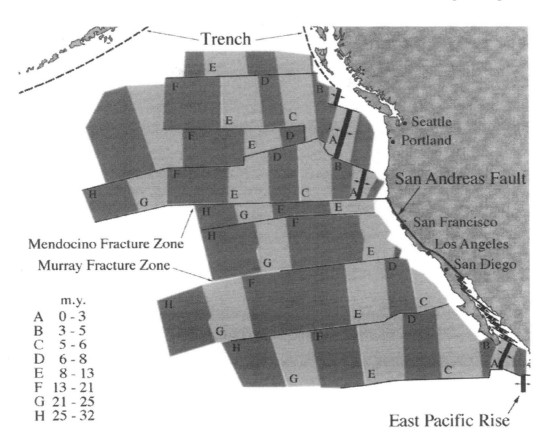

When a magnetometer was trailed behind a ship on a US Navy survey off the west coast of the United States 'magnetic anomalies' were discovered. Intriguingly, the 'stripes' were offset by the fractures in the ocean floor. Significantly, there were east–west oriented step-like scarps on the ocean floor. It looked as if the ocean floor had been sliced up and differentially offset. Once it was realised that the ocean floor is made by spreading at the mid-ocean ridge system and that the Earth's magnetic field reverses its polarity intermittently, it became evident that the magnetic pattern had been made by the East Pacific Rise.

the same age having the same field irrespective of their location. Not only was the magnetic field capable of reversing, it had been doing so intermittently for millions of years.[48,49,50,51,52,53] The deep sea sediment cores from Antarctica provided an independent proof.[54] Unlike the lava flows, ocean sediments yielded a continuous record and the microfauna present meant that the timescale of field reversals could be calibrated. Vine used this timescale to *predict* what the pattern ought to be over the crest of the Reykjanes Ridge (which forms the mid-Atlantic ridge's median to the south of Iceland) on the assumption of a spreading rate of 1 centimetre per year, and this pattern was confirmed by an airborne magnetometer survey.[55] The island is split by the rift, which is open to inspection. In the late 1960s, Ronald Mason of Imperial College London monitored a network of survey sites on either side of the rift and confirmed Vine's assumption about the rate of opening. The island is growing by spreading to either side of the rift. The high degree of volcanic activity is believed to be due to the fact that it is sited directly over a mantle plume. The existence of this plume explained why the mid-Atlantic ridge is so extensively built-up in this location. Once it was realised that magnetic reversals had been occurring for such a long time it became evident that the stripes imprinted on the Murray fracture zone had been made by the East Pacific Rise which (as the most active part of the mid-ocean ridge system) is producing ocean floor at the astonishingly rapid rate of 16 centimetres per year.

Initially, it was assumed that the mid-ocean rift had begun as a continuous feature and that shear forces had torn the crust and progressively offset the short segments of the rift. In 1965, however, J.T. Wilson at Toronto University suggested that the rift had never formed a single continuous feature and had always been fragmented by faults. Furthermore, he argued that the process of sea floor spreading required that only the section of a fault stretching between adjacent rifts would be active because the newly created ocean floor on either side of the fault would be travelling in opposite directions; the continuations beyond the rifts would be moving in the same direction. It had already been noted that the mid-ocean ridge formed a locus for seismic activity, and a thorough analysis by Lynn Sykes at the Lamont-Doherty Observatory in 1966 found that the seismic activity on such faults was limited to between adjacent rift sections, just as Wilson had predicted.[56,57] At that time the significance of strike-slip faults on land had not been appreciated. The predominant fault action was believed to involve vertical displacement; therefore the evidence of extensive lateral displacements on the ocean floor was encouraging for those seeking a *mechanism* for continental drift because it established that segments of the ocean floor were highly mobile.

Global seismic studies established several interesting facts. Firstly, the loci of earthquakes were not randomly distributed but were concentrated in well-defined zones which, although often narrow, stretched for thousands of kilometres. In the 1950s, Hugo Benioff had studied earthquakes near island arcs in the Pacific,[58] and found a linear relationship between the depths of the epicentres and their distances from adjacent trenches. Of course, the idea of subduction had yet to be conceived but these plots traced slabs of ocean floor diving at angles that varied between a shallow 30 degrees and an almost vertical 80 degrees to depths of 700 kilometres. In contrast, the epicentres of earthquakes in the mid-ocean ridges are shallower than 20 kilometres. Given that the mantle is so viscous, it is remarkable that a subducting slab can penetrate so deeply. It may be fragmented as it is thermally eroded, but it will eventually be totally reassimilated into the mantle and be recycled.

In 1967, independent studies of systems of ridges, trenches and faults by Jason Morgan at Princeton[59] and Dan McKenzie of Cambridge in England, working with Robert Parker of the Scripps Oceanographic Institute,[60] discovered that the Earth's surface is broken into a number of rigid lithospheric 'plates' whose motions are governed by spherical geometry.

The most far-reaching insight was that virtually all seismic, volcanic and tectonic activity occurs at the *margins* of the plates. This was a powerful unifying idea that revolutionised the study of the Earth, but the new theory also introduced new terminology. *Constructive* margins are places where lithosphere is produced, and include not just the mid-ocean rifts, but also the rifts onshore where continents are being torn apart. *Destructive* margins are trenches where lithosphere is being subducted. *Conservative* margins are places where plates slip past each other without producing or destroying lithosphere, and the (strike-slip) boundaries are referred to as 'transform' faults.

The basis of the new theory was Euler's analysis of motion on the surface of a sphere.[61] This stated that every plate displacement could be considered as a simple rotation about an appropriate axis which passed through the planet's centre and intersected the surface at a 'pole'. Every point on the moving plate therefore traced out a section of a 'small circle' (corresponding to a line of latitude) about this axis. The opening of the Atlantic involved the Americas rotating around an Eulerian pole situated at about 58 degrees north and 36 degrees west, which lies on the mid-Atlantic ridge just south of Iceland. On a spherical surface the amount of sea floor spreading would increase with angular distance from the Eulerian pole – and this not only convinced J.T. Wilson that the spreading was real but also explained why the fracture zones subtended longer arcs in the South Atlantic than in the North Atlantic. This differential spreading *requires* that the rift take the form of short sections producing narrow strips of sea floor. The theory also predicted that the faults would trace small circles – as indeed they do. Each section of the rift traces a meridional line radiating from the Eulerian pole. Furthermore, the form of any boundary between plates was dictated by the geometry – boundaries which coincided with small circles formed conservative margins and those that did not conform to this pattern were obliged to form either constructive or destructive margins.

The theory of 'plate tectonics' provided the context to interpret the San Andreas strike-slip fault. A trench parallels most of the western coast of the Americas and the diving oceanic floor induces the chain of volcanoes inland. Generalising somewhat, the East Pacific Rise currently intersects the Americas in the Gulf of California and the boundary to the north is a strike-slip fault that runs onshore to just north of San Francisco, where it heads offshore and rejoins the next segment of ridge. It took a long time to work out the fine detail,[62] but the basic idea is now readily understood and, indeed, serves as an exemplar.[63,64]

The first point of the East Pacific Rise to meet the trench was at the eastern extremity of the Mendocino fracture zone, some 28 million years ago. This started the San Andreas strike-slip fault extending to the south. Subduction off California ended when the trench met the eastern extremity of the Murray fracture zone and the last of the segment of spreading ridge between the two zones was consumed. The length of the San Andreas remained fixed while the trench consumed the eastern half of the Murray fracture zone, but once the trench started to consume the spreading ridge to the south it resumed its southerly growth. The American plate had been advancing west throughout. During the last 25 million years or so, the section of continent on the margin of the oceanic plate has slipped several hundred kilometres north. Los Angeles will eventually pass San Franciso. In fact, the boundary marked by the San Andreas is not a clean interface. The plate margin is badly fragmented,

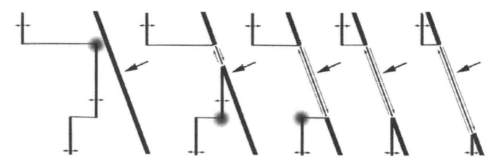

The theory of 'plate tectonics' provided the context to interpret the San Andreas strike-slip fault. As the American plate drifts westward, it overrides the ocean floor, which is subducted. The first point of the East Pacific Rise to start to subduct was at the eastern extremity of the Mendocino fracture zone, some 28 million years ago. This started the San Andreas strike-slip fault extending to the south. Subduction off California ended when the trench met the eastern extremity of the Murray fracture zone and the last of the segment of spreading ridge between the two zones was consumed. The length of the San Andreas remained fixed while the trench consumed the eastern half of the Murray fracture zone, but once the trench started to consume the spreading ridge to the south it resumed its southerly growth. The American plate had been advancing throughout. During the past 25 million years the section of continent on the margin of the oceanic plate has slipped several hundred kilometres north. Volcanic activity was rife in California until subduction off the coast ceased. It still occurs in Mexico to the south, and in Oregon and Washington to the north, because subduction is still occurring there.

and there is an extensive system of faults. As the two plates slip past one another, their action tends to rotate the fragments that are caught in this vise, so the seismic zone extends far inland. Volcanic activity used to be rife in California until subduction off the coast ceased. It still occurs in Mexico to the south, and in Oregon and Washington (and indeed into Canada) to the north because subduction is still occurring there.

With the insight provided by plate tectonics, oceanic spreading ridges, fracture zones, trenches, onshore strike-slip faults and volcanic island arcs were shown to form an integrated global system. Initially, Hawaii posed a slight problem: the theory of plate tectonics posited that volcanic activity would occur at plate margins, whereas Hawaii is located in the middle of the Pacific plate. In 1963, in assessing the implications of sea floor spreading, Wilson suggested that Hawaii is a lithospheric hot spot created by a mantle plume and, furthermore, pointed out that because the progressively older islands of the chain rise from the Hawaiian Ridge they serve to indicate the direction and the rate of motion of the sea floor over the hot spot.[65] Also the north–south Emperor Seamount chain appeared to adjoin the western end of the Hawaiian Ridge. Dating the volcanoes on the individual islands showed that the entire chain recorded 80 million years of sea floor motion, and that the direction of motion had abruptly changed about 35 million years ago.[66] The answer to the dilemma for plate tectonics lay in the existence of the plume. Where one burns through an oceanic plate, a vast basaltic shield is constructed on the ocean floor.

Sea floor spreading 29

The San Andreas fault can be discerned transecting the right-hand side of this overhead image of the Los Angeles area.

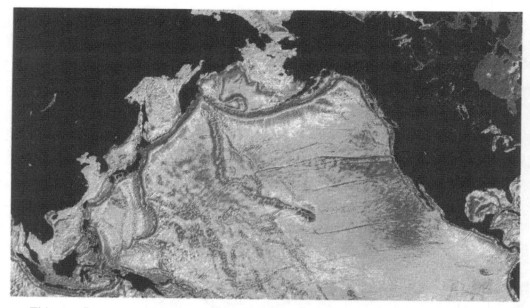

This sea floor map of the north Pacific strikingly illustrates the peripheral trenches around the northern continental margins, the back-arc basins (such as Japan's Inland Sea), the manner in which the western Pacific comprises a number of distinct slabs of oceanic plate and the island arcs above where the main plate is subducted, and the dog-leg of the Hawaiian/Emperor island chain which indicates a change in direction in the main plate's motion.

Once a specific island moves away from the hot spot and its magma source is cut off, its volcano will become extinct. The hot spot will eventually activate a new vent, thereby extending the chain. In the case of the Hawaiian islands, the new edifice has still a kilometre or so to rise before it broaches the surface.[67,68] Thus, although plate tectonics tends to concentrate volcanic activity at plate margins, volcanism can also be prompted in other circumstances – an important lesson for the generalists.

In the 1960s, following a pioneering study of the propagation of seismic energy by Beno Gutenberg at the California Institute of Technology, it was found that there is a zone in which the velocity of shear waves decreases to a minimum. The velocity of shear waves depends on the compressibility, rigidity and density of the medium through which it travels. The velocity of such a wave increases with increasing pressure, and decreases with increasing temperature. Since the pressure should increase progressively with depth, the observed 'low-velocity zone' must correspond to a steep thermal gradient in which the temperature rises. Calculations show that whereas rock in the lithosphere is rigid, the conditions in the zone (which has been called the 'asthenosphere') are precisely those required to melt the rock, so the lithospheric is decoupled at a depth of 100 to 200 kilometres. The geochemical and petrological evidence shows that the 'upper' mantle (above this layer) has the same ultramafic composition as the convecting 'lower' mantle, with the primary minerals being pyroxene and olivine, but the lower temperatures and pressures permit it to solidify.

The key to the impasse which had faced Wegener's theory was that the granitic continents do not drift over a basaltic basement but ride on mobile lithospheric plates which float on the asthenosphere. The ocean floor between continents that have been split apart was created by the act of splitting, so it is *new*. The ocean floor over which a continental plate advances is being *subducted*. The criticism of Wegener's theory that had addressed the impossibility of a buoyant continent drifting over its basaltic basement was rendered obsolete. The mantle convection is by solid-state crystalline flow (a phenomenon that was not known when continental drift was proposed). The force driving the motions of the plates is believed to be plumes rising, flattening out on the underside of the lithosphere, and descending again, in the process applying traction to the underside of the plates they impinge upon (as Arthur Holmes had suggested in 1930).

When Hess first conceived of sea floor spreading, it was in the context of interpreting the mid-Atlantic ridge. At first, he supposed that the line of the rift was where the crust had been weakened by a curvilinear riser of a mantle convection cell. However, when the rift was found to be discontinuous it was realised that this simplistic view was naive. The ridge system is *not* static with respect to the mantle. This can be inferred from the fact that the African plate is bounded on three sides by spreading ridges. The ridge system is therefore receding from the African plate and, as a result, is moving with respect to the mantle. Also, because the rate of spreading is not uniform along the ridge's length, its shape is continuously evolving.

The ocean floor is composed of several distinct layers whose structure is similar in all the world's oceans, so oceanic plates display a remarkable uniformity. The lava erupted from the rift is an olivine tholeiite, which is a basalt derived from partial melting of the mantle at shallow depth where the temperature in the rising magma is high and the pressure is relatively low. It is rich in orthopyroxene and calcic plagioclase feldspar, but is deficient in the potassic and sodic varieties of plagioclase.[69]

Sea water at the typical ocean rift depth of 3 kilometres is 2 °C and at very high pressure, so lava is promptly chilled. Because it has no time to crystallise, the lava develops a glassy rind a few centimetres thick. As it cools it contracts, shatters and implodes and more lava bursts out to form another protrusion. The result is rather like squeezing toothpaste from a tube. However, a lavaform of this type has little structural strength and readily breaks into fragments which pile up. The topmost layer of the ocean floor is therefore a stack of these 'pillow' lavas. Furthermore, the hydrostatic pressure is sufficient to inhibit the exsolution of volatiles. Indeed, the lava will readily absorb sea water, and this hydration modifies the lava chemistry.[70,71]

Pillow lavas had long been encountered on land, but their unusual characteristics had remained something of a mystery. One early theory suggested that they formed when lava intruded into sediments saturated with water.[72] It was only when they were seen forming as a lava flow ran offshore that the reason for their characteristic shape was understood. A submersible that was surveying the mid-Atlantic ridge in 1973 found the first pillow lavas *in situ* in the rift.[73]

Beneath the pillow lavas and the intrusive dykes, the rest of the oceanic plate consists of a thick layer of olivine peridotite that cooled sufficiently slowly to enable large crystals to form, and so is described as gabbro. In his original formulation, Hess had thought that this layer was a serpentised peridotite but studies of seismic propagation contradicted this and established that the material is gabbroic.[74,75] The uniform gabbro yields at depth to a cumulate gabbro tracing a progression of increasingly ultramafic pyroxenites, olivinites and peridotites which formed as the magma became stagnant and underwent chemical sedimentation.

In the late 1950s, the US National Academy of Sciences funded a project to drill through the ocean floor to sample the rock below the Mohorovicic Discontinuity, but in 1966, after the budget had escalated twenty-fold, this effort was abandoned.[76] In retrospect, it turned out to have been misguided.

By the end of the 1960s, geologists knew enough about the ocean floor to recognise parts of it that had somehow become integrated into the rocks that were now exposed onshore. In 1963 Ian Gass and David Masson-Smith in the UK had noted that the Troodos massif of western Cyprus appeared to be composed of the kinds of rocks that could be expected to have formed at a depth of several kilometres beneath the floor of the Tethys Ocean (the ocean which separated Africa from Eurasia, and has almost completely been 'pinched out' by Africa's northward motion). In 1971 a team detonated explosives in holes left by miners seeking copper veins, and verified that the seismic velocities of these rocks matched those of the rocks deep within the ocean floor. In fact, the Troodos massif is a cross-section which transitions from pillow lavas, down through dykes, gabbro and cumulates into the olivine-rich 'moho' mantle. Such outcrops are called 'ophiolites' (snake-like, reflecting the texture of their surfaces). The distribution of ophiolites highlights where old ocean floors have been subducted.[77] As an oceanic plate slides into a trench the sediment is scraped off and the initially horizontal layers are intensely folded and ultimately accumulated into the margin of the plate beyond the trench. If the trench is irregular, pieces of ocean plate can break off and become incorporated into the folded sediment.[78] The shallow depth above a trench in which this material is compressed is a high-pressure but low-temperature environment which induces a mild form of metamorphism. A geochemical study found that the Troodos ophiolites came from a 'marginal basin' behind an offshore island arc where the 'back-arc' spreading* yielded less mafic magma than the tholeiitic magma formed at an ocean rift,[79,80] a scenario which was consistent with their having formed as the Tethys Ocean was subducted beneath southern Europe.

Cambridge geophysicist Dan McKenzie later filled in the details of why the mantle erupts through the mid-ocean rift.[81] The mantle is a viscous fluid which is convecting by solid-state flow. As the mantle rises the diminishing pressure causes melting, in the process creating the asthenosphere. In effect, therefore, the state of the mantle is controlled by the quantity of rock above it. As the overburden reduces, the mantle rock adopts less dense crystalline phases and expands. McKenzie's calculations revealed that the ocean floor is 7 kilometres thick precisely *because* this

---

* 'Back-arc' spreading in 'marginal basins' is explained in Chapter 7.

is the amount of basaltic rock required to 'suppress' the mantle from 'decompressing' and inhibit it from bursting out in a massive eruption. However, once an opening has been created by a hot plume, the opening is self-sustaining. The asthenosphere rises almost to the surface beneath the mid-ocean ridge. The only way to stem the rising tide is to cap it, and because the ocean floor is continuously transported from the rift the rift can only be sealed if a continental plate floats over it – as happened when the North American plate encroached upon the northern part of the East Pacific Rise. Thus, once an intense mantle plume has split a continent and sea floor spreading has been initiated, the rift is not only self-sustaining, it is also free to drift. The Earth's surface is therefore much more dynamic than even Wegener had envisaged. In expanding a suggestion by J.T. Wilson that mantle plumes might serve as static references, in 1972 W.J. Morgan worked out an 'absolute' coordinate system for tracking the motions of plates world wide.[82]

SUPERCONTINENT CYCLES

Perhaps the most significant insight into the evolution of the continents provided by plate tectonics was the role of mountain building, a process that in 1890 American geologist G.K. Gilbert had called 'orogeny'.

It was evident that ranges such as the Alps of southern Europe were sedimentary strata which had been intensely folded – but what force could have applied such tremendous compression? Once it was observed that fold belts occurred on plate margins, it was realised that they were the inevitable result of continental plates colliding with one another. In terms of the geological timescale, the rate at which mountain belts rise is rapid, with several kilometres of uplift being produced in only a few million years. As mountain belts are eroded by weathering, their deep roots rise isostatically to compensate. In an ancient range such as the Appalachians in the eastern United States, there has been so much erosion from the top that the root has been virtually eliminated. Mountain belts are therefore transitory. Over time, the deformations introduced by continental collisions are erased.

When South America broke away from Africa, the mountain belt that ran from Africa to Brazil was already badly eroded. If mountain belts form when continents collide, then plate tectonics *did not start* with the break up of Pangea; the supercontinent was a transitionary state. The insight that orogeny marked the collisions of continents enabled the *assembly* of Pangea to be inferred from old eroded mountain ranges.

The Caledonian belt extends from Spitsbergen through Greenland, Norway, northwest Scotland, northwest Ireland, Newfoundland and the northern section of the Appalachians. The Hercynian belt which extends across northern Africa, southern Europe, southern Britain, Newfoundland and the southern section of the Appalachians is rather younger. They derive from two continental collisions. A remarkable conclusion that could be inferred from these 'sutures' was that when continents collide they *completely* pinch out the intervening oceanic plates. Their previous margins were unlikely to have been mirror images, but no gaps were left

A map showing the world's mountain ranges. The significance of such large-scale orogenic structures (in conjunction with oceanic trenches) was not fully appreciated until the theory of plate tectonics.

with segments of ocean floor. The forces driving the continents were so great that their peripheries were crumpled into mountain belts (so the Mediterranean is destined to be transformed into an even higher mountain range than the Alps as Africa continues northward). This renders it impracticable to 'undo' ancient sutures and reconstruct continental margins as they would have been prior to a collision, so maps depicting the continents that came together to form Pangea are at best only 'illustrative'. In fact, as a continental plate overrides a subducting plate it accretes material from it. By way of an example, much of the northwestern periphery of the United States is composed of a hotchpotch of 'terranes' derived from islands swept up from the Pacific, so maps purporting to depict North America even several tens of millions of years ago are necessarily speculative.[83]

Evidently, the Caledonian belt was formed when North America integrated with northern Europe. The Urals were created some time later when Asia was added to complete Laurasia.[84] The Hercynian belt was formed when Laurasia collided with Gondwanaland's northern edge. In 'broad brush' terms, therefore, Gondwanaland formed about 500 million years ago and Laurasia 370 million years ago. They combined to form Pangea 250 million years ago and 50 million years later this began to break up. The Atlantic and Indian oceans were formed by the break-up of Pangea, the Tethys Ocean was destroyed, and the Pacific is in the process of being consumed. The 'widest' segment of oceanic plate is that between the East Pacific Rise and the trench off Japan. The northwestern part of this plate is the oldest piece of ocean floor but it is only 200 million years old, so oceanic plates, like mountain belts and supercontinents, are also transitory artefacts of plate tectonics. As the continental landmasses are torn apart by rifting and battered by collisions, their shapes and topography are constantly evolving. Since mantle convection is the force that drives

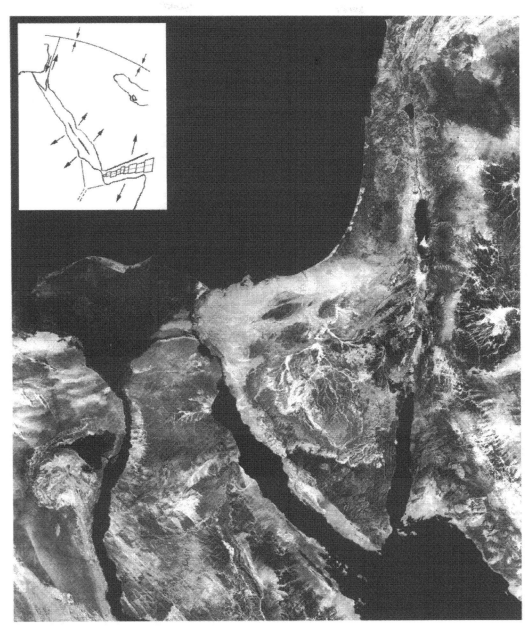

Africa is being torn apart. The 'triple junction' at the northern end of the system of grabens of the East African Rift marks where the continental rift meets the spreading ridge on the floor of the Red Sea and the stepwise spreading ridge which runs out into the Arabian Sea and links up with the Carlsberg Ridge in the Indian Ocean. At the north of the Red Sea the rift runs up the Gulf of Aqaba as a transform fault and continues inland through the Dead Sea and the Sea of Galilee. (Courtesy of NASA/EOSAT.)

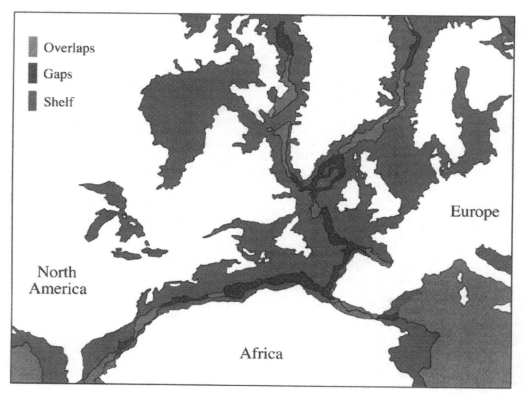

A map of the relative positions of the continents prior to the opening of the North Atlantic with the break-up of Pangea.

plate motions, it may be that the process of supercontinent aggregation and break-up is *cyclic* over a period of several hundred million years.[85,86]

Compared to the Earth's highly dynamic crust, the basaltic mare plain at Hadley–Apennine on the Moon is strikingly passive. While exploring the site in 1971, Dave Scott drilled a core sample. Only the upper half-metre of the regolith had been 'gardened' by the impact process. The state of preservation of the fine layering in the next metre showed that it had lain undisturbed for 400 million years.[87] On the Earth during this same period Pangea had been formed and torn apart.

Supercontinent cycles 37

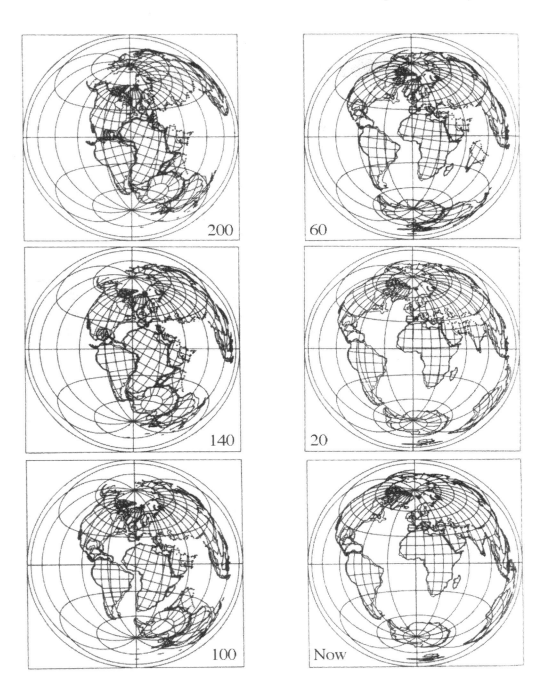

A palaeomagnetic reconstruction of the continents illustrating the break up of Pangea in Lambert equal-area south polar stereographic projection. (From *Mesozoic and Cenozoic Palaeocontinental Maps*, A.G. Smith and J.C. Briden, Cambridge University Press 1977, with permission.)

NOTES

1. *The annals of the world*, J. Ussher. London, 1658.
2. *Asimov's new guide to science*, I. Asimov. Penguin Books, 1984.
3. *A guide to Earth history*, Richard Carrington. Chatto & Windus, p. 22, 1967.
4. 'Bishop Ussher, John Lightfoot and the age of creation', W.R. Brice. *J. Geological Education*, vol. 30, p. 18, 1982.
5. *Epochs in nature*, G.L.L de Buffon. Paris, 1778.
6. *A theory of the Earth*, James Hutton. Royal Society, Edinburgh, 1785.
7. *Earth story: the shaping of our world*, S. Lamb and D. Sington. BBC, p. 14, 1998.
8. *Principles of geology, being an attempt to explain the former changes of the Earth's surface by causes now in operation*, C. Lyell. London, 3 volumes, 1830–1833.
9. 'An estmate of the geological age of the Earth', J. Joly. *Ann. Rep. Smithsonian Inst.*, p. 247, 1899.
10. 'On the age of the Sun's heat', W. Thomson. *Macmillan's Magazine*, p. 388, March 1862.
11. 'On the secular cooling of the Earth', W. Thomson. *Trans. Roy. Soc. Edinburgh*, vol. 23, p. 157, 1862.
12. 'On geological time', W. Thomson. *Trans. Geol. Soc. Glasgow*, vol. 3, p. 1, 1871.
13. *Lord Kelvin and the age of the Earth*, J. Burchfield. Macmillan Co., 1975.
14. Despite the evidence that the Sun is powered by nuclear fusion and its surface is hot, alternative theories are still being proposed. The vicar of Stoughton in Sussex, England, for example, recently offered a mathematical treatise that argued that the Sun is not hot: *The temperate Sun*, P.H. Francis (privately published) 1970. For a lighthearted review of the theories offered by Independent Thinkers, see *Can you speak Venusian: a trip though the mysteries of the Cosmos*, P. Moore. Wyndham, 1976.
15. For a review of radiometric dating methods, see *The age of the Earth*, G.B. Dalrymple. Stanford University Press, 1991.
16. *Thunderstones and shooting stars: the meaning of meteorites*, R.T. Dodd. Harvard University Press, p. 49, 1986.
17. *On the connection of geology with terrestrial magnetism*, E. Hopkins. London, 1844.
18. *Key to the geology of the globe*, R. Owen. New York, 1857.
19. *The creation and its mysteries unveiled*, A. Snider. Paris, 1858.
20. 'On the physical cause of the ocean basins', O. Fisher. *Nature*, vol. 25, p. 242, 1882.
21. Gondwanaland was named after Gondwana in India.
22. *The origin of continents and oceans*, A.L. Wegener. Methuen, 1915.
23. The shoreline has fluctuated greatly in the last few million years, so although the current level dominates the human environment, it has no special significance.
24. 'The fit of the continents around the Atlantic', E.C. Bullard, J.E. Everett and G.A. Smith. In *A symposium on continental drift*, Blackett, Bullard and Runcorn. The Royal Society, 1965.
25. *Our wandering continents*, A.L. du Toit. Oliver & Boyd, 1937.
26. 'The confirmation of continental drift', P.M. Hurley. *Scient. Am.*, April 1968.
27. 'The carboniferous glaciation of South Africa', A.L. du Toit. *Trans. Geol. Soc. South Africa*, vol. 24, p. 188, 1921.
28. The Curie point varies for different minerals, but for magnetite it is 580 °C and for hematite is 675 °C.
29. 'The remanent magnetism of some sedimentary rocks in Britain', J.A. Clegg, M. Almond and P.H.S Stubbs. *Philosophical Magazine, Series 7*, vol. 45, p. 583, 1954.

Notes

30 'The direction of the geomagnetic field in remote epochs in Great Britain', K.M Creer, E. Irving and S.K. Runcorn. *J. Geomagnetism Geoelectricity*, vol. 6, p. 163, 1954.
31 'Paleomagnetic comparisons between Europe and North America', S.K. Runcorn. *Proc. Geol. Soc.Canada*, vol. 8, p. 77, 1956.
32 'Rock magnetism in India', J.A. Clegg, E.R. Deutsch and D.H. Griffiths. *Phil. Mag.*, Series 8, vol. 1, p. 419, 1956.
33 'Analysis of rock magnetic data', P.M.S. Blackett, J.A. Clegg and P.H.S. Stubbs. *Proc. Roy. Soc.*, Series A, vol. 256, p. 291, 1960.
34 *Principles of physical geology*, A. Holmes. Thomas Nelson & Sons Ltd (first edition), p. 319, 1944.
35 *The origin of continents and oceans*, A.L. Wegener. Methuen (third edition), p. 40, 1922.
36 'The thermal history of the Earth', A. Holmes. *J. Washington Academy of Science*, vol. 23, p. 169, 1933.
37 *Principles of physical geology*, A. Holmes. Thomas Nelson & Sons Ltd (first edition), p. 373, 1944.
38 The heat flow is typically 66×10^{-2} joules per square metre per second, but it varies geographically. In the case of the ocean floor, the heat imparted at the spreading ridge is retained for tens of millions of years so it remains 'warm' even far from the ridge. Continental crust is self-warming because it is rich in radioactives. It transpired that there is a relationship between the heat flow and the time since the last major tectonic event. That is, the heat in a continental shield is low and that from sites of recent orogeny is higher. See: 'The heat flow through oceanic and continental crust and the heat loss of the Earth', J.G. Sclater, C. Jaupart and D. Galson. *Earth Rev. Geophys. Space Phys.*, vol. 18, p. 269, 1980.
39 'History of ocean basins', H.H. Hess. In *Petrologic studies*, Geological Society of America, 1962.
40 'Continental ocean basin evolution by spreading of the sea floor', R.S. Dietz. *Nature*, vol. 190, p. 854, 1961.
41 It is the leakage of heat from the core which induces the convection within the mantle that in turn drives plate tectonics.
42 'Spin control', M. Carlowicz. *Earth Mag.*, p. 21, December 1996.
43 'Magnetic survey off the west coast of North America', R.G. Mason and A.D. Raff. *Bull. Geol. Soc. Am.*, vol. 72, p. 1259, 1961.
44 'Magnetic anomalies over the Reykjanes ridge', J.R. Heirtzler, X. Le Pichon and J.G. Baron. *Deep Sea Res.*, vol. 13, p. 427, 1966.
45 'Magnetic anomalies over oceanic ridges', F.J. Vine and D.H. Matthews. *Nature*, vol. 199, p. 947, 1963.
46 'Reversals of the Earth's magnetic field', A.V. Cox, R.R. Doell and G.B. Dalrymple. *Science*, vol. 144, p. 1537, 1964.
47 'Reversals of the Earth's magnetic field', A.V. Cox, G.B. Dalrymple and R.R. Doell. *Scient. Am.*, vol. 216, p. 44, 1968.
48 'Reversals of the Earth's magnetic field', E.C. Bullard. *Phil. Trans. Roy. Soc. Series A*, vol. 263, p. 481, 1968.
49 'The origin of the oceans', E.C. Bullard. *Scient. Am.* (Special Issue 'The Ocean'), p. 16, 1969.
50 *History of the Earth's magnetic field*, D.W. Strangway. McGraw-Hill, 1970.
51 'Sea floor spreading', F.J. Vine. In *Understanding the Earth: a reader in the Earth sciences*, I.G. Gass, P.J. Smith and R.C.L. Wilson (Eds). Artemis Press, p. 233, 1971.
52 *Plate tectonics and geomagnetic reversals*, A. Cox (Ed.). Freeman Co., 1973.

53 *The Earth's magnetic field: its strength, origin and planetary perspective*, R.T. Merrill and M.W. McElhinny. Academic Press, 1983.
54 'Paleomagnetic study of Antarctic deep sea cores', N.D. Opdyke, B. Glass, J.D. Hays and J. Foster. *Science*, vol. 154, p. 349, 1968.
55 'Spreading of the sea floor: new evidence', F.J. Vine. *Science*, vol. 154, p. 1405, 1966.
56 'A new class of faults and their bearing upon continental drift', J.T. Wilson. *Nature*, vol. 207, p. 343, 1965.
57 'Seismology and the new global tectonics', B. Isacks, J. Oliver and L.R. Sykes. *J. Geophys. Res.*, vol. 73, p. 5855, 1968.
58 'Orogenesis and deep crustal structure: additional evidence from seismology', H. Benioff. *Bull. Geol. Soc. Am.*, vol. 66, p. 358, 1954.
59 'Rises, trenches, great faults and crustal blocks', W.J. Morgan. *J. Geophys. Res.*, vol. 73, p. 1959, 1968.
60 'The North Pacific: an example of plate tectonics on a sphere', D.P. McKenzie and R.L. Parker. *Nature*, vol. 216, p. 1276, 1967.
61 In the 18th century, the Swiss mathematician Leonard Euler had worked out motion on a spherical surface.
62 *Plate tectonics: how it works*, A. Cox and R.B. Hart. Blackwell Scientific, 1986.
63 *Continental drift: the evolution of a concept*, U.B. Marvin. The Smithsonian Press, 1973.
64 *Drifting continents and shifting theories*, H.E. Le Grand. Cambridge University Press, 1998.
65 'Evidence from islands on the spreading of ocean floors', J.T. Wilson. *Nature*, vol. 197, p. 536, 1963.
66 Lines of islands and seamounts in French Polynesia, and in the Austral Islands that linked to the Gilbert and Ellice Islands showed similar orientations.
67 'Exploring Loihi: the next Hawaiian island', N. Parks. *Earth Mag.*, p. 56, September 1994.
68 'Loihi rumbles to life', N. Parks. *Earth Mag.*, p. 42, April 1997.
69 'Geochemical characteristics of mid-ocean ridge basalts', S.S. Sun, R.W. Nesbitt and A.Y. Sharashin. *Earth Planet. Sci. Lett.*, vol. 44, p. 119, 1979.
70 When an ultramafic basalt is hydrated under pressure the calcic plagioclase feldspar yields to albite and epidote and the olivine and pyroxene are transformed into serpentine and chlorite.
71 Although pillows formed in shallow water will have glassy shells due to the thermal shock suffered upon meeting sea water, they will not be significantly chemically altered because the pressure is insufficient to force the water into the rock.
72 'Origin of pillow lavas', J.V. Lewis. *Bull. Geol. Soc. Am.*, vol. 25, p. 591, 1914.
73 *Photographic atlas of the mid-Atlantic ridge rift valley*, R.D. Ballard and J.G. Moore. Springer-Verlag, 1977.
74 'Sea floor spreading, progressive alteration of layer 2 basalts, and associated changes in seismic velocities', N.I. Christensen and M.H. Salisbury. *Earth Planet. Sci. Lett.*, vol. 15, p. 367, 1972.
75 'Structure and constitution of the lower ocean crust', N.I. Christensen and M.H. Salisbury. *Rev. Geophys. Space Phys.*, vol. 13, p. 57, 1975.
76 'Mohole: geopolitical fiasco', D.S. Greenberg. In *Understanding the Earth: a reader in the Earth sciences*, I.G. Gass, P.J. Smith and R.C.L. Wilson (Eds). Artemis Press, p. 343, 1971.
77 'Tectonic setting of basic volcanic rocks determined using trace element analysis', J.A. Pearce and J.R. Cann. *Earth Planet. Sci. Lett.*, vol. 19, p. 290, 1973.

78 'Geochemical evidence for the genesis and eruptive setting of lavas from Tethys ophiolites', J.A. Pearce. In *Ophiolites*, A. Panayiotou (Ed.), Geol. Surv. Cyprus, p. 261, 1980.
79 'The Troodos massif: its role in the unravelling of the ophiolite problem and its significance in the understanding of constructive plate margin processes', I.G. Gass. In *Ophiolites: Proceedings of the International Ophiolite Symposium*, Nicosia, Cyprus, 1979, A. Panayiotou (Ed.). Geol. Surv. Cyprus, p. 23, 1980.
80 'The emplacement of ophiolites by collision', Z. Ben-Avraham, A. Nur and D. Jones. *J. Geophys. Res.*, vol. 87, p. 3861, 1982.
81 'The Earth's mantle', D.P. McKenzie. *Scient. Am.*, p. 25, 1983.
82 'Deep mantle convection plumes and plate motions', W.J. Morgan. *Bull. Am. Assoc. Petrol. Geol.*, vol. 56, p. 203, 1972.
83 'Growth of western North America', D.L. Jones, A. Cox, P.J. Coney and M. Beck. *Scient. Am.*, vol. 247, p. 70, 1982.
84 'Assembling Asia', J. Shurkin and T. Yulsman. *Earth Mag.*, p. 52, June 1995.
85 'Large-scale mantle convection and the aggregation and dispersal of supercontinents', M. Gurnis. *Nature*, vol. 332, p. 695, 1988.
86 'The big flush', S. Vogel. *Earth Mag.*, p. 39, March 1994.
87 *Exploring the Moon: the Apollo expeditions*, D.M. Harland. Springer-Praxis, p. 176, 1999.

3

The lure of the Moon

THE ASTRONOMERS

In 1608 the Dutch spectacle-maker Hans Lippershey invented an optical device employing two lenses. This 'looker' made distant objects appear closer.[1] In May 1609, Galileo Galilei, a scientist living in Padua, Italy, heard of this invention and after improving its design he found that he could identify ships coming over the horizon a few hours earlier than was possible by the unaided eye, which provided his Venetian merchant friends with a significant commercial advantage. Early in 1610 Galileo indulged his long-standing interest in astronomy and aimed his telescope at the Moon, which he saw in unprecedented detail.[2] His stature as an observer is confirmed by his reluctance to follow his contemporaries in classifying the large dark areas as seas.[3] Unlike his contemporaries who accepted the Pythagorean view that the Moon is another Earth, Galileo did not "believe that the body of the Moon is composed of earth and water". He noted that the bright regions were much more rugged than the dark plains. The most intriguing features, however, were the profusion of ringed structures which were clearly large holes. He was also able to resolve individual mountains, and by measuring their shadows when the Sun was low on the lunar horizon he was able to infer that they were up to 10 kilometres high. His drawings of the surface features were only rough sketches, however, and they are difficult to correlate with features we see today.

The first reasonable rendition of the Moon was drawn in 1645, by Flemish cartographer M.F. Langrenus whose lunar observations were under the aegis of the King and Queen of Spain in an effort to determine whether sunrise and sunset of isolated mountains could serve as timing references for measuring longitude at sea. In 1647, Johannes Hevelius, having established an observatory in Danzig, included a map in a book.[4] Neither nomenclature resulted in the lunar names in use today, but Hevelius – who was of the belief that the lunar surface consisted of land and water – set the precedent of naming features after terrestrial equivalents. However, the lack of glint from sunlight soon convinced most observers that there were no expanses of open water on the Moon.

44 The lure of the Moon

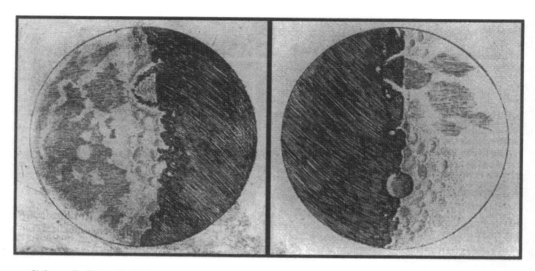

When Galileo Galilei turned his first telescope towards the Moon in 1610, he was astonished to see its surface scarred with large holes. His drawings were only rough sketches however, and are difficult to correlate with features we see today.

An Italian Jesuit priest Giovanni Riccioli produced a map in 1651 based on observations made by his student Francesco Grimaldi, and introduced most of the crater names that are used today. In 1665, Robert Hooke dropped balls of 'lead-shot' into viscous clays, and observed that the impacts made craters resembling those on the Moon. However, it had yet to be realised that meteorites fall from space, so Hooke could not explain why the Moon had been struck by projectiles. In another experiment, Hooke heated dry gypsum powder and noted that when it released water vapour the bubbles rose through the material and left craters as they broached the surface. He eventually came to the conclusion that the lunar craters represented some form of endogenic igneous process,[5] and the implication that the Moon's surface had once been molten was more readily accepted.

A more accurate map was produced in 1775 by the German astronomer Tobias Mayer. In 1779 J.H. Schroter, a magistrate in Lilienthal near Bremen with a passion for astronomy in general and the Moon in particular, established his own observatory equipped with a telescope built by William Herschel. Whilst Schroter did not produce a formal map, over the next thirty years he made hundreds of annotated drawings of specific features under various illumination conditions recording the finer detail.[6] Having discovered a curvilinear depression in the Moon in 1787, which he dubbed a 'rille', he instigated a search, and by 1801 had mapped ten more. The larger picture was recorded by W.G. Lohrmann, a surveyor in Dresden, who published a map on a larger scale than Mayer's. However, it was the book that was published in 1838 by the team of Wilhelm Beer and Johann Madler in Berlin that became the standard reference.[7] Julius Schmidt, a German astronomer who was appointed director of the Athens Observatory in 1858, published a map that amplified the detail on Beer and Madler's map and listed 425 rilles.[8] A few kilometres in width, up to 200 kilometres in length, and 100 to 500 metres deep, rilles were presumed to be fissures in the rocky surface induced by rapid cooling.[9] One of the largest rilles, located on a plateau near the crater Aristarchus, was after Schroter.

Although the lunar craters were believed to be volcanic calderas, R.A. Proctor published a book in 1876 in which he argued that they marked impacts.[10] However, in the absence of any terrestrial counterparts it was difficult to believe that the Moon could have been bombarded so intensely. That same year, Edmund Neison-Neville published a book that described every feature that had so far been named. With W.R. Birt, he founded the Selenographical Society (later transformed into the British Astronomical Association) to coordinate lunar studies.

The invention of photography transformed the study of the Moon. The first daguerreotype image was captured in 1840 by J.W. Draper in New York, but it was just an inch in diameter.[11] A 5-inch image secured by J.A. Whipple was displayed at London's Great Exhibition in 1851, where it impressed members of the public who had never viewed the Moon telescopically. An even better result was obtained the following year by Warren de la Rue using the wet collodian technique. As a single image recorded a wide area in an instant, photography soon became the routine way of studying the Moon, and a photographic atlas of the Moon, displaying the lunar surface under different illumination phases, was published by W.H. Pickering in 1904 using photographs secured by the Harvard College Observatory. However, as the shimmering effect of the Earth's atmosphere blurs photographs, observers wishing to resolve the finer detail did so visually and awaited a brief moment of clear 'seeing'.

In 1919, while undergoing tests, the 100-inch Hooker reflector on Mount Wilson near Los Angeles photographed the Moon, but once the telescope (at that time the world's largest) was commissioned the astronomers were too busy pursuing 'real' astronomy to devote time to the 'dead' Moon in the Earth's backyard. For nearly half a century, therefore, the Moon remained the province of the amateur astronomers and in the 1960s, when the Moon became the focus of professional attention once again, it was claimed by the geologists.

THE PHOTOGEOLOGISTS

In 1892, G.K. Gilbert, an American businessman, interpreted detailed observations of the Moon as signifying that the lunar craters are the result of impacts.[12] He performed experiments by firing lead-shot into clay, and made craters that strongly resembled those on the Moon. But firing lead-shot into clay does not accurately mimic the tremendous energy released during a large impact, so experimental investigations were of limited value. Specifically, it had been believed that the impactor's momentum would stretch a crater into an ellipse. In fact, energy is a scalar quantity so the energy derived from the impactor's momentum is released symmetrically and a 'primary' impact will make a circular crater irrespective of the angle of incidence. The elliptical craters observed on the Moon are 'secondaries' created by *ejecta* from primary events, flying on shallow ballistic trajectories with lower energies. Initially, Gilbert argued that a crater 1 kilometre wide near Winslow

G.K. Gilbert in 1893 was the first to recognise that the ejecta from massive impacts 'sculpted' the surrounding terrain in a radial pattern extending for thousands of kilometres.

in Arizona was an impact crater, but after a magnetic study failed to find the nickel–iron meteorite that he thought should lie beneath its floor, he concluded that the pit must be a maar created by an explosive release of steam. Impact craters, it appeared, were unique to the lunar surface.

Gilbert was the first to recognise that the 'mountain rings' that enclose the 'circular maria' are craters representing catastrophic impacts; it was clear that a 'basin' large enough to contain Mare Imbrium could not be a volcanic caldera. Furthermore, he pointed out a radial pattern in the surrounding terrain which had evidently been gouged by ejecta thrown out on very low-angle trajectories. He traced this Imbrium 'sculpture' all across the near side of the Moon.

In 1935, geologists John Boon and Claude Albritton suggested that several large circular structures on the Earth might mark terrestrial impacts, and they called them astroblemes ('star wounds'). After expressing his belief that lunar craters were made by impacts,[13] Robert Dietz of the Scripps Oceanographic Institute discovered 'shatter cones' in the nickel-rich Sudbury structure in Ontario, Canada, and thereby proved its impact origin.[14] Evidently, the Moon was *not* unique in having been battered by cosmic debris.

Ralph Baldwin drew all the evidence together in a book published in 1949, and effectively proved the case for the Moon's craters being due to impact.[15] A 'crater curve' which plotted the frequency of craters of different sizes displayed the same characteristics as a curve based on a study of holes made during intensive bombing. It was evidence that the dynamic process was the same; volcanism would not have produced craters with such a characteristic distribution of sizes.

After proving that the crater in Arizona which Gilbert had concluded was a maar was really an impact, Gene Shoemaker established the US Geologic Survey's Branch of Astrogeology in 1961. He then proceeded to demonstrate that the history of the lunar surface could be studied using conventional stratigraphic analysis techniques.[16]

This enabled the geologists to 'steal' the Moon from the astronomers. Their first task concerned the relationship of the circular maria to the enclosing rings of mountains.[17] 'Photogeologists' had interpreted what seemed to be small volcanic domes and vents, flow fronts, lava tubes and compressional ridges as indicating that the maria were lava flows. Low-angle illumination established that the maria had regional slopes typically not exceeding 1 in 2,000. This implied that the lava was an extremely low-viscosity variety. The flows had erupted from fissures and literally flooded all the low-lying terrain they could reach, creating vast plains. High points on features which had not been completely subsumed could be seen protruding through the flows. It was possible to see where the flows had embayed surrounding rough terrain. The nearest terrestrial equivalent seemed to be the flood basalts that erupted from fissures and then buried their sources beneath vast plateaux of lava. However, eruptions required the Moon's interior to have once been hot enough to melt rock, and not everyone agreed that this was the case.

After having read Baldwin's book, Harold Urey developed an interest in the origin of the Moon. As far as he was concerned the Moon was a 'pristine' object that was uniform throughout.[18] This became known as the 'cold' Moon hypothesis. Because the Moon had not undergone thermal differentiation, it could not have given rise to volcanism. He believed the 'lava' of the maria was created by a particularly massive impact that melted crustal rock and 'splashed' it across low-lying terrain. He suggested, in effect, that the maria were a side-effect of the formation of the Imbrium basin. However, the crater curves showed that the maria are not all of the same age – an observation that was at odds with their having been splashed out by a single impact.

THE FIELD GEOLOGISTS

The first probe to operate on the lunar surface was the Soviet Union's Luna 9, which returned a panoramic image of the Oceanus Procellarum in January 1966. Over the next few years, in the run up to Apollo, NASA dispatched several Surveyor landers to test the surface's physical and chemical characteristics.

During their brief excursion on the lunar surface Neil Armstrong and Buzz Aldrin sampled Mare Tranquillitatis. Its rock was a dense titanium-rich basalt. Given the Moon's low bulk density, it was clear that it could not be composed entirely of such rock – it was thermally differentiated and the lightweight aluminous silicates had migrated to the surface. This effectively killed Urey's 'cold' Moon hypothesis. The absence of oxidised iron implied that the lava formed in a 'reducing' environment, so oxygen had been absent. The most striking fact was the total absence of hydrous minerals. The lunar basalt was deficient in volatile metals such as sodium. The plagioclase was of the calcic type. Such a low-alkali (that is, sodium-poor) lava would have had an extremely low viscosity, which is why it flowed so far and left so few 'positive-relief' flow fronts. The discovery of basalt lent support to Gerard Kuiper's proposal that the maria were volcanically extruded; however, he had also argued that it would have a 'frothy' texture due to having crystallised in vacuum, which was not the case, and although some of the lavas were vesicular they did not show any peculiar 'vacuum' texture. As Shoemaker had predicted, the actual lava flow was masked by a seriate regolith of weathered debris several metres deep.

The Tranquillitatis lavas accumulated by episodic volcanism over *several hundred million years* – from 3.84 to 3.57 billion years ago. Furthermore, the fact that *two* forms of basalt were present implied either that there had been several chemically distinct sources, or that one source had undergone chemical evolution over time.

What a difference one brief field trip had made. Its 'ground truth' had scythed through the theories without consideration for the professional standing of their proponents. Furthermore, a generation of previously minor players – such as Anthony Turkevich – were thrust into the limelight by being proved correct. Turkevich had led the team that developed an instrument for carriage on a Surveyor lander to determine the composition of the lunar surface by measuring the way that alpha particles were backscattered. Two weeks prior to Apollo 11 he had published a paper in which he predicted that Mare Tranquillitatis would be titanium-enriched basalt.[19]

Not only was there a variation in the lavas of Mare Tranquillitatis, but the lavas that were recovered by Pete Conrad and Al Bean at Apollo 12's sampling site in Procellarum had crystallised a few hundred million years later still. Also, the fact that they were rather less rich in titanium than the Tranquillitatis lavas confirmed that the maria were derived from different sources. In fact, four types of basalt were identified at Procellarum. These could be characterised by their terrestrial equivalents as olivine basalt, pyroxene basalt, ilmenite basalt and feldspathic basalt. The crystallisation dates, however, clustered within a fairly narrow time margin, indicating that several distinct flows had flooded this area in rapid succession. This suggested that the extrusions resulted from partial melting of distinct pockets of rock, locally and near the surface, rather than a succession of flows fed by a single reservoir deep in the mantle.

Having determined that the maria were formed at different times, the next step was to date the excavation of the Imbrium basin as this was the most significant event in the relative time sequence derived from Shoemaker's stratigraphic mapping.

Apollo 14 was sent to sample the Fra Mauro Formation, which was believed to be a thick blanket of crustal rock ejected by this impact. Al Shepard and Ed Mitchell struggled up a ridge to sample rocks on the rim of Cone crater in order to obtain a sample from as deep as possible within this deposit. As expected, there was little mare lava at this site; the rocks were complex breccias made of fragments of intensely shattered rocks in a fine-grained matrix of powdered rock. Analysis was more complicated than at a mare site because the breccias had to be picked apart and their clasts analysed individually. After a thorough analysis, the date of the Imbrium impact was determined as being between 3.85 and 3.82 billion years ago.[20]

What type of lava lay within the Imbrium basin? Dave Scott and Jim Irwin of Apollo 15 were assigned to investigate a site on the basin's eastern rim. It was an adventurous mission involving a steep descent after passing low over the Apennines (some of the tallest mountains on the Moon) in order to land on a patch of mare that had embayed the mountains. The landing site was close to Hadley Rille, which was

The field geologists 49

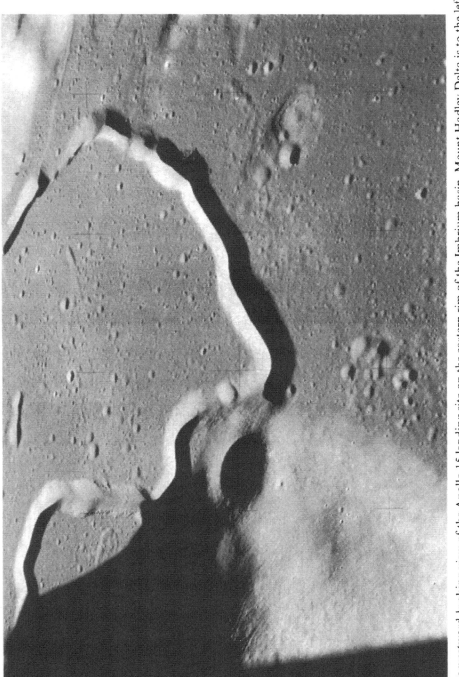

In this westward-looking view of the Apollo 15 landing site on the eastern rim of the Imbrium basin, Mount Hadley Delta is to the left with St George on its flank and the craters of the South Cluster on the mare plain at its base; Hadley Rille cuts across the mouth of the embayed valley; and the hilly-cratered North Complex is right of centre. The landing site was this side of the rille, midway between the North Complex and South Cluster.

The Apollo 15 astronauts gained a unique perspetive of the 1.5 kilometre-wide, 300 metre-deep Hadley Rille.

believed to be a channel through which lava had flowed from a vent in the mountains down to the mare floor.[21] The samples indicated that lava had last flowed down the rille half a billion years after the basin formed. Was the mare a single monolithic slab, or an accumulation of separate flows? Stratification in the wall of the rille indicated a series of fairly thick flows. The primary objective, however, was to find a piece of the original aluminous anorthositic crust. On the assumption that the Apennines were deeply faulted blocks of anorthosite that were forced up by the impact, Scott and Irwin sampled on the flank of Mount Hadley Delta, and on the

GEOLOGIC CROSS-SECTION OF THE APOLLO 15 LANDING SITE

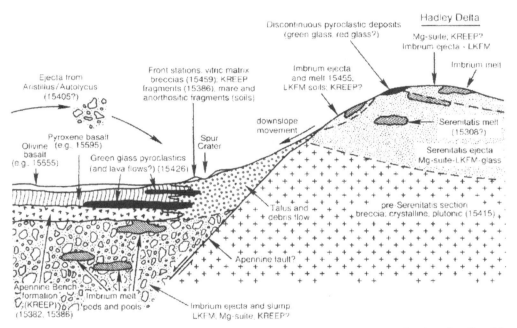

A schematic northwest–southeast geological cross-section through the Apollo 15 landing site on the eastern rim of the Imbrium basin showing the complex geological transition between mare and highlands (modified from Spudis and Ryder, 1985; Swann et al, 1972; Swann, 1986). In the mare (left) post-Imbrian basalt lavas overlie a thick deposit of Imbrium ejecta. In the highland area (Mount Hadley Delta, right) older (pre-Imbrian) ejecta from the Serenitatis basin overlies ancient crust. Numbers refer to specific collected samples that are representative of the various units inferred to be present. (Courtesy the Lunar and Planetary Institute and Cambridge University Press.) *References:* Spudis, P.D. and Ryder, G. (1985) 'Geology and petrology of the Apollo 15 landing site: past, present and future understanding', *EOS Trans. AGU*, vol. 66, p. 721. Swann, G.A. et al. (1972) 'Preliminary geological investigation of the Apollo 15 landing site', in *Apollo 15 Preliminary Science Report*, NASA SP-289, 1972. Swann, G.A. (1986) 'Some observations on the geology of the Apollo 15 landing site', in *Workshop on the Geology and Petrology of the Apollo 15 landing site*, G. Ryder and P.D. Spudis (Eds). Lunar & Planetary Institute Tech. Report 86-03, p. 108.

rim of Spur crater they found a piece of the underlying massif that had been excavated and was literally sitting atop a pedestal awaiting collection.[22,23,24]

On Apollo 16, John Young and Charlie Dule went to sample a patch of gently undulating terrain in the Central Highlands, called the Cayley Formation. Its light-tone had prompted the suggestion that it was a flow of a more aluminous lava with fewer magnesium- and iron-bearing minerals (that is, a less mafic composition) than the dark mare basalts. It embayed the Descartes hills which appeared to be volcanic domes of an even more silicic lava approaching rhyolite.[25] However, the Cayley turned out to be predominantly 'fluidised ejecta' that a basin-forming impact had 'splashed' across the surface.[26] Apollo 17 landed in a valley in the Taurus Mountains on the southeastern rim of the Serenitatis basin to sample what appeared to be 'recent' volcanism – that is, vents believed to have been active in the past billion years. Although Gene Cernan and Jack Schmitt were delighted to find 'orange soil' on the rim of Shorty crater, the fire fountain that vented this pyroclastic had done so 3.64 billion years ago. The crater had been mistaken for a volcanic vent because the impact had excavated the buried deposit.[27]

Although the samples retrieved during the Apollo programme implied that the lunar 'heat engine' had been shut down for about 3.0 billion years, there had been so few landings that the sampling could not be considered comprehensive. In fact, some photogeologists cite evidence that volcanism continued on a local scale in Procellarum.[28] The lavas that almost engulfed the Flamsteed crater (where Surveyor 1 landed) appear recent. The Gruithuisen and Marius Hills in the far western Procellarum are reminiscent of clusters of silicic volcanic domes.[29] However, the ages of these sites will have to remain conjectural until further samples are collected.

In 1959, when the Soviet Union's Lunik 3 flew an extended orbit to photograph what lay beyond the Moon's eastern limb, it revealed a dichotomy between the near side and far side hemispheres.[30] While the maria occupy nearly 30 per cent of the near side, they account for no more than 2 per cent of the far side. For some reason they formed preferentially on the side of the Moon that permanently faces the Earth.

One outstanding question concerned the *thickness* of the maria. Hadley Rille established that they were composed of individual flows, but did they stack up 25 kilometres thick as one interpretation of the seismic evidence suggested? It took many years to gather sufficient data to show that the basalt flows are no more than a veneer on a thick deposit of intensely brecciated rock.[31] In fact, the maria comprise a mere 1 per cent of the lunar crust, and the crust is only 12 per cent of the volume of the Moon. Hence, the 10 million cubic kilometres of lava that have been extruded are volumetrically insignificant.

The largest meteor impact recorded by the Apollo seismometer network was a rock with a mass of about 5 tonnes. The propagation of the shock showed not only that the crust is 60–65 kilometres thick at the Procellarum and Fra Mauro sites, and slightly thicker (75 kilometres) at the Apollo 16 site in the highlands, but also that there is a sharp transition in density – a lunar Mohorovicic Discontinuity. By good fortune this meteor struck the far side, so analysis of the manner in which the energy was propagated through the interior indicated that the Moon has a molten iron-rich core. While it was difficult to estimate the size of the core from such limited data, it was evident that it could not have a radius of more than 500 kilometres. Gravity mapping by Lunar Prospector in the 1990s showed that its radius is about 300 kilometres,[32] accounting for no more than 1 per cent of the Moon's mass.[33] The Apollo seismometers also revealed that as it travels around the Earth the Moon undergoes episodes of 'moonquakes' with deep epicentres, some of which are 800–1,000 kilometres beneath the surface, probably due to 'tidal' magma movements. Remarkably, having expended so much effort to emplace it, NASA switched off the network of scientific instruments in 1977.[34]

The Apollo measurements of the magnetic field on the lunar surface found great variability, suggesting that the field was remanent, within the surface rocks. The overall field is negligible because the crust has been so heavily brecciated, and the rocks tossed around by impacts, that the randomly aligned fields in the individual rocks cancel out. Lunar Orbiter imagery showed a bright 'swirl' pattern near the Reiner crater in Procellarum, and a magnetometer survey found that this site also has an unusually intense (at least in terms of the Moon) magnetic field. There are more extensive swirls on the eastern limb north of Mare Marginis, and on the far side near the Van de Graaff crater. Significantly, these sites are antipodal to the Orientale and Imbrium basins, respectively, and show a 'jumbled' terrain comprising hills and grooves. Initially, it was thought that this was evidence of silicic volcanism, but the antipodal alignments raised the possibility that the surface had been disrupted by the 'focusing' of seismic shock waves which had travelled around the globe.[35,36] Furthermore, it was proposed that a transient magnetic field could have been sustained long enough to have become fossilised in the cooling rock. When Lunar Prospector made its remote-sensing survey its magnetometer and electron reflectometer instrument found that crustal magnetic fields fill most of the zones antipodal to the Imbrium and Serenitatis basins. This supported the theory that basin-forming impacts result in magnetisation of the antipodal crust. The field antipodal to Imbrium is strong enough to deflect the solar wind and form a miniature magnetosphere, magnetosheath and bow shock system several hundred kilometres in diameter.[37] These localised magnetospheres may have shielded the initially light regolith from the darkening effect of the plasma in the solar wind creating the swirl patterns.[38,39]

CRUSTAL FORMATION

When Surveyor 7 landed in the Southern Highlands in the ejecta blanket around the crater Tycho, its alpha-particle backscattering instrument measured the elemental composition of the regolith. The geological community's first reaction upon finding that the site was very rich in aluminium was to reason that it was an alumina-rich basalt, but Gene Shoemaker argued that it was anorthositic rock.[40] Normally, rocks are composed of several major minerals and a range of minor constituents, but anorthosite is at least 90 per cent calcic plagioclase feldspar, which is an aluminous silicate mineral. Some of the tiny fragments in the Tranquillitatis regolith consisted of more than 70 per cent plagioclase. While insufficiently pure to be called

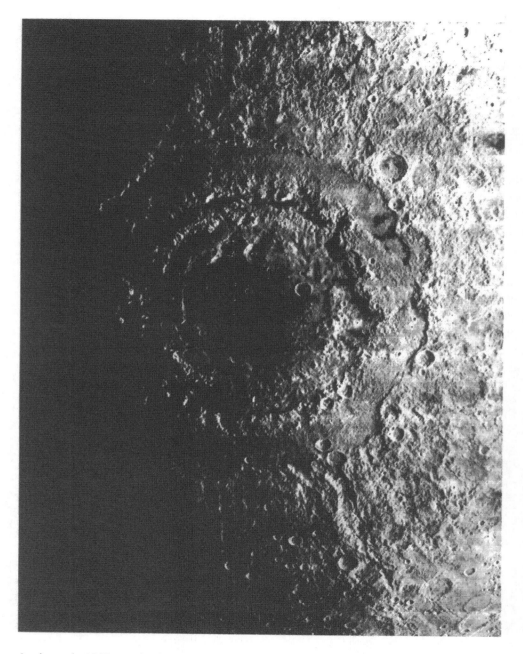

In the early 1960s, geologists realised that the Moon had suffered massive impacts early in its history, but it was not until Lunar Orbiter 4 returned this mosaic in 1966 showing the multiple-ringed Orientale basin on the far side that they fully appreciated the awesome scale of such an impact. The mare-filled basin is ringed by the 650-kilometre-diameter Rook Range and the 1,000-kilometre-diameter Cordillera Range. Note the pools of mare between the rings, and the ejecta deposited on the highlands far beyond the outer structure. All the lunar basins would initially have looked like this.

anorthosite, these were definitely an anorthositic gabbro with pyroxene and olivine as their mafic constituents. Apollo 15 recovered fragments of anorthosite. All the anorthosites from the various sites visited by the Apollo missions proved to be strikingly homogeneous and all were at least 4.4 billion years old, indicating that they crystallised *very* soon after the Moon accreted.

Orbital multispectral surveys, calibrated by the 'ground truth' of the Apollo samples, have shown that anorthosite is the predominant rock type of the highlands and, since the highlands account for almost 70 per cent of the surface, it is actually the dominant rock in the lunar crust. In fact, a study of variations in the strength of the Moon's gravitational field has revealed that the highlands comprise a layer of anorthosite at least 20 kilometres thick.[41] It probably forms the basement beneath the maria, and so may well be global in extent. So how did the Moon's crust become anorthositic?

Anorthosite contains about 35 per cent alumina by mass, which is more than any other kind of rock, so it has a low density of just 2.9 g/cm^3, and as crystals form in a mafic melt they rise and float on the surface. For anorthosite to have formed a 20-kilometres thick crust on a global basis, the entire Moon must have been molten to a depth of several hundred kilometres. This has been dubbed the 'magma ocean'.[42,43]

Although predominant, anorthosite is not ubiquitous because there are also rocks in which the plagioclase is significantly diluted by pyroxene and olivine (in some cases, with these being about half of the rock). If the primary mafic mineral is pyroxene, the rock is called norite, and if the mafic mineral is olivine it is called troctolite. Because these two mafic minerals are rich in magnesium, norite and troctolite are together referred to as the 'magnesian suite' and the rocks that predominate in the highlands are often characterised as 'ANT'.[44]

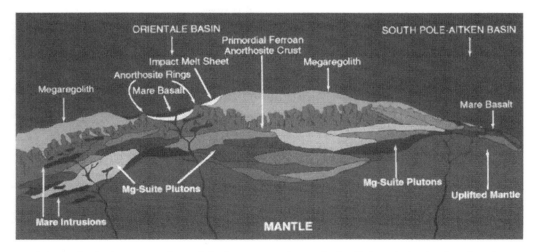

A hypothetical cross-section of the lunar crust beneath the Orientale basin showing the extent to which the impact has influenced the crust. (Courtesy of Paul Spudis of the Lunar and Planetary Institute in Houston.)

Whereas the anorthosite crystallised very early on, the magnesian suite rocks formed over a period of several hundred million years. Their local variation in chemistry indicates that they did *not* crystallise from a homogeneous magma, but from a number of pockets of magma that had undergone varying degrees of chemical differentiation. In fact, it is now believed that they formed the *lower* crust. It is thought that the magma ocean was very convective, and that the early scum of plagioclase was locally augmented by the formation of 'rockbergs' when anorthosite was concentrated above the descending flow of a convection cell. Some of the mafic minerals would have become incorporated into the underside of such blocks. Later, plutons rising from the mantle would have remelted and assimilated the plagioclase and crystallised to produce the troctolite and norite in the lower crust. This process lasted several hundred million years, and produced the local variations in the chemistry of the magnesian suite rocks.

The removal of large sections of the outer crust during basin-forming impacts exposed the deeper magnesian suite, some of which was excavated and distributed all across the Moon by the late bombardment, which was when the thick brecciated 'megaregolith' was formed. The eruptions of aluminous 'non-mare' basalt which began about 4.2 billion years ago indicate that some of the modified plutons broached the surface. The more mafic basalt that erupted for half a billion years after the bombardment dwindled was derived from the mantle, and exploited the deep faults in the basin floors to reach the surface.

The Apollo samples revealed some surprises, one of which was a rock type rich in potassium, some of the rare earth elements (such as uranium and thorium)

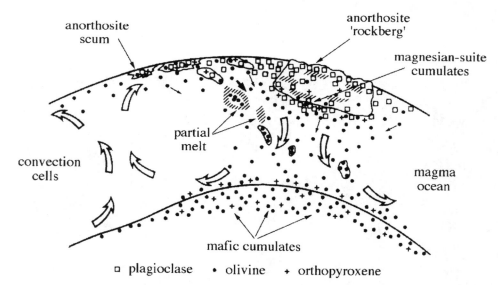

An illustration of how large blocks of anorthositic crust (rockbergs) may have formed above downwelling flows in the convecting magma ocean. Magnesian suite minerals accrete on the undersides of the blocks. (Based on 'Pyroxene stability and the composition of the lunar magma ocean', J. Longhi. Proc. Lunar Planet. Sci. Conf., p. 285, 1978.)

and phosporus. The chemical symbols for these elements gave rise to the name 'KREEP'. Small fragments of a light-toned basalt rich in KREEP recovered by the Apollo 15 crew were revealed to have crystallised 3.84 billion years ago. Remote sensing from orbit revealed that this was common in the Apennine Bench Formation which forms the light-toned plain within the Imbrium basin between the Apennines and the Archimedes Plateau (from which these Apollo 15 samples probably originated). This light-toned plain had been believed to be impact-melt, but it is an aluminous lava rich in KREEP that lined the basin floor almost immediately after its formation. It may well continue as a substrate beneath the dark mafic mare lavas which later flooded the basin.[45]

For a while, the origin of this highly radioactive constituent was a mystery, but once the implications of the magma ocean model were explored it was realised that such a rock would inevitably form. Because the KREEP elements did not 'fit' into the minerals that crystallised from the magma (they are referred to as 'incompatibles' for this reason) they remained in the melt and became ever more concentrated. As the anorthosite rose and the mafics sank, a layer of incompatibles remained sandwiched between, and when the dregs crystallised the result was this highly radioactive rock. This rock could reach the surface either by being excavated by a massive impact or by being remelted by an intrusion of mafic magma and carried to the surface as an exotic constituent.[46] In addition, Lunar Prospector's gamma-ray spectrometer established that the KREEPy radioactives are most concentrated in and around the near side western maria. There is some in the Aitken basin, too. As the largest of the Moon's basins, Aitken excavated the deepest into the magnesian suite of the lower crust.[47] However, possibly because it lies on the far side, it has not flooded with lava.

Considering that Urey had been arguing — reasonably, from his point of view — that the Moon was undifferentiated, finding that it had once been molten to a considerable depth came as a surprise. How could a body as *small* as the Moon have been so hot? The Earth is fully 81 times more massive. Did it, too, once possess a magma ocean?

WHENCE CAME THE MOON?

The Moon's gravity induces tides in the Earth's seas. In open ocean, the Moon's presence raises the sea level by about 75 centimetres. In fact, the Moon also distends the rocky surface, but only by about 11 centimetres.[48] Immanuel Kant realised in 1754 that the friction of tidal water against the shore ought to slow down the Earth's rotation.

Two centuries of astronomical observations established that the Earth's day is lengthening by about 1.8×10^{-3} seconds per century,[49] corresponding to 22 arcseconds per century.[50,51] The length of the Earth's day is therefore increasing by approximately 1 second every 50,000 years and, in response, the Moon is receding at a rate of about 2.5 centimetres per year in order to conserve the overall angular momentum of the two-body system.[52] This has been confirmed utilising the laser reflectors emplaced on the lunar surface by the Apollo astronauts. Early in its

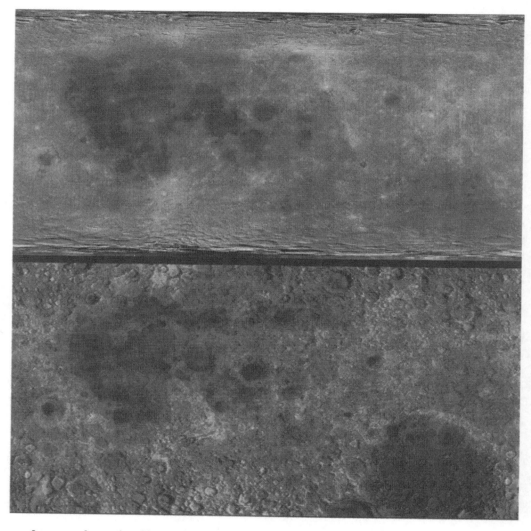

Imagery from the Clementine spacecraft as it orbited the Moon in 1994 has been mosaicked to produce the albedo map (top). The mare-dominated near side is on the left. The Orientale basin is at the extreme left. The 'false-colour' topographic map (bottom) has been arranged to match the albedo map, although without the distortion of perspective in the polar regions. The 'dark' region at the lower right is the massive Aitken basin, which is barely recognisable as an albedo feature. (Courtesy of Maria Zuber of the Massachusetts Institute of Technology.)

history, the Earth rotated more rapidly, and the Moon was much nearer. At the start of the Archean (3.8 billion years ago), the day was a mere 14 hours.[53] A study of sediments laid down during tidal cycles a billion years ago indicated that the day was 18 hours.[54] A study of annual growth rings in fossilised coral that is 400 million years old revealed that there were about 400 days in the year, which meant that each day lasted 21.5 hours.[55]

In 1879, George Darwin (son of the renowned biologist Charles Darwin) realised that at some point in the remote past the period of the Moon's orbit would have corresponded exactly to the time that the Earth took to rotate on its axis. At that time, he calculated, the centres of the two bodies would have been about 20,000 kilometres apart, the period would have been about 4 hours, their rotations would have synchronised and each would have had a bulge pointed towards the other.

In pondering how this situation might have arisen, Darwin posited that the Moon had actually been 'shed' by the Earth. By way of a mechanism, he cited the tidal effects that would have resulted from the Sun's gravity (the Sun raises weak ocean tides on the Earth today). He suggested that a resonance might have developed between the 2-hour solar tides and the period of free oscillation of the progenitor, whereby once the amplitude of this oscillation increased it detached the mass that formed the Moon. Over time, as its tides slowed the Earth's rotation the Moon withdrew to preserve the angular momentum of the system. This has become known as the fission hypothesis.[56,57]

The theory was well received. The Earth would already have been thermally differentiated, so the Moon would have formed from material drawn from the upper mantle. In fact, the bulk density of the Moon is almost exactly the same as that of the Earth's mantle. In 1882 Osmond Fisher, a geophysicist, suggested that the Pacific marked where the Moon had been torn from the Earth; this had split the continental landmass and its fragments are now drifting 'downhill' into the cavity from which the Moon was torn.[58] Unfortunately, if the Moon formed from the Earth's mantle, it ought to be chemically similar, but it is depleted in water, sodium, lead and other volatiles, and is enriched in refractories such as titanium, uranium and some of the rare earth elements.[59,60,61]

Furthermore, there are also dynamical flaws with this proposal. Although the Earth–Moon system has a very high angular momentum, it would not have been sufficient (by a factor of 3, it has been calculated) to have formed a bulge on the scale required to induce fission of a 'blob' of material. Also, as Harold Jeffreys pointed out, the development of a resonance such as Darwin proposed as a mechanism for fission would have been inhibited by internal friction. In fact, modelling has demonstrated that if this had actually happened it would have ejected about 20 per cent of the progenitor's mass, whereas the Moon is only 0.012 of the Earth's mass. Even if a blob had been ejected, it would have been subjected to disruptive tidal stresses as it would have been within the Earth's Roche limit, and its debris would more likely have been reimpacted than placed in a stable orbit.

Another class of theory envisaged that the Moon originated somewhere else, and was captured by the Earth's gravity during a close encounter. Inevitably, over the years, there has been debate over where the Moon could have formed, with suggestions ranging all the way to the frozen wastes beyond Neptune. However, the temperature of the solar nebula from which the planets condensed would have decreased with distance from the new Sun, and if the Moon had condensed far from the Sun it would incorporate many compounds which would not have been able to condense in the hot inner Solar System. The fact that the Moon is so spectacularly deficient in volatiles implies that it formed in the inner Solar System.[62,63]

Although most supporters of the capture theory agreed that this must have taken place very early in the Earth's history (because there is evidence that ocean tides were present as far back as the Archean[64]) there has nevertheless been a proposal that the Moon was formed beyond the Solar System and was placed in orbit around the Earth by benevolent aliens 572 million years ago.[65] However, the evidence for this is rather thin, to say the least.

The problem with the capture theory is the low probability of a close encounter resulting in a stable orbit. For capture to occur, specific dynamical conditions must be established because a fly-by or a collision is more likely. There must be a mechanism for stealing energy from the intruding body as it passes the Earth in order to slow it sufficiently to enter orbit. Furthermore, the initial orbits must be very similar, for otherwise the energy difference will prevent capture. If the Moon originated in the outer Solar System, then it would have to have been in a highly eccentric orbit to have approached the Earth, in which case the encounter would have been too energetic to result in capture.

H. Gerstenkorn proposed that the Moon initially assumed a retrograde orbit. He envisaged a torquing effect by which the orbital inclination increased as the orbit decayed. Upon reaching a minimum – almost at the Earth's Roche limit – its inclination would have been 90 degrees. Because its inclination exceeded 90 degrees when it began to withdraw once more, its orbital motion would have become prograde.[66] This was a rather contrived model for which there was no direct evidence. In 1963 Harold Urey developed a way for the Moon to have been captured directly into a prograde orbit and then receded in response to its tidal braking of the Earth.[67] A variation of the capture theory argued that an intruder strayed so close to the Earth that it was tidally disrupted, leaving debris in orbit which subsequently accreted to form the Moon.[68] This built upon an early idea by G.K. Gilbert that the Moon had accreted from debris swept up by the Earth.

Another theory posited that the Earth and the Moon condensed from the solar nebula in the same vicinity and soon established a gravitational embrace, in effect forming a double planet.[69] This had the advantage of establishing an encounter geometry with sufficiently low energy for capture to have resulted. However, the subtle difference in chemistry between the two bodies was difficult to account for if they condensed in close proximity. Indeed, there were reasons to think that double planets would *not* condense from the solar nebula: by the time that an object had accreted to a certain mass it would have swept its immediate environs free of raw material, and this would have inhibited the development of a companion. The large angular momentum of the Earth–Moon system was tricky to account for too, as indeed was the fact that the Moon orbits the Earth in a plane that is inclined to that in which the system orbits the Sun. Despite its problems, Kuiper (Urey's eternal rival in all things pertaining to the Moon[70]) was sure that the double planet theory would prove to hold an essence of truth.

As NASA prepared to send the first Apollo crew to the Moon, it was generally recognised that none of the contending theories of the Moon's origin was satisfactory, but it was thought that the issue would be resolved as soon as lunar samples were analysed. There was therefore considerable surprise when the samples seemed to rule against *all* of the competing theories.[71] Frustratingly, Urey's quip that the scientists appeared to have proved that the Moon could not exist, had come true.[72]

For over a decade, the scientific teams mulled over their analyses, and attempted to figure out what the data really meant.[73] The key to the mystery proved to be the role of impacts early in the Solar System. At least a dozen protoplanets may have condensed close to the Sun. The individual objects might have been as large as the planet Mars. But collisions towards the end of the accretionary period diminished their number, as the larger protoplanets were struck by their smaller relatives.

The basis for what became known as the giant impact theory had been laid a decade earlier by W.K. Hartmann and by A.G.W Cameron,[74,75] but the results were announced at the Lunar Science Conference held in Kona, Hawaii, in 1984[76] and followed up in published works and a number of papers in journals.[77,78,79,80,81] Specifically, the theory envisaged a Mars-sized object (which was itself thermally differentiated into a core and a mantle) striking the Earth a glancing blow which, in fact, had been almost a near-miss. Much of their mantle rock was vaporised, and much of the rest was ejected. Most of the impactor's core coalesced with the Earth's core, but some of its iron accompanied the debris, which first formed a disk around the Earth and then accreted to create the Moon. This was a unifying idea that drew in aspects of the collisional theory, the ejection of debris (the fission theory) which, once captured, condensed nearby to form a double planet system. It incorporated many of the 'acceptable' aspects of earlier theories, but did not inherit their flaws. For example, the high angular momentum of the resulting double planet system is explained as a consequence of the impact geometry; as is the inclined orbit. The bulk density of the Moon is the same as that of the mantle because it was derived from mantle rock, and all the volatiles were boiled out of the ejecta by the tremendous heat of the collision. In these circumstances, the accretion of the Moon would have been a *rapid* process taking no more than a few tens of thousands of years, and the energy that this liberated would have kept the surface molten, thereby accounting for the magma ocean.[82] Furthermore, the superheated ejecta would have driven off the volatile sodium causing the scum of aluminous silicate to be of the refractory calcic type, thus accounting for the primitive crust being anorthositic.

While the hand-waving argument was greeted with some scepticism, computer modelling demonstrated that the giant impact theory was physically possible a fact which in itself marked a 'first' for theories of lunar origin.[83,84] To all intents and purposes, therefore, the issue of how the Earth acquired a large satellite is considered to have been resolved.[85,86]

In retrospect, the essence of the idea could be traced back to 1946, when R.A. Daly published a paper proposing that the interloper had been only partially disrupted by the impact: most of it had survived intact, had entered orbit and had promptly been struck by the debris that had been torn free – and this, he had argued, explained how the ancient population of lunar craters had formed.[87,88]

The recent discovery that by 4.4 billion years ago – that is, within 100 million years of its accretion – the Earth had cooled sufficiently to form a crust and a hydrosphere meant that the Moon must have formed a considerable time prior to this.[89] Calculations indicate that the phase of protoplanetary collisions towards the end of the accretionary process should have lasted no more than 25 million years.[90] In fact, simulations suggest that the Earth may have been hit *several* times during this period.[91,92]

Like the other theories of how the Moon became a satellite of the Earth, we are left wondering why there are no large moons circling the other planets of the inner Solar System. The circumstances envisaged by the giant impact theory make it likely that other planets *did* suffer similarly large impacts. Indeed, this may well explain why Mercury has such an anomalously high density.[93] If Venus had suffered a collision which countered rather than augmented its spin and tipped it over, this could explain that planet's unusual rotation.

THE WHOLE STORY

Stratigraphic analysis had provided a relative timescale for the events that characterised the lunar surface and the Apollo moon rocks gave the absolute dates required to calibrate this timescale. The oldest samples, the anorthosites, were about as old as anything on the Moon could possibly be, but most of the radiometric dates for the other samples fell in the age range 4.0 to 3.2 billion years.[94]

So, while geologists were searching for terrestrial rocks exceeding 3.5 billion years*, their lunar counterparts were discovering that almost all rocks on the Moon are at least that old. The surfaces of these two objects tell complementary stories. The Moon records the violent events of the early Solar System. In effect, the intensely battered lunar surface shows what the Earth must have looked like at the start of the Archean.

WHAT NEXT?

One of the most remarkable discoveries about the Moon is that there appears to be water ice in the deeply shadowed craters at the poles. This was hinted at by the Clementine spacecraft in 1994 – by a single observation whose interpretation was intensely debated[95,96] – and then confirmed a few years later by Lunar Prospector's remote-sensing.[97] The next logical step is to dispatch a robotic lander to confirm that there is indeed ice in the regolith.[98]

Beyond the scientifically intriguing aspects, deposits of ice on the Moon would have many practical implications for human space exploration. Since there is no other source of water in near-Earth space, and as shipping water into orbit would be extremely expensive (of the order of $10,000 per kg), lunar water could do much to facilitate the exploration of the Solar System, because it could provide oxygen and hydrogen to serve as rocket fuel.

---

*The search for the Earth's oldest rocks is related in Chapter 7.

What next? 63

John Young stands on the Cayley Plain of the Apollo 16 landing site in 1972. By this point in the programme, lunar explorers were admirably equipped for field geology. In addition to the rack of tools on the rear of the Rover, each man had personal kit including a cuff-checklist, a shin-pocket for a hammer (in this case absent), a Hasselblad camera, a sample-bag dispenser (alongside the camera) and a carrier on the side of his life support backpack.

Paul Spudis, who led the Clementine team who made the original discovery, has reflected that the combination of an uninterruptable power supply, a fairly benign thermal environment, abundant mineral resources, stable communications with Earth and a vast 'reservoir' of water makes the 'Mountain of Eternal Light' near the Moon's south pole 'the most valuable piece of extraterrestrial real estate in the Solar System'.[99]

NOTES

1. The name 'telescope' (which means 'to see at a distance') was not coined until 1612 when Ionnes Dimisiani, a Greek mathematician, suggested it to an Italian cardinal.
2. Galileo Galilei's observations were published in *Sidereus Nuncius* (*Sidereal Messenger* or *Starry Messenger*) in 1610.
3. *Mapping and naming the Moon: a history of lunar cartography and nomenclature* E.A. Whitaker. Cambridge University Press, p. 19, 1999.
4. *Selenographica*, J. Hevelius. Danzig, 1647.
5. *Micrographia*, R. Hooke, 1665; reproduced by Dover Publications in 1961.
6. J.H. Schroter's drawings of the Moon were published in *Selenotopographische Fragmente*, which comprised two volumes published in Gottingen in 1791 and 1802.
7. *Der Mond*, W. Beer and J. Madler. Berlin, 1838.
8. *Uber rillen auf dem Monde*, J. Schmidt. 1866.
9. *A popular history of astronomy*, A.M. Clerke. Black Co., p. 308, Edinburgh: 1885.
10. *The Moon*, R.A. Proctor. 1876.
11. The 'daguerreotype' photographic process was devised by Frenchmen L.J.M. Daguerre and J.N. Niepce in the 1820s, but they did not announce it until 1839.
12. 'The Moon's face: a study of the origin of its surface features', G.K. Gilbert. *Bull. Phil. Soc. Washington*, vol. 12, p. 241, 1893.
13. 'The meteoritic impact origin of the Moon's surface features', R.S. Dietz. *J. Geology*, vol. 54, p. 359, 1946.
14. 'Sudbury structure as an astrobleme', R.S. Dietz. *J. Geology*, vol. 72, p. 412, 1964.
15. *The face of the Moon*, R.B. Baldwin. University of Chicago Press, 1949.
16. 'Stratigraphic basis for a lunar time scale', E.M. Shoemaker and R.J. Hackman. In *The Moon*, Z. Kopal and Z.K. Mikailov (Eds). Academic Press, p. 289, 1962.
17. 'Lunar maria and circular basins: a review', D. Stuart-Alexander and K.A. Howard. *Icarus*, vol. 12, p. 440, 1970.
18. 'Origin and history of the Moon', H.C. Urey. In *Physics and astronomy of the Moon*, Z. Kopal (Ed.). Academic Press, p. 481, 1962.
19. 'Chemical composition of the lunar surface in Mare Tranquillitatis', A.L. Turkevich, E.J. Franzgrote and J.H. Patterson. *Science*, vol. 165, p. 277, 1969.
20. 'Rb-Sr ages of igneous rocks from the Apollo 14 mission and the age of the Fra Mauro Formation', D.A. Papanastassiou and G.J. Wasserburg. *Earth Planet. Sci.*, vol. 12, p. 36, 1971.
21. 'On the origin of lunar sinuous rilles', V.R. Oberbeck, W.L. Quaide and R. Greeley. *Modern Geol.*, vol. 1, p. 75, 1969.
22. *Exploring the Moon: the Apollo expeditions*, D.M. Harland. Springer-Praxis, p. 132, 1999.
23. 'Geology of Hadley Rille', K.A. Howard, J.W. Head and G.A. Swann. *Proc. Lunar Planet. Sci. Conf.*, p. 1, 1972.

24 'The formation of Hadley Rille and implications for the geology of the Apollo 15 region', P.D. Spudis, G.A. Swann and R. Greeley. *Proc. Lunar Planet. Sci. Conf.*, p. 243, 1988.
25 'Descartes region: evidence for Copernican-age volcanism', J.W. Head and A.F. Goetz. *J. Geophys. Res.*, vol. 77, p. 1368, 1972.
26 *Geology of the Apollo 16 area, central lunar highlands*, G.E. Ulrich, C.A. Hodges and W.R. Muehlberger. USGS Professional Paper No. 1048, 1981.
27 *The geologic investigation of the Taurus-Littrow valley, Apollo 17 landing site*, E.W. Wolfe, N.G. Bailey, B.K. Lucchitta, W.R. Muehlberger, D.H. Scott, R.L. Sutton and H.G. Wilshire. USGS Professional Paper No. 1080, 1981.
28 'Beginning and end of lunar mare volcanism', P.H. Schultz and P.D. Spudis. *Nature*, vol. 302, p. 233, 1983.
29 'Imbrian-age highland volcanism on the Moon: the Gruithuisen and Mairan domes', J.W. Head and T.B. McCord. *Science*, vol. 199, p. 1433, 1978.
30 *An atlas of the Moon's far side: the Lunik III reconnaissance*, N.P. Barabashov, A.A. Mikhailov and N.Y. Lipskiy. Reprinted by Sky Publishing, 1961.
31 'Thickness of mare flow fronts' A.W. Gifford and F. El-Baz. *Proc. Lunar Planet. Sci. Conf.*, p. 382, 1978 (abstract); *Moon and Planets*, vol. 24, p. 391, 1981 (paper).
32 'Lunar Prospector overview', A.B. Binder. *Science*, vol. 281, p. 1475, 1998.
33 'Improved gravity field of the Moon from Lunar Prospector', A.S. Konopliv, A.B. Binder, L.L. Hood, A.B. Kucinskas, W.L. Sjogren and J.G. Williams. *Science*, vol. 281, p. 1476, 1998.
34 *ALSEP termination report*, J.R. Bates, W.W. Lauderdale and H. Kernaghan. NASA Reference Publication No. 1036, 1979.
35 *The once and future Moon*, P.D. Spudis. Smithsonian Institute Press, p. 34, 1966.
36 P.H. Schultz and D.E. Gault. *Moon*, vol. 12, p. 159, 1975.
37 'Lunar surface magnetic fields and their interaction with the solar wind: results from Lunar Prospector', R.P. Linn, D.L. Mitchell, D.W. Curtis, K.A. Anderson, C.W. Carlson, J. McFadden, M.H. Acuna, L.L Hood and A.B. Binder. *Science*, vol. 281, p. 1480, 1998.
38 'Lunar nearside magnetic anomalies', L.L. Hood, P.J. Coleman and D.E. Wilhelms. *Science*, vol. 205, p. 53, 1979.
39 L.L. Hood and G. Schubert. *Science*, vol. 208, p. 49, 1980.
40 'Lunar theory and processes', D.E Gault, J.B. Adams, R.J. Collins, T. Gold, G.P. Kuiper, H. Masursky, J.A. O'Keefe, R.A. Phinney and E.M. Shoemaker. In *Surveyor Program Results*, NASA SP-184, p. 351, 1969.
41 *The once and future Moon*, P.D. Spudis. Smithsonian Institute Press, p. 149, 1996.
42 'Lunar anorthosite and a geophysical model of the Moon', J.A. Wood, J.S. Dickey, U.B. Marvin and B.N. Powell. *Proc. Lunar Sci. Conf.*, p. 965, 1970.
43 'Fragments of terra rocks in the Apollo 12 soil samples and a structural model of the Moon', J.A. Wood. *Icarus*, vol. 16, p. 494, 1972.
44 'Mineralogy, petrology and chemistry of ANT-suite rocks from the lunar highlands'. M. Prinz and K. Keil. *Phys. Chem. Earth*, vol. 10, p. 215, 1977.
45 *The once and future Moon*, P.D. Spudis. Smithsonian Institute Press, p. 150, 1966.
46 It is important to realise that KREEP is not a mineral as such, it is more of a chemical pollutant.
47 'Global elemental maps of the Moon: the Lunar Prospector gamma ray spectrometer', D.J. Lawrence, W.C. Feldman, B.L. Barraclough, A.B. Binder, R.C. Elphic, S. Maurice and D.R. Thomsen. *Science*, vol. 281, p. 1484, 1998.
48 W. Thomson. *Phil. Trans.*, vol. 153, p. 573, 1862.

49 'The Earth-Moon system', Z. Kopal. In *Understanding the Earth: a reader in the Earth sciences*, I.G. Gass, P.J. Smith and R.C.L. Wilson (Eds). Artemis Press, (second edition), p. 111, 1971.
50 'The age of the Earth', G.B. Dalrymple. Stanford University Press, p. 48, 1991.
51 In fact the modern value is 24 ± 4 arcseconds per century. See: 'Tidal recession of the Moon from ancient and modern data', F.R. Stephenson. *J. Brit. Astron. Assoc.*, vol. 91, p. 136, 1981.
52 For an account of early research into tidal braking see *A popular history of astronomy*, A.M. Clerke. Black Co., p. 314, 1885.
53 'A sensitivity study of changes in Earth's rotation rate with an atmospheric general circulation model', G.S. Jenkins. *Global and Planetary Change*, vol. 11, p. 141, 1996.
54 See news item on, p. 18 of *Earth Mag.*, October 1996.
55 'Coral growth and geochronology', J.W. Wells. *Nature*, vol. 197, p. 948, 1963.
56 'On the precession of a viscous spheroid on the remote history of the Earth', G.H. Darwin. *Phil. Trans. Roy. Soc.*, vol. 170, p. 447, 1879.
57 *The tides and kindred phenomena in the Solar System*, G.H. Darwin. Freeman & Co., 1989.
58 'On the physical cause of the ocean basins', O. Fisher. *Nature*, vol. 25, p. 242, 1882.
59 The Moon has more iron in the ferrous form (FeO) than does the Earth's upper mantle. Of course, the Earth has more iron overall than does the Moon, because it has an iron core – indeed, the core is far larger than the Moon itself.
60 'Structure and evolution of the Moon', S.R. Taylor. *Nature*, vol. 281, p. 105, 1979.
61 'The lunar core and the origin of the Moon', H.E. Newson. *EOS*, vol. 65, p. 369, 1984.
62 'Genetic relations between the Moon and meteorites', R.N. Clayton and T.K. Mayeda. *Proc. Lunar Sci. Conf.*, p. 1761, 1975.
63 'Isotopic anomalies in the early Solar System', R.N. Clayton. *Ann. Rev. Nuclear Particle Sci.*, vol. 28, p. 501, 1978.
64 The Moon was evidently present 3.5 billion years ago because tidal currents left characteristic patterns in the sediments on Archean beaches. Stromatolites grew in the intertidal zone which, because the Moon was closer, was larger than it is now.
65 *Dark Moon – Apollo and the whistle-blowers*, M. Bennett and D.S. Percy. Aulis Publishers, p. 450, 1999.
66 H. Gerstenkorn. *Z. Astrophys.*, vol. 36, p. 245, 1955.
67 'The capture hypothesis of the origin of the Moon', H.C. Urey. In '*The Earth-Moon system*', B.G. Marsden and A.G.W. Cameron (Eds). Plenum, p. 210, 1966.
68 J.A. Wood and H.E. Mitler.
69 'The Moon's face: a study of the origin of its features', G.K. Gilbert. *Bull. Phil. Soc. Washington*, vol. 12, p. 241, 1893.
70 'Nickel for your thoughts: Urey and the origin of the Moon', S.G. Brush. *Science*, vol. 217, p. 891, 1982.
71 *Lunar science: a post-Apollo view*, S.R. Taylor. Pergamon Press, 1975.
72 *To a rocky Moon: a geologist's history of lunar exploration*, D.E. Wilhelms. University of Arizona Press, p. 352, 1993.
73 'A history of modern selenogony: theoretical origins of the Moon from capture to crash', S.G. Brush. *Space Sci. Rev.*, vol. 47, p. 211, 1988.
74 'Satellite-sized planetesimals and lunar origins', W.K. Hartmann and D.R. Davis. *Icarus*, vol. 24, p. 504, 1975.
75 'On the origin of the Moon', A.G.W. Cameron and W. Ward. *Proc. Lunar Sci. Conf.*, p. 120, 1976.

76 'Lunar origin meeting favours impact theory', G.J. Taylor. *Geotimes*, vol. 30, no.4, p. 16, 1985.
77 'The origin of the Moon', S.R. Taylor. *Am. Scientist*, p. 469, September-October 1984.
78 'The origin of the Moon', A.P. Boss. *Science*, vol. 231, p. 341, 1986.
79 *Origin of the Moon*, W.K. Hartmann, R.J. Phillips and G.J. Taylor (Eds). Lunar and Planetary Institute, 1986.
80 'Origin of the Moon: the collisional hypothesis', D.J. Stevenson. *Ann. Rev. Earth Planet. Sci.*, vol. 15, p. 271, 1987.
81 'Geochemical implications of the formation of the Moon by a single giant impact', H.E. Newsom and S.R. Taylor. *Nature*, vol. 338, p. 29, 1989.
82 'Moon over Mauna Loa: a review of hypotheses of formation of Earth's Moon', J. Wood. In *Origin of the Moon*, W.K. Hartmann, R.J. Phillips and G.J. Taylor (Eds). Lunar and Planetary Institute, p. 17, 1986.
83 'A preliminary numerical study of colliding planets', M.E. Kipp and H.J. Melosh. In *Origin of the Moon*, W.K. Hartmann, R.J. Phillips and G.J. Taylor (Eds). Lunar and Planetary Institute, p. 643, 1986.
84 'Impact model featured on film', L.L. Hood. *Geotimes*, vol. 31, p. 15, 1986.
85 *Lunar sourcebook: a user's guide to the Moon*, G.H. Heiken, D.T. Vaniman and B.M. French (Eds). Lunar and Planetary Institute and Cambridge University Press, 1991.
86 'The scientific legacy of Apollo', G.J. Taylor. *Scient. Am.*, vol. 271, p. 26, July 1994.
87 'Origin of the Moon and its topography', R.A. Daly. *Proc. Am. Phil. Soc.*, vol. 90, p. 104, 1946.
88 'Historical review of a long-overlooked paper by R.A. Daly concerning the origin and early history of the Moon', R.B. Baldwin and D.E. Wilhelms. *J. Geophys. Res.*, vol. 97, p. 3837, 1992.
89 'In the beginning', A.N. Halliday. *Nature*, vol. 409, p. 144, 2001.
90 'Terrestrial accretion rates and the origin of the Moon', A.N. Halliday. *Earth Planet. Sci. Lett.*, vol. 176, p. 17, 2000.
91 'Occurrence of giant impacts during the growth of the terrestrial planets', G.W. Wetherill. *Science*, vol. 228, p. 877, 1985.
92 'Accumulation of the terrestrial planets and implications concerning lunar origin', G.W. Wetherill. In *Origin of the Moon*, W.K. Hartmann, R.J. Phillips and G.J. Taylor (Eds). Lunar and Planetary Institute, 1986.
93 'Accumulation of Mercury from planetesimals', G.W. Wetherill. In *Mercury*, University of Arizona Press, p. 670, 1988.
94 *Geology of the Moon*, J.E. Guest and R. Greeley. Wykeham, 1977.
95 The initial results of the Clementine mission were published together in *Science*, vol. 266, pp. 1835–1862, 1994.
96 S. Nozette *et al. Science*, vol. 274, p. 1495, 1996.
97 'Fluxes of fast and epithermal neutrons from Lunar Prospector: evidence for water ice at the lunar poles', W.C. Feldman, S. Maurice, A.B. Binder, B.L. Barraclough, R.C. Elphic and D.J. Lawrence. *Science*, vol. 281, p. 1496, 1988.
98 A European proposal to do this fell by the wayside, and NASA has no immediate plans to follow up on the discovery of water at the lunar poles.
99 "I made that statement at the Pentagon press conference where we announced the Clementine discovery of ice on the Moon (December, 1996). I've since used it in talks and presentations I've given. The media seemed to pick it up (they LOVE sound-bites)" – Paul Spudis (private communication).

4

The Red Planet

EARLY STUDIES

Soon after discovering in 1610 that Jupiter has a system of satellites, Galileo Galilei took a look at Mars, but his small telescope could not resolve detail on the planetary disk. Francesco Fontana of Naples was the first to glimpse dark markings on the ochre disk but they were not very distinct. Christiaan Huygens was the first to map the surface markings. Prominent on his 1659 sketch was a dark triangular feature.[1] By monitoring its travel across the planet's disk he was able to infer that Mars rotates on its axis in a little over 24 hours. Later, R.A. Proctor was able to measure the rotational period with great accuracy by calculating the number of revolutions that had occurred between two observations some 200 years apart. An error of 0.1 second in his period would have resulted in a difference of two hours over that interval, so these very early observations were of real value.[2]

In 1666 G.D. Cassini saw that the polar regions of Mars were white, and inferred that they were snow-covered. In 1719 G. Maraldi reported that the caps waxed and waned in size. William Herschel noted that the polar caps grew and diminished in phase with winter in each hemisphere. The southern cap grows as far north as 45 degrees of latitude in midwinter, and covers an area of 10 million square kilometres. The northern cap never reaches this size. The range of seasonal temperature variation is greater in the southern hemisphere because the planet's orbit is distinctly elliptical, the planet's rotational axis is tilted and the northern winter occurs near perihelion, whereas the southern winter occurs near aphelion. Furthermore, as Maraldi pointed out, the caps are not centred precisely on the axial poles.[3] Continuing studies identified a pattern in the way that the caps shrank in spring. Along the fringe of the southern cap as it receded, two dark rifts developed and a small patch became isolated. After observing this repeatedly Giovanni Schiaparelli announced in 1877 that the patch was a high plateau that retained its snow cover for some time after it had cleared from the adjacent lower-lying terrain.[4]

As long ago as 1785, William Herschel considered that

> Mars is not without considerable atmosphere, for besides the permanent spots on its surface I have noticed occasional changes of partial bright belts, and also once a darkish one in a pretty high latitude. And we can hardly ascribe such alterations to any other cause than the variable disposition of clouds and vapours floating in the atmosphere of the planet.[5]

He also opined that its inhabitants "probably enjoy a situation in many respects similar to our own". Mars, however, is somewhat smaller than the Earth, and as its gravity is weaker, lightweight molecules will tend to leak into space, diminishing the atmosphere.

How could the air pressure on Mars be measured? One technique was to estimate the 'air mass' as a function of angle from the centre of the disk. A site on the meridian will be viewed through a shorter 'column' than one near the limb. In principle, the manner in which a bright patch varied in brightness as the planet rotated would permit the atmospheric density profile to be measured, and the surface pressure calculated. Of course, the method was subjective, so a variety of results were derived. Nevertheless, by 1960 the results from a variety of techniques had produced general agreement that the surface pressure was 80 to 120 millibars; 10 per cent of that at sea level on Earth. Evidently, the surface pressure of Mars was equivalent to that of our stratosphere at an altitude of 18 kilometres.

Identifying the composition of the Martian atmosphere involved spectroscopic analysis of how it reflected sunlight. However, the fact that the light carrying the signature of the Martian atmosphere passed through the Earth's atmosphere on its way to the instrument meant that the resulting spectrum was a combination of the two. Several ingeneous ways were devised to try to distinguish between them.

In 1867 Jules Janssen set up a spectroscopic telescope on the 9,800-foot summit of Mount Etna on the island of Sicily in the hope that observing from altitude would minimise the local absorption. After analysing reflected moonlight to establish the residual absorption lines from the Earth's atmosphere, he examined Mars and concluded that its atmosphere was quite rich in water vapour. When this was 'confirmed' by William Huggins in the UK and H.C. Vogel in Germany, it reinforced the view that the polar caps were seasonal snow fields that surrounded permanent cores of water ice.

In the first phase of study of the Red Planet telescopes were able to show only the overall albedo features — namely, an ochre planet with some dark areas and white polar caps. By about 1830, however, the art of telescope-building had been refined and genuine cartographic work began in earnest. Maps were produced by William Beer and J.H. Madler in 1840, Frederick Kaiser in 1864, Camille Flammarion in 1876 and Nathaniel Green in 1877 (to name just a few). Apart from the polar regions, 'mapping' involved distinguishing dark features on the ochre background. Individual observers drew them differently and named them variously,[6] but the classical nomenclature developed by Schiaparelli in the 1880s was promptly accepted as standard.

The dark features were initially assumed to be oceans, but in 1860 Emmanuel Liais of the Rio de Janeiro Observatory in Brazil suggested that they may be dense

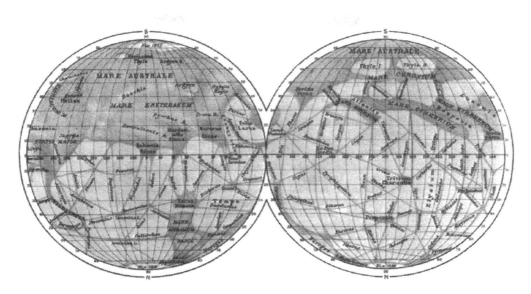

In 1888, after many years of observation, G.V. Schiaparelli published a map of Mars with a large number of dark narrow straight lines criss-crossing the ochre areas. Prior to the 'Space Age', astronomers drew the planets with the disk inverted, so south is at the top.

vegetation. Also, in 1863 Schiaparelli realised that the absence of glint from the Sun indicated that they were not bodies of open water. Further observations highlighted albedo evolutions that seemed to be seasonal growth following one or other of the polar caps infusing water vapour into the atmosphere, which the prevailing winds carried to lower latitudes. Some observers even identified a 'wave of darkening' which radiated away from the polar areas. Indeed, after observing the southern hemisphere in the 1939 opposition, Gerard de Vaucouleurs in France was sure that this was a systematic effect.

> The greater the distance of the dark areas from the south polar cap, the later the darkening begins. We see, in short, a wave of darkening proceeding from the south polar region; a wave which begins at the end of winter at the border of the cap, near the latitude 60 degrees south; reaches the equator by the middle of spring; ... and extends, by the end of the southern spring, as far as latitude 40 degrees north.[7]

He reported, with typical precision, that the wave advanced at a rate of about 46 kilometres per day. Other people noted seasonal albedo variations, but considered such a systematic 'wave' to be illusory.

For those who did not favour vegetation, in 1912 Svante Arrhenius in Sweden suggested that the seasonal albedo variations were due to hygroscopic salts in the soil darkening as they absorbed water vapour with the onset of spring, and D.B. McLaughlin of the University of Michigan proposed in 1954 that Mars was intensely volcanic and the dark areas were ash from continuous eruptions being blown around by the seasonal winds.[8,9] Even though Tsuneo Saheki in Japan had observed a 'flare' on the terminator near Tithonius Lacus in 1951, which he considered to have been a massive volcanic explosion,[10] this theory attracted little support. De Vaucouleurs was insistent that the lack of shadows on the terminator meant that the planet must be essentially flat: there could not be any terrestrial-style chains of mountains and any isolated peaks could scarcely exceed a few thousand feet in height.

A map drawn by Schiaparelli in 1888 marked a milestone in Martian cartography because he drew curvilinear streaks which he described as 'canali', meaning channels. Percival Lowell in America was intrigued and in 1894 produced a map with an elaborate network of lines. In a series of books, he argued that these lines were *canals* built by the inhabitants of Mars in an attempt to irrigate a dying planet by channelling water from the poles to the populated equatorial areas.[11,12,13] At that time, the technologically advanced nations on Earth were in the process of digging networks of canals to form a transportation system, and since it was thought that Mars was 'older' than the Earth it seemed natural that a dying Martian civilisation would be capable of building canals on a *planetary* scale. However, many of Lowell's contemporaries (notably E.E. Bernard) saw no sign of linear features, with the result that by the favourable opposition of 1909 serious observers had dismissed the idea of artefacts produced by a dying civilisation.

In 1933 the 100-inch Hooker reflector at the Mount Wilson Observatory near Los Angeles was used by W.S. Adams and T. Dunham to measure the wavelengths of absorption lines precisely so that over time the lines imprinted by the Martian

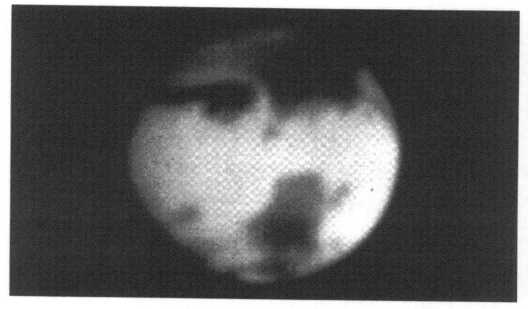

Mars is a difficult object to photograph even with the 200-inch Hale telescope of the Palomar Observatory due to shimmering effects in the Earth's atmosphere. However, during moments of clear 'seeing', visual observers were able to resolve considerable surface detail.

atmosphere would first appear shifted shortward and then longward of their terrestrial counterparts as the planets moved with respect to one another as they orbited the Sun. Among their results, they set an upper limit to Mars's atmospheric oxygen content: it forms 21 per cent of the Earth's atmosphere but is negligible in the Martian atmosphere. At first, this was surprising. Where had all the oxygen gone? Rupert Wildt at Princeton suggested in 1934 that the oxygen had been bound up as iron oxide (that is, rust) in the rocks of the ochre areas. In 1947 Gerard Kuiper reported detecting carbon dioxide at low concentration in the Martian atmosphere. In the early 1960s, Audouin Dollfus, working spectroscopically high in the Swiss Alps in order to weaken terrestrial effects, reported that there was significantly less water vapour than Janssen's result had indicated. A photographic study at Mount Wilson using emulsions sensitive to the infrared showed that the 'precipitable' water on Mars (that is, the thickness of the layer it would form if it all condensed onto the surface) was a mere 14 micrometres.[14] Evidently, Mars was drier than the most arid terrestrial desert. Even so, liquid water would be able to exist at 90 millibars surface pressure as long as the temperature did not exceed 35 °C. Because the pressure would be greatest in the low-lying areas, it was concluded that the dark areas were ancient sea beds which, since they would act as moisture traps for the seasonal infusion of water vapour, were able to support hardy forms of vegetation such as lichen.

Although the Martian polar caps cover a vast area, the discovery that there is so little water vapour in the atmosphere led to the realisation that the caps must be extremely *thin* coverings. At the turn of the 20th century, Cowper Ranyard and Johnstone Stoney had suggested that the caps were frozen carbon dioxide instead of water ice,[15] but this was 'relegated to the limbo of forgotten things' as Patrick Moore has delightfully expressed it.[16] In 1949 Kuiper was adamant that the caps were not carbon dioxide, and de Vaucouleurs calculated in 1954 that they could be no more than a few centimetres thick. Others likened them to a coating of hoar frost rather than to a field of snow. At 90 millibars surface pressure at the onset of winter, the water vapour would freeze directly from the atmosphere and the snow/frost would sublime to vapour in the spring, so there would be no liquid water. However, a spectroscopic analysis in 1964 cut the upper limit of the surface pressure to 25 millibars,[17] and the *partial pressure* of carbon dioxide was estimated at about 4.2 millibars.[18]

As the first spacecraft approached the planet in 1965, the standard map was that drawn by E.M. Antoniadi in 1930, who had extended Schiaparelli's nomenclature,[19] and it was believed that life had a foothold on the planet.[20]

THE FIRST CLOSE LOOK

The 'minimum energy' trajectory to Mars is a Hohmann 'transfer' which traces out half of an ellipse with its perihelion tangential to the Earth's orbit and its aphelion, 250 days or so later, tangential to Mars's orbit.[21] Naturally, the planet must be present when a spacecraft arrives, so calculating back from the interception and allowing for Mars's motion gives the date of launch. In fact, Mars will travel about 130 degrees around the Sun in that interval. Opposition occurs when the Earth is directly between a planet and the Sun, so launch opportunities to Mars arise about fifty days prior to opposition. In fact, there is a 'window' of several weeks around this optimum date during which a near-minimal energy launch can take place, but it is a fine balance of energy against time. Also, Mars's orbit is eccentric, so the key factor that controls the mass that a specific rocket can dispatch to Mars is whether Mars's opposition occurs near perihelion or near aphelion – a perihelic opposition transfer requires less energy allowing a heavier spacecraft to be dispatched at that time. The last opposition of Mars before spacecraft started to visit the planet was in March 1965, but it was an aphelic opposition and the planet was not well placed for telescopic observers.[22]

On 14 July 1965, Mariner 4 snapped a series of 22 images as it slipped by the trailing limb, closing to within 9,850 kilometres of the surface.[23] This sequence traced a discontinuous track inclined at an angle of 40 degrees to the equator, and ran about a quarter of the way around the planet from 180 to 90 degrees longitude.[24] The first image was a view of the leading limb from 17,000 kilometres. It was an oblique perspective of Amazonis in the vicinity of the dark patch, Trivium Charontis. The sequence ran southeastward across Zephyria, Atlantis, Phaetontis and Memnonia, and the final three, which were beyond the terminator, were featureless. Considered in today's terms, Mariner 4's harvest was a small stack of rather fuzzy pictures, but when they had been 'cleaned up' they revealed detail that was much finer than that visible by telescope.[25] What they revealed was startling – craters. Although barely 1 per cent of the surface was recorded, some 300 craters were in evidence. Surprisingly, there was no obvious difference between the classical albedo features, and there were craters on both. The largest, in Mare Sirenum, an area which had been expected to be thick with vegetation, was starkly depicted on the 11th image in the sequence.[26] It is 120 kilometres in diameter. The profusion of craters implied that the planet still bore the scars of its early history, which in turn meant that it had not undergone significant geological activity, at least not for hundreds of millions of years, and perhaps even not for several billion years.[27]

Interestingly, around 1950 Clyde Tombaugh had considered the possibility that Mars had suffered massive impacts. He reasoned that the linear 'canals' which appeared to radiate from 'oases' on Lowell's map might be vast cracks in the crust created by the shock of the impacts that excavated the nodes.[28] Ralph Baldwin and E.J. Opik had independently developed similar ideas.[29,30] At that time, however, there had not seemed to be any likelihood of ever finding out whether the oases really were craters.

Although the Moon was very heavily cratered, their process of formation was still a matter of debate in 1965 with some people saying that they were impacts and others arguing that they were of endogenic origin. Few terrestrial impact craters had been recognised, because they are rapidly eroded. The case for a 1-kilometre-wide crater near Winslow in Arizona being the result of an impact had only just been demonstrated.[31] Overnight, therefore, our impression of Mars was transformed into an ancient cratered planet upon which, in all likelihood, life had *failed* to gain a foothold.

The final nail in the coffin for Martian life was delivered by the occultation experiment. The strength of the spacecraft's radio signal was carefully monitored as it flew behind the planet's limb as viewed from Earth some two hours after the encounter, and again an hour or so later when it emerged from the far side. Knowing the spacecraft's trajectory, a 'refractivity profile' could be measured. Different mixtures of gases have different refractive effects on a signal, so the observed profile could be used to refine the model of the atmosphere in terms of chemical composition, temperature and pressure as functions of altitude for the region above which the signal passed. The first sounding was made between Electris and Mare Chronium at about 55 degrees south, and the second was in darkness over Mare Acidalium at 60 degrees north. The surface pressures were in the range 4.0 to 6.1 millibars.[32] Liquid water would certainly not be stable at such a low pressure, and this also meant that the surface was much colder than expected, because after sunset the surface would be able to radiate the heat it had absorbed from the Sun during the day. Although there is little water vapour in absolute terms, the fact that Mars is so cold means that at night the atmosphere is nevertheless close to its saturation point. The realisation that the polar regions fall below 128°C resurrected the long-discarded idea that the deposit is frozen carbon dioxide rather than water ice.[33] Considering the discovery by Mount Wilson that the partial pressure of carbon dioxide on Mars's surface is a mere 4 millibars, the spacecraft's occultation data implied that the atmosphere is at least 95 per cent carbon dioxide. Based on a weak analogy with the Earth, it had been presumed that the rest of the '25 millibar atmosphere' was predominantly nitrogen; in fact, the nitrogen concentration is only 2.7 per cent. Argon is the third most common gas at 1.6 per cent, and oxygen is negligible, as Mount Wilson had found. Mariner 4 also carried instrumentation designed to sense whether Mars had a magnetic field and, if so, whether charged particles were circulating within it, corresponding to the Earth's Van Allen Belts. No such field was found, so the upper atmosphere is exposed to the blustery plasma of the solar wind. The ancient surface and the thin atmosphere prompted scepticism that Mars could *ever* have possessed a hydrosphere. While there was no evidence of fluvial erosion in the imagery, it was acknowledged that the resolution was too low to show river valleys unless they were on a considerably larger scale than those on Earth.[34]

FOLLOW UP

After skipping the 1967 launch opportunity, NASA resumed its exploration of Mars with Mariners 6 and 7, which were launched in February and March 1969 on fly-by missions. The spacecraft were similar to Mariner 4, but had an improved imaging system in which the standard lens was augmented with a 'telephoto'. As they closed to within a million kilometres of Mars, they started taking pictures to record its rotation. The resolution was better than the telescopic view and the albedo features and the southern polar cap were seen with unprecedented clarity, and although very little additional detail could be seen, this enabled improved maps to be drawn. However, the ring-like appearance of the bright spot which, in 1879, Giovanni

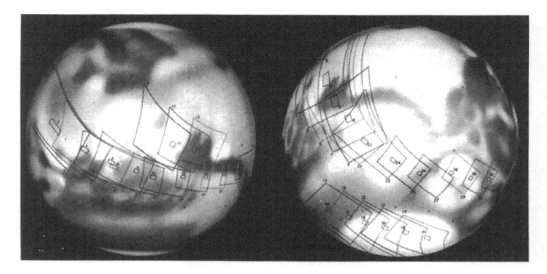

The positions of the images taken of Mars during the Mariner 6/7 fly-by missions of 1969 are shown in relation to the major albedo features. In this 'Space Age' presentation, north is at the top.

Schiaparelli had named Nix Olympica (the 'Snows of Olympus') prompted its interpretation as a large impact crater.[35] Not surprisingly, although certain linearities corresponded to some of the canals that Percival Lowell had drawn, there was no evidence that they formed an artificial irrigation system.

When Mariner 6 passed Mars on 31 July it took 25 near-encounter images of a tract in the equatorial zone from 320 degrees to 60 degrees longitude, providing a look at Aurorae Sinus, Pyrrhae Regio and Deucalionis Regio, with its closest approach some 3,500 kilometres above Sinus Meridiani. A week later, Mariner 7 returned 33 images in two swaths, one crossing the equator and running from 100 to 20 degrees and the other in the southern hemisphere near the meridian that yielded a look at Thymiata, Deucalionis Regio, Hellespontus, Hellas and Mare Hadriaticum. The occultations confirmed that since Mars's gravity is weaker than the Earth's the pressure *profile* is shallower. At 6.5 millibars, the surface pressure at Sinus Meridiani was typical, but the 3.5-millibar pressure at Hellespontica Depressio indicated that this feature had been misnamed because it was actually an elevated area. It was decided to utilise 6.2 millibars as a convenient 'sea level' datum on a dry planet.[36] This pressure is the 'triple point' of water. Liquid water is unstable at lower pressures. In fact, given geographical and seasonal pressure variations, the mean surface pressure is close to this value. The suggestion that the polar caps were not water ice and were completely composed of frozen carbon dioxide was strengthened by Mariner 7's measurement of the temperature at the south pole of $-123\,°C$, because this was close to the carbon dioxide 'frost point'. Interestingly, the floors of some of the craters on the fringe of the cap were coated with frost.

When mapping a planet, photogeologists start by identifying the various different surface units in terms of physiography. Overall, the three fly-by missions had imaged only 10 per cent of Mars's surface at moderate resolution, and three types of terrain were discerned: 'cratered', 'chaotic' and 'featureless' (in order of most-common to least-common). Astonishingly, given the historical focus on albedo, there seemed to be no correlation between the type of terrain and its albedo. The morphology of the chaotic terrain just south of the equator at about 40 degrees longitude suggested a general collapse of the subsurface. If so, this implied that at least some endogenic process had been active; but what could have eroded the subsurface on such a scale? A bright circular feature called Hellas, which had sometimes brightened to such a degree that it resembled an offset polar cap, had been suspected of being a plateau that attracted snow in winter, but it was found to be a depression and – in striking contrast to the adjacent heavily cratered terrain – appeared to be featureless. It was all very puzzling.

A GLOBAL VIEW

The fly-bys of Mariners 4, 6 and 7 had overturned our soundly reasoned impression of the Red Planet, transforming the anticipated cold dry world that nevertheless harboured vegetation into a planet which still bore the scars of ancient impacts and had an atmosphere far too thin to sustain free water, icy caps, or life. As a result, the tax-paying public and many scientists lost interest, but the geologists were fascinated by the forbidding surface and were eager to map it. Global coverage would require a spacecraft to be placed into orbit around the planet. In fact, it would be better to use two spacecraft. When we observe the Moon at its 'full' phase we see only albedo variations. At other phases, shadows indicate the topography. Mars orbits the Sun beyond the Earth, so we view it only at near-full phases. No telescope had ever even hinted at topography on the terminator line, which is why early studies focused on the albedo features. A spacecraft in orbit would be able to view the surface under a variety of illuminations, but whereas a high point over the daylit hemisphere would favour albedo studies, a low point over the terminator was required for topographic mapping. NASA therefore decided that Mariner 8 would adopt a highly inclined orbit that dipped low near the terminator to map 70 per cent of the planet at high resolution, while Mariner 9 adopted an almost equatorial orbit at high altitude in order to monitor variations in surface albedo over time.[37] Unfortunately, its Centaur stage failed and Mariner 8 was dumped into the Atlantic Ocean. The mission was redesigned to attempt as many observations as possible using a single spacecraft, but the compromise of an orbital inclination of 65 degrees would provide illumination at shallower angles than the ideal for studying albedo and higher than the ideal for studying topography. Accordingly, after reprogramming, Mariner 9 was successfully dispatched on 30 May 1971.

It had long been evident that the Martian weather system did not involve cyclonic systems of the type that prevail on the Earth, but from time to time white clouds had been reported that were presumed to be composed of crystals of water ice at high altitudes. 'Yellow' clouds also occasionally appeared and rapidly expanded to mask large areas, and these were thought to be dust storms stirred up by strong seasonal winds. A particularly extensive dust storm was seen by E.M. Antionadi at the 1909 perihelic opposition.[38] When somewhat lesser ones appeared in 1924, 1929, 1941 and 1956 it became evident that they tended to coincide with summer in the southern hemisphere, when the planet was near perihelion and the effects of differential solar heating were most significant.[39,40]

In February 1971, Charles Capen of the Lowell Observatory had predicted that during that year's opposition a major dust storm would likely develop over Hellespontus, and he warned that this might interfere with the forthcoming Mariner mapping.[41] On 21 September this region did indeed become obscured. It was first photographed by Gregory Roberts of Johannesburg, South Africa, and by 27 September it had obscured a wide area west of Hellas. The dark features were still visible at the end of the month but within a week the mid-southern latitude zone was masked and, at the end of October, the entire planet was obscured.[42] When Mariner 9 began its far-encounter imaging on 10 November, the only features on display were the south polar cap and a cluster of fuzzy dark spots, three of which were on a line extending northeast from near the equator and spaced about 700 kilometres apart, and another on its own off to the northwest. Although the trio were promptly dubbed 'North Spot', 'Middle Spot' and 'South Spot' it was soon realised that these positions corresponded to Ascraeus Lacus, Pavonis Lacus and Nodus Gordii respectively. The fourth dark spot corresponded to Nix Olympica, which the Mariner 6/7 far-encounter views had suggested was a large crater.[43] How could depressed areas be visible through the dust storm?

When Mariner 9 fired its braking rocket on 14 November, it became the first spacecraft to adopt Martian orbit. Luckily, it was able to be reprogrammed to postpone its mapping activity for a few months until the dust storm had abated. Meanwhile, whenever the spacecraft crossed the planet's limb its radio signal was monitored so as to measure the physical properties of the dust-laden atmosphere. The Soviets had dispatched a pair of spacecraft on ambitious missions to Mars, but they did not fare very well. Shortly before entering orbit on 27 November, Mars 2 released a small probe as planned, but this fell silent soon after penetrating the atmosphere. The other spacecraft seemingly reached the surface, but failed as it was preparing to return a panoramic image of the landing site.[44] Perhaps they fell victim to the dust storm? Frustratingly, the orbiters were incapable of being reprogrammed, so they set about mapping oblivious to the dust storm that rendered their efforts futile. On 27 October 1972, after returning a total of 7,329 images and having turned Mars back into an *interesting* planet, Mariner 9 exhausted its supply of attitude control propellant and was switched off. It had revealed a planetary landscape with striking variety, and by placing high-resolution close-up views into the wider context it had given the geologists the long-sought global perspective.[45,46] This enabled the photogeologists to analyse the surface using tried and tested methods based upon the principle of superposition.

Although interferometry by terrestrial radio telescopes operating as radars had enabled the surface of Venus to be discerned through the planet's veil of cloud, this was impracticable in the case of Mars because the rapid rotation frustrated the synthesis technique. But when Mars was at opposition it was possible to process the time delays in the reflections and 'profile' the surface by radar altimetry. The initial test in 1967 by the Haystack antenna had shown that the dark albedo features were 'radar bright', implying that they were rough on the submetre-scale. In addition, there was poor correlation between the measured elevation and radar reflectivity.[47] Over successive oppositions the Goldstone antenna built up profiles to help to interpret the features charted by Mariner 9 by adding the 'third dimension' to the imaging dataset.[48,49,50] For centuries astronomers had been limited to classifying the albedo variations, but their efforts to infer correlations with topography had been in vain because it turned out that there is no such systematic relationship. When this was realised, the International Astronomical Union revised the nomenclature, classifying the surface features in terms of their morphological characteristics.[51,52]

When the two Viking Orbiters arrived in the summer of 1976, they mapped the planet at a resolution of 200 metres per pixel and studied selected areas at resolutions as fine as 6 metres per pixel. Furthermore, because they functioned for several years they were able to monitor the seasonal variations for two orbits of Mars around the Sun.

VOLCANOES!

Mariner 9's first discovery was startling. Having recently come to terms with the implications of ancient cratered terrain revealed by the fly-by missions, the fact that mountains were protruding through the dust storm came as a surprise, if only because telescopic observers were sure that there could be *no* large mountains on the planet. Furthermore, as the dust started to clear, it became evident that each edifice had a complex of summit craters which could only be calderas: the mountains were volcanoes!

Olympus Mons (as Nix Olympica was subsequently renamed) is 600 kilometres across its base, so it is a volcanic 'shield'. Nevertheless, its summit stands some 25 kilometres above the surrounding plain. In fact, it is by far the tallest volcano in the Solar System.[53] The other three 'spots', Ascraeus, Pavonis and Arsia Mons were atop an enormous crustal bulge in the Tharsis region. By a stroke of bad luck, the 10 per cent of the Martian surface observed by the fly-bys had been almost entirely cratered terrain. Mars is a world of remarkable contrasts.

LINE OF DICHOTOMY

To a first approximation, the Martian surface consists of two morphologically distinctive hemispheres. The transitional 'line of dichotomy' has its most northerly incursion near longitude 330 degrees at about 50 degrees latitude, north of Sabaeus Terra and Meridiani Terra. The terrain to the south is heavily cratered and evidently ancient. The smooth plains to the north are only lightly cratered, so they are relatively young. They could be either sedimentation or effusive volcanism on a vast scale. The ghostly rims of craters that protrude in some places indicate that these flows, whatever their form, were thin and overran a heavily cratered terrain. In some places large areas of this terrain survived – for example, Tempe Terra is an elevated patch of ancient terrain northeast of Tharsis and the Phlegra Montes stand north of the line.

Significantly, the northern plains lie several kilometres below the datum. The boundary is characterised by an irregular but shallow scarp that runs around the planet, except for where it has been masked by the rampant volcanism in the Tharsis region. It is scalloped where major impacts left basins that were later flooded from the north. The Isidis basin is a striking example. The Chryse basin is less obvious, because it has been masked by the sediment from outflow channels. The scarp is an erosional feature that has progressively migrated southward. It has been dissected by faults and subjected to erosion that has left distinctive terrains. The fracturing of the highlands near the scarp has created linear trench-like valleys, and mass wastage has eroded their walls, widening them. This disruption is more pronounced closer to the line. The high ground is so broken into angular blocks that it consists of buttes isolated by 'fretted' channels.[54] Across the line of dichotomy, these blocks become progressively smaller knobby hills. Continuing erosion is evident by the scalloped debris aprons at the foot of their steep slopes.[55] This transitionary terrain ultimately blends into the northern plains. Apart from where outflow channels debouch onto the low-lying northern plains, this fretting process has been controlled by fracturing.[56] The line is embayed in places.[57] Clearly, an event early in the planet's history had for some reason resurfaced the low-lying terrain to the north, but there has been considerable debate over whether the hemispheric dichotomy is due to endogenic or exogenic processes.

In 1979 it was suggested that an intense mantle plume in the northern hemisphere scraped the bottom off the lithosphere causing it to settle isostatically,[58] and in 1981 it was posited that this heat may have been released by post-accretional core separation,[59] but this was contradicted by evidence from the SNC meteorites that the core formed early in the process of accretion. As an exogenic alternative, in 1984 it was suggested that widely separated massifs within the line of dichotomy trace the rim of a vast basin which formed early in the planet's history, possibly as long as 4.2 billion years ago.[60] Measured around the curve of the globe, this basin would have been over 7,500 kilometres in diameter, and would have included 80 per cent of the surface of the northern hemisphere.[61] However, the ejecta blanket and the 'sculpture' that such a massive impact would have left is lacking. In 1988 it was proposed that a concentration of overlapping basins so weakened the northern lithosphere that deep fractures provided conduits for lava to spill over from one basin to another to create the Vastitas Borealis in the same way as the maria spread across much of the Moon's nearside.[62] A variety of endogenic models have been devised over the years,[63,64,65] but if the dichotomy really is the result of an impact then the key discovery was made while modelling the process of accretion by computer.[66] If a giant impact occurred at the tail-end of the planet's accretion, when the crust was in the process of forming but before it began to record the impacts that we see preserved on its ancient surface, then the ejecta sculpture would not be expected to be evident. A great deal of material would have been blasted into space, and although much of this would have

Mars Global Surveyor provided a crucial insight into the nature of the northern lowlands, and the line of dichotomy, in the form of a slice of the crustal structure along the meridian derived from gravity and topography data. The crust is about 40 kilometres thick beneath the northern plains and the adjacent terrain, but 70 kilometres thick at high southern latitudes. Although the line of dichotomy defines the transition from the lowlands to Arabia Terra, the fact that the thin crust extends southwards with uniform thickness under Arabia Terra indicates that the lowlands and Arabia Terra form a *single* crustal province, suggesting that the line of dichotomy's geological manifestation is primarily due to surficial processes. (Courtesy of MGS/RS/MOLA Science Teams.)

escaped, some would have entered orbit. Mars has a strangely high number of elliptical craters[67] and perhaps these were excavated by the low-energy impacts as this debris later rained down. The main legacy of this giant impact *should* have been a significant thinning of the northern lithosphere that was readily punctured by the continuing bombardment. But how could this theory be tested? Doppler tracking of Mariner 9 and the Viking Orbiters found no evidence of a gravitational transition over the line.[68] However, the gravitational and topographical data from Mars Global Surveyor enabled the crustal thickness to be inferred, and this revealed a thickness of about 40 kilometres beneath the northern plains and 70 kilometres beneath the high southern latitudes. Although the line of dichotomy defines the transition from the lowlands to Arabia Terra, the fact that the thin crust extends southwards with uniform thickness under Arabia Terra indicates that the lowlands and Arabia Terra form a *single* crustal province, distinct from the southern terrain. The fact that the boundary with that thicker southern crust does not correspond to the line of dichotomy implies that the geological manifestation is primarily due to surficial rather than internal processes.[69] Nevertheless, early in the planet's history, heat would have been lost more effectively through this area than through the southern highlands, particularly if the mantle was vigorously convecting, and the heat loss could either have stimulated intense volcanism or have devolatised the crust and prompted vast gaseous or fluid release.[70] Perhaps the safest position to take on the issue, however, is to admit that as yet we simply do not know how the crustal dichotomy occurred.[71]

VALLEY NETWORKS

After all the evidence against Mars ever having possessed a hydrosphere, the discovery by Mariner 9 of what appeared to be dry riverbeds came as a welcome surprise.[72] Their classification has been divided into 'runoff' and 'sapping' channels.[73]

The runoff channels so resemble terrestrial drainage systems that the case for their having been progressively etched by slowly running surface water is compelling. They are typically less than a kilometre wide, seldom exceed 100 kilometres in length, start small and increase in size downstream, have tributaries that form dendritic inflow systems and, intriguingly, tend to end abruptly as though the flow had disappeared underground.[74] On Earth, the erosion which forms karst in permeable rock often creates such 'blind alleys'.[75] There are also features which look as if they formed by collapse after permafrost melted and water erupted from the ground. Such valleys run more or less straight and have steep walls. The tributaries are very short and there are large areas of pristine surface between them, indicating that they did not drain the surrounding plain. Furthermore, the tributaries do not make contact with the valley floor, they break out high on the walls, so the water would have cascaded by waterfall to the valley floor. It is thought that such a valley would have been progressively extended by sapping at its head, and that the tributaries formed in the same manner after sapping had eroded the cliff-like wall. Indeed, the fact that the tributaries 'start' in amphitheatre-like cavities which show no sign of having served as runoff collectors supports the sapping interpretation. The sapping channels are generally longer than the runoff channels. At almost 800 kilometres long, Ma'adim Vallis, which drains north into the crater Gusev just south of the line of dichotomy, is one of the longest. Almost all the channels occur in the cratered highlands; they appear to be an integral part of this terrain, and are therefore contemporary with the cratering. Significantly, the runoff channels did not link up to establish rivers that drained into depressed areas and formed seas. Compared to terrestrial river systems, relatively little material would have been eroded and redistributed by these 'immature' drainage systems. The northern flank of Alba Patera contains one of the few that are not in the southern highlands, and it includes the best-developed fluvial valleys on Mars. It is also a younger network than its southern counterparts.[76] Although in this case the water seems to have been produced by sapping, as the water flowed downslope it developed a drainage system with a degree of integration comparable to that of the terrestrial Hawaiian shields.[77]

The valley networks spoke profoundly of Mars's early development. As the primordial atmosphere would have been enhanced by volcanic outgassing, it may have been an intense hydrological cycle in this early atmosphere that created the valley networks in the southern hemisphere. In any case, the fact that the atmosphere is now extremely thin indicates that the planet has undergone a major climatic change. It may even alternate between 'warm and wet' and 'cold and dry' periods, in response to dynamical instabilities.[78,79,80,81] Recent studies have suggested that the vast majority of the valleys derive their morphologies from sapping,[82,83] but there is also evidence to suggest that in many cases a V-shaped valley etched by runoff was reactivated and extended with a U-shaped profile by headward sapping.[84] The timescales of these phases of activity are not clear, however.

CRATERS

When the fly-by missions returned pictures of the southern cratered terrain, it was noticed that the depth-to-diameter ratios of the craters are significantly different to the lunar highlands. Craters less than about 20 kilometres in diameter are simple bowl-shaped cavities with 'sharp' profiles and raised rims, but the larger craters appear 'shallower' than their lunar counterparts, with terraced walls, complex central peaks, degraded rims and ejecta blankets, and floors that have been levelled by various in-fill materials.[85] The paucity of 'ray' craters and chains of secondaries was also striking.

As a rule, Martian craters do not have such extensive blankets of hummocky ejecta. On the Moon, the continuous ejecta deposit is typically confined within 0.7 radius of a crater's rim. On Mercury, the confinement is 0.5 radius. But the continuous ejecta for craters on Mars extends out 2 radius. Debris on ballistic trajectories would be expected to travel further on the Moon than on Mercury, because the Moon is smaller. However, Mars is comparable in size to Mercury, so why should the blankets extend farther on Mars? The answer seems to involve the form of the ejecta. Whereas ejecta on the Moon and Mercury is usually blocky close to the rim and grades progressively to finer debris that ultimately blends into the surrounding terrain, many Martian craters exceeding 5 kilometres in diameter are surrounded by overlapping 'sheets' with lobate margins. There is considerable variety in the form of such ejecta.[86] In some cases it is thin and the underlying topography projects through. In other cases, the ejecta is thick enough to mask the terrain and there is frequently a distinct radial pattern. Sometimes the ejecta is concentrated in a thick annulus. Such impacts evidently 'splashed out' slushy ejecta which then flowed over the surface. In some cases, the mudflow has either been deflected around a nearby crater's rim or sloshed over and flooded it. Larger impacts excavate to deeper levels than do smaller ones, so it was suggested that only impacts exceeding a certain energy had been able to penetrate deep into the permafrost.[87,88] The variation in crater types may indicate that the excavated rock had differing amounts of water and ice. There is a geographical correlation. The fluidised ejecta runs farther from the rim of craters at higher latitudes.[89] It has been calculated that the subsurface in which water would freeze varies from 1 kilometre thick in the equatorial zone to 3 kilometres thick in the polar regions.[90,91] The intensely brecciated outer crust (the 'megaregolith') would have been porous, and would have soaked up a great deal of water when the hydrological cycle was active. In fact, it has been calculated that the megaregolith could have absorbed *90 per cent* of the total volume of water originally on the planet.[92]

BASINS

Martian craters exceeding 100 kilometres in diameter are classified as impact basins. On the Moon, this distinction has been drawn at 200 kilometres. Some two dozen Martian basins have been identified, but most of them have been significantly modified by processes or erosion and deposition. Basins greater than 400 kilometres wide possess multiple-ring structures (as on the Moon) but their ringforms are fragmentary.

Hellas in the southern hemisphere is the largest of the well-preserved Martian basins; it is about 2,000 kilometres in diameter.[93] Degraded and fragmentary rings are marked by an arc of mountains to the north and west. Although individual massifs are up to 180 kilometres across, there is no steep 'front' like that of the Apennines facing into the Moon's Imbrium basin: the southeastern rim is degraded and the southwestern rim is completely masked by a volcanic plain associated with a pair of calderas. The northeastern rim is extensively etched by two fluvial channels that have drained down the fairly shallow slope and deposited material on the floor which, although filled in, is still some 7 kilometres below the planetary datum. Although the floor is often masked by frost or cloud and appeared 'featureless' to Mariner 7, one of the Vikings was able to image it on a clear day and showed albedo variations suggestive of a series of massive flows.[94] The Hellespontus Montes to the west are all that remains of one of the outer rings. At 3.3 kilometres elevation above the datum, Hellas's ejecta blanket is the highest terrain in the southern hemisphere. As with the lunar basins, the rim massifs are crustal blocks up-thrusted by the shock of the impact. However, there is little 'sculpture' from flying debris.

The Argyre basin, about half the size of Hellas,[95] is more rugged, with prominent rings in the form of the Nereidum and Chartitum Montes. Like Hellas, channels run into the basin and the floor (which is 4 kilometres below the datum) is masked, but the rims of craters protrude through this in-fill so it is evidently shallow. Although there are no large volcanic constructs around the basin's periphery, there are ridged plains on the floor that might be lava flows.

Many other basins (particularly those north of the line of dichotomy) are so degraded that their presence is only hinted at by isolated tell-tale tectonic features.[96] Isidis, for example, was recognised as a basin only once Mariner 9 imaged it. It straddles the line of dichotomy. Only the southern rim remains as Libya Montes, because whatever process resurfaced the lowland terrain has totally erased the rest of the rim and submerged its floor. And the volcanism on the Syrtis Major Planum to the west of Isidis appears to have exploited deep fractures that trace a 1,900-kilometre-diameter ring centred on Isidis.[97] The Chryse basin, which also straddles the line of dichotomy, has been totally erased by the flows that formed Chryse Planitia. There are several basins beneath the northern plains. The craters which poke through Utopia Planitia indicate the presence of a 3,000-kilometre-wide basin.[98,99] The extremely flat Daedalia Planum to the southwest of Tharsis appears to be an ash deposit masking an 1,800-kilometre-wide basin.[100] The evidence of bombardment is masked. Even when a Martian basin has been eroded to the point where the primary ringform is no longer recognisable, its existence can be inferred from tell-tale effects on its surroundings. A large impact produces regularly spaced ring faults and, for large basins, these fractures probably penetrate 15 kilometres into the lithosphere. These faults are often the only indication of an old basin.[101] The zones of weakness that deep faults create influence later regional stress patterns. For example, wrinkle ridges tend to be deflected by the rings, and their formation is inhibited inside the rings, so a study of the texture of the plains flooding a basin can reveal the presence of the smothered ring structure.[102] It was such an analysis which demonstrated the presence of an impact basin beneath Chryse Planitia.

The largest basins on Mars are considerably larger than Imbrium on the Moon and Caloris on Mercury. Interestingly, whereas the surficial coverage of Martian basins is 1 per 10 million square kilometres, the coverage is 8 per 10 million square kilometres on the Moon and 14 per 10 million square kilometres on Mercury.[103,104] Even if the resurfacing of the northern lowlands removed half of Mars's basins, this figure would still be low. This might not be an 'anomaly', however; it could indicate an inverse relationship between a planet's distance from the Sun and its basin coverage, perhaps because the population of large impactors in the early Solar System was highly concentrated. A 'scaling' effect might have to be applied to account for different populations in the inner Solar System.[105,106] An analysis of the morphological features that are characteristic of large impacts tentatively identified an additional two dozen severely degraded basins.[107] Clearly, attempting to infer conditions in the early Solar System from the Martian cratering record is fraught with uncertainties.

Some 'alternative thinkers' have argued that Mars's northern hemisphere is a vast volcanic plain set below the datum because the crust was ejected into space as spall by the shockwaves from the Hellas and Argyre impacts.[108] They also claim that this occurred just 12,000 years ago, at which time Martian life was wiped out, but this interpretation of the geological evidence is not widely accepted.

CRATERING RATES

The rate of cratering of Mars is not known, but given assumptions about the population of impactors in the inner Solar System the exposure ages of the various Martian surface units can be extrapolated from the Moon, for which we have firm information.[109,110]

The cratered highlands of Mars differ from those of the Moon in that the curve plotting the distribution of crater frequency to diameter for the lunar highlands is 'saturated' down to sizes of several tens of metres. On Mars, in contrast, the saturation point occurs at a crater diameter of 30 kilometres. In fact, the frequency of 10-kilometre-sized craters is 10 per cent of that for the Moon. Because the Martian highlands are far from saturation point, there is a 'shortage' of smaller craters. Interestingly, while most of the large craters are degraded, the smaller craters are fresher looking.

In 1973, after analysing the Mariner 9 imagery, William Hartmann argued that the smaller craters constituted a systematically younger population.[111] He proposed that an erosive process was active during the bombardment and that this degraded the large craters and eliminated most of the smaller ones. The larger craters are etched by valley networks, indicating that there was water erosion at that time. Also, as large areas of the southern highlands have been masked by volcanogenic deposits, volcanism played a key role in eroding the ancient structures. Since the highlands are not saturated, the craters are more widely spaced than their lunar counterparts. The intercrater material has clearly been superposed because it has masked ejecta blankets and has encroached on some of the craters themselves. Most of this material is believed to be lava erupted from large fissures.[112] Hartmann proposed that the smaller craters appear under-represented because they formed *after* the intense bombardment, at a time when the impact rate had declined and when the initially intense erosional processes had diminished. This 'late' population of small craters is superimposed upon the plains between the larger craters and many have produced fluidised ejecta blankets. Furthermore, Hartmann has recently noted that erosion is still at work, smoothing out craters up to 100 metres in diameter by filling them with dust, sedimentary deposits and lava flows.[113]

HIGHLAND VOLCANISM

In addition to the enormous shields in the Tharsis and Elysium regions north of the line of dichotomy, there are some volcanoes in the southern highlands. This 'highland volcanism' is concentrated around the Hellas basin.[114] The reduction in pressure on the lower lithosphere by the removal of crustal material during such a massive impact would have stimulated localised decompressional melting,[115] and deep faults would have accommodated magmatic intrusions and provided lava with access to the surface, where it produced extremely low-profile structures with complex summit calderas.[116]

Situated 1,500 kilometres northeast of Hellas's rim, Tyrrhena Patera is highly degraded.[117] Its irregular caldera measures 12 by 40 kilometres and is transected by an oval discontinuous graben 80 kilometres long. A series of explosive eruptions suggest that the magma intrusions rose through water-saturated megaregolith and the water flashed to steam.[118] The flanks are deeply eroded with radial channels which suggest that pyroclastic flows had laid down thick blankets of ash which became 'welded' into ignimbrites as they settled.[119] The wide base and very shallow profile suggest that it may be similar to a terrestrial 'ash shield'. The complex caldera indicates episodic volcanism, and it has been proposed that once the vent had exhausted its supply of volatiles it switched to erupting lava, and that it was this, rather than water, that eroded the deep channels in the ash on its flank.[120] At about 20 kilometres across, the channels are wider than the lava channels on the flanks of the Tharsis and Elysium shields. They are also significantly broader than the erosional valley networks. One of the most prominent channels emerges from the caldera and runs westwards for more than 200 kilometres, to where it merges with the surrounding volcanic ridged plain. The floors of both the caldera and this channel are covered with lava, which supports the case for the channels having been

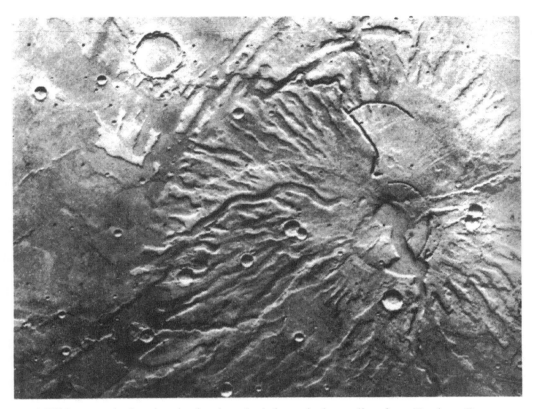

A Viking mosaic showing the deeply etched channels that radiate from Tyrrhena Patera, a low shield volcano in the southern highlands near the Hellas basin. The irregular caldera measures 12 by 40 kilometres and is transected by an oval discontinuous graben 80 kilometres long. The deeply eroded valleys that radiate from the summit suggest thick blankets of welded-ash deposited during early phraeto-magmatic eruptions when the magma rose through a water-saturated crust and the water flashed to steam.

formed by lava. If the ash was etched by water, however, this is more likely to have been by sapping induced by the local heat flow through the crust melting subsurface ice rather than by the runoff of prodigious precipitation. The cratering implies that Tyrrhena Patera is at least 3 billion years old, so all of this activity took place early in Mars's history. Hadriaca Patera is near Tyrrhena, and Amphitrites and Peneus Paterae are southwest of Hellas's rim. They also underwent explosive eruptions that deposited thick blankets of ash. Indeed Hadriaca's subdued caldera produced an apron 300 kilometres across that has been so thoroughly etched by radiating channels that it is now strikingly *ridged*. Given its location on the regional slope, Hadriaca's lavas formed channels 400 kilometres long to the southwest. With slopes of just a fraction of a degree, Amphitrites and Peneus are so subdued that their calderas resemble impact craters,[121] and the concentric fractures around Peneus suggest that it has suffered general collapse. Their lavas embayed the ancient cratered terrain, in the process creating the Malea Planum ridged plain. The most explosive terrestrial volcanoes are the 'resurgent calderas' which produce

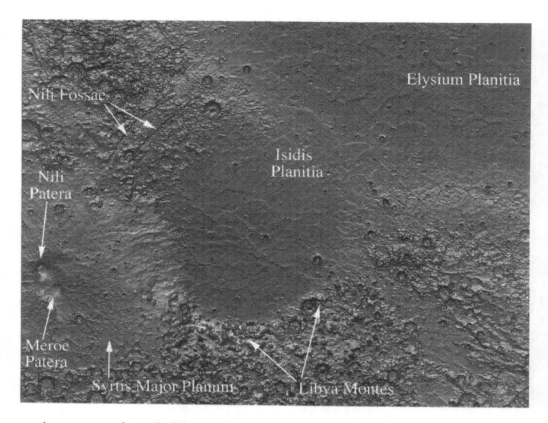

A computer-enhanced Viking view of the line of dichotomy as it passes between Syrtis Major Planum with its two volcanic calderas (left) and the low-lying floor of the Isidis basin (centre) and Elysium Planitia beyond. The only relics of the Isidis impact are a short section of the rim of mountains (Libya Montes) and arcs of circumferential ring faults (Nili Fossae). The Syrtis magma may have exploited these zones of weakness to gain access to the surface. Notice that the Syrtis lavas have drained onto the adjacent basin's floor.

shallow edifices with very large calderas. The most recent eruption of Yellowstone in Wyoming blanketed much of North America with ash, so it is probably a reasonable analogue for a Martian highland patera.

Apollinaris Patera is in the transitional terrain just beyond the line of dichotomy south of Elysium Planitia. Although it may have initially blasted out ash, it is believed to be the earliest of Mars's lava shields. The western flank is surprisingly steep. The final phase of activity left the 70-kilometre-diameter caldera full of lava, and spilled a massive lava extrusion down the southern flank which is etched by a dendritic pattern which, if it is fluvial erosion, would date this volcano to the early period with a 'wet' climate.[122]

Nili and Meroe Paterae on Syrtis Major Planum's dark plateau represent another form of highland volcanism.[123] They may actually be ultra-low-profile shields on a volcanic plateau exceeding 1,000 kilometres across.[124] Arcuate grabens in the central area trace out a shallow depression 280 kilometres across which has been interpreted as being the result of a magma chamber 'foundering'. This volcanism appears to be related to the Isidis basin in the same way that Tyrrhena Patera is related to Hellas, because the vents lie on a partial ring of fractures concentric to the basin. Eastward lava flows have drained down onto Isidis Planitia. The rest of the plateau's periphery transitions into the intercrater plains of the southern highlands. Tempe Patera is a similarly subdued caldera-like feature north of the line of dichotomy, but it is on the isolated patch of cratered terrain northeast of Tharsis. It too is ancient, possibly the oldest such feature on the planet.

There are no such low-profile central-vent structures on the low-lying northern plains. The conditions that gave rise to their formation evidently applied only in the ancient cratered terrain. Yet, given the expanse of such terrain, it is remarkable that there are so few of them.

VOLCANIC PLAINS

The presence of flow fronts with distinct contacts with pre-existing terrain, encroachment of flow fronts upon craters, rims that protrude through the plains material, and small cones with summit craters, suggest that about half of the Martian plains are of volcanic origin. Many lava flows can be traced to specific vents. Most are only several tens of kilometres long, but some run for hundreds of kilometres. Lobate flow fronts terminate at scarps several tens of metres high. A few flows have leveed lava channels (in the case of Ceraunius Tholus on Tharsis, the channel starts at the broken wall of a caldera and runs down the flank). Some flows appear to have been fed by lava tubes, because they emerge some distance from their likely sources.[125]

Other lightly cratered plains form 'ridged plains' which are characterised by widely spaced low ridges. These are found primarily in four areas: within the Isidis basin, two areas close to the Hellas basin (one of which has masked its the southern rim) and Lunae Planum to the east of Tharsis. These plains are believed to be lava flows far more extensive than those produced by central vents. As they were large eruptions of low-viscosity lavas, such plains resemble the lunar maria. Like terrestrial flood basalts, they probably erupted from fractures and left no trace of their sources. Curvilinear wrinkles span several tens of kilometres and have slopes of only a few degrees leading up to a narrower relief feature on the crest. These plains are clearly very thick, because they completely mask the previous topography. It was once thought that the lunar wrinkles were extrusions of lava from fissures,[126,127] but they are actually the result of mild tectonism induced by compressional stresses as the lava cooled, densified and settled. This also imparted extensional forces which caused the periphery to withdraw and open up an arcuate crack.[128,129] Although compressional ridges formed on the Martian volcanic plains, there is no evidence of withdrawal. The fact that many of the ridges on Lunae Planum are circumferential to Tharsis suggested that the stresses arose because the plain is now on a slight slope.[130,131] The cross-cutting relationships also indicate that Lunae Planum formed

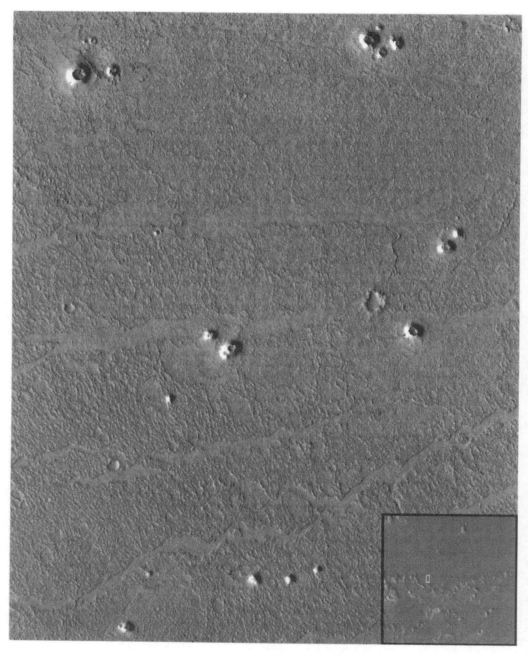

Mars Global Surveyor discovered a field of small cones superimposed on fresh lava flows on northwestern Amazonis Planitia. The most likely interpretation is that they are volcanic features known as 'pseudocraters' or 'rootless cones' formed by explosions when the hot lava crossed a water-rich surface. Possible Martian pseudocraters are of interest because they may mark the locations of shallow water or ice at the time the lava was emplaced.

before the Valles Marineris canyon system opened, otherwise the lava would have flowed down into the canyons that transect the southern part of the plain.[132] While the ridged plains appear pristine apart from continuing cratering, other plains, particularly those at high northern latitudes, are more complex. Furthermore, large areas of these plains have been masked by erosional deposits, particularly where an outflow channel debouched onto them and built up sediment. In fact, it has recently been argued that many of these plains are in fact *primarily* sedimentary in form.

The northern lowland plains are dotted with a variety of topographical features, including polygonal fractures, closely packed dimple depressions and irregular hills, and there are literally thousands of sub-kilometre-sized domes.[133] There are also cones with summit craters which may be piles of tephra from localised phreatic explosions induced by a lava flow running over an area rich in volatiles.[134] Both Tempe Terra and Syria Planum are dotted with 'breached' cones which may indicate where the walls of cones were undermined by lava extrusions.[135,136] The Snake River Volcanic Plain in Idaho may be a reasonable analogue for such rich fields of volcanoes.[137,138] A few rough-looking edifices with sharp radial flank ridges look as if interleaved ash and lava flows have built up steep cones which seem to be the equivalent of terrestrial stratovolcanoes.[139] One such structure lies in the highlands south of Apollonius. Its intensely ridged 20-degree slopes rise 2 kilometres to a caldera with a lava lake that has spilled over the rim and sent a flow down the flank onto the plain.[140,141]

Cratering indicates that most of the fluid lavas that were erupted to form the plains units occurred in the half billion years after the tail-end of the bombardment, essentially at the same time as the maria were forming on the Moon, but since then volcanism has been more or less confined to the elevated Elysium and Tharsis provinces, where a dozen vast shields and some smaller volcanoes have been constructed.

THARSIS

Measured in terms of how it rises from the surrounding terrain, the Tharsis bulge has an asymmetric profile extending for 4,000 kilometres north–south, ranging from the plains of Vastitas Borealis south over the equator and 3,000 kilometres from Lunae Planum in the east over to Amazonis and Arcadia Planitia in the west. Its steep northwestern flank rises from the low-lying northern plains and forms a series of young lava flows, but the shallow eastern flank, which is an old ridged plain, transitions into the southern highlands. Syria Planum on the broad domical summit stands 8 to 10 kilometres above the planetary datum.

Alba Patera would appear to have been the first volcano to develop in the Tharsis province and lies on the northern flank of the crustal bulge. Although its base is 1,500 kilometres across, it is an extremely shallow structure whose summit rises just a few kilometres.[142] Its periphery is defined by a system of concentric fractures which imply that the entire structure is in an advanced state of collapse.

Early in its history, Alba Patera might have been an explosive vent which

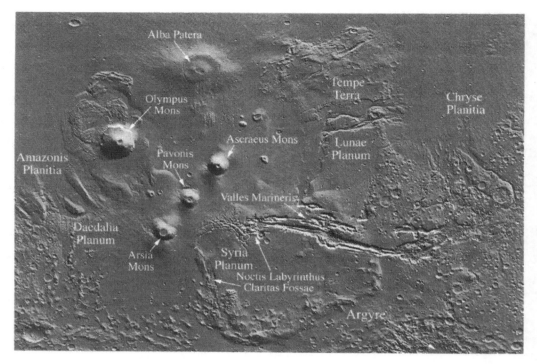

A section of the topographic map by the Mars Global Surveyor's laser altimeter showing the Tharsis province. The true scale of Alba Patera is strikingly revealed. A significant discovery was broad channels draining onto the lowlands immediately west of Tharsis, larger than those to the east which drain onto Chryse Planitia. Their sources seem to have been overrun by lava flows from Arsia Mons. (Courtesy of MGS/MOLA Science Team.)

blanketed its environs with pyroclastics.[143] In Mars's low gravity, an ash cloud would have been blasted to an altitude of at least 100 kilometres.[144] The low air pressure would not have been able to support it and the plume would have collapsed and *surged* far across the surface. In fact, this ash is still visible in a ring extending from 250 to 450 kilometres from the caldera. The volcano later switched to effusive eruptions and sheets of lava were fed by channels and tubes which can be traced up to 1,000 kilometres westward from vents near the caldera complex on the summit.[145] The 100-kilometre-wide main caldera is rather irregular and extremely subdued, with only its western rim being marked by a scarp, but there is a rather better preserved subsequent caldera in its southeastern quadrant. Some of Alba's flows are contemporary to those on the Tharsis ridge, but most of the activity seems to have occurred about 3 billion years ago. The small mounds on its western edge could be cinder cones formed by a more viscous siliceous lava, but they may also be 'rootless volcanoes'.[146]

There are numerous fractures radiating out from the summit of the Tharsis bulge, some of which run for 3,000 kilometres. Most of the faults radiate down its flank but many form short concentric arcs on the lower slopes. In many cases, slabs of crust have dropped between pairs of faults to form grabens a few kilometres wide which extend for 1,000 kilometres. Cratering indicates that the fracturing ended several billion years ago – before the large shields on the summit formed and much of the terrain was hidden by lava flows. The faults are most evident where they have fractured outcrops of older terrain but they must extend beneath the younger volcanic flows too. The angle of the graben walls indicate that the faults penetrate to depths of only a few kilometres.[147] The linear faults which prominently swarm over Alba Patera are part of the radial system of the Tharsis bulge,[148] as are the faults that run northeast across Tempe Terra (an isolated patch of cratered terrain to the east of Alba). The faults of Claritas Fossae, which run southeast of Syria Planum into the southern highlands, are also radial to Tharsis but are more complex, having intensely faulted floors littered with tilted blocks.[149] The Valles Marineris canyons run east from Syria Planum. In effect, the extensional tectonism that created Noctis Labyrinthus on the summit of the Tharsis bulge spawned two sets of rifts, only one of which developed to the extent of forming canyon-scale grabens with faults which go deep into the lithosphere.[150,151]

One early suggestion for how the Tharsis bulge developed posited that it began with broad lithospheric 'doming' by a rising mantle plume, and that this plume stimulated the volcanism that built up the bulge with flows and populated it with volcanoes.[152] An alternative proposal for the uplift argued that the base of the lithosphere was 'underplated', and the uplift resulted from isostatic adjustment.[153] Recent studies of lithospheric thicknesses strongly imply isostatic support.[154] However, although it was initially believed that the bulge was the result of a single uplift event centred on Syria Planum, an analysis of the radial faults indicated several distinct episodes of uplift.[155,156] The Tharsis tectonism appears to have influenced 30 per cent of the planet's surface. Modelling of the stresses derived from the topography and the extant gravity field indicated that the radial fracture system resulted from the bulge's existence, rather than being a side-effect of the act of uplift.[157] A study of how the fracture systems relate to the surrounding terrain and to one another established that the earliest fractures are located on Tempe Terra, implying that the processes which produced the bulge were active early in the planet's history.[158] In fact, it was a *defining* event in the planet's evolution. Most of the volcanism in the central section of Tharsis occurred after most of the fractures had formed, but the role of volcanism is still debated. Some have argued that Tharsis is primarily the result of uplift and that it is thinly veneered by volcanics,[159] while others consider it to be a massive volcanic construct.[160] A stratigraphic study established that the central-vent edifices were constructed concurrently with the emplacement of the lava plains on the flanks of the bulge.[161] It appears that a deep fracture transected the northwestern flank, and gave rise to the volcanism which built up the line of large shield volcanoes, together referred to as the Tharsis Montes, which are spaced about 700 kilometres apart and lie more or less on the line of dichotomy's projection through the bulge. The sequence is clear: firstly, the lithosphere was uplifted, the crust was radially fractured and then, as the trio of shields were being erected on its summit, floods of lava buried some of the radial fractures.

A map showing the distribution of the fractures, grabens and wrinkle ridges associated with the flanks of the vast Tharsis rise, dominating the hemisphere. The randomly intersecting troughs of Noctis Labyrinthus mark the summit. Swarms of grabens cross the older Alba Patera shield and the ancient cratered terrain of Tempe Terra (to the northeast). Concentric patterns of ridges wrinkle the eastern flank. (Courtesy of M.H. Carr, *The Surface of Mars*, 1981, Yale University Press.)

Arsia Mons is the southernmost of the trio. Lava has spilled from its 120-kilometre-wide caldera and run down the flank in leveed channels several hundred metres wide. The 5-degree slope near the summit rapidly diminishes to less than 1 degree on the lower flank.[162] A system of concentric fractures on its lower flanks have been transected to the northeast and southwest (that is, along the line of the chain) by a series of major fractures from which lava erupted and flowed for up to 300 kilometres.[163] Since the discharge rate is essentially independent of flow parameters, and is directly proportional to the maximum extent of the flow, an analysis of the flows demonstrated that these flank vents extruded lava at a rate of 1 million cubic metres per second,[164] which is a far greater rate than the flood basalts that constructed the Columbia River Plateau on Earth.[165]

All of the large shields atop Tharsis display similar growth cycles,[166] in that they were built up by repeated extrusions of fluidic lava. To have flowed so far, this lava would need to have been a low-silica type, which is consistent with the chemical analyses by the Viking landers of the fine-grained material from eolian erosion. The low potassium and aluminium and high iron and magnesium indicated that the source rocks were mainly mafic basalts.[167] In the later stages of the development of the shields, parasitic vents formed on the northeastern and southwestern flanks and extruded vast amounts of lava that spilled onto the surrounding bulge. Only when the shallow reservoirs had drained did the summits collapse to open calderas.[168] However, the complex cavities indicate that the volcanoes were reinvigorated several times. Wrinkle ridges on the floors of some of the calderas indicate that lava lakes solidified *in situ* in the final phase of activity. The crispness of the rims and the absence of slumping and other forms of erosion indicate that the calderas were formed relatively recently.[169] The local free-air gravity anomalies indicate that these edifices do not have individual 'roots', they are not in isostatic equilibrium but are supported by the underlying crustal bulge.[170]

The lavas which spilled down the bulge appear to have engulfed a number of much smaller volcanic cones which had already formed. For example, an arc of Ulysses Patera's caldera wall has been partly destroyed by an impact. The downhill ejecta of another crater on its lower flank has been overrun by the adjacent lava plain, indicating that the volcano is older than the surrounding terrain. Since its flanks have such distinct contacts with the plain and its caldera is disproportionately large, it has been proposed that Ulysses is actually the top of a larger volcano which is projecting through the lava from Pavonis Mons. However, Biblis Patera is a few hundred miles further west and is comparable in size, so their proximity would have constrained the scale of their bases. Although they could have been two peaks on a single edifice, there are no multiple-summit shields elsewhere. Tharsis Tholus on the northeast flank of the Tharsis rise is similar: it is cut by a wide northeast-trending graben that has very abrupt contacts with the plain, and is undoubtedly a short section of one of the Tharsis radials which has since been buried by the lava flows that have run down from Ascraeus Mons.[171] Nevertheless, these smaller volcanoes rise between 2 and 6 kilometres above their surroundings, so they are fairly substantial in their own right.

OLYMPUS

Olympus Mons, Mars's largest volcanic shield, is not located on Tharsis's summit but on the northwestern flank of the bulge. Its base is slightly elliptical, varying between 500 and 600 kilometres wide with its main axis aligned radial to the bulge. Its flank is truncated by a scarp which, at 8 kilometres tall, matches the elevation of the bulge's summit. The entire edifice has been evidently built up by individual lava flows, some of which, particularly in the northeast and the southwest, have 'draped' over the scarp. In some places the scarp is clean, and forms an almost sheer cliff. It follows an irregular line, however, and short sections are radial to the structure. The slope immediately above the scarp is 2 or 3 degrees, so although the summit is 17 kilometres higher it is 200 kilometres distant. About a third of the way up, the mountain's profile becomes a succession of 10-degree ramps leading to broad terraces that were probably created by shallow thrust-faults as the edifice inflated and deflated in response to magma first rising into and then draining from the upper reservoir – a process referred to as 'breathing'.[172] However, although it is far more massive than the shields on the ridge, it does not possess the largest caldera and it has not developed to the same degree as its neighbours as it does not have parasitic flank vents. The 80-kilometre-wide caldera complex is marked by a sheer cliff that falls 2 kilometres to the multifaceted floor, which records at least six phases of activity.[173] An analysis of the stresses which made the prominent concentric grabens near the base of this cliff, and of the wrinkle ridges towards the centre of the main caldera, found that the shallow magma chamber was 10 to 15 kilometres below the summit, so it was far above the elevation of the scarp and at about the level at which the flank yielded to terracing.[174,175,176,177] Although it does not have concentric fractures around its periphery, the edifice is clearly collapsing under its own mass. In fact, it has been suggested that the scarp was formed by thrust-faulting at the base of the mountain which resulted in fault-propagation folding.[178,179]

There are intensely textured lobate features several hundred kilometres out from the scarp, but the origin of these 'aureole' is disputed. One aspect which any model must accommodate is the fact that they were not formed simultaneously. As the large northwestern one partially masks the somewhat smaller northern ones, they are deposits of some sort. Since they cover hundreds of thousands of square kilometres, they represent a vast quantity of material. The striking ridges and grooves imply that the deposits are several kilometres thick. The lobate character strongly suggests that the material was emplaced by surface flows.

It has been suggested that they *are* the material that slumped from the mountain to expose the scarp.[180,181,182] If the aureole represent the structural collapse of the lower flank, the material must have swept outward in a 'base surge'. Interestingly, the three shields atop Tharsis have neither scarps nor aureole. Another proposal is that enormous pyroclastic flows swept down from Olympus, in which case the grooves were readily eroded ash deposits and the ridges were welded ash tuff or ignimbrite.[183] Yet another suggestion drew an analogy with what happens when a terrestrial volcano forms beneath a thick ice sheet: its periphery is confined by the ice, and so it forms a scarp until it broaches the surface of the ice, at which time it starts to build a shield, but this presupposes that when Olympus Mons first erupted, the northwestern flank of Tharsis was under 8 kilometres of ice.[184]

The summits of the trio of shields atop Tharsis are 27 kilometres above the planetary datum, but they are only 17 kilometres tall because they are on the 10-kilometre-tall bulge. Olympus Mons's summit is at the same altitude, but it is actually a much larger edifice because it stands on terrain that is just 2 kilometres above the datum.[185] The fact that they all peaked at the *same* altitude means that growth from their summit vents ceased when the pressure within the crust was no longer able to force magma up through the feed pipes. Once the summits of the shields on the bulge fell dormant, the prevailing pressure opened the parasitic vents

A three-dimensional rendition of the northern flank of Olympus Mons derived from Viking data showing the enigmatic lobate aureole structures.

on the lower flanks on the line of the deep-seated fault, which offers a clue into the deep lava source. Magma rising through a volcano's feed pipe is driven by the hydrostatic pressure induced by the difference in density between the magma and the rock through which it passes. In the case that these densities were comparable to terrestrial values, the magma feeding these volcanoes would have been drawn from a depth of about 250 kilometres.[186] This is much deeper than the source of the magma of the Hawaiian shields, which is at a depth of 60 kilometres. The scale of the Martian shields derives from their broad bases. Their profiles are no steeper than those of the Hawaiian shields. However, their summit calderas are vastly larger than their terrestrial counterparts. The Hawaiian and Emperor volcanic islands form a single chain in the northwest Pacific Ocean and were drawn from a single deep source. Even if the Pacific plate had not been in motion, and *all* of this magma had contributed to a single edifice, it would not have rivalled Olympus. On the other hand it has to be borne in mind that the Hawaiian and Emperor chains were built up over only 80-million years, whereas Olympus has been under construction for several billion years. Most of the volcanoes on Earth today are extremely young by this standard. The Mauna Loa and Kilauea complex on the Big Island of Hawaii is only 1 million years old, and Mount Etna (Europe's largest shield) has been built up over less than half a million years. The vast bulk of the Martian shields, which are striking illustrations of the ability of volcanism to build edifices, derives from the fact that the lithosphere is immobile.[187]

In the last billion years, only Olympus Mons and the trio of large shields on Tharsis have been active. This tail-off of volcanic activity may have reflected mantle cooling and associated lithospheric thickening. However, the thickness of the Martian lithosphere has been a bone of contention. For such a massive uplift to have survived for so long, either the lithosphere must be sufficiently thick to support it or, if the lithosphere is thin, it must be actively supported by the persistent mantle plume. The fact that the Tharsis volcanism persisted long after activity in other areas had ceased implies that the plume is still present. The cratering of the flows which originate in the peripheral vents of the three large shields have produced a wide range of ages, depending upon assumptions. For example, the stratigraphically youngest surfaces have been variously estimated as old as 2.5 billion years,[188] or as recent as 100 million years.[189] A study suggested that activity on Arsia Mons ended some 700 million years ago, Pavonis Mons 300 million years ago, and Ascraeus Mons 100 million years ago.[190] This implied a progression of activity from the summit of the bulge towards the northeast. Olympus's freshest-looking lava flows have been estimated at a mere 30 million years old,[191,192,193] so it is probably dormant rather than extinct.

In addition to lava, the Tharsis volcanoes would appear to have blasted out large clouds of ash. Amazonis Planitia is west of Tharsis, and its radar reflectance is so low that it must constitute a blanket of clay-sized particles with no protruding boulders or even cobbles, so it may be a blanket of volcanic ash that was carried by prevailing winds,[194,195] possibly as much as 15 metres thick.[196]

ELYSIUM

Located north of the line of dichotomy, the Elysium rise is only 1,500 to 2,000 kilometres across, its summit rises only 5 kilometres above the low-lying plain, and its three large shields are somewhat smaller than those atop Tharsis.

The first volcano to form in this area was Hecates Tholus, which is a 180-kilometre-wide shield rising about 6 kilometres. The centre of activity then shifted 850 kilometres to the south to make Albor Tholus, which is a 30-kilometre-wide dome that is topped by a 7-kilometre caldera. The 10-kilometre-tall shield of Elysium Mons was added between them at a later period and is the largest of the group. The flanks of these edifices are much steeper than Tharsis's shields. Both Elysium Mons and Hecates Tholus have comparatively small calderas, just 12 kilometres across. The smooth appearance of their upper flanks has been interpreted as being blankets of pyroclastics.[197,198] The fluted flanks imply fluvial erosion, probably by sapping, as heat leaked out through the crust. The explosive eruptions by magma enriched by volatiles and the fluvial erosion features suggest that these volcanoes probably date back to the time of a 'wet' climate. The lava from Elysium Mons has encircled both of its neighbours and the flows that have run 1,300 kilometres to the east have transitioned into an intercrater plain in the older terrain. The concentric fractures some distance out from the base of Elysium Mons indicate that the edifice has subsided significantly.[199] In places, these fractures have been buried by the flows of Elysium Planitia, indicating that the plains formed after this subsidence.

A system of faults radiating northwest and southeast have created the grabens of Elysium Fossae, but the rise's presence has not imparted sufficient stress on the lithosphere to create a canyon system comparable to Valles Marineris. The fluid that erupted from these fossae, and excavated outflow channels as it drained northwest onto Elysium Planitia, seems to have been permafrost that was melted by the heat of

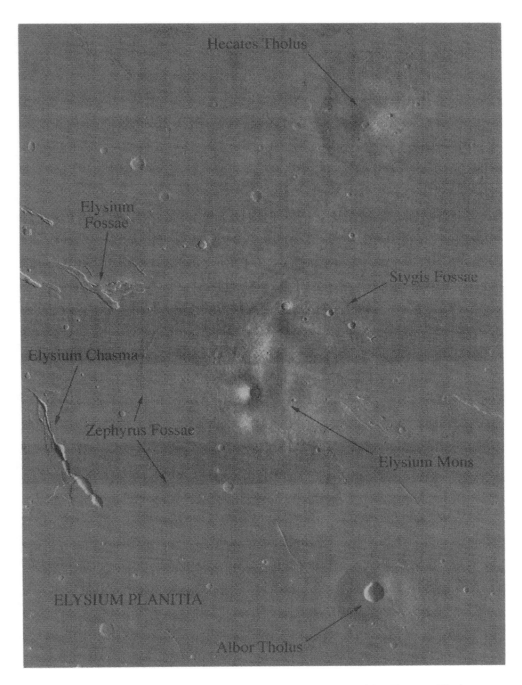

A Viking mosaic of the cluster of volcanoes on Elysium Planitia. Hecates Tholus was the first to form. The centre of activity then shifted 850 kilometres to the south to Albor Tholus, which is a 30-kilometre-wide dome which is topped by a 7-kilometre caldera. Later, Elysium Mons formed between them. Its lava encircled its neighbours. A system of faults radiating northwest and southeast have created the grabens of Elysium Fossae.

magma intrusions,[200] just as glaciers atop terrestrial volcanoes melt as magma rises into the edifice prior to an eruption.[201] The erosional deposits, which spread over an area of 1 million square kilometres on Elysium Planitia, are so sparsely cratered that this flooding must have occurred long after the climate had turned 'dry'.[202]

CANYONS

The Valles Marineris canyon system starts in Noctis Labyrinthus, a network of short deep randomly intersecting gashes incised into Syria Planum. These are not erosional features; they formed when the crust was cracked by extensional stresses.[203] As the atmospheric pressure atop Tharsis is only a few millibars, mists form in the grabens of Noctis Labyrinthus during the night.[204]

As Valles Marineris descends the bulge's eastern flank, it becomes a series of essentially parallel chasms separated by ridges spanning 700 kilometres from one ridge crest to the next. The widest section is Ophir, Candor and Melas Chasmata.[205] These merge to the east to form Coprates Chasma, the primary canyon, which comprises a single trough 600 kilometres wide with a floor lying as much as 7 kilometres below the level of the plain.[206] Coprates opens into the considerably wider Eos Chasma – a mosaic of angular slabs of crust separated by narrow trenches approximately 1 kilometre deep.[207] A number of outflow channels emerge from it, run north down the regional slope, and debouch onto Chryse Planitia. The fault lines in the eastern section of Valles Marineris are less evident, and the canyons are shallower with hummocky floors, so this section of the system was evidently shaped more by erosion than by drop-faulting. At 4,000 kilometres long, the system runs 25 per cent of the way around the planet's equator.[208] In the process it descends from the 10-kilometre-tall Tharsis bulge to terrain that is only 1 kilometre above the datum. The point where Coprates opens into Eos is at an elevation of 4 kilometres and it is the contact between the ridged plain that forms Tharsis's southeastern flank and the ancient cratered terrain to the south.

The canyon walls are fault scarps aligned radially to Tharsis, so they are evidently tectonic in origin. Indeed, the entire structure seems to be a rift where the crust was split asunder. The individual canyons formed when crustal blocks dropped between pairs of faults. The fact that the canyon floor is a depressed section of the plain can be inferred from the preserved craters, but the act of down-faulting appears to have triggered landslides which have masked parts of the floor with aprons of debris up to 100 kilometres across.[209] In one case, a large block of the north wall has slipped to form a terrace. Gullies etch the upper part of some walls but it is not apparent whether this was due to fluvial erosion or mass wastage. Although shallow grabens and lines of pits run parallel to the main canyons, the surface of the plain is strikingly pristine right to the edge of a canyon. The fault lines stand out clearly where there has been little mass wastage, as evidenced by the characteristic triangular terminations of the ridges that run down the wall, forming a line along the foot of the wall.

The canyon system provides a unique opportunity to study a vertical exposure that is vast in both horizontal and vertical scope. Evidence of layering in the walls might record the series of lava flows which built up the plain.[210] Fine patterns on the floor of Coprates Chasma may be the water marks left by lakes that pooled in the deepest parts of the canyon. As the build up of water breached a succession of barriers the lakes would have drained downhill, and ultimately debouched onto the hummocky floor of Eos Chasma.[211]

There is therefore a causal link between the formation of the tectonically induced canyons, the crustal collapse that created the patches of chaotic terrain and the flooding of the low-lying northern plains.

OUTFLOW CHANNELS

The most startling of Mariner 9's discoveries were what appeared to be enormous outflow channels. Apart from two in the southern highlands on Hellas's eastern rim, and some on the flank of Elysium Mons, the outflow channels drain across the line of dichotomy onto the low-lying northern plains. The largest channels debouch into the southern part of Chryse Planitia. This vast drainage system comprises Ares, Tiu and Simud (which emerge from the chaotic terrain at the eastern end of Valles Marineris), Shalbatana (which is derived from a small patch of chaos just north of Eos and crosses Xanthe Terra), Maja (drawn from Juventae Chasma on the eastern side of the ridged lava flow of Lunae Planum) and Kasei (which emerges from Echus Chasma to the west of Lunae Planum, and then runs north before swinging east into Chryse).[212] By way of an example, the Tiu outflow channel – which exceeds 600 kilometres in length, in places spans 25 kilometres across and is over 2 kilometres deep – is just one part of a larger system with Ares to the east and Simud to the west. Significantly, these three channels emerge fully formed from elongated chaotic depressions which are typically 100 kilometres wide. Although some of the major outflow channels have tributaries, these are fairly short and emerge from their own patches of chaos. It has been suggested that these sources formed on ring fractures that are concentric to the impact basin which underlies Chryse Planitia, in which case the permafrost may have been melted by magma that exploited these deep faults.[213] Once the water found an outlet and the 'pore space' was vacated the crust above it collapsed. On the other hand, it has been argued that the flooding from these chaotic zones is on such a massive scale because it was driven by a 'break out' of an aquifer that had been 'pumped up' by gravity-driven artesian flow within the Tharsis bulge.[214] The zone of weakness created by the ring faults of the adjacent impact basin would have provided outlets. The pressure must have been renewed periodically, however, because the channels appear to have flooded repeatedly over an extended period of time. Early estimates of the cratering rate implied that the last Chryse flow occurred only a billion years ago,[215] but a more recent study had pushed this date back to 3 billion years, so the spasm of outflow activity was in the Hesperian.[216,217]

These floods were extremely erosive. They 'sculpted' isolated obstacles on the open plain into teardrop-shaped forms. As they flowed around craters, they swept

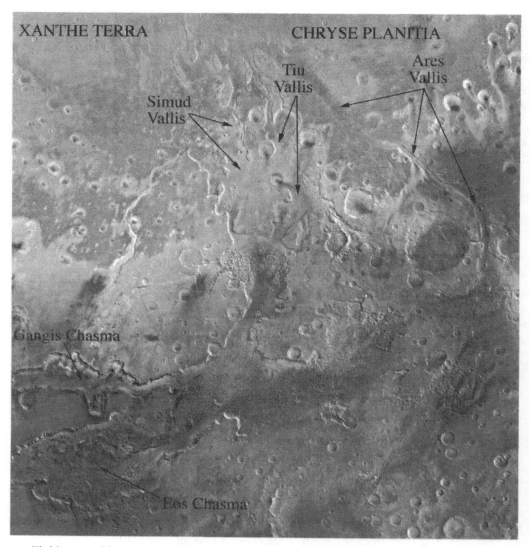

Fluid erupted by the chaotic terrains at the eastern end of the Valles Marineris etched a number of enormous channels which drain north over the line of dichotomy onto the low-lying Chryse Planitia.

away segments of the ejecta blankets and incised the flanks, leaving them standing on a pedestal. Terracing on some of these pedestals could either indicate layering in the rock, or erosion by a succession of flows. Certainly the height of the pedestals provides a dramatic indication of how much material was eroded. Upon being obstructed by a ridge, they accumulated until the level rose sufficiently to overflow the ridge crest, then the flow etched a deep channel as the dam drained. A channel's depth is inversely proportional to its width; that is, the deepest sections mark where a channel sliced a narrow route through a dam. After debouching onto the low-lying plains, they spread out and deposited the suspended debris forming gravel bars

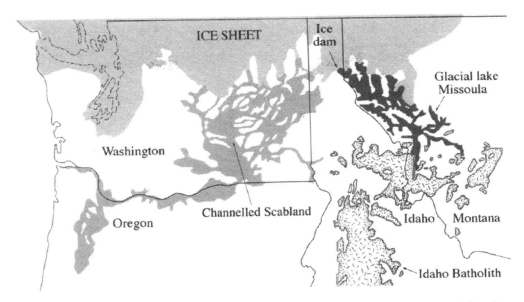

As the polar cap retreated at the end of the last Ice Age, 10,000 years ago, a large lake of melt water (Glacial Lake Missoula) formed in northwestern Montana behind an ice dam in northern Idaho. When this dam burst, the lake rapidly drained in a vast flood. In so doing, it etched the 'channelled scablands' into the Columbia Plateau, creating a landscape strikingly similar to the outflow channels which drain onto Chryse Planitia on Mars.

through which subsequent flows eroded 'braided' channels.[218] As such enormous volumes flowed down these channels, losses to sublimation in the thin atmosphere would have been insignificant even in a 'dry' climate.[219]

The Scablands of eastern Washington State were carved out when an ice dam in northern Idaho broke and released a large glacial lake with a volume of water equivalent to Lakes Erie and Ontario (known as 'Lake Missoula') in Montana. It seems likely that the Martian outflow was similar to this terrestrial flood in terms of how it carved the landscape, if not actually in terms of its origin.[220,221]

On the other hand, it has been argued that some of the features in the outflow channels are the result of slow erosion by artesian-fed glaciers. A comprehensive study found examples of U-shaped valleys, anastomising valleys, hanging valleys, and scouring marks that resemble terrestrial ice-sculpted terrains.[222,223,224] It has recently been suggested that a cryogenic form of 'debris flow' sustained by degassing of carbon dioxide-rich ices etched these channels.[225]

ON MARS

The Soviet Union exploited the very favourable 1971 launch window to dispatch a pair of spacecraft to Mars which released descent probes 4 hours prior to themselves entering orbit to conduct mapping missions.[226] Mars 2's probe entered the atmosphere on 27 November, but it suffered a malfunction and crashed some 500 kilometres southwest of Hellas. Mars 3's probe successfully soft-landed between Electris and Phaetontis on 2 December, but as it prepared to transmit its first view of the landing site it fell silent. Although the consensus is that the probe failed, some of the engineers believe that their lander worked and it was the spacecraft's relay that failed.[227] Although Mars 2 entered the intended 1,280 x 24,900-kilometre orbit, its radio link was so poor that little usable data was received. After suffering a propellant leak, Mars 3 was able to limp into an orbit with a 190,000-kilometre apoapsis, thus severely limiting its opportunities for observing the planet.[228] In any case, however, the dust storm meant that their preprogrammed imaging sequences were futile. Some of the remote-sensing instruments were able to provide useful information — in particular, a radio 'sounder' established that the ochre tracts were probably covered with 'fines'; that is, sand.

The Soviets exploited the 1973 window, but the planetary alignment was not as favourable as in 1971 and the landers had to be dispatched separately from the orbiters. However, the flotilla produced few results. A propellant leak prevented Mars 4 from braking but it was able to take some useful pictures as it made a 2,200-kilometre fly-by. Mars 5 successfully entered an orbit that was almost synchronised with the planet's rotation, but no sooner had it returned images near the Argyre basin than its instrument compartment lost pressure. The landers fared no better. Mars 6 transmitted telemetry all the way down to the surface but then fell silent, and after being prematurely released by its bus, Mars 7 missed the planet. With these débâcles, the Soviets decided to redesign their Mars spacecraft, and meanwhile focus their effort on Venus, where they were having greater success.

When NASA initiated its Viking programme in 1969 to search for life on Mars, it had hoped to use the 1973 launch window, but the cost of developing such a complex mission obliged a two-year postponement.[229] When Viking 1 settled into Martian orbit on 19 June 1976 it took a look at the nominal landing site where the outflow channels debouched onto Chryse Planitia, where it was hoped that there might be subsurface water. When this site was judged to be too rough, a survey of the general area was performed from orbit. This was the advantage of making the landing from orbit, rather than prior to entering orbit. After much debate a smooth-looking sedimentary deposit 60 kilometres beyond the mouth of an outflow channel was chosen and Viking 1 successfully touched down on 20 July.[230] With the historic first landing achieved, the safety criteria for the second lander were relaxed somewhat, and on 3 September Viking 2 landed on Utopia Planitia, which forms part of the Elysium volcanic province.

Viking 1's first image from the surface was a startlingly crisp monochrome view of one of its footpads. It then assembled a panorama that revealed Chryse Planitia to be a landscape reminiscent of the eroded lava flows of the American Southwest. Although it had appeared to be smooth in orbital imagery, it was strewn with rocks ranging from pebbles to boulders, and the vehicle had set down within 8 metres of a boulder 3 metres across.[231] A later colour image showed an orange-tinted vista of rusty rocks and sand dunes, and a salmon-pink sky.

The Viking surface mission was to determine whether there was organic life, and

Table 4.1 Viking surface chemistry

Component	Percentage
Silica	44
Iron oxide	17.5
Aluminium oxide	7
Sulphur oxide	7
Magnesium oxide	6
Calcium oxide	6
Titanium oxide	0.5
Potassium	<0.1

Note: Several significant light elements, including oxygen, were not measured. The total of all measured components is typically only 70 per cent of the sample mass. To get the ratios shown it was *assumed* that everything was fully oxidised as individual fully-stoichiometric oxides; the total was then added up and percentages were normalised to suit.

most of the instruments were designed to test this, but one was to determine the chemical composition of the regolith. After the scoop had retrieved a sample of soil (Table 4.1) the instrument irradiated it with X-rays in order to identify the elements by their fluorescence.[232,233]

Knowing which elements are present is one thing, but inferring the mineral composition is something else. One study concluded that the Martian regolith was probably ferromagnesian clays that contained (in decreasing amounts) a hydrated sulphate of magnesium, calcite, and a trace of titanium oxide.[234,235] Calcite is calcium carbonate, and its formation would have drawn carbon dioxide from the atmosphere. A sulphate would account for the high sulphur ratio. On Earth, carbonates and sulphates are found both in evaporites which are formed when water rich in dissolved ions precipitates minerals to form 'salt flats' and in hydrothermal vents in which volcanically heated water precipitates minerals underground to form rich veins.

Although the analyses were performed on soil, this is thought to be derived from the rocks by erosion so the homogenised samples enabled general conclusions to be drawn regarding the composition of the rocks. Firstly, the low silica and high iron and magnesium abundances implied a high-temperature low-viscosity mafic lava which would have flowed far and wide and formed open plains. The sulphur abundance was higher than in terrestrial rocks, but sulphur is a siderophile and its abundance is consistent with the high levels of iron. The potassium abundance (which was below the instruments' detection limit) indicated a lack of volatiles, which implied that the eruption drew upon a deep source and that the magma was not contaminated by partial melting as it rose through the lithosphere.[236,237]

Studies of the erosion of the volcanic plains of the American Southwest indicated that lava tends to erode by fragmenting along fractures whose spacing is proportional to depth within a flow, so the larger rocks come from greater depths. An analysis of the rocks scattered around the two Viking landing sites concluded that the upper metre or so of these plains had been eaten away, with the rocks being the debris. Clearly, therefore, the erosional rate on Mars is considerably slower than on Earth, where a metre's worth of erosion could be expected per million years – in effect, erosion on Mars proceeds at a rate of only 1/1000th that on the Earth.[238,239] The only sign of ongoing activity was the detection by Viking 2's seismometer of a 'marsquake' which measured 2.8 on the Richter Scale.[240]

The nominal design lifetime of the Viking landers was 90 days, but Viking 1, which was renamed the Mutch Memorial Station in honour of the late Tim Mutch, the leader of the lander imaging team, monitored seasonal variations for fully three Martian years until it fell silent on 13 November 1982.[241,242]

The Viking landers finally established the composition of the atmosphere at the surface to be 95.3 per cent carbon dioxide, 2.7 per cent nitrogen, 1.6 per cent argon, and tiny amounts of oxygen, carbon monoxide and water vapour.[243] This information proved very insightful a few years later.

PIECES OF MARS

Although the ultimate objective for the robotic exploration of Mars is to return a sample of the crust to the Earth for comprehensive analysis, we apparently already possess pieces of the planet in the shape of a dozen or so meteorites which have been classified into three types, the first of which to be found in each case fell on Chassigny in France in 1815, Shergotty in India in 1865, and Nakhla in Egypt in 1911. These are known as the SNC meteorites. However, the significance of their difference to other stony meteorites was not realised until the early 1970s, when a radiological analysis of one of the Nakhlites revealed that it crystallised just 1.3 billion years ago.[244,245] Elemental abundances implied that the parent body of the Nakhlites was a geologically evolved planet-sized object.[246] The case for Mars was initially circumstantial,[247] but the discovery of shock-generated glassy nodules in Shergottites which had trapped noble gases with isotopic ratios precisely matching the Viking atmospheric data strengthened the case.[248,249,250,251] When spectroscopy revealed that the surface of 4 Vesta is basaltic, it looked as if the SNCs might derive from a chemically evolved asteroid.[252,253,254,255] However, when it was discovered that the high isotopic content of heavy-nitrogen that distinguishes the atmosphere of Mars from virtually all other 'volatile reservoirs' in the Solar System was also present in the SNCs, the case was considered solid.[256,257] It was deemed to be proved in 1997 when the Sojourner rover on the Pathfinder mission analysed rocks *in situ* and found that key indicators, such as the ratio of iron to manganese, closely match those of the SNC meteories.[258]

Although Mars's escape velocity is only 5 kilometres per second, this is too high for these rocks to have been blasted into space by explosive volcanic activity, and they could only have been ejected by a large impact. The complex radiometric chronologies of the SNC meteorites, particularly the Shergottites, have been variously interpreted as either the time that they crystallised[259] or the shock of the impact that ejected them into space.[260] Some have argued that a major impact, most likely in the southern highlands, excavated them from the crust and ejected the shocked rocks into space 200 million years ago on trajectories that eventually encountered the Earth.[261,262,263] It has been calculated that such an impact would have left a crater 200 kilometres in diameter, so it should be possible to identify it.

The SNCs and the chemical analysis of the soils on Chryse and Utopia Planitia suggest that the composition of Mars's mantle is 50 per cent olivine, 25 per cent orthopyroxene, 15 per cent calcic clinopyroxene and, 10 per cent garnet, and thus is similar to the Earth's mantle (but with more iron).[264]

On Earth, the lava which erupts onto the surface from deep magma sources is basaltic, but the lower lithosphere is composed primarily of pyroxene and olivine.[265] Partial melting within the lithosphere will yield an alkaline lava rich in sodic and potassic volatiles. As the temperature increases, the iron- and magnesium-based refractories will join the melt, and yield progressively more mafic lavas. Temperatures of 1,700 to 2,000 K are sufficient to induce a partial melt of 40 per cent and create an ulramafic lava called komatiite. On Earth, this lava was produced in the early Precambrian, when the mantle was hotter, the heat flow was greater and the lithosphere was thinner. Interestingly, komatiitic volcanism appears to predominate on Io, which is incredibly hot for its size because its interior is heated by gravitational tidal stresses. Even though Mars's mantle would have been cooler than the Earth's, if it was enriched by iron then ultramafic lavas would have been created at lower melting ratios, and komatiite may have been common. It has a lower viscosity than basalt, and so would have flowed even more readily than the fluid basalt that pours from vents on Hawaii and so formed the Martian plains. The Chassignites are sufficiently rich in olivine to be called dunite; the Nakhlites are pyroxene-rich and the Shergottites are basalts combining feldspar, pyroxene and olivine. They probably formed by a process of sedimentation in which denser minerals progressively settled within a magma chamber, forming differentiated strata of cumulates.

PATHFINDER

In July 1997, NASA's Mars Pathfinder settled upon the mouth of the Ares Vallis outflow channel on Chryse Planitia.[266] Although the site was rougher than that on which Viking 1 had set down, the use of inflatable bags to cushion a 'hard' impact had made it accessible. Once it was safely on the surface it was renamed the Carl Sagan Memorial Station.[267] It then released a rover, called Sojourner, equipped for field geology. The light-travel time meant that although the robot was directed from Earth it could not be controlled in real time, so its smart processor was capable of autonomous activity within this overall direction.

Sojourner's 'baseline' mission was only one week but it survived for almost three months. Its Alpha particle, Proton and X-ray Spectrometer (APXS) performed ten chemical analyses of eight rocks and two soil samples. It had been hoped that by landing on the flood plain beyond the mouth of a major channel believed to have been carved by a flood it would be possible to sample rocks from a wide variety of sources, and this did in fact happen. It was a rolling plain, strewn with rocks ranging in size from pebbles to boulders. Some of the rounded rocks resembled those on Earth that have been swept along by water, and there was also a stack of large boulders, partially embedded in the ground, leaning upon one another as if they had been deposited by flowing water. The scale of the litter was certainly suggestive of a massive flood.[268,269,270,271] Nevertheless, the case is not conclusive. It has been proposed that the channel was carved by a (dry) debris flow,[272] or by glacial action.[273] Working on the assumption that it was liquid water, Michael Malin speculated that it flowed at a rate of 100 million cubic metres per second. Although exceedingly erosive, such an enormous outpouring may have lasted just a few days.

The first rock to be examined by APXS was 'Barnacle Bill', so-named because of the big knobs on its surface. The result was surprising.[274,275] It had a higher silica ratio than had been expected on the basis of the compositions of the SNC meteorites and of the soils measured by the Vikings. In fact, this composition was similar to a terrestrial 'dry' andesite.[276] 'Yogi', the second rock, had a composition similar to terrestrial basalt. The variety of silica-rich compositions meant that the crust had undergone considerable chemical differentiation. Indeed, on Earth, free silica is characteristic of continental crust. Mars is therefore a rather more geologically evolved planet than had been believed. The significance was the *heat* needed to drive this igneous evolution, but when did the 'heat engine' shut down?

A detailed visual inspection of the rock textures[277] confirmed the igneous form of many of the rocks, and revealed that some of the boulders are conglomerates in which pebbles are bound up in a fine matrix. The sedimentary rocks were evidently formed during a flood when a suspension settled out. Some of these boulders had small surficial cavities, as if pebbles had been eroded out, and indeed, there were pebbles lying around that may have been shed in this manner. The rounded form of these pebbles implies that they had undergone considerable erosion prior to being incorporated into the conglomerate. Furthermore, the fact that the conglomerates were delivered to this site intact implied that they had formed either in a river bed or in the aftermath of an *earlier* flood. Other conglomerates, particularly the angular ones, may be ejecta from the impact that formed 'Big Crater', several kilometres to the south of the landing site.

With the benefit of hindsight, the Sojourner scientists belatedly wished that they had fitted the robot with a brush so that it could clean the rocks of 'pollution' prior to sampling. Indeed, it has been argued that the readings suggesting andesitic rocks were misleading: if these rocks had become coated with a silica-rich 'rind', then the chemical analyses would have measured this surficial layer rather than the underlying rock.[278]

The high sulphur content in some of the rocks may have been due to a thin coating of dust as the Vikings had shown the soil to be rich in sulphur. This was confirmed by Sojourner's soil analysis which thereby strengthened the theory that the global dust storms homogenised the Martian dust so that it is now the same everywhere. Significantly, despite the discovery of andesitic rocks, the analyses implied that the dust is derived from weathered basalt.[279] In addition to confirming the Viking discovery that there is a highly magnetic mineral in the dust, the Pathfinder lander's magnetic properties experiment showed that this is at least partly

These images were taken by Mars Pathfinder (the Carl Sagan Memorial Station) on the 'flood plain' of the Ares Vallis. The 'Twin Peaks' on the horizon stand 30 metres tall. The northern peak (on the right) is about 860 metres west of the landing site and the southern peak is 1,000 metres away. 'Barnacle Bill', the first rock to be examined by Sojourner's APXS, has a silica-rich composition resembling 'dry' andesite from Iceland. 'Yogi', the second rock, is similar to terrestrial basalt.

due to maghemite. On Earth, maghemite is found in iron-rich aqueous solutions that have been freeze-dried. The maghemite in the dust may be evidence of a warm and wet environment early in Mars's history.

The Pathfinder lander greatly exceeded its one-month baseline mission and continued to report on the local weather while slowly assembling a stereoscopic 'super panorama' of the landing site at maximum imaging resolution and at a dozen wavelengths. It had been hoped to monitor the onset of the northern winter, but the lander fell silent on 27 September 1997 perhaps because its battery malfunctioned and so the transmitter froze during the night. Nevertheless, its meteorological sensors provided 100 times better temporal resolution than those on the Viking landers, so its data represented a significant advance.

THE CORE

Throughout the time it was operating on the surface, the doppler on Pathfinder's radio signal was monitored to establish the planet's rotational state. When integrated with the Viking results to form a 21-year baseline, the data revealed that Mars's axis precesses over a period of 170,000 years (the corresponding cycle in the case of the Earth is 26,000 years). From this, it was possible to infer the planet's moment of inertia, and hence deduce how mass is distributed within the planet. If the mantle really is similar in composition to the SNC meteorites, and the core is iron, then the core must be of the order of 2,600 kilometres in diameter. If the core contains a significant amount of sulphur (which is less dense than iron) then it will be slightly larger, at 4,000 kilometres. The core is therefore relatively large, spanning 40 to 60 per cent of the planet's diameter.[280]

Until the early 1980s, it was believed that Mars's core formed subsequent to the period during which (possibly as much as several billion years after) the planet accreted, once radioactive heating had raised the temperature of the interior to facilitate thermal differentiation via gravitational separation.[281,282] This was deemed to be consistent with the evidence of extensional tectonism which was assumed to have been the result of expansion as the planet's interior slowly warmed.[283]

However, contradictory insight into Mars's early development has been provided by the SNC meteorites. They indicate that as the planet accreted, the heat of the massive early impacts penetrated deeply and induced the chemical differentiation of the interior into a core, a mantle and a crust.[284] Ratios of radioactive elements indicated[285] that the core formed within a few hundred million years of the planet's accretion, just as in the case of the Earth.[286]

While Mars Global Surveyor was repeatedly dipping into Mars's upper atmosphere to use controlled aerobraking to adjust the apoapsis of its orbit, its periapsis was temporarily lowered below the planet's ionosphere and the magnetometer was able to detect a magnetic field with a strength of 0.00125 of that at the Earth's surface. This had not been detected by earlier orbiters because it had been masked by the ionosphere. The spatial variations in the field implied that it is not a global dipole but a remanent field preserved in the crust, which implied that

Mars once had a dipole 5 to 10 times stronger than the Earth's present-day dynamo. The magnetic pattern in the crust indicated that the field had undergone intermittent polarity reversals.[287,288,289] At each low pass, the spacecraft's magnetometer produced a magnetic profile for a narrow strip of the surface. However, this survey was pre-empted in early 1999 when the periapsis was raised to initiate the primary mapping mission, so the survey is frustratingly coarse.[290] Nevertheless, if it were not for the structural problems with the solar panels which had necessitated a longer than planned period of aerobraking, the magnetic anomalies would not have been detected, so this important discovery was a serependitious bonus.

In the northern hemisphere, the remanent field is weak and is randomly aligned, but in the ancient southern terrain the anomalies form stripes of alternating polarity oriented east-to-west. Without plate tectonics, this pattern cannot be due to lithospheric spreading. Furthermore, the well-preserved craters of the ancient terrain rule out horizontal crustal displacements after the tail-end of the bombardment. One theory is that as lava that was erupted from long fissures crystallised, it 'fossilised' the prevailing field. However, there is no correlation between these anomalies and the pattern of volcanism in the intercrater plains. Furthermore, the large basins in the south do *not* show distinct magnetic patterns. As the dipole field must have diminished by the tail-end of the bombardment, the conditions required for the core to generate a magnetic field would appear to have *lasted* only a few hundred million years.[291] Hence, this magnetisation predates most of the craters on show, which would seem to have struck an intensely volcanic crust.[292] The fact that this ancient cratered surface is preserved in the Martian highlands means that the shock-metamorphosed rocks from the tail-end of the bombardment are still *in situ* and so are available for sampling by future field geologists.

Mars's post-accretional history is one of secular cooling: being small, the planet has a high ratio of surface area to volume, so vigorous early mantle convection would have rapidly drawn off the heat from the core and prompted the earliest phase of volcanic activity, which produced the explosive highland paterae and the floods of lava that formed the plains. The plains were ridged by compressional tectonic forces later on, but as the crust thickened this activity gradually abated. The remaining plume activity sustained the Elysium and Tharsis uplifts, and their associated volcanism.

PLATE TECTONICS?

While the Valles Marineris canyon system shows that Mars has undergone tectonic rifting this does not imply incipient plate tectonics, because the vast rift is due to the existence of the Tharsis bulge.

A terrestrial analogue for Valles Marineris, although considerably smaller, might be the East African and Red Sea rift valleys which mark where the forces of plate tectonics are dissecting a continental lithospheric plate. Although these fractures formed when a mantle plume impinged upon the underside of the lithosphere and 'domed' it, this deformation is considerably smaller than Mars's Tharsis bulge.

In this topographic map by the Mars Global Surveyor's laser altimeter, the line of dichotomy between the smooth lowland plains to the north and the rough southern highlands is apparent. (Courtesy of the MGS/MOLA Science Team.)

In fact, the north shore of the Horn of Africa, where the East African rift meets the rift that runs up the Red Sea, is actually a 'triple junction' because a mature submarine spreading ridge runs out through the Arabian Sea and on into the Indian Ocean. A system of rifts spaced at 120 degrees is a characteristic of doming the Earth's lithosphere. In the case of Tharsis, however, this did not occur. Nevertheless, the stress on Tharsis's summit is evident from the fact that it opened not only Valles Marineris to the east but also the grabens of Claritas Fossae to the south.

We can state with confidence that the Martian highlands are not divided into mobile plates that jostle one another, because the large craters are neither dissected by faults nor distorted by regional compressional forces. This conclusion is consistent with the absence of belts of 'fold' mountains and transform faults crossing the highlands. It had been suggested that the northwest-trending Gordii Dorsum escarpment in the transition zone southwest of Tharsis could be the surface expression of a lithospheric fault that underwent about 35 kilometres of left-lateral motion,[293] but this was recently discovered to be a ridge between two previously unrecognised outflow channels. Some have suggested that the Phlegra Montes on the northern plains east of Elysium are the surface expression of such a fault.[294]

One study explored the possibility that the hemisphere north of the line of dichotomy is the result of *vigorous* plate tectonics.[295] In this theory, the southern highlands remained as a single unbroken plate and the northern region was created by a spreading margin which had its Eulerian rotational pole in the volcanic province of Syrtis Major Planum. It was also speculated that the Tharsis bulge marks where this plate was consumed by a subduction trench. By analogy with terrestrial processes (such as in Washington State where the subducting oceanic plate produced the volcanoes of the Cascade Range spaced out in a line 50 kilometres apart and 50 kilometres inland) it was suggested that the large shields in line on Tharsis might be the result of melting stimulated by subduction. All of this, of course, is speculative, but it makes the point that we ought not to dismiss the possibility that plate tectonics started on Mars, only to be stifled, as it may well explain some otherwise mysterious features.

Surveyor's analysis of the planet's thermal emission established that two distinct surface compositions are present in dark albedo regions. Their distributions are divided more or less along the planetary dichotomy, with the ancient southern highlands and Syrtis Major Planum being basaltic and the younger northern lowlands being dominated by a more silicic lava, suggesting that Martian volcanism may have become more evolved over time. There are no widespread komatiite or SNC compositions however, suggesting that these types of rock are present only at depth.[296]

THE SEARCH FOR WATER

As Surveyor conducted its mapping mission, it revealed striking evidence of sedimentary rock formations.[297,298] Outcrops occur in a variety of locations around the planet: inside craters, between craters, and within chasms. Three main outcrop

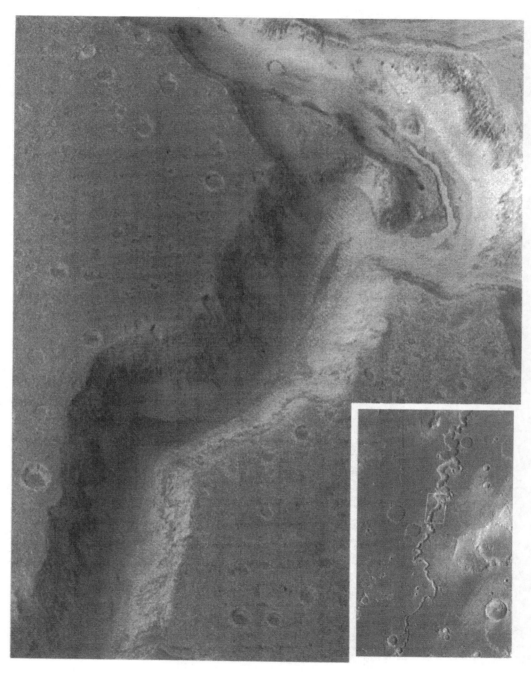

As one of Mars Global Surveyor's earliest images, this view of Nanedi Valles was a startling improvement over the Viking imagery (insert). Not only did it show horizontal layering in the valley wall, and a bench on the floor of the valley on the inside of the bend, but there was also a narrow channel etched in the adjacent valley floor that looked as if it had been cut by running surface water.

types were discerned. Layered units are relatively thin rock beds, some just a few metres thick, which are neatly stacked on top of one another in distinct groups. Massive units are bulky with no clearly defined bedding. In some cases, massive units were found together with a layered unit, although in every case the massive unit was found on top. The caps of eroded massive or layered units are shallow mesas with surfaces ranging from smooth to pitted, ridged and grooved. All of these outcrops appear to be fine-grained materials in horizontal layers. In some places, hundreds of

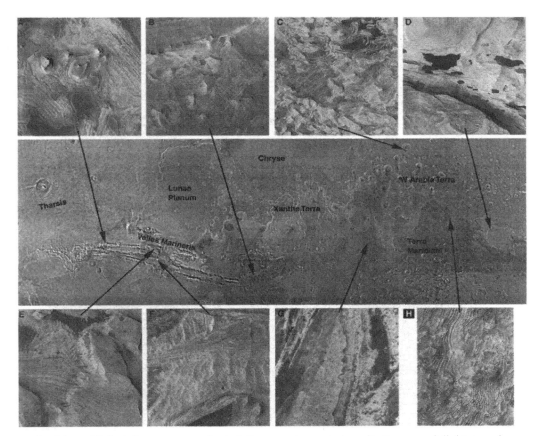

As Mars Global Surveyor conducted its mapping mission, it discovered light-toned, layered, cliff-forming material exposed at a variety of locations which some researchers consider to be ancient sedimentary rock formations. The outcrops were found inside craters, between craters, and within chasms, suggesting that they were deposited in lakes confined within these areas, and the sediments formed as the lakes evaporated. The lack of channels where water flowed into these depressions has been interpreted as meaning that subterranean water was driven to the surface by artesian pressure as the local 'water table' rose. Their contexts indicate that the layered deposits were formed very early in Martian history, between 4.3 and 3.5 billion years ago. If life developed on Mars during this time period, then fossils or other indications of that life may be contained within these sediments, as they are on Earth. (Courtesy of M.C. Malin and K.S. Edgett of Malin Space Science Systems.)

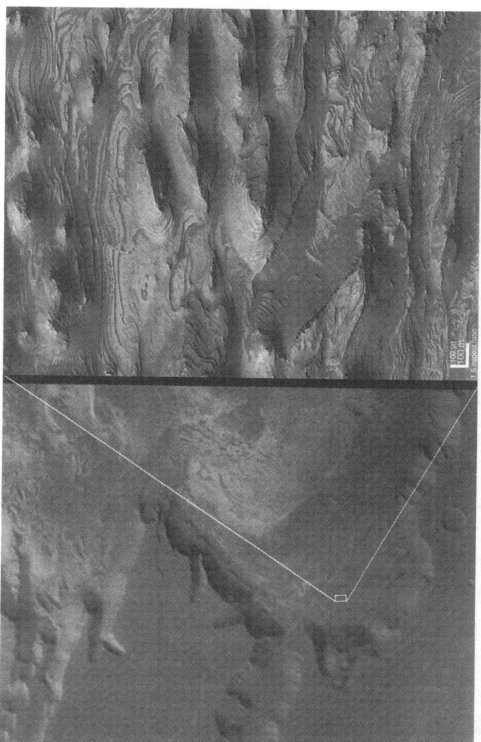

Mars Global Surveyor found layered deposits on the floor of West Candor Chasma in Valles Marineris suggestive of sedimentary deposition in ancient lakes.

individual beds are exposed. In a few cases there are stacks of alternating light and dark deposits several kilometres thick.

Although it is *possible* that these deposits were accumulated during intense dust storms in ancient times when the atmosphere was denser, it seems more *probable* that they were formed in standing water because although the exposures occur at a wide variety of locations, they are preferentially found in confined locations such as canyons and deep craters where water could be expected to have pooled. The lack of channels where water flowed *into* these depressions has been interpreted as meaning that subterranean water was driven to the surface by artesian pressure as the local 'water table' rose. The fact that the pools were confined implies that they stood as lakes and eventually evaporated. The surrounding contexts indicate that these layered deposits were formed *very early* in Martian history, between 4.3 and 3.5 billion years ago.

OCEANUS BOREALIS?

As soon as the Viking Orbiter imagery began to stream in, it was noted that there were vast deposits of sediments and low-lying terrains 'patterned' with giant polygons. Both these features were interpreted as evidence that there was once a large body of standing water north of the line of dichotomy.[299] The Mars Orbiter Laser Altimeter (MOLA) aboard Surveyor could measure elevations to within a metre, which finally facilitated global topographic mapping. As Surveyor performed its aerobraking, MOLA indicated that the northern plains are *remarkably* flat.[300] The first topographic map, which was issued in early 1999 when the spacecraft finally started its mapping mission,[301,302,303] revealed that over distances of many hundreds of kilometres the lowlands vary in elevation by just a hundred metres or so. Furthermore, the elevation profiles along the line of dichotomy were strikingly reminiscent of the terrestrial continental slopes that run down onto the abyssal sedimentary plains. Intriguingly, benches of uniform elevation are situated around the edges of some of the northern plains suggestive of beaches at sea level. These discoveries revived speculation that the northern plains are the floor of an ancient ocean.

As Timothy Parker studied the Viking imagery of the northern plains he noticed subtle arcs, stripes and ripple-marks that bore a remarkable resemblance to the margins of Lake Bonneville, a now-dry lake that flooded parts of Utah, Nevada and Idaho some 100,000 years ago.[304] When a thorough study identified several such features, it was suggested that a series of oceans with differing amounts of water had built up and retreated.[305] It was speculated that these oceans had formed from permafrost that melted and rose to the surface when magma intruded into the crust.[306] As Surveyor pursued its main mapping mission, it became possible to construct a detailed topographic map of the northern plains to determine whether these features *could* be shorelines on the premise that they would have to maintain a fairly even elevation.[307] Unfortunately, the more expansive of Parker's two candidates, the so-called Arabian shoreline, followed an undulating line. However, the 'shallower' Deuteronilus shoreline followed one contour remarkably well along its *entire* circumference. It does not deviate by more than 280 metres from its mean, and where the 'fit' is poor it is obvious why this is so. For example, on Elysium Planitia, the shoreline has been overrun by lava flows. To James Head the data provided four types of quantitative evidence in favour of there having been an ancient ocean within this putative shoreline.[308] As noted, the feature is nearly a level surface. The topography is more subdued below this level than above it, which consistent with smoothing by sedimentation. A series of interior terraces are highly suggestive of a receding shoreline. Finally, the regional topography shows that the volume contained within the shoreline is consistent with estimates of the amount of water available on the planet.[309] Micheal Carr has argued that when the outflow channels debouched their floods onto Chryse Planitia the planet was cold and dry, but Head has concluded from the fact that over a distance of 2,200 kilometres six major outflows debouched within 180 metres of the elevation of the Deuteronilus shoreline that the outflows drained into an ocean whose depth was about 600 metres on average and 1,500 metres at its deepest point.[310]

However, Michael Malin did not find the high-resolution imagery of selected parts of the putative shoreline very convincing.[311] Perhaps on a planet that is dominated by eolian erosion, ancient shorelines may be best observed from a distance because they are so badly eroded that they become difficult to identify close up. Rather more disconcerting is the lack of evaporites from when the ocean eventually evaporated. However, the next year a study of the interior of a Nakhlite established the presence of water-soluble ions, either from a hydrothermal system or from an evaporating brine pool. Sodium and chlorine were the most abundant elements so the Martian oceans seem to have been similar to the Earth's in terms of chemical composition and ionic concentration.[312] Despite this circumstantial support, there is a serious problem: if there was an ocean, even one that was capped by a sheet of ice, then there *ought* to be a substantial deposit of carbonate rock lining the now-dry sea bed, but even where impact craters excavated the covering layers of eolian sediment, Surveyor's Thermal Emission Spectrometer did not find *any* carbonates. The case for Oceanus Borealis has ebbed and flowed over the years, with the same data being interpreted both 'for' and 'against'.[313] For example, the features in the MOLA data that Head believes to be beaches left by receding shorelines are thought by others to be of tectonic form.[314] They display the characteristic wrinkle ridge profile of compressive stress, and appear to be related to regional stress centres, such as the Tharsis rise.[315] Furthermore, as Surveyor's documention of the northern lowlands inexorably increased, Head was obliged to acknowledge that "comprehensive analysis of hundreds of high-resolution ... images shows little compelling evidence for features that can confidently be interpreted to represent shorelines",[316] although it is unclear exactly what billion-year-old shorelines would look like in any event. At this juncture, notwithstanding the recent 'negative' findings, the ultimate fate of the Oceanus Borealis hypothesis is uncertain.

Recently, Nick Hoffman of La Trobe University in Australia has not only argued that there never was an ocean but also that the outflow channels were not made by floods of water. In his view, Mars has been cold and dry for most of its history and the channels were eroded by gravity flows of supercold carbon dioxide gas erupted from shallow subsurface 'liquifers' of carbon dioxide. In this 'White Mars' theory, it was 'cryoclastic' flows which accumulated the vast sedimentary deposits on the northern lowlands.[317,318] A cryoclastic flow would differ from a pyroclastic flow in one key respect. The energy that drives a pyroclastic flow derives from volcanically heated rock, which rapidly cools. As a result, even with supersonic flow speeds, this cooling limits the range to at most several hundred kilometres. The energy source for the Martian cryoclastic flows, however, would be the degassing of carbon dioxide-rich ices. This process is constrained by the rate of energy supply to the crystals,

A depiction of the hypothetical 'Oceanus Borealis' according to the 'inner contact' which some researchers have interpreted as an ancient shoreline. (Courtesy of MGS/MOLA Science Team. Rendering by Peter Neivert of Brown University.)

which in turn is controlled by internal friction. If the flow began to fade, the friction would increase and the grains would be compressed and warmed, and so would release more carbon dioxide that would sustain the flow and enable it to travel considerably further.

On Earth, pyroclastic flows can be produced in a variety of ways. Some are generated by the collapse of an eruption column above an explosive volcanic vent, and others are generated by the collapse of 'pasty' lava domes, or as secondary flows when thick ash falls slide off the edge of steep hillsides. This latter mode of origin is similar to how the cryoclastic flows might have formed on Mars. The collapse of layered terrain comprising dust, thin layers of rock, and volatile-rich ice would generate the turbulent cloud of the density flow. As a result such flows would emerge essentially fully formed from collapse zones (precisely as the outflow channels emerge from the chaotic zones) and build up a thick sediment of debris in the low-lying terrain to the north.[319] In all likelihood, such activity would have been

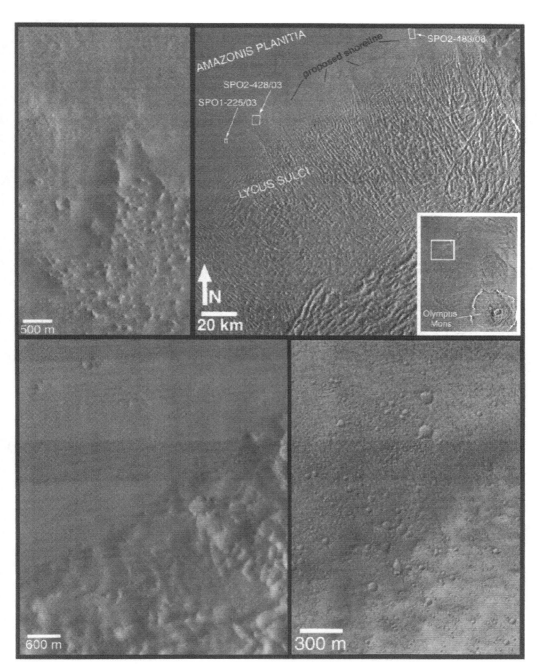

On the basis of Viking imagery of the surface north of where Olympus Mons's Lycus Sulci aureole deposit meets Amazonis Planitia, some researchers discerned what appeared to be an ancient shoreline, but when Mars Global Surveyor inspected it in detail the case became very subjective.

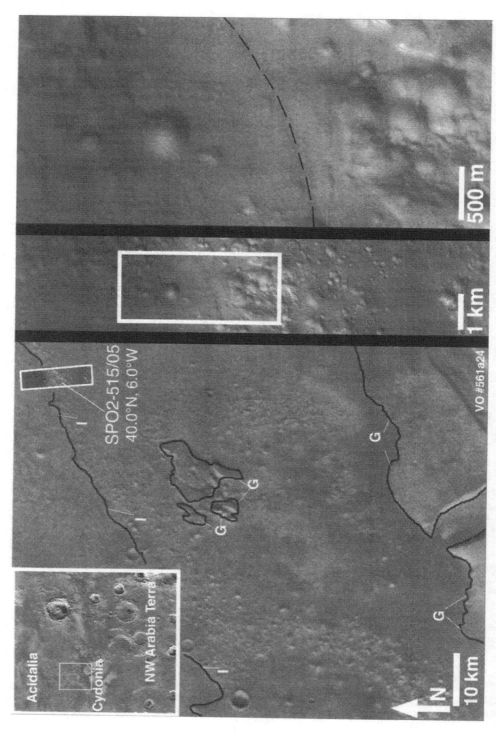

When Mars Global Surveyor inspected the putative 'shoreline' in the Acidalia Planitia–Cydonia Mensae region, the evidence was ambiguous.

stimulated by magmatic intrusions.[320] It has even been argued by the 'White Mars' proponents that the line of dichotomy (and the northern lowlands that it delimits) is due to a single hemispheric-scale episode of catastrophic venting of carbon dioxide and associated crustal collapse.[321] The great appeal of this theory is that because cryoclastic flows require Mars to have been cold and dry, it avoids the need to explain how the climate transformed from an early warm and wet environment into the current cold and dry one.[322]

Surveyor's spectrometer found evidence that Mars has been cold and dry for a very long time. It found what appears to be a sulphate in the pervasive dust and many bright soils. Sulphates had been expected because the Viking landers found the soils to be rich in sulphur, but they had not identified the host mineral. It could have been a sulphide (sulphur attached to a metal) or a sulphate (sulphur attached to a trio of oxygen atoms). In addition, the spectrometer detected olivine and pyroxene. These minerals were distributed predominantly in the darker basaltic volcanic rocks in the southern hemisphere, but also in the Valles Marineris canyon system.[323] The dust includes trace amounts of fine-grained hematite that is probably the result of mechanical erosion of the coarser grained rocks. The spectrometer found a coarse-grained hematite in a 300-kilometre-wide deposit in the equatorial zone near the meridian. It may be significant that this correlates to a large 'unconfined' ancient sedimentary deposit.[324,325] This coarse-grained hematite is different from the fine-grained hematite in the dust. The much larger grains may have developed when geothermally heated water rich in dissolved iron minerals pooled on the surface for some time before finally leaving behind the minerals when it evaporated.[326] However, this process could also be expected to have left other minerals which have *not* been seen.[327] Furthermore, the presence of water and chemical weathering would also have made clay minerals, but these have *not* been seen.[328] On the other hand, the lack of minerals from chemical weathering is consistent with the olivine abundance, in that it implies that the rate of chemical weathering has been very slow, which in turn implies that the surface has been cold and dry for a long time. Since olivine readily weathers in a warm and wet environment, Mars must have been cold and dry when the olivine was emplaced, presumably in the form of lava eruptions billions of years ago. If water surged down the outflow channels onto the northern plains as recently as some people have suggested, then the planet must have been cold and dry at the time. Thus, even if water was flowing down these channels, this fact in and of itself ought not to be interpreted as evidence of the climate being warm and wet at that time.

RECENT ACTIVITY?

Early in aerobraking, Surveyor had spotted indications that liquid water may have flowed on Mars's surface in the *recent* past.[329] Gullies etched into the walls of a few southern hemisphere impact craters and Nirgal Vallis and Dao Vallis (two of the larger channels) suggested ground-water seepage and runoff. More were found as the spacecraft began to methodically map the planet, and in late 2000 Michael Malin

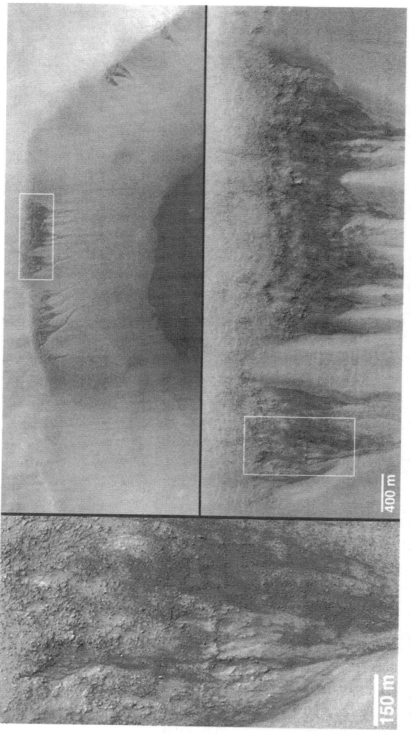

The first clue that there might be places on Mars where liquid groundwater seeps out onto the surface was revealed by the Mars Global Surveyor even as it was undertaking aerobraking, in the form of dark V shapes that taper downslope as narrow, somewhat curved channels on the wall of a 50-kilometre-wide impact crater in southern Noachis Terra. This was interpreted as seepage where springs emerge on a slope and water runs downhill. Later in the mission, new imagery showed that the dark features host many small channels that coalesce downslope at the apex (or point) of each 'V'. This interpretation was supported by a large number of boulders, some of them the size of houses, that have eroded out of the crater wall.

Mars Global Surveyor revealed massive stratification in the wall of Coprates Chasma.

and Ken Edgett published a list of over 100 tiny channels and gullies.[330] The fresh appearance and the absence of eolian sedimentation or erosion meant that they were formed within the last few million years. It was tempting to assume that they were etched by running water, but many of the sites are far from the equator, where the ground is likely to be frozen to a depth of several kilometres, whereas the gully sources are only a few hundred metres downslope. Nevertheless, if water *was* being driven towards the surface and was bursting from strata exposed in steep slopes, the distribution of sites suggested that there are regional subterranean aquifers.

William Hartmann has reported crater counts which suggest that lava flows on Amazonis and Elysium Planitia were erupted within the last 30 million years, so perhaps the associated geothermal energy sporadically melts the underside of the permafrost to form these aquifers.[331,332] On the other hand, Nick Hoffman has said that the gullies in the polar regions could have been made by localised carbon dioxide venting, with the flows being dry cryoclastics.[333] However, the fact that these sources did not prompt extensive collapse, and the flow features are so suggestive of liquid water could mean that the gullies were etched by clathrates, which would erupt far less violently. Michael Carr has argued that if a carbon dioxide clathrate was vented from a permafrost exposure, the volume would contain about 1 per cent of water, so a muddy apron would form downslope of the source. Unfortunately, the spatial resolution of Surveyor's spectrometer was insufficient for it to be able to isolate the chemical composition of these small aprons from their surroundings.

AN OPEN BOOK

In an early image of the wall of Tithonium Chasma in Valles Marineris, Surveyor spotted a 1,000-metre-high triangular cliff at the end of a ridge displaying 80 layers in outcrop.[334] It had been thought that, apart from a thin veneer, most of the Martian crust was undifferentiated, but this observation established that it is intensely stratified. As imagery from other sections of the canyon system was secured, it was realised that this layering is present across a broad area.[335] Because there has evidently been no lithospheric recycling, this layering may well represent a cross-section dating back to the earliest times. One day, field geologists *in situ* will be able to 'read' the planet's history just like turning the pages of a book.[336]

NOTES

1 This dark feature was subsequently named the Kaiser Sea, and is now known as Syrtis Major Planum.
2 *Mon. Not.*, vol. 28, p. 37.
3 It is now known that the centre of the residual cap (the fraction which remains during summer) in the southern hemisphere is displaced 400 kilometres from the rotational pole; in the north, the offset is only 65 kilometres.

4 Schiaparelli named the dark rifts Rima Australis and Rima Angusta, and the patch of the cap that they isolated Novissima Thyle.
5 *Patrick Moore on Mars*, P. Moore. Cassell, p. 68, 1998.
6 For an excellent account of these early mapping efforts, see *The planet Mars: a history of observation and discovery*, W. Sheehan, University of Arizona Press, 1996.
7 *Patrick Moore on Mars*, P. Moore. Cassell, p. 75, 1998.
8 'Volcanism and aeolian deposition on Mars', D.B. McLaughlin. *Geol. Soc. Am. Bull.*, vol. 65, p. 715, 1954.
9 'The volcano-aeolian hypothesis of Martian features', D.B. McLaughlin. *Pub. Astron. Soc. Pacific.*, vol. 68, p. 211, 1956.
10 'Martian phenomena suggesting volcanic activity', T. Saheki. *Sky & Telescope*, vol. 14, p. 144, 1955.
11 *Mars*, P. Lowell. Longmans & Green, 1895.
12 *Mars and its canals*, P. Lowell. Macmillan Co., 1906.
13 *Mars as the abode of life*, P. Lowell. Macmillan Co., 1908.
14 'The detection of water vapour on Mars', H. Spinrad, G. Munch and L.D. Kaplan. *Ap. J.*, vol. 137, p. 1319, 1963.
15 *The planet Mars: a history of observation and discovery*, W. Sheehan, University of Arizona Press, p. 128, 1996.
16 *Patrick Moore On Mars*, P. Moore. Cassell, p. 81, 1998.
17 'An analysis of the spectrum of Mars', L.D. Kaplan, G. Munch and H. Spinrad. *Ap. J.*, vol. 139, p. 1, 1964.
18 Nitrogen did not offer any convenient absorption lines with which to measure its concentration in the Martian atmosphere, but it was difficult to see what else the predominant gas could be. This was a case of reasoning by analogy with the Earth – always a risky proposition in the field of comparative planetology.
19 *The planet Mars*, E.M. Antionadi. Herman Co., 1930; English translation by P. Moore. Reid Co., 1975.
20 *Life on Mars*, F.L. Jackson and P. Moore. The Scientific Book Club, Routledge & Kegan Paul, 1965.
21 The 'minimum energy' transfer orbit was explored mathematically by W. Hohmann in Germany in 1925.
22 The Mariner spacecraft shared systems with the Ranger lunar probes. At this early stage of space exploration NASA built pairs of vehicles for its deep space missions in case either a booster failed or a spacecraft was lost. The first two Mariners were assigned to Venus in 1962. Mariner 1 was lost at launch. Mariner 2 made the first Venus fly-by. The Soviet Union had attempted to dispatch a probe to Mars in October 1960 but it had failed to attain Earth orbit and went undeclared. Two further probes were launched in November 1962. The first was lost but the other made it into orbit and was named Mars 1 once it was safely on its way. On 21 March 1963 it fell silent a hundred million miles from Earth. Although when it flew by at a range of 200,000 kilometres in June it became the first human artefact to reach Mars, it was inert. The first Zond was dispatched to Venus in April 1964 but it too fell silent before reaching its objective. Mariner 3 was intended for Mars, but its aerodynamic shroud failed to release following launch on 5 November 1964. Zond 2 set off for Mars two days after Mariner 4 on 28 November 1964, but it fell silent in May 1965. In those days, the reliability of Soviet spacecraft systems had much to be desired.
23 Mariner 4's primary instrument was its 'TV camera', more formally referred to as the imaging experiment. It focused its image on a vidicon tube with a 200 by 200 pixel array.

For each image, this array was scanned by a TV camera and the signal stored on magnetic tape. At 8 bits per second, it took 4 days to replay the 22 images from the tape.

24 Planetary longitudes are measured westward of the meridian, so all longitudes are westerly. This obviates the confusion over west and east. A small dark spot on the equator was chosen by Camille Flammarion in 1876 to serve as the meridian, which he appropriately named Sinus Meridiani (Meridian Bay).

25 The images were broken down into an array of greyscale readings, and as soon as the raw data streamed in the scientists 'filled in the squares' on a large sheet of paper to gain some idea of what the spacecraft had seen. This manually reconstructed image is now framed and on display in JPL. The raw imagery was low-contrast, but the data was in digital form so computers were used to expand the contrast in order to 'tease out' detail. 'Computer enhancement' techniques had only just been developed. The surface resolution of the best image was slightly in excess of 1 kilometre per pixel.

26 This 120-kilometre-diameter crater was subsequently named 'Mariner Crater' in honour of the spacecraft that discovered it, Mariner 4.

27 *Mariner Mars 1964 Project Report: TV experiment*, R.B. Leighton. JPL Technical Report 32-884, 1967.

28 *The planet Mars: a history of observation and discovery*, W. Sheehan, University of Arizona Press, p. 253, 1996.

29 *The face of the Moon*, R. Baldwin. University of Chicago Press, 1949.

30 'Collision probabilities with the planets', E.J. Opik. *Proc. Irish Academy*, vol. 54A, p. 165, 1951.

31 'Impact mechanics at Meteor Crater', E.M. Shoemaker. A chapter in *The Moon, meteorites and comets*, B.M. Middlehurst and G.P. Kuiper (Eds). University of Chicago Press, 1963.

32 'Occultation experiment: Results of the first direct measurement of Mars' atmosphere and ionosphere', A.J. Kliore, D.L. Cain, G.S. Levy, V.R. Eshelman, G. Fjeldbo and F.D. Drake. *Science*, vol. 149, p. 1243, 1965.

33 'Behaviour of carbon dioxide and other volatiles on Mars', R.B. Leighton and B.C. Murray. *Science*, vol. 153, p. 136, 1966.

34 *An analysis of the Mariner 4 photographs of Mars*, C.R. Chapman, J.B. Pollack and C.E. Sagan. SAO Special Report 268, 1968.

35 'The surface of Mars: the cratered terrains', B.C. Murray, L.A. Soderblom, R.P. Sharp and J.A. Cutts. *J. Geophys. Res.*, vol. 76, p. 313, 1971.

36 Nowadays, Mars has been completely topographically mapped by Mars Global Surveyor's laser altimeter, so the early elevation scale based on atmospheric pressure has been superseded, but an equivalent value is still used for continuity.

37 'Spacecraft exploration of Mars', C.W. Snyder and V.I. Moroz. In *Mars*, H.H. Kieffer *et al.* (Eds). Arizona University Press, p. 71, 1992.

38 *The planet Mars*, E.M. Antoniadi. Herman Co., p. 54, 1930; English translation by P. Moore. Reid Co., 1975.

39 The eccentricity of its orbit means that Mars receives 40 per cent more solar energy when it is at perihelion than when at aphelion.

40 The runaway dust storms on Mars develop because there is a feedback mechanism between the amount of dust in the atmosphere and the magnitude of 'tidal' winds, and the process by which dust is drawn into the tenuous atmosphere is called 'saltation'.

41 'Martian yellow clouds — past and future', C.F. Capen, *Sky & Telescope*, vol. 41, p. 117, 1971.

42 *Patrick Moore on Mars*, P. Moore. Cassell, p. 77, 1998.

43 In fact, even though we now know that these features are volcanoes, and they have been assigned appropriate names, some of the 'old hands' still refer to them as 'North Spot', 'Middle Spot' and 'South Spot'.
44 *Solar System log*, A. Wilson. *Jane's*, p. 66, 1987.
45 'Preliminary Mariner 9 report on the geology of Mars', J.F. McCauley, M.H. Carr, J.A. Cutts, W.K. Hartmann, H. Masursky, D.J. Milton, R.P. Sharp and D.E. Wilhelms. *Icarus*, vol. 17, p. 289, 1972.
46 *The geology of Mars*, T.A. Mutch, R.E. Arvidson, J.W. Head, K.L. Jones and R.S. Saunders. Princeton University Press, 1976.
47 'Radar measurements of Martian topography', G.H. Pettengill, C.C. Counselan, L.P. Rainville and I.I. Shapiro. *Astron. J.*, vol. 74, p. 461, 1969.
48 'Mars radar observations: a preliminary report', G.S. Downs, R.M. Goldtein. R.R. Green and G.A. Morris. *Science*, vol. 174, p. 1324, 1971.
49 'Martian topography and surface properties as seen by radar: the 1971 opposition', G.S. Downs, R.M. Goldtein. R.R. Green, G.A. Morris and P.E. Reichley. *Icarus*, vol. 18, p. 8, 1973.
50 'Radar measurements of Martian topography and surfaces: the 1971 and 1973 oppositions', G.S. Downs, P.E. Reichley and R.R. Green. *Icarus*, vol. 26, p. 273, 1975.
51 'The new Martian nomenclature of the IAU', G. de Vaucouleurs, J. Blunck, M. Davies, A. Dollfus, I.K. Koval, G.P. Kuiper, H. Masursky, S. Miyamoto, V.I. Moroz, C.E. Sagan and B. Smith. *Icarus*, vol. 26, p. 85, 1975.
52 *Mars and its satellites: a detailed commentary on the nomenclature*, J. Blunck. Exposition Press, 1977.
53 Mauna Loa on the Big Island of Hawaii is the largest shield on Earth. Its base is 120 kilometres wide, and its summit is 9 kilometres above the floor of the Pacific. Nevertheless, it contains barely 10 per cent of the bulk of Olympus Mons.
54 'Channels on Mars', R.P. Sharp and M.C. Malin. *Geol. Soc. Am. Bull.*, vol. 86, p. 593, 1975.
55 'The distribution of lobate debris aprons and similar flows on Mars', S.W. Squyres. *J. Geophys. Res.*, vol. 84, p. 8087, 1979.
56 'Mars: south polar pits and etched terrains', R.P. Sharp. *J. Geophys. Res.*, vol. 78, p. 4222, 1973.
57 'A widespread common age resurfacing event in the highland-lowland transition zone in eastern Mars', H.V. Frey, A.M. Semeniuk, J.A. Semeniuk and S. Tokarcik. *Proc. Lunar Planet. Sci. Conf.*, p. 679, 1988.
58 'Tectonic evolution of Mars', D.U. Wise, M.P. Golombek and G.E. McGill. *J. Geophys. Res.*, vol. 84, p. 7934, 1979.
59 'Martian thermal history, core segregration, and tectonics', G.F. Davies and R.E. Arvidson. *Icarus*, vol. 45, p. 339, 1981.
60 'The Martian hemispheric dichotomy may be due to a giant impact', D.E. Wilhelms and S.W. Squyres. *Nature*, vol. 309, p. 138, 1984.
61 The Borealis basin would have been centred at 50 degrees north and 190 degrees longitude, just north of the Elysium bulge.
62 'Large impact basins and the mega-impact origin for the crustal dichotomy on Mars', H.V. Frey and R.A. Schultz. *Geophys. Res. Lett.*, vol. 15, p. 229, 1988.
63 'Geologic evidence supporting an endogenic origin for the Martian crustal dichotomy' G.E. McGill. *Proc. Lunar Planet. Sci. Conf.*, p. 667, 1989.
64 'Origin of the Martian global dichotomy by crustal thinning in the Noachian or early Hesperian', G.E. McGill and A.M. Dimitriou. *J. Geophys. Res.*, vol. 95, p. 12595, 1990.

65 'Origin of the Martian crustal dichotomy: evaluating hypotheses', G.E. McGill and S.W. Squyres. *Icarus*, vol. 93, p. 386, 1991.
66 'Accumulation of the terrestrial planets and implications concerning lunar origin', G.W. Wetherill. In *Origin of the Moon*, W.K. Hartmann *et al.* (Eds). Lunar and Planetary Institute, 1986.
67 'Grazing impacts on Mars: a record of lost satellites', P.H. Schultz and A.B. Lutz-Garihan. *J. Geophys. Res. Suppl.*, vol. 87, p. A84, 1982.
68 'Mars gravity field based on a short-arc technique', W.L. Sjogren, J. Lorell, L. Wong and W. Downs. *J. Geophys. Res.*, vol. 80, p. 2899, 1975.
69 'Internal structure and early thermal evolution of Mars from Mars Global Surveyor topography and gravity', M.T. Zuber, S.C. Solomon, R.J. Phillips, D.E. Smith, G.L. Tyler, O. Aharonson, G. Balmino, W.B. Banerdt, J.W. Head, F.G. Lemoine, P.J. McGovern, G.A. Neumann, D.D Rowlands and S. Zhong. *Science*, vol. 287, p. 1788, 2000.
70 'Evidence for magmatically driven catastrophic erosion on Mars', K.L. Tanaka, J.S. Kargel and N. Hoffman. *Proc. Lunar Planet. Sci. Conf.*, p. 1989, 2001.
71 I am grateful to Alex Blackwell of the University of Hawaii for pointing this out.
72 'The ancient rivers of Mars', D.C. Pieri. *The Planetary Report*, vol. 3, p. 4, 1983.
73 'Channels on Mars', R.P. Sharp and M.C. Malin. *Geol. Soc. Am. Bull.*, vol. 85, p. 593, 1975.
74 'The geology of Mars', M.H. Carr. *Am. Scientist*, vol. 68, p. 626, 1980.
75 *Planetary landscapes*, R. Greeley. Allen & Unwin, p. 183, 1987.
76 'Fluvial valleys and Martian paleoclimates', V.C. Gulick and V.R. Baker. *Nature*, vol. 341, p. 514, 1989.
77 'Origin and evolution of valleys on Martian volcanoes', V.C. Gulick and V.R. Baker. *J. Geophys. Res.*, vol. 95, p. 14325, 1990.
78 'Long-term orbital and spin dynamics of Mars', W.R. Ward. In *Mars*, H.H. Kieffer *et al.* (Eds). Arizona University Press, p. 298, 1992.
79 'Quasi-periodic climate change on Mars', H.H. Kieffer and A.P. Zent. In *Mars*, H.H. Kieffer *et al.* (Eds). Arizona University Press, p. 1180, 1992.
80 'Periodic insolation variations on Mars', B.C. Murray, W.R. Ward and S.C. Young. *Science*, vol. 180, p. 638, 1973.
81 'Past obliquity oscillations of Mars: the role of the Tharsis uplift', W.R. Ward, J.A. Burns and O.B. Toon. *J. Geophys. Res.*, vol. 84, p. 243, 1979.
82 'Groundwater formation of Martian valleys', M.C. Malin and M.H. Carr. *Nature*, vol. 397, p. 589, 1999.
83 'Metre-scale characteristics of Martian channels and valleys', M.H. Carr and M.C. Malin. *Icarus*, vol. 146, p. 366, 2000.
84 'Morphometric measurements of Martian valley networks from Mars Orbiter Laser Altimeter (MOLA) data', R.M. Williams and R.J. Phillips. *J. Geophys. Res.*, 2001.
85 'The Martian impact cratering record', R.G. Strom, S.K. Croft and N.G. Barlow. In *Mars*, H.H. Kieffer *et al.* (Eds). Arizona University Press, p. 383, 1992.
86 These have been variously called 'rampart', 'splash', 'splosh', 'ejecta flow' and 'fluidised ejecta' craters.
87 '*A method for measuring heat flow in the Martian crust using impact crater morphology*', J.M. Boyce. NASA TM-80339, 1979.
88 'Martian permafrost features', M.H. Carr and G.G. Schaber. *J. Geophys. Res.*, vol. 82, p. 4039, 1977.
89 'Martian fluidised crater morphology: variations with crater size, latitude, altitude and target material', P. Mouginis-Mark. *J. Geophys. Res.*, vol. 84, p. 8011, 1977.

90 'Ground ice on Mars: inventory, distribution and resulting landforms', L.A. Rossbacher and S. Judson. *Icarus*, vol. 45, p. 39, 1981.
91 By way of a caveat on this point, it should be noted that the calculations for the depth of the permafrost were made on the assumpton of pure water. If the ice is a hyperconcentrated brine then the permafrost is likely to be thinner and so liquid water will occur at a shallower depth.
92 'Ground ice on Mars: inventory, distribution and resulting landforms', L.A. Rossbacher and S. Judson. *Icarus*, vol. 45, p. 39, 1981.
93 Estimates as to the diameters of the Martian basins vary depending upon what is being measured. In the case of Hellas, which is slightly elliptical, the estimates ranged from 1,600 to 2,300 kilometres. If inner rings have been submerged by the flows which filled the cavity, then the visible rim will not mark the impact's excavation radius. In fact, a map by Mars Global Surveyor's laser altimeter revealed that what had been assumed to be the outermost ring is actually located on the inner slope of a ring 4,000 kilometres across, so Hellas is the largest impact basin in the Solar System.
94 *The Cambridge photographic atlas of the Solar System*. G.A. Briggs and F.W. Taylor. Cambridge University Press, p. 141, 1982.
95 Estimates of Argyre's diameter range from 1,000 to 2,000 kilometres depending upon whether the inner ring or the periphery of the continuous ejecta is being measured. It is comparable in size to Imbrium on the Moon.
96 *Planetary landscapes*, R. Greeley. Allen & Unwin, p. 153, 1987.
97 'Syrtis Major: a low-relief volcanic shield', G.G. Schaber. *J. Geophys. Res.*, vol. 87, p. 9852, 1982.
98 'Evidence for a very large basin beneath Utopia Planum', G.E McGill. *Proc. Lunar Planet. Sci. Conf.*, p. 752, 1988.
99 'Buried topography of Utopia Planitia: persistence of a giant impact depression', G.E. McGill. *J. Geophys. Res.*, vol. 94, p. 2753, 1989.
100 'Evidence of an ancient impact basin in Daedalia Planum', R.A. Craddock, R. Greeley and P.R. Christensen. *J. Geophys. Res.*, vol. 95, p. 10729, 1990.
101 'The impacted Martian crust: structure, hydrology and some geological implications', D.J. MacKinnon and K.L. Tanaka. *J. Geophys. Res.*, vol. 94, p. 17359, 1989.
102 'Global and regional ridge patterns on Mars', A.F. Chicarro, P.H. Schultz and P. Masson. *Icarus*, vol. 63, p. 153, 1985.
103 *Exploring the planets*, W.K. Hamblin and E.H. Christiansen. Macmillan Company, p. 131, 1990.
104 'Comparison of impact basins on Mercury, Mars and the Moon', C.A. Wood and J.W. Head. *Proc. Lunar Sciences Conf.*, p. 3629, 1976.
105 'Mars: a standard crater curve and possible new time scale', G. Neukum and D.U. Wise. *Science*, vol. 194, p. 1381, 1976.
106 'Problems associated with estimating the relative impact rates on Mars and the Moon', G.W. Wetherill. *Moon*, vol. 9, p. 227, 1974.
107 'The structure and evolution of ancient impact basins on Mars', P.H. Schultz, R.A. Schultz and J. Rogers. *J. Geophys. Res.*, vol. 87, p. 9803, 1982.
108 *The Mars mystery – a warning from history that could save life on Earth*, G. Hancock, R. Bauval and J. Grigsby. Penguin Books, 1998.
109 'Martian ages', G. Neukum and K. Hiller. *J. Geophys. Res.*, vol. 86, p. 3097, 1981.
110 'Cratering and obliteration history of Mars', C.R. Chapman and K.L. Jones. *Ann. Rev. Earth Planet. Sci.*, vol. 5, p. 515, 1977.

111 'Martian cratering: Mariner 9 initial analysis of cratering chronology', W.K. Hartmann. *J. Geophys. Res.*, vol. 78, p. 4096, 1973.
112 'Volcanism in the cratered terrain hemisphere of Mars', R. Greeley and P.D. Spudis. *Geophys. Res. Lett.*, vol. 5, p. 453, 1978.
113 'Evidence for recent volcanism on Mars from crater counts', W.K. Hartmann *et al. Nature*, vol. 397, p. 586, 1999.
114 'Antipodal effects of major basin forming impacts on Mars', J.E. Peterson. *Proc. Lunar Planet. Sci. Conf.*, p. 885, 1978.
115 'The impacted Martian crust: structure, hydrology and some geological implications', D.J. MacKinnon and K.L. Tanaka. *J. Geophys. Res.*, vol. 94, p. 17359, 1989.
116 'The chronology of Martian volcanoes', J.B. Plescia and R.S. Saunders. *Proc. Lunar Planet. Sci. Conf.*, p. 2841, 1979.
117 'Volcanic geology of Tyrrhena Patera', R. Greeley and D.A. Crown. *J. Geophys. Res.*, vol. 95, p. 7133, 1990.
118 'Volcanism in the Noachis-Hellas region of Mars', J.E. Peterson. *Proc. Lunar Planet. Sci. Conf.*, p. 3411, 1978.
119 'Evolution and erosion of Tyrrhena and Hadriaca Paterae, Mars: new insights from MOC and MOLA', T.K.P. Gregg, D.A. Crown and S.E.H. Sakimoto. *Proc. Lunar Planet. Sci. Conf.*, p. 1628, 2001.
120 'Volcanoes on Mars', R. Greeley and P.D. Spudis. *Rev. Geophys. Space Phys.*, vol. 19, p. 13, 1981.
121 In *Mars: the story of the Red Planet*, P. Cattermole (Chapman & Hall, 1992), Amphitrites and Peneus are referred to as 'volcanic ring structures'.
122 This raises the issue of how the volcano survived the retreat of the line of dichotomy's scarp without being 'fretted'.
123 'Syria Planum, Mars: a major volcanic construct in the early history of Tharsis', J.W. Head, B.M. Webb, B.E. Kortz and S. Pratt. *32nd Vernadsky-Brown Microsymposium on Comparative Planetology (Moscow)*, p. 58, October 2000.
124 'Syrtis Major: a low-relief volcanic shield', G.G. Schaber. *J. Geophys. Res.*, vol. 87, p. 9852, 1982.
125 'Volcanism on Mars', R. Greeley and P.D. Spudis. *Rev. Geophys. Space. Phys.*, vol. 19, p. 13, 1981.
126 'Rilles, ridges and domes: clues to maria history', W.L. Quaide. *Icarus*, vol. 4, p. 374, 1965.
127 'Lunar mare ridges, rings and volanic ring complexes', R.G. Strom. In IAU Symposium No.47 *The Moon*, 1972.
128 'Wrinkle ridges as deformed surface crusts on ponded mare lava'. W.B. Bryan. *Proc. Lunar Sci. Conf.*, p. 93, 1973.
129 'Origin of planetary wrinkle ridges based on the study of terrestrial analogs', J.B. Plescia and M.P. Holombek. *Geol. Soc. Am. Bull.*, vol. 97, p. 1289, 1986.
130 'Chronology and global distribution of fault and ridge systems on Mars', D.H. Scott and J.M. Dohm. *Proc. Lunar Planet. Sci. Conf.*, p. 487, 1990.
131 'Origin of the periodically spaced wrinkle ridges on the Tharsis region of Mars', T.R. Watters. *J. Geophys. Res.*, vol. 96, p. 15599, 1991.
132 'Cross-cutting relations and relative ages of ridges and faults in the Tharsis region of Mars', T.R. Watters and T.A. Maxwell. *Icarus*, vol. 56, p. 278, 1983.
133 'Sub-kilometre Martian volcanoes: properties and possible terrestrial analogs', H.V. Frey and M. Jarosewich. *J. Geophys. Res.*, vol. 87, p. 9867, 1982.

Notes 133

134 'Small volcanic constructs in the Chryse Planitia region of Mars', R. Greeley and E. Theilig. *Rept. Planet. Geol. Prog.* NASA TM-79729, p. 202, 1978.

135 'Tectonic history of the Syria Planum province of Mars', K.L. Tanaka and P.A. Davis. *J. Geophys. Res.*, vol. 93, p. 14893, 1988.

136 'The Tempe volcanic province on Mars and comparison with the Snake River Plains of Idaho', J.B. Plescia. *Icarus*, vol. 45, p. 586, 1981.

137 'The Snake River Plain, Idaho: representative of a new category of volcanism', R. Greeley. *J. Geophys. Res.*, 1982.

138 'Road log from American Falls to Split Butte', R. Greeley and J.S. King. In *Volcanism of the Eastern Snake River Plain, Idaho: a comparative planetary geology guidebook*. NASA CR 15554621, p. 295, 1977.

139 *Volcanoes of the Solar System*, C. Frankel. Cambridge University Press, p. 108, 1996.

140 'Volcanism in the cratered terrain hemisphere of Mars', R. Greeley and P.D. Spudis. *Geophys. Res. Lett.*, vol. 5, p453, 1978.

141 *Planetary landscapes*, R. Greeley. Allen & Unwin, p. 165, 1987.

142 Alba Patera's ultra-low-profile shield is unique on Mars. Considered in terms of its base, it is said to be the largest single volcanic construct in the Solar System. A close relative may be the low shield Ra Patera on Io.

143 'The physical volcanology of Mars', P.J. Mouginis-Mark, L. Wilson and M.T. Zuber. In *Mars*, H.H. Kieffer *et al.* (Eds). University of Arizona Press, p. 424, 1992.

144 'Validity of convective plume rise models for volcanic eruptions on Mars'. L.S. Glaze and S.M. Baloga. *Proc. Lunar Planet. Sci. Conf.*, p. 1209, 2001.

145 'Polygenic eruptions on Alba Patera', P.J. Mouginis-Mark, L. Wilson and J.R. Zimbelman. *Bull. Volcanology*, vol. 50, p. 361, 1988.

146 *Planetary landscapes*, Ron Greeley. Allen & Unwin, p. 165, 1987.

147 'Discontinuities in the shallow Martian crust', P.A. Davis and M.P. Golombek. *Proc. Lunar Planet. Sci. Conf.*, p. 224, 1989.

148 'Tectonic history of the Alba Patera–Ceraunius Fossae region of Mars'. K.L. Tanaka. *Proc. Lunar Planet. Sci. Conf.*, p. 515, 1990.

149 'Stress and tectonics on Mars', W.B. Banerdt, M.P. Golombek and K.L. Tanaka. In *Mars*, H.H. Kieffer *et al.* (Eds). University of Arizona Press, p. 249, 1992.

150 'The Valles Marineris–Noctis Labyrinthus–Claritas Fossae region of Mars', P. Masson. *Moons and Planets*, vol. 22, p. 211, 1980.

151 'Martian tension fractures and the formation of grabens and collapse features at Valles Marineris', K.L. Tanaka and M.P. Golombek. *Proc. Lunar Planet. Sci. Conf.*, p. 383, 1989.

152 'Tectonism and volcanism of the Tharsis region of Mars', M.H. Carr. *J. Geophys. Res.*, vol. 79, p. 3943, 1974.

153 'Tharsis province of Mars: geologic sequence, geometry, and a deformation mechanism', D.U. Wise, M.P. Golombek and G.E. McGill. *Icarus*, vol. 38, p. 456, 1979.

154 'The global topography of Mars and implications for surface evolution', D.E. Smith, M.T. Zuber, S.C. Solomon, R.J. Phillips, J.W. Head, J.B. Garvin, W.B. Banerdt, D.O. Muhleman, G.H. Pettengill, G.A. Neumann, F.G. Lemoine, J.B. Abshire, O. Aharonson, C.D. Brown, S.A. Hauck, A.B. Ivanov, P.J. McGovern, H.J. Zwally and T.C. Duxbury. *Science*, vol. 284, p. 1495, 1999.

155 'Mars Tharsis region: volcano-tectonic events in the stratigraphic record'. D.H. Scott and K.L. Tanaka. *Proc. Lunar Planet. Sci. Conf.*, p. 2403, 1980.

156 'Tectonic history of the Tharsis region', J.B. Plescia and R.S. Saunders. *J. Geophys. Res.*, vol. 87, p. 9775, 1982.

157 'Geophysical observations pertaining to solid-state convection in terrestrial planets', R.J. Phillips and E.R. Ivins. *Phys. Earth Planet. Interiors*, vol. 19, p. 107, 1979.
158 'Tectonic evolution of Mars', D.U. Wise, M.P. Golombek and G.E. McGill. *J. Geophys. Res.*, vol. 84, p. 7934, 1979.
159 'Tectonic history of the Tharsis region', J.B. Plescia and R.S. Saunders. *Proc. Lunar Planet. Sci. Conf.*, p. 2841, 1979.
160 'Evolution of the Tharsis province on Mars: the importance of heterogeneous lithospheric thickness and volcanic construction', S.C. Soloman and J.W. Head. *J. Geophys. Res.*, vol. 87, p. 9755, 1982.
161 'Mars: palaeostratigraphic restoration of buried surfaces in Tharsis Montes', D.H. Scott and K.L. Tanaka. *Icarus*, vol. 45, p. 304, 1981.
162 'Radar altimetry of south Tharsis', L.E. Roth, G.S. Downs, R.S. Saunders and G. Schubert. *Icarus*, vol. 42, p. 287, 1980.
163 These parasitic vent flows are very prominent on the Mars Global Surveyor's MOLA elevation map.
164 'Calderas on Mars: models of formation for the Arsia type', L.S. Crumpler, J.C. Aubele and J.W. Head. *Proc. Lunar Planet. Sci. Conf.*, p. 269, 1991.
165 To put this into context, the flank vent of Arsia Mons which produced the 300 kilometres long flow erupted lava at a rate comparable to that at which water flows from the mouth of the Amazon, the Earth's largest river.
166 'Structural evolution of Arsia Mons, Pavonis Mons and Ascraeus Mons in the Tharsis region of Mars', L.S. Crumpler and J.C. Aubele. *Icarus*, vol. 34, p. 496, 1978.
167 'Geochemical and mineralogical interpretation of the Viking inorganic chemical results', P. Toulmin, A.K. Baird, B.C. Clark, K. Keil, H.P. Rose, R.P. Christian, P.H. Evans and W.C. Kelliher. *J. Geophys. Res.*, vol. 82, p. 4625, 1977.
168 One study indicated that the shallow reservoir of Olympus Mons had resided at a level in the mountain which was several kilometres above the base. If this was the case for the shields on the ridge, the vents on their lower flanks would have rapildy drained their reservoirs and induced summit collapse. However, Olympus Mons has a caldera complex even though it hasn't developed flank vents.
169 'Structural evolution of Arsia Mons, Pavonis Mons and Ascraeus Mons on the Tharsis region of Mars', L.S. Crumpler and J.C. Aubele. *Icarus*, vol. 35, p. 496, 1978.
170 'Mars gravity: high resolution results from Viking Orbiter 2', W.L. Sjogren. *Science*, vol. 203, p. 1006, 1979.
171 If the domes are really the summits of buried shields, then the introduction of the designation tholus was an unfortunate complication. However, while the physiographic differences form an objective classification, their interpretation as buried shields is a subjective judgement which is open to dispute.
172 There is an interesting point of comparison with Olympus Mons and the Hawaiian shields, which are ringed by concentric scarps (called 'pali') where the periphery is progressively falling away on steeply dipping normal faults.
173 This count is based upon the arrangement of the calderas within the main cavity, whose formation obliterated evidence of earlier activity.
174 'Constraints on depth and geometry of the magma chamber of the Olympus Mons volcano', M.T. Zuber and P.J. Mouginis-Mark. *Proc. Lunar Planet. Sci. Conf.*, p. 1387, 1990.
175 *Constraints on magma chamber depth of the Olympus Mons volcano*, M.T. Zuber and P.J. Mouginis-Mark. LPI Tech. Rep. 90-04, p. 58, 1990.
176 'Distribution of strain in the floor of Olympus Mons caldera', T.R. Watters and D.J. Chadwick. *Proc. Lunar Planet. Sci. Conf.*, p. 1310, 1990.

177 *Distribution of strain in the floor of Olympus Mons caldera*, T.R. Watters, D.J. Chadwick and M.C. Liu. LPI Tech. Rep. 90-04, p. 293, 1990.
178 *The basal scarp of Olympus Mons*, E.C. Morris. NASA TM-84211, p. 389, 1981.
179 'Flank tectonics of Martian volcanoes', P.J. Thomas, S.W. Squyres and M.H. Carr. *J. Geophys. Res.*, vol. 95, p. 14345, 1990.
180 'The aureole of Olympus Mons', S.A. Harris. *J. Geophys. Res.*, vol. 82, p. 3099, 1977.
181 'Origin of the Olympus Mons aureole and perimeter scarp', R.M.C. Lopes, J.E. Guest and C.J. Wilson. *Moon and the Planets*, vol. 22, p. 221, 1980.
182 'Further evidence for a mass movement origin of the Olympus Mons aureole', R.M.C. Lopes, J.E. Guest, K. Hiller and G. Neukum. *J. Geophys. Res.*, vol. 87, p. 9917, 1981.
183 'Aureole deposits of the Martian volcano Olympus Mons', E.C. Morris. *J. Geophys. Res.*, vol. 87, p. 1164, 1982.
184 'The subglacial birth of Olympus Mons and its aureoles', C.A. Hodges and H.J. Moore. *J. Geophys. Res.*, vol. 84, p. 8061, 1979.
185 The summit of Olympus Mons is at a slightly higher elevation than the shields atop Tharsis.
186 'The geology of Mars', M.H. Carr. *Am. Scientist*, vol. 68, p. 626, 1980.
187 'Volcanism on Mars', M.H. Carr. *J. Geophys. Res.*, vol. 78, p. 4049, 1973.
188 'Mars: a standard crater curve and possible new time scale', G. Neukum and D.U. Wise. *Science*, vol. 194, p. 1381, 1976.
189 'Relative crater production rates on planets', W.K. Hartmann. *Icarus*, vol. 31, p. 260, 1977.
190 *Volcanoes of the Solar System*, C. Frankel. Cambridge University Press, p. 103, 1996.
191 *Volcanoes of the Solar System*, C. Frankel. Cambridge University Press, p. 138, 1996.
192 *Planetary volcanism: a study of volcanic activity in the Solar System*, P. Cattermole. WileyPraxis, p. 265, 1996.
193 'Mars: chronological studies of the large volcanoes in Tharsis'. J. Grier, W. Bottke, W.K. Hartmann and D.C. Berman. *Proc. Lunar Planet. Sci. Conf.*, p. 1823, 2001.
194 *Volcanoes of the Solar System*, C. Frankel. Cambridge University Press, p. 104, 1996.
195 K.S. Edgett *et al. J. Geophys. Res.*, vol. 102, p. 21545, 1997.
196 D.O. Muhleman *et al. Ann. Rev. Earth Planet. Sci.*, vol. 23, p. 337, 1995.
197 *Volcanoes of the Solar System*, C. Frankel, Cambridge University Press, p. 107, 1996.
198 'Explosive volcanism on Hecates Tholus: investigation of eruption conditions', P.J. Mouginis-Mark. L. Wilson and J.W. Head. *J. Geophys Res.*, vol. 87, p. 9890, 1982.
199 Mars Global Surveyor's data suggests that the circumferential fracture zones around some of the large shields may have been due to collapse following 'delamination' of the mountain's 'root' by later mantle convection, which would explain the gravitational anomalies specifically associated with these structures. Alba Patera may be an extreme case of such undermining.
200 'Formation of Martian channels', H. Masursky, J.M. Boyce, A.L. Dial, G.G. Schaber and M.E. Strobell. *J. Geophys. Res.*, vol. 82, p. 4016, 1977.
201 'Classification and time of formation of Martian channels based on Viking data', H. Masursky, J.M. Boyce, A.L. Dial, G.G. Schaber and M.E. Strobell. *J. Geophys. Res.*, vol. 82, p. 4016, 1977.
202 'Global stratigraphy', K.L. Tanaka, D.H. Scott and R. Greeley. In *Mars*, University of Arizona Press, p. 345, 1992.
203 'Structure pattern analysis of the Noctis Labyrinthus–Valles Marineris region of Mars', P. Masson. *Icarus*, vol. 30, p. 49, 1977.
204 'Noctis Labyrinthus' means 'Labyrinth of the Night'. It was the maze-like appearance of the fracture system which prompted the use of this designation.

205 The Martian canyons are considerably larger than the largest terrestrial canyon, the Grand Canyon, which was etched by the Colorado River as the surrounding plain was elevated. In fact, the Grand Canyon is comparable in size to the tributary canyons of Valles Marineris.
206 Coprates Chasma's 7-kilometre cliffs are three times higher than those of the Grand Canyon. As it happens, Coprates corresponds to one of Lowell's 'canals', but while it may indeed once have contained water it is not an artificial irrigation channel.
207 'Geology of Valles Marineris – first analysis of imagery from the Viking Orbiter primary mission', K.R. Blasius, J.A. Cutts, J.E. Guest and H. Masursky. *J. Geophys. Res.*, vol. 78, p. 4049, 1977.
208 Valles Marineris was so-named in honour of Mariner 9, which discovered it.
209 'Landslides in Valles Marineris', B.K. Lucchitta. *J. Geophys. Res.*, vol. 84, p. 8097, 1979.
210 *Planetary landscapes*, R. Greeley. Allen & Unwin, p. 166, 1987.
211 'Geological map of the Coprates quadrangle of Mars', J.F. McCauley. USGS Map I-897, 1978.
212 The 3,000 kilometres long dog-leg around Lunae Planum makes Kasei one of the longest outflow channels, and part of it has been masked by lavas from Tharsis.
213 'The structure and evolution of ancient impact basins on Mars', P.H. Schultz, R.A. Schultz and J. Rogers. *J. Geophys. Res.* vol. 87, p. 9803, 1982.
214 'Formation of Martian flood features by release of water from confined aquifers', M.H. Carr. *J. Geophys. Res.*, vol. 84, p. 2995, 1979.
215 'Ages of rocks and channels in prospective Martian landing sites of Mangala Valles region', H. Masursky, M.G. Chapman, A.L. Dial and M.E Strobell. *Proc. Lunar Planet. Sci. Conf.*, p. 520, 1986.
216 'Crater production function for Mars', G. Neukum and B.A. Ivanov. *Proc. Lunar Planet. Sci. Conf.*, p. 1757, 2001.
217 W.K. Hartmann and G. Neukum. Presented at the Intl. Space Sci. Inst. in Bern and published in the associated 'special issue' of *Space Sci. Rev.* 2001.
218 *Water on Mars*, M.H. Carr. Oxford University Press, 1996.
219 The flow rates of the Martian outflow channels have been computed to have been around ten-thousand times that of the Earth's largest river, the Amazon.
220 'An overview of geological results from Mariner 9', H. Masursky. *J. Geophys. Res.*, vol. 78, p. 4009, 1973.
221 'Erosion by catastrophic floods on Mars and Earth', V.R. Baker and D.J. Milton. *Icarus*, vol. 23, p. 27, 1974.
222 'Mars and Earth: comparison of cold-climate features', B.K. Lucchitta. *Icarus*, vol. 45, p. 262, 1981.
223 'Martian outflow channels sculpted by glaciers', B.K. Lucchitta. *Proc. Lunar Planet. Sci. Conf.*, p. 634, 1980.
224 'Ice sculpture in the Martian outflow channels', B.K. Lucchitta. *J. Geophys. Res.*, vol. 87, p. 9951, 1982.
225 'Huge CO_2-charged debris flow deposits and tectonic sagging in the northern plains of Mars', K.L. Tanaka, W.B. Banerdt, J.S. Kargel and N. Hoffman. *Geology*, vol. 29, p. 427, 2001.
226 The exceptionally favourable launch window of 1971 meant that the standard rocket had sufficient capacity to send an integrated orbiter/lander to Mars, but the rather less favourable window in 1973 precluded this and the landers had to be dispatched separately.
227 *Solar System log*, Andrew Wilson. *Jane's*, p. 78, 1987.

Notes 137

228 'Spacecraft exploration of Mars', C.W. Snyder and V.I. Moroz. In *Mars*, H.H. Kieffer *et al.* (Eds). Arizona University Press, p. 71, 1992.
229 *On Mars: exploration of the Red Planet 1958–1978*, E.C. Ezell and L.N. Ezell. NASA SP-4212, p. 185, 1984.
230 'The Viking landing sites: selection and certification', H. Masursky and N.L. Crabill. *Science*, vol. 193, p. 809, 1976.
231 'The geology of the Viking Lander 1 site', A.B. Binder, R.E. Arvidson, E.A. Guinness, K.L. Jones. E.C. Morris, T.A. Mutch, D.C. Pieri and C.E. Sagan. *J. Geophys. Res.*, vol. 82, p. 4439, 1977.
232 A similar device had been used by some of the Surveyor spacecraft which had landed on the Moon prior to the Apollo programme, and some of the Apollo motherships had carried X-ray spectrometers to scan the surface by using the Sun as the irradiator.
233 'The Viking X-ray fluorescence experiment: analytical methods and early results', B.C. Clark, A.K. Baird, H.J. Rose, P. Toulmin, R.P. Christian, W.C. Kelliher, A.J. Castro, C.D. Rowe, K. Kiel and G. Huss. *J. Geophys. Res.*, vol. 82, p. 4577, 1977.
234 Magnesium sulphate is also known as epsom salts.
235 'Geochemical and mineralogical interpretation of the Viking inorganic chemical results', P. Toulmin, A.K. Baird, B.C. Clark, K. Keil, H.J. Rose, R.P. Christian, P.H. Evans and W.C. Kelliher. *J. Geophys. Res.*, vol. 82, p. 4625, 1977.
236 On Earth, it was the build up of potassium (a large-ion lithophile) which drove the formation of the buoyant crustal rock. The rate of continental development 'took off' when activity switched from sodic to potassic lava.
237 'Absence of silicic volcanism on Mars: implications for the crustal composition and volatile abundance', P.W. Francis and C.A. Wood. *J. Geophys. Res.*, vol. 87, p. 9881, 1982.
238 *Volcanoes of the Solar System*, C. Frankel. Cambridge University Press, p. 123, 1996.
239 'The geology of the Viking Lander 2 site', T.A. Mutch, R.E. Arvidson, A.B. Binder, E.A. Guinness and E.C. Morris. *J. Geophys. Res.*, vol. 82, p. 4452, 1977.
240 'Seismology on Mars', D.L. Anderson, W.F. Miller, G.V. Latham, Y. Nakamura, R.L. Kovach and T.C.D. Knight. *J. Geophys. Res.*, vol. 82, p. 4524, 1977.
241 Viking 2 was switched off on 11 April 1980.
242 *To Utopia and back*, N.H. Horowitz. W.H. Freeman, 1986.
243 'The composition of the atmosphere at the surface of Mars', T. Owen, K. Biermann, D.R. Rushneck, J.E. Biller, D.W. Howarth and A.L. Lafleur. *J. Geophys. Res.*, vol. 82, p. 4635, 1977.
244 'Evidence for the late formation and young metamorphism in the achondrite Nakhla', D.A. Papanastassiou and G.J. Wasserburg. *Geophys. Res. Lett.*, vol. 1, p. 23, 1974.
245 'SNC meteorites: clues to Martian petrologic evolution?', H.Y. McSween. *Rev. Geophys.*, vol. 23, p. 391, 1985.
246 'A petrological model of the relationships among achondritic meteorites', E. Stolper, H.Y. McSween and J.F. Hayes. *Geochimica Cosmochimica Acta*, vol. 43, p. 589, 1979.
247 'Dynamical, chemical and isotopic evidence regarding the formation locations of asteroids and meteorites', J.T. Wasson and G.W. Wetherill. In *'Asteroids'*, T. Gehrels (Ed.). University of Arizona Press, 1979.
248 Allan Hills (ALH 77005) and Elephant Moraine (EET 79001) recovered from the Antarctic ice sheet were the first two Shergottites to be found to contain 'Martian gases'.
249 'Trapped noble gases in the EETA 79001 shergottite', D.D. Bogard. *Meteoritics*, vol. 17, p. 185, 1982.
250 'Martian gases in an Antarctic meteorite?', D.D. Bogard and P. Johnson. *Science*, vol. 221, p. 651, 1983.

251 'Noble gas contents of Shergottites and implications for the Martian origin of the SNC meteorites', D.D. Bogard, L.E. Nyquist and P. Johnson. *Geochimica Cosmochimica Acta*, vol. 48, p. 1723, 1984.
252 'The origin of the SNC meteorites: an alternative to Mars', A.V. Singer and H.J. Melosh. *Am. Geophys. Union EOS Trans.*, vol. 62, p. 941, 1981.
253 'Possible asteroidal origin of SNC meteorites', A.V. Singer and H.J. Melosh. *Proc. Lunar Planet. Sci Conf.*, p. 742, 1982.
254 'SNC meteorites: evidence against an asteroidal origin', L.D. Ashwal, J.L. Warner and C.A. Wood. *J. Geophys. Res.* Supp., vol. 87, p. A393, 1982.
255 'Are all the "Martian" meteorites from Mars?' U. Ott and F. Begemann. *Nature*, vol. 317, p. 509, 1985.
256 'The case for a Martian origin of the Shergottites: nitrogen and noble gases in EETA 79001', R.H Becker and R.O. Pepin. *Earth Planet. Sci. Lett.*, vol. 69, p. 225, 1984.
257 'Evidence of Martian origins', R.O. Pepin. *Nature*, vol. 317, p. 473, 1985.
258 'Mars Pathfinder update', R. Naeye. *Astronomy*, p. 30, November 1997.
259 'A discussion of isotopic systematics and mineral zoning in the Shergottites: evidence for a 180 m.y. igneous crystallisation age'. J.H. Jones. *Geochimica Cosmochimica Acta*, vol. 50, p. 969, 1986.
260 'Formation ages and evolution of shergotty and its parent planet from U–Th–Pb systematics', J.H. Chen and G.J. Wasserburg. *Geochimica Cosmochimica Acta*, vol. 50, p. 955, 1986.
261 'The large crater origin of SNC meteorites', A.M. Vickery and H.J. Melosh. *Science*, vol. 237, p. 738, 1987.
262 'SNC meteorites: igneous rocks from Mars?', C.A. Wood and L.D. Ashwal. *Proc. Lunar Planet. Sci. Conf.*, p. 1359, 1981.
263 'Orbital evolution of impact ejecta from Mars', G.W. Wetherill. *Meteoritics*, vol. 19, p. 1, 1984.
264 *Volcanoes of the Solar System*, C. Frankel. Cambridge University Press, p. 127, 1996.
265 It is important to bear in mind that the terms 'lithosphere' and 'crust' are not synonymous. The lithosphere is the rigid outer shell above the fluidic mantle. The crust is the chemically distinct outer lithosphere. There is no chemical differentiation across the mantle lithosphere boundary, only a change of phase.
266 Subsequent analysis of the flood sculpture indicated that the Pathfinder landing site was dominated by the Tiu Vallis outflow which had cut across the mouth of Ares Vallis, 'So what did Mars Pathfinder land on?', K.L. Tanaka and J.A. Skinner. *Proc. Lunar Planet. Sci. Conf.*, p. 2189, 2001.
267 'Welcome to Mars!', .C. Petersen. *Sky & Telescope*, p. 34, October 1997.
268 'Overview of the Mars Pathfinder mission and assessment of landing site predictions', M.P. Golombek *et al. Science*, vol. 278, p. 1743, 1997.
269 'Results from the Mars Pathfinder camera', P.H. Smith *et al. Science*, vol. 278, p. 1758, 1997.
270 'Observations at the Mars Pathfinder landing site: do they provide "unequivocal" evidence of catastrophic flooding?', M.G. Chapman and J.S. Kargel. *J. Geophys. Res.*, vol. 104, p. 8671, 1999.
271 'Geology of Xanthe Terra outflow channels and the Mars Pathfinder landing site', D.M. Nelson and R. Greeley. *J. Geophys. Res.*, vol. 104, p. 8653, 1999.
272 'Mars Pathfinder landing site: regional geology and mass-flow interpretation', K.L. Tanaka. *Proc. Lunar Planet. Sci. Conf.*, p. 1800, 1998.
273 'Pathfinder landing site: alternatives to catastrophic floods and an antarctic ice flow

analog for outflow channels on Mars', B.K. Lucchitta. *Proc. Lunar Planet. Sci. Conf.*, p. 1287, 1998.
274 'The chemical composition of Martian soil and rocks returned by the mobile APXS: preliminary results from the X-ray mode', R. Rieder *et al. Science*, vol. 278, p. 1771, 1997.
275 In fact, uncertainties over the calibration of Sojourner's APXS complicated the analysis of the results. While the X-ray data was satisfactory, the alpha particle and the proton readings could not be accepted until follow-up laboratory work verified the instrument's calibration in a simulated Martian environment. For the results of this effort see 'Revised data of the Mars Pathfinder Alpha Proton X-ray Spectrometer: geochemical behaviour of major and minor elements', J. Brückner, G. Dreibus, R. Rieder and H. Wänke. *Proc. Lunar Planet. Sci. Conf.*, p. 1293, 2001.
276 H.Y. McSween. *J. Geophys. Res.*, vol. 104, p. 8679, 1999.
277 'Petrologic interpretation of rock textures at the Pathfinder landing site', T.J. Parker, H.J. Moore, J.A. Crisp and M.P. Golombek. *Proc. Lunar Planet. Sci. Conf.*, p. 1829, 1998.
278 'The effects of thin coatings of dust or soil onthe bulk APXS composition of the underlying rocks at the Pathfinder landing site', J.A. Crisp. *Proc. Lunar Planet. Sci. Conf.*, p. 1962, 1998.
279 'Chemical constraints on the origin of Martian global dust and its processing into soil', H.Y. McSween and K. Keil. *Proc. Lunar Planet. Sci. Conf.*, p. 1022, 2000.
280 'Mars Pathfinder: better science?', J. Bell. *Sky & Telescope*, p. 36, July 1998.
281 'The thermal state and internal structure of Mars', D.H. Johnston, T.R. McGetchin and M.N. Toksoz. *J. Geophys. Res*, vol. 79, p. 3959, 1974.
282 'Formation, history and energetics of cores in the terrestrial planets', S.C. Solomon. *Phys. Earth Planet. Interiors*, vol. 19, p. 168, 1979.
283 'On volcanism and thermal tectonics on one-plate planets', S.C. Solomon. *Geophys. Res. Lett.*, vol. 5, p. 461, 1978.
284 'Occurrence of giant impacts during the growth of terrestrial planets', G.W. Wetherill. *Science*, vol. 222, p. 877, 1985.
285 'Formation ages and evolution of shergotty and its parent planet from U–Th–Pb systematics', J.H. Chen and G.J. Wasserburg. *Geochimica Cosmochimica Acta*, vol. 50, p. 955, 1986.
286 'Thermal evolution of Earth and Moon growing by planetesimal impacts', W.M. Kaula. *J. Geophys. Res.*, vol. 84, p. 999, 1979.
287 'Mixed news from Mars', W.K. Hartmann. *Astronomy*, p. 24, February 1998.
288 'Magnetic lineations in the ancient crust of Mars', J.E.P. Connerney *et al. Science*, vol. 284, p. 794, 1999.
289 'Early Martian magnetism tape-recorded in rock'. *Astronomy*, p. 30, August 1999.
290 'Global distribution of crustal magnetisation discovered by the Mars Global Surveyor MAG/ER experiment', M.H. Acuna *et al. Science*, vol. 284, p. 790, 1999.
291 'Timing of the Martian dynamo', G. Schubert, C.T. Russell and W.B. Moore. *Nature*, vol. 408, p. 667, 2000.
292 'Mars's magnetic mystery'. News feature in *Sky & Telescope*, p. 18, August 1999.
293 'Is the Gordii Dorsum escarpment on Mars an exhumed transcurrent fault?', R.D. Forsythe and Z.R. Zimbelman. *Nature*, vol. 336, p. 143, 1988.
294 *Volcanoes of the Solar System*, C. Frankel. Cambridge University Press, p. 133, 1996.
295 'Martian plate tectonics', N.H. Sleep. *J. Geophys. Res.*, vol. 99, p. 5639, 1994.
296 'Martian volcanism: a global dichotomy of basaltic and andesitic materials', J.L. Bandfield, P.R. Christensen and V.E. Hamilton. *Proc. Lunar Planet. Sci. Conf.*, p. 1099, 2000.

297 'Sedimentary rocks of early Mars', M.C. Malin and K.S. Edgett. *Science*, vol. 290, p. 1927, 2000.
298 'Ancient Martian lakes? Perhaps', D. Tytell. *Sky & Telescope*, p. 20, March 2001.
299 'Sedimentary deposits in the northern lowland plains, Mars', B.K. Lucchitta, H.M. Ferguson and C. Summers. *J. Geophys. Res.* Supp., vol. 91, p. E166-E174, 1986.
300 'Topography of the Northern Hemisphere of Mars from the Mars Orbiter Laser Altimeter', D.E. Smith *et al. Science*, vol. 279, p. 1686, 1998.
301 'New topographic map may explain Martian history', R. Graham. *Astronomy*, p. 24, September 1999.
302 'The global topography of Mars and implications for surface evolution', D.E. Smith *et al. Science*, vol. 284, p. 1495, 1999.
303 'Mars: northern hemisphere slopes and slope distributions', O. Aharonson *et al. Geophys. Res. Lett.*, vol. 25, p. 4413, 1998.
304 'Transitional morphology in west Deuteronilus Mensae, Mars: implications for modification of the lowland/upland boundary', T.J. Parker, R.S. Saunders and D.M. Schneeberger. *Icarus*, vol. 82, p. 111, 1989.
305 'Coastal geomorphology of the Martian northern plains', T.J. Parker, D.S. Gorsline, R.S. Saunders, D.C. Pieri and D.M. Schneeberger. *J. Geophys. Res.*, vol. 98, p. 11061, 1993.
306 'Ancient oceans, ice sheets and the hydrological cycle on Mars', V.R. Baker, R.G. Strom, V.C. Gulick, J.S. Kargel, G. Komatsu and V.S. Kale. *Nature*, vol. 352, p. 589, 1991.
307 'Oceans in the past history of Mars: tests for their presence using Mars Orbiter Laser Altimeter (MOLA) data', J.W. Head III, M. Kreslavsky, H. Hiesinger, M. Ivanov, D.E. Smith and M.T. Zuber. *Geophys. Res. Lett.*, vol. 24, p. 4401, 1998.
308 'Possible ancient oceans on Mars: evidence from Mars Orbiter Laser Altimeter data', J.W. Head, H. Hiesinger, M.A. Ivanov, M.A. Kreslavsky, S. Pratt and B.J. Thomson. *Science*, vol. 286, p. 2134, 1999.
309 The three-dimensional study of the polar caps established that they contain somewhat less water-ice than had been thought, but still enough water to form to a layer over the planet about 25 metres deep. In practice, of course, the water would not form an even cover, it would pool in the low-lying northern plains.
310 'Chryse Planitia, Mars: topographic configuration and tests for hypothesised ancient bodies of water using Mars Orbiter Laser Altimeter (MOLA) data', M.A. Ivanov and J.W. Head. *32nd Vernadsky-Brown Microsymposium on Comparative Planetology (Moscow)*, p. 66, October 2000.
311 'Oceans or seas in the Martian northern lowlands: high-resolution imaging tests of proposed coastlines', M.C. Malin and K.S. Edgett. *Geophys. Res. Lett.*, vol. 26, p. 3049, 1999.
312 C. Moore, D. Sawyer, M. McGehee and J. Canepa. *Meteoritics and Planetary Science*, July 2000.
313 'Mapping of possible "Oceanus Borealis" shorelines on Mars: a status report', T.J. Parker. *Proc. Lunar Planet. Sci. Conf.*, p. 1965, 1998.
314 'Enigmatic northern plains of Mars', P. Withers, and G.A. Neumann. *Nature*, vol. 410, p. 651, 2001.
315 R.A. Schultz. *J. Geophys. Res.*, vol. 105, p. 12035, 2000.
316 'Oceans in the northern lowlands of Mars: further tests using MGS data', J.W. Head, M.A. Ivanov, H. Hiesinger, M. Kreslavsky, S. Pratt and B.J. Thomson. *Proc. Lunar Planet. Sci. Conf.*, p. 1064, 2001.
317 'White Mars: a new model for Mars' surface and atmosphere based on CO_2', N. Hoffman. *Icarus*, vol. 146, p. 326, 2000.

Notes 141

318 'CO$_2$ phase changes and flow mechanisms for non-aqueous "floods" on Mars', N. Hoffman. *Proc. Lunar Planet. Sci. Conf.* p. 1288, 2001.
319 'Emplacement of a debris ocean on Mars by regional-scale collapse and flow at the crustal dichotomy', N. Hoffman, K.L. Tanaka, J.S. Kargel and W.B. Banerdt. *Proc. Lunar Planet. Sci. Conf.*, p. 1584, 2001.
320 'Evidence for magmatically driven catastrophic erosion on Mars', K.L. Tanaka, J.S. Kargel and N. Hoffman. *Proc. Lunar Planet. Sci. Conf.*, p. 1989, 2001.
321 'Huge CO$_2$-charged debris-flow deposit and tectonic sagging in the northern plains of Mars', K.L. Tanaka, W.B. Banerdt, J.S. Kargel and N. Hoffman. *Geology*, vol. 29, p. 427, 2001.
322 'The origin of pervasive layering on early Mars through impact/atmosphere feedback mechanisms', N. Hoffman. *Proc. Lunar Planet. Sci. Conf.*, p. 1582, 2001.
323 'Compositional variability of Martian olivines using Mars Global Surveyor Thermal Emission Spectra', T.M. Hoefen and R.N. Clark. *Proc. Lunar Planet. Sci. Conf.*, p. 2049, 2001.
324 'The new Mars of MGS MOC: ridged layered geologic units (they're not dunes)', K.S. Edgett and M.C. Malin. *Proc. Lunar Planet. Sci. Conf.*, p. 1057, 2000.
325 'Preliminary stratigraphy of Terra Meridiani, Mars', B.M. Hynek, R.E. Arvidson, R.J. Phillips and F.P. Seelos. *Proc. Lunar. Planet. Sci. Conf.*, p. 1179, 2001.
326 'The distribution of crystalline hematite on Mars from the Thermal Emission Spectrometer: evidence for liquid water', P.R. Christensen, M.C. Malin, D. Morris, J. Bandfield, M. Land and K.S. Edgett. *Proc. Lunar Planet. Sci. Conf.*, p. 1627, 2000.
327 Surveyor's early TES measurements were obscured by the amount of dust in the atmosphere. The emissions from the airborne dust masked the subtle spectral signatures of the surface minerals, so the search for evaporites was frustrated. Nevertheless, the project scientists are confident that they will eventually be able to deconvolve the dust signature from the data.
328 In July 1997, before Surveyor reached Mars, infrared observations by the Hubble Space Telescope detected a spectral signature which suggested the presence of water-bearing minerals in the soil of Acidalia Planitia (to the north of Chryse Planitia where Pathfinder landed that month). One interpretation was that these were clays, but Surveyor's TES, despite having higher spatial and spectral resolution, did not confirm this.
329 'Evidence for recent groundwater seepage and surface runoff on Mars', M.C. Malin and K.S. Edgett. *Science*, vol. 288, p. 2330, 2000.
330 M.C. Malin and K.S. Edgett. *Science*, vol. 288, p. 2330, 2000.
331 'Young volcanism on Mars and implications for planetary exploration', W.K. Hartmann. *32nd Vernadsky-Brown Microsymposium on Comparative Planetology (Moscow)*, p. 50, October 2000.
332 'Recent volcanism on Mars'. *Sky & Telescope*, p. 34, October 2000.
333 'Fresh polar channels on Mars as evidence of continuing CO$_2$ vapour-supported density flows', N. Hoffman. *Proc. Lunar Planet. Sci. Conf.*, p. 1271, 2001.
334 'Red Planet rendezvous', W.K. Hartmann. *Astronomy*, p. 50, March 1998.
335 A.S. McEwen *et al*. *Nature*, vol. 397, p. 584, 1999.
336 'Invading Martian territory: Mars Global Surveyor looks for clues to the origin of life', W.K. Hartmann. *Astronomy*, p. 46, April 1999.

5

Mysterious Venus

EARLY OBSERVATIONS

Galileo Galilei's interest in astronomy was prompted by the appearance of a supernova in 1604 which, at its peak, rivalled Venus in brightness. In the autumn of 1609, armed with the newly invented telescope, he observed that Venus showed lunar-like illumination phases.[1] This contradicted the Ptolemaic concept of a 'celestial sphere',[2] and proved the proposition of Polish astronomer Nicolaus Copernicus that, with the exception of the Moon, the 'wandering bodies' moved around the Sun rather than around the Earth.[3]

When Venus is closest to the Earth, it is at inferior conjunction. However, it then presents its darkened hemisphere to the Earth and is lost in the solar glare. When its disk is fully illuminated it is at superior conjunction, and is once again lost in the Sun. The planet is therefore best placed for observation when it is at elongation.

In 1650, James Gregory pointed out that simultaneously observing transits of Venus from widely spaced locations across the Earth would measure the solar parallax which, in turn, would yield the Earth's distance from the Sun. Transits would be commonplace if Venus's orbit was aligned with the ecliptic but the 3-degree divergence is sufficient to make them very rare events – in fact, transits occur in pairs, 8 years apart, every hundred years or so.[4] Johann Kepler predicted in 1627 that Venus would cross in front of the Sun on 6 December 1631, and Pierre Gassendi in Paris watched for it all day but was disappointed because it did not start until after sunset. The first transit to be observed was on 23 November 1639 by William Crabtree in England, after J. Horrocks had realised that they occur in pairs.[5]

After unsatisfactory results from the 1761 transit, Captain James Cook led an epic voyage to Tahiti in the South Pacific to enable Charles Green, an assistant to Nevil Maskelyne (then the Astronomer Royal), to observe the 1769 transit. The results allowed J.F. Encke to calculate the distance from the Earth to the Sun. This provided a yardstick for the scale of the Solar System, whose relative size was known from Kepler's laws of orbital motion. The value of what has been called the

'Astronomical Unit' was refined by subsequent transits. Venus's next transit is in 2004. Observations will have little scientific value, but it should be an interesting event for amateur astronomers.

When Francesco Fontana observed Venus in 1643, he drew light and dark patches that he assumed to be continents and oceans. In 1667 G.D. Cassini noted bright dusky markings and estimated that the planet rotated in 23 hours 21 minutes, but he was subsequently unable to see anything to confirm this,[6] and his son J.J. Cassini was unsuccessful in this regard during his many years of observing the planet.[7] In 1725 Francesco Bianchini drew another map showing surface detail and naming the features,[8] and from these sightings he concluded that the planet's period of rotation was some 24 *days*.

When the fuzzy edge of Venus's disk was viewed against the Sun during the 1761 transit, M.V. Lomonosov in Russia concluded that Venus possessed a dense atmosphere. In 1769 R. Rittenhouse noted that there was light refracting around that part of the planet that had yet to encroach on the solar disk. The extension of the 'cusps' when the planet posed a thin crescent phase suggested to J.H. Madler in 1849 that sunlight was being refracted over the poles. C.S. Lyman confirmed this in 1866. Indeed, in 1898 H.N. Russell noted that when the planet was only a degree or so from the Sun these cusps extended almost all the way around the planet's disk.

Having seen nothing more than a few dusky markings over many years of observing the planet, William Herschel concluded that Venus's surface was masked by impenetrable cloud.[9] During a decade of observing the planet J.H. Schroter saw no detail on Venus's disk but in 1788 (and again in 1811) he noticed a pattern of 'filmy streaks' from which he inferred that Cassini had been correct in reporting a period of just under 24 hours.[10] In 1841, F. de Vico supported this conclusion.

On three occasions between 1789 and 1793, Schroter observed that one of the cusps was truncated and there was an isolated point of light shining near where the tip would have been. He inferred that there were transitory gaps in the otherwise complete cloud cover, and that he was observing a tall mountain catching the sunlight.[11] In 1813 Franz von Paula Gruithuisen of Munich Observatory noted that Venus displayed bright 'hoods' and, because they were where he expected the poles to be if the planet rotated in an upright manner, he presumed that they were bright polar ice caps, like those on Mars.[12] In 1878 E.L. Trouvelot saw the southern hood as a series of distinct spots, which he believed were mountains. He also speculated that they might have been seen as dark projections beyond the limb during the 1882 transit.[13,14] As recently as 1952, J.C. Bartlett argued that despite being closer to the Sun, and therefore generally warmer, Venus had ice caps at the poles which could be viewed from time to time through gaps in the clouds.[15]

G.B. Riccioli noticed that the darkened hemisphere 'glowed' when Venus was a cresent in 1643. This 'ashen light' was also observed by Derham in 1715,[16] Kirch in 1721, and Schroter and Harding in 1806.[17] A similar 'secondary' illumination is readily observed when the Moon is a crescent, and the dark hemisphere is faintly illuminated by sunlight reflected by the Earth, a condition known as 'the old moon in the new moon's arms'. Upon finding the ashen light to be particularly striking in 1883, Zenger suggested that Venus was being similarly illuminated, but the distances were too great for 'Earthshine' to be significant.[18] After Zenger advanced the theory that Venus showed so little surface detail because it had a near-global ocean, Schafarik proposed that the glow was due to phosphorescence in this ocean.[19] In the early 19th century, Gruithuisen speculated that the glow might be due to fires lit by its inhabitants.[20] In 1872 P. de Heen suggested that the glow was an auroral display. Nowadays, the ashen light is presumed to be an electrical phenomenon in the tenuous upper atmosphere, but the details are not known.

Various reports were made of a Venusian satellite. The first report of an object close to the planet showing identical phases was made by G.D. Cassini in 1672. Similar reports were made by James Short in 1740[21] and Montaigne in 1761. However, there is little doubt that what they observed was a 'ghost' image of the exceedingly bright planet induced by minor optical flaws in their telescopes. One observer is reputed to have remarked dryly that he had a telescope that could show a companion alongside any bright star!

G.V. Schiaparelli observed Venus methodically from 1877 to 1890 and, based upon his observations of the south polar hood, concluded that the planet's axial rotation was synchronised with its orbital period of 225 days.[22,23,24] Percival Lowell agreed that it rotated synchronously,[25] but thought that the atmosphere was transparent.[26] In 1897 he drew a map that showed a series of linear streaks (which he named) radiating from a central dark hub.[27] In contrast to the 'canals' that Lowell perceived on Mars and believed to be of artificial construction, he thought that the linearities on Venus were of natural origin. Although his contemporaries dismissed his reports of linearities (he also drew them on Mercury and Jupiter's satellites), R. Barker reported linearities in 1932,[28] as did R.M. Baum in 1951,[29] but they had utilised small telescopes. Such renowned planetary observers as E.E. Barnard of the Lick Observatory, E.M. Antoniadi at the Meudon Observatory, and A. Dollfus at the Pic du Midi Observatory reported only faint fuzzy markings of a transitory nature. Henry McEwen (for half a century the director of the British Astronomical Association's Mercury and Venus Section) concluded that Venus's atmosphere was 1,600 kilometres deep,[30] which was remarkable.

In 1921 W.H. Pickering departed from the two 'camps' that favoured Venus's rotation as being either essentially 24 hours or 225 days, by offering a figure of 68 hours.[31] Furthermore, he made the radical proposal that its spin axis was tipped over at 85 degrees, much like Uranus. Later spectroscopic efforts to establish Venus's rotation by observing the doppler effect across its disk found that the limb velocities were negligible.[32] A photographic study suggested a period of at least a few weeks.[33] In 1924 W.H. Steavenson derived a period of 8 days based on observations of a conspicuous marking, and agreed with Pickering concerning the axial tilt, but he was using a small telescope.[34] Thermocouple measurements in the late 1920s by S.B. Nicholson and E. Pettit using the 100-inch Hooker reflector on Mount Wilson, which was then the largest telescope in the world, revealed that the atmosphere on the dark hemisphere was not as cold as expected if the planet's rotation had been synchronous. It was concluded from this that the period was several months.[35] In 1953, Gunter Roth, in Germany, measured the doppler shift across the planet's disk and decided that it rotated in 15 days.[36] At this same time, J.C. Bartlett concluded that the obliquity was near-zero.[37] It was evident from the lack of agreement between observers over the years that Venus was a very tricky object to study. Apart from those who discerned surface detail, visual telescopic observers had spotted only dusky features of, at best, a transitory nature, but in 1927 F.E. Ross experimented with photographing the planet through a variety of filters and discovered that ultraviolet showed detail that was not visible to the naked eye.[38,39] As ultraviolet light does not penetrate far into Venus's atmosphere, these features had to be at high altitude. That same year at the Lick Observatory, W.H. Wright showed that there were no markings in the infrared.

In the 1950s, following up on Ross's discovery of ultraviolet atmospheric features, G.P. Kuiper at the McDonald Observatory, N.A. Kozyrev in the USSR, A. Dollfus at Pic du Midi, and R.S. Richardson at Mount Wilson noted a 'banded' atmospheric structure, which Kuiper interpreted as indicative of ascending (dark) airflow and descending (bright) airflow, and he noted that the symmetry implied that the axial tilt was 32 degrees and the rotational period was 'a few weeks'.[40,41] Richardson suspected a *retrograde* rotation. A study by C. Boyer at Pic du Midi identified a stable Y-shaped pattern with a retrograde rotation of 4 days. Furthermore, the orientation of this pattern established that the axis was *not* inclined at 85 degrees as Pickering had proposed, and not at 32 degrees as Kuiper had proposed, but at 180 degrees – in effect, the planet was upside down. However, Boyer's analysis was lost in the mass of conflicting evidence.

In 1915, C.E. Housden, who believed that Venus's rotation was synchronous, proposed that as a result of keeping one hemisphere facing the Sun and the other in perpetual darkness the atmospheric circulation would be very different to that of the Earth. Specifically, air would rise at the subsolar point and then flow at high altitude around to the dark hemisphere, whence it would be chilled, descend, and be drawn back onto the daylit hemisphere at low level. Housden also believed that the atmosphere was wet, and predicted that there would be an accumulation of ice on the frozen hemisphere. Furthermore, favouring Lowell's view that the atmosphere was fairly transparent, he suggested that the network of streaks Lowell had drawn were an irrigation system built by the inhabitants in order to draw water from the far side onto the desert-like near side.

Many of those who believed that Venus's rotation was not synchronised subscribed to the popular notion that Mars was an 'ancient' world that had already dried up and 'died' and that Venus was a 'younger' version of the Earth.[42] Indeed, in 1918 Svante Arrhenius in Sweden proposed that Venus was a lush environment strikingly similar to the Earth as it was some 200 million years ago during the Carboniferous period when much of its surface was a vast swamp thick with luxuriant vegetation. In 1924, W.H. Pickering ventured that Venus might possess oceans.[43] Water vapour is difficult to identify unambiguously in a planet's atmosphere by telescopic spectroscopy because the signal is swamped by the water vapour in the Earth's atmosphere, but in 1897 Johnstone Stoney had reported Venus's atmosphere to be laden with water vapour, so a hot and wet Venus seemed plausible. A polarisation study by B. Lyot in 1929 determined that the clouds were so 'brilliant' because the sunlight was reflecting from a dense suspension of tiny droplets of liquid. Rupert Wildt at Princeton suggested in 1937 that the clouds were droplets of formaldehyde, a compound of carbon, hydrogen and oxygen whose formation might be induced by ultraviolet and which is white in the presence of water vapour, but a search for appropriate absorption bands was inconclusive.[44] H. Suess allowed that Venus may have started out that way, but he believed the surface to be hot and arid and he argued that the clouds contained chloride salts left over from when the oceans evaporated.[45]

In 1932, W.S. Adams and T. Dunham of the Mount Wilson Observatory detected three absorption bands in the infrared.[46] These were later identified as being due to carbon dioxide. The presence of such a 'heavy' gas in the upper atmosphere meant that the lower atmosphere would be rich in carbon dioxide, and since this is a 'greenhouse' gas it followed that the surface would be warmer than the planet's location close to the Sun would imply. In 1937 Arthur Adel of the Lowell Observatory argued that the surface temperature probably exceeded 50 °C due to the carbon dioxide.[47] In 1939 Rupert Wildt argued that the temperature likely exceeded the boiling point of water, in which case there could be no oceans.[48] However, V.A. Firsoff proposed a means by which the lower atmosphere might be predominantly oxygen. If Venus possessed a strong magnetic field, then (he said) its atmosphere would be stratified with the gases being sorted at levels according to their magnetic susceptibilities. Being strongly paramagnetic, oxygen would be retained at the surface. Carbon dioxide, being strongly diamagnetic, would be 'repelled' to the upper atmosphere.[49] But there was no independent evidence that Venus had such an intense magnetic field. Despite the suggestion that Venus was extremely hot, in 1955 F.L. Whipple and D.H. Menzel, accepting Lyot's conclusion the clouds were composed of droplets and taking this to be water vapour, envisaged such a vigorous water cycle within the predominantly carbon dioxide atmosphere that chemical erosion by carbonic acid (formed when the water absorbed carbon dioxide) would mean that there ought not to be *any* exposed land – it was a *global ocean*.[50]

Also in 1955, the ever imaginative Fred Hoyle claimed that Venus's ocean held hydrocarbons – oil – rather than water. It was his view that there could not be *any* water left. The clouds (he argued) were a dense smog made of dust motes and tiny droplets of oil. His logic was that if hydrocarbons that accumulated at shallow depth in the crust eventually broached the surface to form lakes, the hydrocarbons would evaporate and be oxidised, in the process drawing the oxygen from the atmosphere. The process would continue (he said) until either all of the oil was oxidised or until the oxygen liberated by the dissociation of water vapour in the upper atmosphere by solar ultraviolet ceased because all of the surface water had evaporated. The presence of the smoggy clouds meant (he said) that the oxygen had run out first.

The development by radio astronomers of a sensitive radiometer in 1956 enabled Venus's *surface* to be investigated directly by its microwave emissions.[51] However, a temperature exceeding 300 °C was puzzling because even though the insolation impinging upon Venus's atmosphere is twice that at the Earth's distance from the Sun, some 85 percent of it is reflected by the clouds. Why should the surface be *that* hot? Nevertheless, such a high temperature fit with Harold Spencer Jones's 'dust bowl' theory.[52]

Clearly, therefore, as the 'Space Age' dawned, it was not known whether Venus was wet and rich in vegetation, awash with oil, or an arid desert. There was even a strange suggestion that the planet was a dormant comet.[53]

RADAR

When large radio telescopes were configured as radars, they were able to study the surface of Venus by the way in which it reflected microwaves.[54] In radar imaging, areas of high reflectivity appeared bright, and areas of lower reflectivity appeared dark. However, because a radar provided its own illumination and the angle at which the surface was viewed was an important factor in how it appeared, the resulting images were not really photographs of the surface. Surface reflectivity depended upon a number of factors, including the characteristics of the surface and the chemistry of the rocks.

The doppler on the radar reflection indicated that Venus rotated extremely slowly, and in a retrograde manner, as Boyer had suggested. Interestingly, at 243

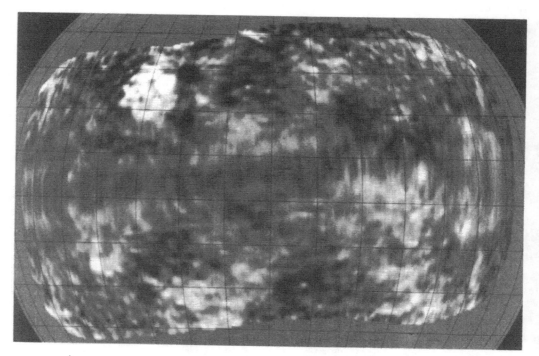

Radio telescopes acting as radars provided the first glimpse of Venus's surface. The technique was limited to the equatorial zone on the hemisphere which faced the Earth when Venus was at inferior conjunction. This map was produced by integrating the results from the Goldstone and Arecibo observatories.

days the planet takes longer to turn on its axis than it does to orbit the Sun. If there were no obscuring clouds, an observer on its surface would see the Sun rise in the west, creep across the sky, and then set in the east 118 days later. This rotation creates a surface equatorial velocity of 4 metres per second. This contrasts with 1,500 metres per second for the rapidly rotating Earth. The Y-shaped pattern in the upper atmosphere circled the globe in just 4 days. If the lower atmosphere shared this rate, the winds at the surface would be intense.

The extremely slow spin meant that Venus had a very low angular momentum.[55] However, local motions within the solar nebula from which the planet condensed should have left it with a significant prograde angular momentum. An early idea was that the Earth's presence served to brake the planet's spin.[56,57] A more recent theory proposed that a giant impact early in the planet's accretion countered rather than augmented the spin and actually tipped the planet right over. It has been suggested that – in contrast to the Earth and the Moon – the ejecta did not coalesce to create a satellite because such a collision would have placed the debris into a close orbit which rapidly decayed, and the material was reaccreted.[58] The issue is far from resolved, though.

The fact that the planet rotated so slowly made it possible to time the signals and derive surface elevations. However, this could be done only when the planet was fairly near, and so the surface coverage was limited. It was not possible to study the polar latitudes at all. The 85-kilometre resolution of this topographic survey was poor, but by establishing the presence of mountainous and low-lying terrains it gave us our first direct view of the planet's surface.

CLOSE LOOKS

In 1961, the Soviet Union was the first to dispatch a spacecraft – named Venera 1 – to study Venus. Although contact was lost with the probe several weeks later, when it was some 7.5 million kilometres from Earth, the mission represented a partial success in that the 'escape stage' had set up a trajectory that, in May, took the probe within 100,000 kilometres of Venus, thereby demonstrating that interplanetary fly-by missions were feasible.

In 1962, NASA's first attempt was frustrated when the launch vehicle malfunctioned and dumped the Jet Propulsion Laboratory's Mariner 1 into the Atlantic. However, Mariner 2 was dispatched successfully. A 16,000-kilometre fly-by had been planned, but the trajectory was slightly off and its closest approach on 14 December 1962 was 35,000 kilometres. The probe did not carry a camera, but it aimed a battery of 'remote sensing' instruments at the planet. Its main objective was to establish whether the high surface temperature measured by microwave radiometry was real, and it confirmed that the surface really is hot.[59] The surface pressure was not known, but liquid water would not be stable at a temperature of 480 °C. This observation disproved both the 'global ocean' and 'carboniferous' theories. The fact that Mariner 2 did not detect any planetary magnetic field was consistent with the extremely slow rotation, because such a field was believed to be induced by electric currents circulating in a rapidly rotating iron core. The absence of a magnetic field ruled out Firsoff's stratified atmosphere theory.

As the first successful interplanetary probe, Mariner 2 demonstrated that a spacecraft with appropriate instruments performing a brief fly-by could yield evidence to sort the wheat from the chaff in the basket of competing theories.

Launch windows for 'fast' trajectories to Venus occur every 19 months and the Soviet Union exploited each opportunity, but continued to suffer from reliability problems with a string of spacecraft either being lost in launch accidents or being stranded in 'parking orbit' by faulty escape stages. In April 1964 a heavier probe – named Zond 1 – was successfully dispatched. However, after making several manoeuvres to refine its trajectory, its radio failed. An even heavier probe (designated Venera 2 once safely on its way) fell silent shortly prior to making a 24,000-kilometre fly-by in February 1966. Although its running mate (Venera 3) was able to refine its trajectory in an attempt to strike the planet the following month, it too had by that time succumbed to some sort of systems failure.

This spate of missions established that the Soviets had advanced beyond the fly-by phase and their goal was to penetrate the planet's atmosphere. This was duly accomplished by Venera 4 on 18 October 1967. An hour after it released the probe, the spacecraft 'bus' was destroyed upon striking the atmosphere. The trajectory (and hence the entry point) was constrained by the requirement that the probe must strike the atmosphere at precisely the right angle (too steep and it would burn up; too shallow and it would bounce off), and also that the probe must have a clear line of sight to the Earth in order to transmit its results. It penetrated the equatorial zone 1,500 kilometres beyond the dark side of the terminator. The 1-metre-diameter spherical probe was designed to survive extreme deceleration during aerobraking. Once slowed to 300 metres per second, it deployed its parachute and started to transmit. By the time the transmission ceased the ambient pressure had risen to 22 bars. Although the Soviets announced that their probe had been disabled upon hitting the surface, this was inconsistent with the final temperature reading of 280 °C because previous indications had implied that the surface was twice as hot. When NASA's Mariner 5 flew by Venus the following day its trajectory took it behind the planet as seen by the Earth.[60] Measurements of the way in which the radio signal was attenuated by the planet's atmosphere yielded a profile from which extrapolation indicated that the pressure at the surface was in the range 75 to 100 bars, so Venera 4 had been 27 kilometres above the planet's surface when it ceased transmitting. Interestingly, the probe was designed to float just in case it should splash down!

Although it had not reached the surface, Venera 4 provided the first *in situ* measurements of the Venusian atmosphere. A nitrogen atmosphere with up to 10 per cent carbon dioxide had been expected, but the carbon dioxide concentration was found to be at least 90 per cent. This discovery went a long way towards explaining the extreme conditions on the surface.[61]

The surface would remain out of reach until a stronger probe could be developed, but with every launch window being exploited it was decided to reduce the size of the parachute so that the probe would descend more rapidly, and penetrate more deeply before succumbing. On 16 May 1969 Veneras 5's final readings at an altitude of 25 kilometres were 320 °C and 27 bars, from which the pressure at the surface was extrapolated to be 140 bars. The final readings from Venera 6 the following day were similar, but its radio altimeter showed that it had reached an altitude of only 12 kilometres, which implied a relatively mild 60 bars and 400 °C at the surface. Clearly something was amiss. Either the altimeter had malfunctioned or the probe had been above an extremely mountainous area. On balance, it was evident that the altimeter had failed. Chemical analyses confirmed the high carbon dioxide content: fully 95 per cent. Most of the rest was an inert gas (later established to be nitrogen). There was at most 0.4 per cent oxygen. Water vapour was present in the cooler upper atmosphere, but it was by no means saturated and the clouds were not water droplets. The hot lower atmosphere was arid.

Having discovered that Venus's atmosphere was denser than expected, the Soviets rebuilt their probe to withstand an incredible 180 bars.

After a parachute descent of 26 minutes on 15 December 1970, Venera 7 appeared to fall silent, but later analysis established that the signal had continued at 1 per cent of its previous strength for a further 23 minutes, from which it was concluded that the probe had rolled over upon touching down, aiming its antenna in the wrong direction. It was also found that the telemetry system had malfunctioned and fixated on the temperature sensor. However, when the profile was combined with the doppler on the radio signal (showing how the probe had been slowed) the pressure could be inferred. Finally, it was established that the pressure at the surface was almost 100 bars.

With the conditions on the surface known, it was possible to design a probe to carry out a detailed programme of surface studies. This development would take time, so another interim mission was ordered for the next window. The mass saved by trimming the pressure shell to cope with 'only' 100 bars was reassigned to improve the thermal shielding. A second antenna was added so as to maintain contact with the probe no matter how it rolled around. Venera 8's arrival on 22 July 1972 was just after local dawn, as far into daylight as it was possible to set down and still have a line of sight to the Earth. After 50 minutes on the surface it was disabled by the thermal stress. This time all the instruments functioned perfectly, and the conditions were unambiguously demonstrated to be 470 (± 8) °C and 90 (± 2) bars. The chemical analysis firmed up the atmospheric composition as 96 per cent carbon dioxide, 3 per cent nitrogen and at most 0.1 per cent oxygen. During the descent, the wind speeds decreased from 100 metres per second at an altitude of 48 kilometres, which marked the base of the cloud deck, to less that 1 metre per second at an altitude of 10 kilometres, so the dense lower atmosphere was *stagnant*. Confident that this probe would reach the surface and survive, a gamma-ray spectrometer had been included to characterise the natural radiation from the surface, and its results suggested a potassium-rich granitic rock.[62] This success brought the first generation of Soviet missions to a conclusion.

When Mariner 10 flew by Venus in 1974, it was using the planet's gravity for a 'slingshot' to Mercury.[63] Science operations were conducted over an 8-day encounter sequence. Although there was no detail to be seen on the planet's disk in the visible spectrum (even from the 5,750-kilometre point of closest approach) the camera had

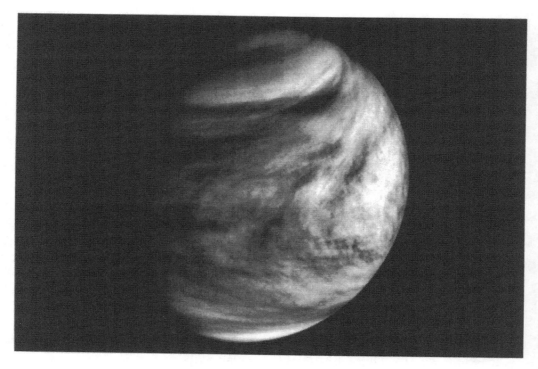

This Mariner 10 ultraviolet image shows the distinctive Y-shaped circulation pattern of Venus's upper atmosphere.

an ultraviolet filter available. The Pic du Midi photographs taken in the mid-1960s (the best terrestrial ultraviolet photographs) showed the Y-shaped pattern only fuzzily, but Mariner 10 saw it in unprecedented detail and resolved structures in the air flow only a few kilometres across. Time-lapse imagery confirmed that this pattern circles the planet in 4 days. This rapid circulation prevents the atmosphere on the night side from cooling down – indeed, Mariner 10's infrared radiometer found the temperature of the cloud tops to be a uniform -23 °C across both hemispheres.

Having demonstrated that their probes could survive the descent, the Soviets were eager to see what the surface looked like. The requirement to have a clear line of sight with the Earth in order that the probe could transmit its data to the Earth had effectively restricted activities to the night side. Admittedly Venera 8 came down just over the terminator but the Sun had been only a few degrees above the horizon. If a probe was to land in full daylight, it would be necessary to provide a relay for its transmission.

After releasing its probe about two days from the planet on a trajectory that would intercept the planet's trailing hemisphere, the new spacecraft would perform a deflection manoeuvre to pass 1,600 kilometres ahead of the planet, at which point it would manoeuvre into an eccentric orbit. As it rose towards its initial apoapsis, it would relay its probe's signal to Earth, and then undertake remote sensing over an extended period. After initial aerobraking, the new robust probes were to use parachutes in the altitude range between 65 and 50 kilometres in order to provide

ВЕНЕРА-9 ОБРАБОТАННОЕ ИЗОБРАЖЕНИЕ

The 180-degree panoramic image returned by Venera 9 was our first direct view of the surface of the planet. The spectacular distortion derives from the imaging system. (The Venera imagery is the copyright of the Soviet Academy of Sciences, USSR, and is courtesy of James Head of Brown University.)

ВЕНЕРА-10 ОБРАБОТАННОЕ ИЗОБРАЖЕНИЕ

The Venera 10 landing site.

time to sample the cloud layer, and then they were to shed their chutes and 'fall' the rest of the way in order to pass through the hot atmosphere as rapidly as possible and thereby maximise their surface time; nevertheless, the air was so thick that they hit the surface at a mere 5 metres per second. The spherical instrument unit was set on a ring-shaped shock absorber that was wide enough to ensure that the probe would stay upright even if it came to rest on a slope.

On 22 October 1975, Venera 9 transmitted the first image of the planet's surface. It was a 180-degree monochrome panoramic view composed over 20 minutes by a line-scan facsimile camera similar to that used by the first probe to soft land on the lunar surface. The probe was on a 20-degree slope on Beta Regio's eastern flank. There was a litter of angular rocks several tens of centimetres across, with small dark particles between them. Because the camera was facing downslope, its foreshortened horizon was just 50 metres away. It appeared that the probe was in among scree on a hillside. Venera 10, which set down three days later, 2,500 kilometres away in the lowlands to the south, found a rather different landscape. In contrast to rocks sitting on the surface, there were fewer and rather slabbier rocks which either formed a remarkably level outcrop or were a fragmented crust laid down as a thin sheet of lava. A dark loosely consolidated material lie in the cracks between the light-toned slabs. According to the gamma-ray spectrometers, both sites were basaltic in nature.

The atmosphere at the surface was transparent, the air was stagnant and there was no dust blowing around. There was no hint of the extreme refraction that some had predicted whereby the view forward would be wildly distorted. Despite the complete overcast, the illumination at the surface was comparable to "a cloudy winter's day in Moscow", as one of the scientists put it.[64] In fact, some 2.5 per cent of the incident insolation penetrated to the surface. However, the light was so effectively diffused that there was no hint at the Sun's position in the sky.

In 1978, NASA's Ames Research Center placed a Pioneer spacecraft into orbit for remote sensing of the planet's atmosphere and radar mapping of its surface, and dropped a flotilla of probes into the atmosphere to undertake simultaneous *in situ* sampling across a variety of sites. The bus released its main probe when 11 million kilometres from the planet, aimed at an entry site near the equator on the dawn side of the terminator.[65] A million kilometres nearer, springs scattered the trio of smaller probes to follow slightly diverging trajectories with arrivals a few minutes apart. One penetrated the atmosphere far to the north, on the night side. A second was just south of the equator on the day side. A third was closer to the equator near local midnight. After aerobraking, the large probe deployed a parachute at 68 kilometres to slow its fall and permit improved sampling as it passed through the clouds, and then this was released at 47 kilometres to enable the probe to fall free. The smaller probes did not have parachutes. Because the probes could not investigate their environment while aerobraking, mass spectrometers on the spacecraft's bus reported on the upper atmosphere prior to burning up. Although not expected to survive the impact, the 'day' probe transmitted on the surface for over an hour, ceasing only when its battery was exhausted. Their results indicated a fine haze (smog) layer at 70 to 90 kilometres, then the main cloud deck which extends down to 48 kilometres, then clear air down to a thin layer of haze at 31 kilometres, below which the air is clear down to the surface. The maximum opacity was at an altitude of 50 kilometres, near the base of the main cloud deck. The convection was limited to the main cloud layer and above; there was little circulation below it. Unfortunately, the inlet to the mass spectrometer on the large probe ingested a cloud droplet early on, so the instruments readings were anomalous until the increasing temperature boiled the droplet away. Nevertheless, it was able to show that the ratio of deuterium to hydrogen is 100-to-1, which is considerably higher than that for the Earth's atmosphere. This was to provide a fascinating clue as to the fate of Venus's water. By the time the orbiter fell into the atmosphere in 1992, having far exceeded its design life, it had essentially 'written the book' on the Venusian atmosphere.[66,67]

By the mid-1970s, terrestrial radar systems had improved to the degree that they were able to provide high resolution imagery for the areas that they could view favourably. Although the radar on the Pioneer orbiter had a surface resolution of only 75 kilometres (which was similar to that of the early terrestrial studies) a mapper in orbit had the advantage of being able to chart the entire globe at the same resolution and hence provide a sense of perspective. The altimetry showed that Venus's overall figure is less oblate than that for the Earth. The absence of polar flattening and an equatorial bulge is undoubtedly a consequence of the exceedingly slow rotation rate.

The Soviets suffered a setback with Veneras 11 and 12 in December 1978. In order

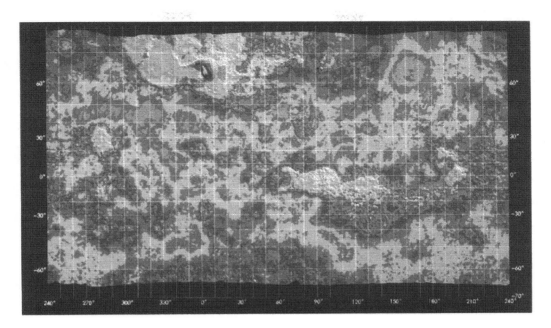

The Pioneer Venus radio altimeter produced a topographic map covering 93 percent of Venus's surface (from 73 degrees north to 63 degrees south) with a surface resolution of 75 kilometres and a vertical resolution of 200 metres (each such measurement was integrated over an area of 100 square kilometres). This revealed that Venus has a monomodal elevation distribution, and that most of the surface is a 'rolling plain' within a few hundred metres of the planetary mean. (Courtesy of NASA and USGS.)

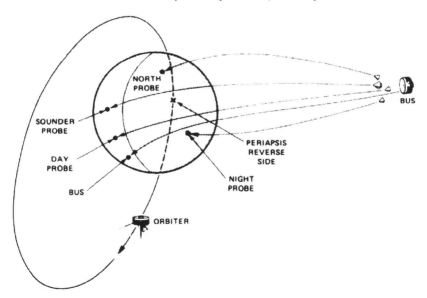

The trajectories of the flotilla of Pioneer Venus probes (including the spacecraft bus) is shown in relationship to the terminator and the plane of the orbit of the companion spacecraft in orbit. (Courtesy of NASA.)

that the probes could carry more instruments, it had been decided not to have the buses perform orbital manoeuvres. Consequently, they reverted to fly-by trajectories that would place each departing bus above its probe's horizon in order to relay its signal back to Earth.[68] Unfortunately, the atmospheric results were ambiguous and the surface imaging was frustrated by the failure of the lens caps to release!

When Harold Masursky of the US Geologic Survey realised that the next pair of Veneras would both sample the low-lying plain to the east of Phoebe Regio, he suggested that one be redirected slightly westward, onto what the Pioneer map indicated was an upland plain that seemed to be considerably older and possibly granitic in composition.[69] In March 1982, Venera 13 set down on this rolling upland plain, and a pair of improved cameras sent colour panoramas which documented the entire horizon. The craft had come to rest on an exposure of slabby rock that was intensely fractured and covered to a varying extent by small fragments of rock and fines. The rocks appeared to be orangey, but this was because the blue and green sections of the spectrum were absorbed by the dense atmosphere, leaving only the red wavelengths to reach the surface. The colour-calibrated images established that the rocks were grey. For the first time, an X-ray fluorescence spectrometer was swung down onto the surface. The chemical analysis suggested a composition similar to a terrestrial potassium-rich alkaline basalt. Venera 14's landing site, on the lowland plain 1,000 kilometres to the southeast, was also a slabby outcrop but with fewer fines. The less potassium-rich chemistry at this site resembled tholeiitic basalt, a primary magma that is erupted on Earth at mid-ocean rifts. The slabs appeared to have a layered structure, implying that the surface was a succession of thin sheets of lava. To have made such a smooth surface on such a scale, the lava would have to have been very fluid and to have had the opportunity to flow far and wide to form a level plain.

Venera 13 introduced a twin-camera system in order to be able to scan almost the entire horizon. The objects in the foreground are the jettisoned lens caps. A new instrument was an armature with a penetrometer to measure the surface's mechanical and electrical properties.

Unfortunately, when Venera 14's penetrometer was rotated down (top) it came to rest upon the lens cap!

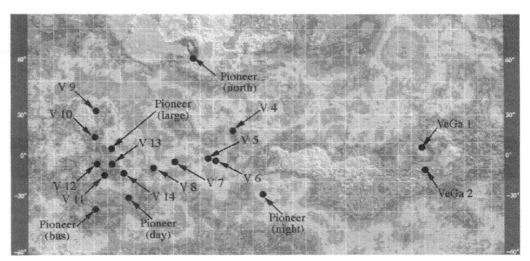

A map showing where the various Soviet and American probes penetrated the atmosphere of Venus, and in most cases, reached the surface.

The varying amounts of fines suggested that some process was actively eroding the rocks, but it was not uniform across the planet. The general sparsity of fines on Venus was marked. Either Venus's slabby rocks are relatively 'fresh', or the weathering processes are very slow.[70] Although the surface is hot and there is a steep thermal gradient, the temperature is uniform at any given elevation and the rocks are not chilled during the long period when the Sun is not in the sky, so they are evidently not eroded by thermal stresses. Fines produced by chemical weathering are likely to remain local because the air is stagnant. This is in contrast to Mars, where the global winds have homogenised the surficial dust. On Venus, the differences in the chemistries of the fines are likely to indicate that there are distinctive volcanic provinces.

With the exception of the Venera 9 site where boulders have accumulated on a slope, the surface of Venus looked as if it would be easy to hike across (for an astronaut in an appropriate environmental suit, of course). Certainly a 'rover' would have little problem in the slabby terrain, and perhaps one day a 'Sojourner' will explore the Venusian plains.

Having profiled the atmosphere and sampled the surface at several locations, the Soviets decided to switch to radar mapping to improve on the Pioneer map. The bus of the Venera spacecraft was modified to deploy a side-looking radar antenna and was fitted with another pair of solar panels to provide the power to operate the radar. Two of these mappers entered orbit in October 1983 as Veneras 15 and 16. They assumed highly elliptical polar orbits with their periapses 60 degrees north, undertook mapping as they dipped down to the 1,000-kilometre peripasis and then, during the 12-hour climb to apoapsis, transmitted the radar data and recharged their batteries. Each pass produced a strip 150 kilometres in width and 7,000 kilometres in length, running from 30 degrees north over the pole, and as the planet slowly rotated beneath them they were able to extend their map around the full range of longitudes. The new 2-kilometre resolution imagery matched the best from the Arecibo radio telescope, whose geographic coverage was severely limited.[71,72] Although the map focused on the northern hemisphere, it provided several significant insights – for example, it was finally possible to unambiguously identify impact craters, and the map showed that there was a minimum size for craters. Evidently, small objects burned up in the dense atmosphere. The cratering record hinted that the plains are less than 1 billion years old. Because these form the majority of the surface, it became evident that Venus had been extensively resurfaced in relatively recent times. It had removed the craters that must have been formed during the bombardment early in the planet's history. At last, therefore, it was possible to infer something significant concerning the nature of the surface of this mysterious cloud-enshrouded planet.

The Soviet Union's final assault on Venus was something of an afterthought – a pair of spacecraft were dispatched to rendezvous with Halley's Comet, and while exploiting Venus's gravity for slingshots in June 1985 they released probes. Reflecting their dual missions, these spacecraft were called VeGa (for Venera-Gallei – the Russian language does not have an 'H', so 'G' is used). In a novel addition to the usual mission, as they dropped through the middle layer of cloud each lander ejected an instruments package which inflated a helium balloon. At an altitude of about 60 kilometres, the pressure was a mere 0.5 bar and the temperature was a moderate 40 °C. An international network of antennae provided continuous reception of their transmissions, using the doppler on the signals to track their motions as they drifted with the prevailing wind. The probes arrived four days apart, and each balloon's battery could sustain two days of operation – during which they would be carried halfway around the planet, and out of contact. The balloons were released at local midnight, the first just north of the equator and the second just south of it. Both drifted some 11,500 kilometres over the terminator, into daylight. Whereas the first had a fairly smooth flight, the second, which passed directly over

Aphrodite Terra, had a rougher ride and fell several kilometres at one point in a strong down-draught. The need for a line of sight to receive the balloon data obliged the probes to land in the dark, so this time the landers did not carry cameras. However they had drills to recover samples for their X-ray spectrometers. Unfortunately, as the first lander headed for a low-lying plain its drill deployed prematurely. VeGa 2's lander set down in the northeastern part of Aphrodite Terra, functioned properly, and found an alumina-rich rock chemistry suggestive of ancient crust.[73]

The great question was whether Venus's lithosphere underwent global plate tectonics? It was difficult to say. The Pioneer radar map had wide coverage but only moderate resolution. The Venera map was more detailed, but was of limited coverage. There were tantalising indications, but the evidence was disputed. A high resolution map of the entire planet was needed. After a false start with a plan for a comprehensive sensor platform designed to study both the surface and the atmosphere, JPL had to settle for a simpler mission with the objectives of mapping the planet by radar and (by detailed tracking) inferring something of its internal structure. After dispatch by the Space Shuttle, the Magellan spacecraft entered into an eccentric polar orbit about Venus on 10 August 1989, and over the next four 243-day 'cycles' it mapped 99 per cent of the planet at a resolution of 120 metres, some of it in stereo, and then, by lowering its orbit, undertook a gravity survey to probe the distribution of mass within the planet.[74,75]

Table 5.1 Landing sites on Venus

Probe	Longitude (deg)	Latitude (deg)	Hemisphere
Venera 4	38	19 N	Night
Venera 5	18	3 S	Night
Venera 6	23	5 S	Night
Venera 7	351	5 S	Night
Venera 8	335	10 S	Dawn
Venera 9	291	32 N	Day
Venera 10	291	16 N	Day
Venera 11	299	14 S	Day
Venera 12	294	7 S	Day
Venera 13	304	7 S	Day
Venera 14	312	13 S	Day
VeGa 1	178	7 N	Night
VeGa 2	181	6 S	Night
Pioneer large probe	304	4 N	Day
Pioneer north probe	5	60 N	Night
Pioneer Night probe	57	29 S	Midnight
Pioneer Day probe	317	31 S	Day
Pioneer bus	291	38 S	Day

Note: Venera 8 was the last of the spherical atmospheric probes, the others were more robust vehicles designed to operate on the surface.

A STRANGE ATMOSPHERE

Once described as the Earth's non-identical twin,[76] Venus is slightly smaller than the Earth: its diameter is 95 per cent, its volume is 86 per cent, and its mass is 82 per cent.[77] The inordinate amount of carbon dioxide in the atmosphere is responsible for the extreme surface conditions. The total amount of carbon dioxide in Venus's atmosphere is probably comparable to the amount the Earth would have if all the carbon dioxide in its carbonate rock could be liberated (or, as it may once have been before the carbonate rock formed). If the carbon dioxide in the Venusian atmosphere could be drawn down into the rock, then it would leave a residual atmosphere of nitrogen with a surface pressure of a comparatively mild 3 bars.

Venus's overall atmospheric circulation system has a single Hadley 'cell' on either side of the equator. The air rises in the equatorial zone, flows at high altitude into the polar zone, then descends. The returning air flow occurs in the middle atmosphere, and the stagnant surface is essentially isolated from the weather system. Such a simplistic circulation system is possible because the planet rotates so slowly. On the rapidly rotating Earth, the Coriolis effect disrupts this simple circulation system by inducing swirling airflows and producing tropical, temperate and polar components.

In Venus's upper atmosphere, the winds of 100 metres per second race from east to west, with the planet's retrograde rotation, and the Y-shape is created by air flows towards the polar regions at barely one-tenth of this rate. This pattern is observable in the ultraviolet because some chemical whose concentration varies across the upper atmosphere absorbs the ultraviolet. In addition to sulphur dioxide, elemental sulphur may also be contributing to this absorption. The atmosphere appears bright where the ultraviolet is reflected, and dark where it is absorbed.

In fact, the weather system is more complex than this, because there is a turbulent zone at the subsolar point where sunlight penetrates more deeply into the atmosphere, prompting vigorous isolated convection cells which cause hot gas to 'bubble out' of the top of the cloud deck, and forming turbulent eddies in the zonal airflow as it passes through this point on its race around the planet.

In 1969, based on measurements of the index of refractivity of the Venusian atmosphere, it was proposed that the cloud condensates were acid-laden water droplets.[78,79,80,81] Although some 75 to 85 per cent of the cloud is composed of aerosols of sulphuric acid, some sulphur particles were detected by the X-ray fluorescence instruments of Veneras 12 and 14 and by the large Pioneer probe's mass spectrometer.[82,83] Sulphuric acid aerosols form during photochemical oxidation at altitudes exceeding 60 kilometres. The dissociation of carbon dioxide or sulphur dioxide in the upper atmosphere produces atomic oxygen which oxidises SO_2 to SO_3, which hydration turns into sulphuric acid (H_2SO_4) droplets. As these droplets 'rain out', they are thermally disrupted upon reaching the 95 °C temperature at an altitude of about 48 kilometres. They produce SO_3 which, upon encountering carbon monoxide, regenerates sulphur dioxide and carbon dioxide, thereby completing the cycle. This precipitation cycle operates in the upper atmosphere, where the temperatures are moderate; the hot air beneath the clouds is essentially clear – for a depth of almost 50 kilometres.

A strange atmosphere 161

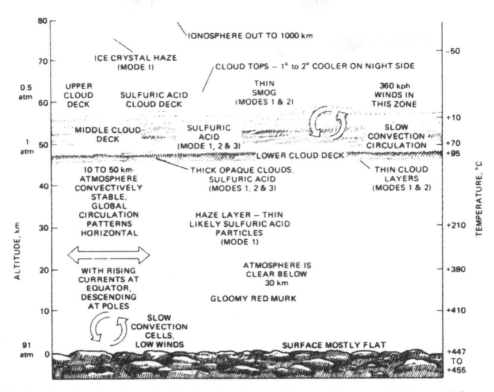

The flotilla of probes dispatched during the Pioneer Venus mission provided a model for the dense Venusian atmosphere. Courtesy NASA.

Venus's surface temperature is uniform, both from pole to pole and from daylight into the darkness. However, there is a very steep thermal profile, with the ambient temperature falling by some 8 °C per kilometre of elevation, so the temperature of the surface depends upon its elevation, and there is a 13-kilometre range between the tallest mountain peak and the floor of the deepest depression.[84] Such a high surface temperature means that water cannot exists. The planet's remaining water is in the cooler upper atmosphere where it participates in the creation of droplets of sulphuric acid, but the total amount must be progressively *diminishing* because the hydrogen atoms released by photodissociation will tend to leak away into space. Liberated hydroxyl radicals will readily combine with rising sulphur dioxide to enhance the process that creates sulphuric acid and any free oxygen that reaches the surface will oxidise the hot rocks and thus be removed from the atmosphere.

Where did the sulphuric acid originate? Sulphur dioxide from volcanoes? Remote sensing by the Pioneer orbiter revealed that the amount of sulphur dioxide in the Venusian atmosphere at an altitude of 80 kilometres progressively diminished between 1978 and 1986 by a factor of ten.[85] This has been interpreted (by some) as evidence of a major volcanic eruption shortly prior to the spacecraft's arrival.[86] There seems to have been a global 'pulse' of volcanic resurfacing about 500 million years ago.[87,88] Was the atmosphere 'pumped up' with sulphurous chemicals at that time? Sulphurous gases are present in the Earth's atmosphere only in trace amounts. They are released by volcanic activity, but unless this is on an enormous scale the sulphur is either soon dissolved in the water droplets in clouds or, in arid regions, bound with oxygen-rich radicals, water vapour or other trace gases to form particles that fall to the ground. When an explosive volcano sends a sulphurous plume into the stratosphere, droplets of sulphuric acid are created but are soon distributed around the globe by the prevailing winds, and by reflecting sunlight they cool the lower atmosphere. Even though most of the sunlight reaching Venus is reflected, the carbon dioxide maintains the high temperature of the lower atmosphere. Could catastrophic volcanism have triggered a 'runaway' evolution? The dense lower atmosphere will inhibit the plumes from volcanoes. Consequently, pyroclastic eruptions must be rare.[89,90] Because the hot lower atmosphere is arid, sulphur dioxide cannot be washed out of the atmosphere. Once it is injected into the atmosphere it will remain there, and provide a ready reservoir for the chemical reactions far above that create the sulphuric acid clouds. Does Venus owe its nasty atmosphere to recent geological activity? Was it at any time more Earth-like?

ANCIENT OCEANS?

As Venus accreted from the solar nebula, it should have acquired as much water as did the Earth. Although Venus is closer to the Sun, the early Sun was 30 per cent less radiant than it is today. If Venus's surface temperature was close to the triple point of water, it may once have had a temperate climate with oceans and a vigorous water cycle in which the water vapour combined with carbon dioxide as carbonic acid that chemically weathered the rock to produce carbonates which were washed into the oceans to settle out as sediment – just as occurs on the Earth.[91]

However, when the Sun heated up, Venus's oceans would have evaporated, producing a super-saturated atmosphere in which the greenhouse effect of the water vapour further increased the temperature. The water vapour in the upper atmosphere would have been dissociated by solar ultraviolet, the hydrogen would have escaped and the oxygen would have been drawn into the rocks as oxides. As this liberated hydrogen escaped into space, the heavier deuterium isotope would have been preferentially retained. The anomalously large 100-to-1 ratio of deuterium to hydrogen is a powerful piece of evidence that Venus has lost a great deal of water.[92] In fact, it has 1/100,000th the water on the Earth. Once the water cycle ceased, there would have been no way to remove carbon dioxide from the atmosphere, and volcanism would have increased its concentration and added sulphurous compounds. Was this how Venus was transformed?

AN ACTIVE SURFACE

On Venus, volcanic eruptions will be influenced by (1) the extremely high temperature of the surface, because it is only marginally less than the melting point of some minerals, and (2) the high air pressure, which will inhibit volatiles in the magma from exsolving and thereby reduce the violence of eruptions.[93] However, the key to understanding how the planet has evolved is its *heat flow*. Venus is not as large as the Earth, so its central pressure is not as great. Consequently, the solidification of the inner core will not have progressed as rapidly and, resultantly, less latent heat will have been released to create convection in the surrounding fluid core (and since this rotates so slowly there is no magnetic field[94]). Nevertheless, this internal heat drives convection in the mantle and stimulates volcanism.

In the case of the Earth, plate tectonics facilitates efficient heat transfer, firstly, by creating ocean floor at spreading rifts, by having this slowly cool as it is transported away from the ridge, and then by having these slabs of cold rock sink back into the mantle to cool the interior. This lithospheric recycling process cools the interior without concentrating the heat flow into localised hot spots. It also gives rise to distinctive landforms. Would Venus be found to have such tell-tale topographic features? Its physiography is 70 per cent rolling uplands, 20 per cent lowland plains and 10 per cent highlands. The Pioneer map, which covered 93 per cent of the surface, established that 60 per cent of it lies within about 500 metres of the mean planetary radius, and only 5 per cent of the surface exceeds 2 kilometres in altitude.[95]

The majority of the Earth's crust is divided into the stable continental platforms which lie a few hundred metres above sea level, and the abyssal plains of the ocean floor 6 kilometres below sea level, with mountain ranges and ocean trenches representing only small proportions of the surface. Venus does not have such a bimodal distribution. The majority of its surface lies at a single level. However, the elevation range of 13 kilometres from the tallest peak to the lowest lying plain is comparable to the range from the tallest terrestrial mountain to the bottom of the deepest ocean trench. To some researchers, Venus's unimodal surface distribution suggested that plate tectonics was *not* active, for otherwise – by terrestrial analogy – there ought to be a bimodal distribution.

The Venusian highlands differ from those of the Earth, which occur in long, narrow fold belts thrust up by interactions between tectonic plates. On Venus, Ishtar Terra, which stands 3 or 4 kilometres above the planetary mean, spans an area larger than the continent of Australia on Earth. The western section, Lakshmi Planum, forms a plateau that is almost completely ringed by belts of mountains which rise another 3 kilometres or so, beyond which steep slopes descend to the surrounding rolling uplands. The steepest of these slopes is Vesta Rupes, on the southern margin. To the east of Lakshmi Planum, Maxwell Montes is the central part of Ishtar Terra and contains the highest peaks on the planet.[96] The fractured plain of Fortuna Tessera further to the east completes the Ishtar Terra highlands, and covers an area of 1,000 by 2,000 kilometres.[97]

To a certain extent, Lakshmi Planum is comparable to the Tibetan Plateau on Earth, which is the direct result of a particularly energetic collision between two continental landmasses (namely, the impact of India with the southern edge of Asia) and its southern flank is buttressed by the Himalayan Range which forms the line of the suture. It took a long time for geologists to fully appreciate the scale of the deformations that occur during plate interactions, but it would seem that Ishtar

164 **Mysterious Venus**

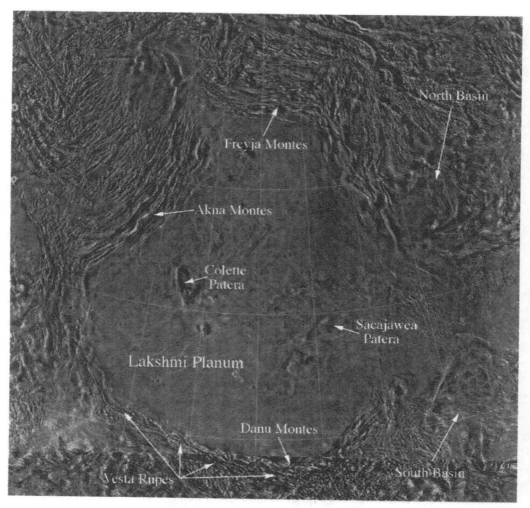

The Venera radar map of Lakshmi Planum, the high plateau on Ishtar Terra. Maxwell Montes is off-frame to the east.

Terra was *not* created in this way.[98] To some researchers, the mountainous terrains are evidence that Venus's lithosphere is too thick to undergo plate tectonics. However, the fact that the Venusian lithosphere is dry means that the diabase substrate is very rigid and can therefore support large-scale elevated structures for much longer than would be the case on the Earth.[99] Indeed, a study established that Ishtar could have been standing for 500 million years.[100] Such a massive structure would undergo gravitational collapsed in only 10 million years in a lithosphere with 'wet' diabase such as on Earth. The fact that Ishtar Terra is so well preserved is a striking indication that Venus suffers very little surface erosion. On Earth, even major mountain belts are severely eroded in only a few tens of millions of years.

Significantly, there are two calderas on Lakshmi Planum – Colette Patera in the west and Sacajawea Patera in the southeast.[101,102,103] Colette's outline is defined by

An active surface

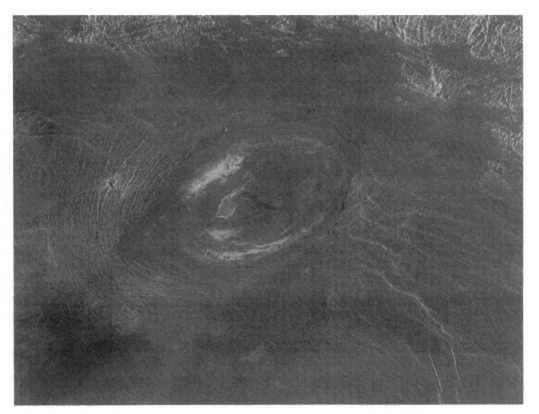

The vast volcanic caldera of Sacajawea Patera on Lakshmi Planum.

arcuate faults that trace out a double ellipse, the outer of which spans 200 kilometres at its widest section.[104] The floor lies about 2 kilometres below its surroundings. Long narrow lava flows radiate out across the plain. To the west, these lava flows have encroached upon and buried the tectonic ridges that bound the plateau. The presence of small patches of tesserae-type terrain on Lakshmi Planum indicates that the smooth surface derives from a succession of eruptions which left only a few highstanding sections of the pre-existing terrain exposed, so the original surface of the plateau was evidently intensely fractured.[105,106] Elsewhere, however, volcanic centres are rare in the tesserae highlands. The tesserae constitute no more than about 8 per cent of the planet's surface.[107] It is thought that they are the oldest preserved features, and date back to when the lithosphere was sufficiently weak (on a global basis) to be so extensively deformed. They therefore provide a glimpse of the state of the surface prior to the pulse of volcanism which resurfaced most of the planet.[108] However, although the stratigraphic relationships show that the tesserae predate the volcanism, subtle differences in their form indicate that they are not all the *same* age.[109]

Is the tallest peak in Maxwell Montes a volcano? The Venera radar map showed a crater on its summit and what appeared to be lava flows on its flanks. Although the identification is still disputed, the higher resolution Magellan map persuaded the majority that this crater (known as Cleopatra) was created by an impact.

A mosaic of Magellan imagery presenting a broad view of Venus's north pole, showing Ishtar Terra and the ridge belt running from the North Basin over the pole and down beside Atalanta Planitia.

Atalanta is a large and roughly circular northern lowland plain that lies some 2 kilometres below the planetary mean. Its smooth appearance on the Venera map prompted the suggestion that it is a depression flooded with lava – Venus's equivalent of a maria. A striking 'fan' of ridges several hundred metres high runs for thousands of kilometres across the rolling upland plain from just north of Ishtar Terra across the pole and down to mid-latitudes east of Atalanta Planitia. The individual ridges are sinuous and bear comparison with the wrinkle ridges on the maria of the Moon and Mars, but they are 'braided' into belts 100 to 200 kilometres in width, with broad relatively smooth valleys between.[110] It is not clear whether the smooth valleys are unmodified crust, or whether they are the result of some settling process. Some people have drawn an analogy with North America's Basin and Range province and argued that this terrain derives from the extensional stress accompanying crustal thinning, while others have interpreted it as a fold belt formed by horizontal compression; the sinuous and overlapping form of some of the ridges support this and suggests thrusting along low-angle faults. Could such compression be related to spreading elsewhere?

Aphrodite Terra (covering an area approaching that of Africa) extends along the equatorial zone for some 10,000 kilometres. It is structurally more complex, and more rugged than Ishtar Terra. In part, it is transected by a number of straight parallel troughs and adjacent ridges with steep walls rising 4 kilometres. The longest troughs can be traced for thousands of kilometres. In some places they are 150 kilometres across and as much as 5 kilometres deep. At the resolution of the Pioneer map, western Aphrodite was strikingly reminiscent of the fracture zones that occur on the terrestrial mid-ocean ridge, with short sections of ridge with a median trough laterally offset from one another and a general sense of bilateral symmetry.[111,112] Was Aphrodite a spreading ridge? The troughs implied that the terrain had been torn apart by extensional stress but this did not in itself indicate plate tectonics.[113] In fact, a later study revealed that the troughs were part of a much larger system that stretched halfway around the planet.[114]

Sapas Mons in Atla Regio at the eastern end of Aphrodite is a volcanic shield 400 kilometres across which rises only a kilometre.[115] It has pits on its summit and massive lava flows on its flanks. To have flowed so far down its shallow slopes the lava would necessarily have been very fluid. On the other hand, Volkova to the east of Atalanta Planitia is a prominent conical edifice with a summit caldera and parasitic cones on its flanks, so some vents produced more viscous lavas.[116]

Beta Regio is a large domical upland terrain comprising two shield volcanoes, Rhea Mons in the north and Theia Mons to the south. Theia Mons is about 500 kilometres across its base, and rises almost 4 kilometres above its surroundings. Theia is a gently sloping conical edifice with a summit caldera which has sent lava radiating down its flanks in a complex overlapping pattern. In contrast to the steep margins of Ishtar Terra, the transition to Beta Regio is gradual. It is actually one of a dozen volcanic rises,[117] each of which is 1,000 to 2,500 kilometres wide and rises 1 to 2.5 kilometres above the planetary mean. In fact, it has been suggested that the entire area delimited by Beta–Atla–Themis Regiones marks a broad mantle upwelling, while the periphery marks the associated downwelling.[118] The volcanic form of this

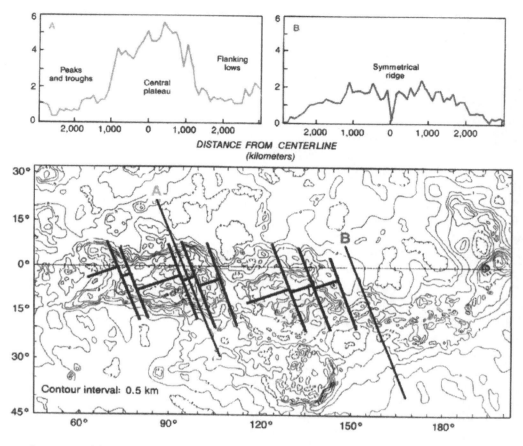

A topographic map of western Aphrodite based on the Pioneer Venus radar altimetry using contours at 500 metres, overlain with an interpretation as a lithospheric spreading ridge. (Courtesy of James Head and Larry Crumpler, with thanks to Kelly Beatty and Andrew Chaikin for permission to reproduce the figure from their *The New Solar System*.)

area derives from the increased heat flow and the associated melting, lithospheric uplift, extension and rifting. The downwelling produced the peripheral lowlands, in which there are ridge belts and ridged plains suggestive of crustal shortening but little lithospheric melting due to the lower heat flow.[119,120] Eistla Regio is a similar complex.[121,122] The volcanic rises are associated with gravitational anomalies,[123] and lithospheric thinning is consistent with a deep mantle plume impinging on the base of the lithosphere.[124,125]

Beta Regio is also transected by Devana Chasma, a trough with a fairly regular system of interior faults spaced 10 to 20 kilometres apart and nested grabens which indicate that it is a tectonic rift.[126] Its width varies, depending upon which specific fault has the most vertical displacement at that point. Theia Mons has encroached upon the chasm and has sent lava flowing down it.[127] Continuing to the south, Devana Chasma runs across the lowlands towards the western end of Aphrodite Terra. Overall, it is

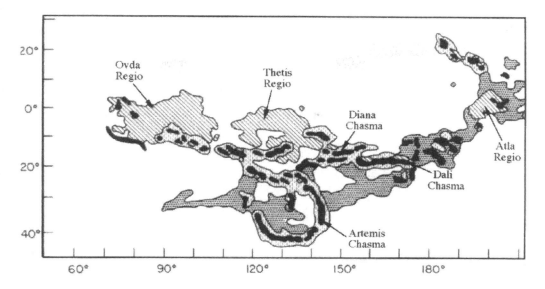

A map of the troughs in Aphrodite Terra by Gerry Schaber. Compare with the interpretation by Head and Crumpler. (Based on a figure in 'Venus: limited extension and volcanism along lines of lithospheric weakness', G.G. Schaber. *Geophys. Res. Lett.* vol. 9, p. 499, 1982.)

strikingly similar to the East African Rift, which marks where a continent is in the process of being torn apart. It would appear that the deep lithospheric fault that created Devana Chasma also provided a route for magma to reach the surface and build up the shields. The association of volcanism and rifting caused by severe extensional stress in the lithosphere reinforced the view that Venus's surface is young in the geological sense. Most of the areas with a low concentration of volcanic centres lie below the planetary mean, and those where volcanism is intense lie on the domical rises.[128,129] Did the lower-lying areas once have a similar population of volcanic centres that were submerged by the enormous volumes of fluid lavas that were extruded in the pulse of volcanism? The absence of any transitionary interface with partially buried structures implies that volcanism was indeed concentrated in the elevated areas, which is consistent with rising plumes producing lithospheric domes and the melting to drive volcanism, whereas the lower heat flow in the surrounding downwelling mantle was not conducive to melting.[130]

The compressional ridge belts and the extensional rift systems suggested lateral forces, but was the lithosphere 'mobile' in the sense of global plate tectonics? If Aphrodite did undergo spreading, when did this take place? Was there subduction elsewhere? A likely candidate for subduction was Artemis Chasma.[131] This horseshoe-shaped feature to the south of central Aphrodite is 2,500 kilometres across and resembles the arcuate trenches that form when one oceanic plate is subducted by another,[132] but this interpretation has been disputed.[133,134] It may be that Ishtar Terra is a block of buoyant continental rock that has been thrust up by the compressional forces imposed by lateral spreading elsewhere, and that

170 **Mysterious Venus**

A mosaic of Magellan imagery presenting a broad view of Beta Regio, with Rhea Mons in the north and Theia Mons to the south, showing how it is transected by Devana Chasma, the vast rift system which extends across the lowlands to the south.

subduction is active at the base of its marginal slopes,[135,136] but there is no evidence of subduction trenches there. Another theory proposed that Ishtar Terra is supported by a mantle plume.[137] Yet another theory posited that the uplift was independent of such spreading as may have occurred elsewhere, and argued that there is a *downwelling* mantle flow beneath Ishtar which has 'drawn in' the surrounding lithosphere to form a 'knot' comprising a plateau above and a 'root' beneath.[138] Such lithospheric thickening would have been accompanied by peripheral thinning, and it was suggested that this produced the ridge belts in the surrounding lowlands (if indeed these were induced by stretching). However, this theory was criticised for requiring an implausible amount of 'drawing in' of the peripheral lithosphere. A refinement of the model proposal that the rigid crust was decoupled from the upper mantle and that although the crust had been drawn in to some extent, most of the thickening was by upper mantle flow accumulating a residuum beneath the structure (effectively creating a 'pond' of melt stationed above the cold spot).[139] By this theory, the volcanism on the plateau derived from remelting in the root. Another interesting hypothesis argued that Ishtar Terra formed when blocks of buoyant granitic crust were drawn together by the convergence over a downwelling plume, so that Akna Montes and Maxwell Montes are separate crustal blocks that are compositionally distinct from Lakshmi Planum, which was built up between them by the compression-induced thickening of the basaltic crust.[140] If Ishtar Terra really is situated above a downwelling plume, then this 'tectonic convergence zone' is of a type that does not (currently) occur on the Earth.

The coronae, so named by the Soviet scientists who first noted them on Venera imagery,[141] initially seemed to be unique to Venus, but there may in fact be similar structures on the Earth and Mars.[142,143] Each corona is a network of concentric ridges enclosing a slightly elevated area typically 250 kilometres in diameter but in a few cases as much as 1,000 kilometres wide.[144,145] Coronae occur in isolation, in clusters associated with volcanic rises and in chains associated with chasmata,[146] and many coronae appear to have spilled lava onto surrounding low-lying plains. It is generally agreed that the coronae were produced when diapirs impinged upon the base of the lithosphere. The inflation of the dome induced the faults which opened the radial troughs. Later, gravitational collapse formed the concentric ridges on the margin. Finally, the interior relaxed as the diapir cooled.[147] However, there was debate concerning the form of the diapirs. Were they thermally driven plumes rising from deep within the mantle,[148] which would indicate that the coronae were genetically related to the volcanic rises,[149] or did the diapirs result from local instabilities in the upper mantle?[150] The large number of coronae (relative to the dozen or so volcanic rises) argued against the diapirs which formed the coronae originating deep in the mantle. Venus's high crustal temperature allows even small, slow magma bodies to rise to the surface in a wide range of thermal regimes, so deep plumes were not a prerequisite to surface eruptions.[151] Coronae are commonly associated with chasmata, which have been classified as symmetric and asymmetric. Symmetric chasmata are commonly associated with corona chains which trend radially from volcanic rises, and their floors are flooded by volcanism. They have been interpreted as rifts.[152] The asymmetric chasmata are bounded on one side by a ridge that is as high as the trough is deep, and since these are the deepest points on the planet the elevation differential from the floor of a trough to the adjacent ridge crest can span 7 kilometres. In light of the similarity to terrestrial subduction zones it was suggested that the asymmetric chasmata on the margins of the largest coronae underwent subduction as the domes settled.[153] There was a twist, however: it was argued that in the case of coronae the trench had migrated away from the interior, such that the diameter increased with time. Since the trench advanced this process has been dubbed 'retrograde' subduction.[154] However, this interpretation has been disputed.[155]

One of the most surprising discoveries was that the volcanic plains which occupy the low-lying terrains are etched by hundreds of narrow channels with a variety of forms ranging from simple sinuous rilles, and channels with flow margins to complex valley networks.[156,157,158] The sinuous rilles are similar in size and morphology to their lunar equivalents. They emanate from a distinct source (typically a depressed feature a few kilometres across), are a few kilometres in width and several tens of kilometres in length, and they narrow and shallow downstream. As on the Moon, most sinuous rilles on Venus do not seem to be associated with specific flow margins. Both cases suggest that the rille-form derives from thermal erosion by flowing silicate lava.[159,160] The sources are concentrated in the uplands, around coronae and related structures in volcanic complexes. Some features *are* clearly associated with large lava flows and some even have levees and lateral spill flow. However, they are much larger than terrestrial lava channels, in some cases flowing for several hundreds of kilometres and their existence indicates vast outpourings of very fluid lavas over sustained timescales.[161,162]

Particularly intriguing were the 'canali' – narrow channels which maintain a constant cross-sectional profile for hundreds of kilometres.[163] Baltis Vallis, located to the west of Atla Regio, is the longest at 6,800 kilometres, but is only about 3 kilometres wide.[164] Its source is a large volcano on the northwestern flank of the Beta–Atla–Themis Regiones and it seems also to have produced one of the widespread plains-forming lavas of the low-lying Rusalka Planitia.[165] The canali occur exclusively on the low-lying planitia. The lava that etched them was so fluid that it even left isolated meander bends.[166] Although some sections of the longest canali appear to have flowed 'up hill', this is a consequence of subsequent deformation to the lava plains across which they ran.

Mafic and ultramafic lavas can flow for distances indicated by the typical canali, but only if they flow beneath an insulating crust.[167] If the canali are silicate lava channels, then they would have been tubes and their current 'open' form is the result of roof-collapse. However, there is no evidence in the Magellan imagery of either short segments of surviving roof or of the litter of roof debris on the channel floors, so they would seem to have been formed as open channels. On Venus, thermal erosion is less efficient due to the smaller temperature difference between a lava flow and the hot rock over which it travels, which would promote longer flows.[168] Unlike the open channels with their levees and spill flows, the canali are clean. The meanders indicate that they were formed by exceptionally fluid lavas. It has been suggested that these might have had exotic (in terrestrial terms) chemistries.[169] Any

An active surface 173

A lava channel meanders its way across a volcanic plain just like a river.

sulphur that was erupted would not solidify in that environment, but it would rapidly evaporate. Nevertheless, an effusive sulphurous lava eruption would flow readily because its viscosity would be comparable to that of water on the Earth's surface, and whilst it would not thermally erode a silicate crust it would mechanically erode the surface. A more likely candidate, however, is an alkali-rich carbonatite. Carbonatites have melting points at or slightly above the ambient surface conditions, and a viscosity orders of magnitude lower than that of a silicate lava, so if extruded they would flow readily for long distances before solidifying.[170] Nevertheless, a carbonatite flow would not be able to thermally erode a silicate surface. The only active terrestrial sulphur flow is on a Hawaiian volcano, and carbonatite volcanism is rare – the best example of carbonatite volcanism is Oldoinyo Lengai located in the East African Rift, which erupts a white 'washing soda' lava.[171] In studying this unusually alkaline volcano, geologists may gain an insight into Venusian volcanism.

Since the Earth and Venus are near twins in terms of physical size, why are their surfaces so different?[172] Water plays a role in magmatism. On Earth, water is *intimately* involved in the processes of plate tectonics. When ocean floor basalt cools it absorbs water. When the plate is subducted back into the mantle the hydrated minerals release this water, which prompts partial melting in the lithosphere above, creating plumes of high silica magmas. If it were not for this water reducing the melting point of the rock, the melt above a subduction zone would create a more mafic magma. However, Venus does not possess oceans. Any water that was outgassed early in the planet's history has been lost. If any water remains within the lithosphere, it might locally influence magma chemistry and give rise to silicic volcanism which would create some buoyant crust immune from subduction. However, 'dry' melting will create magmas that are rich in refractories such as iron, magnesium and calcium. Being compositionally similar to the mantle, such ultramafic basalts would be readily reassimilated. Perhaps Venus's lithosphere is unusually thin, at only 45 kilometres, or unusually thick, at 250 kilometres? Arguments have been advanced for both.[173] A thick crust is implied by the deep root under the high-standing Ishtar Terra,[174] the unrelaxed crater forms,[175] and a variety of gravitational anomalies.[176,177,178] If the lithosphere is dry, it will not be very plastic, in which case it will be able to support large edifices even if it is relatively thin.[179] If the entire lithosphere is constructed of the same kind of rock, and this is chemically similar to the mantle, then vertical recycling may occur as a result of isostatic settling in response to the local accumulation of lava (as occurs on Jupiter's volcanic moon, Io), in which case subduction of the large-scale terrestrial variety might not have initiated. If the lithosphere is 'immobile' then the volcanoes constructed over rising mantle plumes would build up in place. On Mars, such volcanism has created shields almost 30 kilometres high, so Venus's pale into insignificance in comparison.

It has been suggested that plate tectonics was active on Venus until about 500 million years ago, when it was 'stifled' by the rigidisation of the 'dry' crust and the progressive thickening of the lithosphere, possibly to a thickness of 250 kilometres.[180] The termination of lithospheric recycling transformed the heat flow to a 'stagnant lid' regime,[181] and the subsequent heat build-up stimulated the pulse of volcanism that modified most of the surface and submerged the lower-lying terrains in fluid lavas.[182] It has also been suggested that vigorous convection within the mantle is stripping off the underside of the lithosphere – a process which has been dubbed 'delamination'.[183] Furthermore, if Venus lacks a significant asthenosphere, the lower lithosphere will be directly coupled to the mantle flow.[184] In effect, therefore the thickness of the Venusian lithosphere may have attained an equilibrium, with all of the activity occurring on its *underside*. Much work remains to be done, however.[185]

NOTES

1. Galileo Galilei reported his discovery of Venus's phases in a letter to Johann Kepler, another 'Copernican', encoding it, as was the style of that time, in the form of an anagram.
2. *The universe of Galileo and Newton*, W. Bixby. Cassell, p. 59, 1966.
3. Copernicus's notes, written over many years, were published as *De revolutionibus orbium coelestium* upon his death in 1543.
4. *The transits of Venus*, H. Woolf. Princeton University Press, 1959.
5. For a review of Horrocks's work, see Graythorp in *J. Brit. Astron. Assoc.*, vol. 47, p. 60, 1936.
6. *J. des Scavans*, p. 122, December 1667.
7. *Mem. Acad. Paris*. 1732.
8. *Observations concerning the planet Venus*, F. Bianchini 1725; translated by S. Beaumont. Springer-Verlag 1995.
9. W. Herschel. *Collected Scientific Papers, Roy. Soc. and Roy. Astron. Soc.* J.L.E. Dreyer (Ed.). vol. 1, p. 449, 1912.
10. *Aphroditographische Fragmente*, J.H. Schroter. 1796.
11. *Phil. Trans.*, vol. 82, p. 201, 1793.
12. *Nova Acta Acad. Naturae Curiosorum*, Bd. 10, p. 239.
13. *Observatory*, vol. 3, p. 416.
14. *Observatory*, vol. 7, p. 239.
15. *Strolling Astronomer*, vol. 6, p. 17, 1952.
16. *Astro-theology*. 1715.
17. *Astron. Jahrbuch*, p. 164, 1809.
18. *Mon. Not. Roy. Astron. Soc.*, vol. 43, p. 331, 1883.
19. *Report Brit. Assoc.*, p. 407, 1873.
20. *New guide to the planets*, P. Moore. Sidgwick & Jackson, p. 55, 1993.
21. *Phil. Trans*, p. 459, 1741.
22. *Astron. Nach.*, No.3304, 1878.
23. 'Considerazioni sul moto rotatorio del planeta Venere', G.V. Schiaparelli. *Opere*, vol. 5, p. 361, 1890.
24. *Ciel de Terre*, vol. 11, 1891.
25. 'The rotation of Venus', P. Lowell. *Astron. Nach.* No. 306, 1897.
26. 'Detection of Venus's rotation period and fundamental physical features of the planet's surface', P. Lowell. *Popular Astron.*, vol. 4, p. 281, 1896.
27. 'Detection of the rotation period and surface character of the planet Venus', P. Lowell. *Mon. Not. Roy. Astron. Soc.*, vol. 57, p. 148, 1897.

28 *J. Brit. Astron. Assoc.*, vol. 42, p. 216, 1932.
29 *Urania*, p. 229, 1952.
30 *J. Brit. Astron. Assoc.*, vol. 36, p. 191, 1926; see also *Guide to the planets*, P. Moore. The Scientific Book Club, p. 54, 1954.
31 W.H. Pickering. *J. Brit. Astron. Assoc.*, vol. 31, p. 218, 1921.
32 W.W. Wolkow. *Bull. Astron. Soc. USSR* 1949.
33 G.P. Kuiper. *Astron. J.* November 1954.
34 *J. Brit. Astron. Assoc.*, vol. 36, p. 299, 1926.
35 *Ap. J.*, vol. 71, p. 102, 1930.
36 *Die Stern*, p. 163, 1953.
37 *Strolling Astronomer*, vol. 7, p. 32, 1953.
38 *Ap. J.*, vol. 68, p. 57, 1928.
39 *A concise history of astronomy*, P. Doig. Chapman & Hall, p. 229, 1950.
40 G.P. Kuiper. *Sky & Telescope*, vol. 14, p. 141, 1955.
41 *Ap. J.*, November 1954.
42 *Other worlds than ours: the plurality of worlds studied under light of recent scientific researches*, R.A. Proctor. Hurst & Co. 1870.
43 *J. Brit. Astron. Assoc.*, vol. 36, p. 303, 1926.
44 *Ap. J.*, vol. 92, p. 247, 1940.
45 *The planet Venus*, P. Moore. Faber & Faber Ltd., p. 78, 1956.
46 *The atmospheres of the Earth and planets*, G.P. Kuiper (Ed.). Chicago University Press, p. 288, 1952.
47 *Ap. J.*, vol. 86, p. 337, 1937.
48 *Ap. J.*, vol. 91, p. 266, 1940.
49 *Our neighbour world*, V.A. Firsoff. p. 209, 1952.
50 D.H. Menzel and F.L. Whipple. *Publications of the Astronomical Society of the Pacific*, vol. 67, p. 161, 1955.
51 C.H. Mayer, T.P. McCullough and R.M. Sloanaker. *Ap. J.*, vol. 127, 1958.
52 *Life on other worlds*, H. Spencer Jones. p. 170, 1952.
53 *Worlds in collision*, I. Velikovsky. Macmillan, 1950.
54 A spate of papers by the various radar teams were published in *Ap. J.*, vol. 69, 1964.
55 'The spin and inertia of Venus', G. Spada, R. Sabadini and E. Boshi. *Geophys. Res. Lett.*, vol. 23, p. 1997, 1996.
56 'Spin–orbit coupling in the Solar System: the resonant rotation of Venus', P. Goldreich and S.J. Peale. *Astron. J.*, vol. 72, p. 662, 1967.
57 'Atmospheric tides and the resonant rotation of Venus', T. Gold and S. Soter. *Icarus*, vol. 11, p. 356, 1969.
58 'Solar tidal friction and satellite loss', W.R. Ward and M.J. Reid. *Mon. Not. Roy. Astron. Soc.*, vol. 164, p. 21, 1973.
59 *Mariner-Venus 1962 final project report*. NASA 1965.
60 *Mariner-Venus 1967 final project report*. NASA 1971.
61 A.P. Vonogradov *et al.* In *Planetary atmospheres*, C.E. Sagan, T.C. Owen and H.J. Smith (Eds). Reidel Publishing, p. 3, 1971.
62 'Geochemistry of the Venera 8 material demonstrates the presence of continental crust on Venus', O.V. Nikolayeva. *Earth Moon Planets*, vol. 50/51, p. 329, 1990.
63 This was the first time that a 'slingshot' manoeuvre had been attempted, and it was regarded as a rehearsal for using Jupiter's gravity to open the way to the outer Solar System.
64 This remark is attributed to Dr. Arnold Selivanov. *Solar System log*, A. Wilson. *Jane's*, p. 87, 1987.

65 *Pioneer Venus*, R.O. Fimmel, L. Colin and E. Burgess. NASA 1983.
66 The preliminary results from the Pioneer Venus missions were published in a 'special issue' of *J. Geophys. Res.*, vol. 85, A13, 1980.
67 A consolidated assessment of the Pioneer Venus results were published in *Venus*, D.M. Hunten, L. Colin, T.M. Donahue and V.I. Moroz. Arizona University Press, 1983.
68 *Solar System log*, A. Wilson. *Jane's*, p. 107, 1987.
69 *Solar System log*, A. Wilson. *Jane's*, p. 109, 1987.
70 'Rock weathering on the surface of Venus', J.A. Wood. In *Venus II: geology, geophysics, atmosphere, and solar wind environment*, S.W. Bougher, D.M. Hunten and R.J. Phillips (Eds). University of Arizona Press, p. 637, 1997.
71 M.C. Malin and R.S. Saunders. *Science*, vol. 196, p. 987, 1977.
72 R.S. Saunders and M.C. Malin. *Geophys. Res. Lett.*, vol. 4, p. 547, 1977.
73 'The volcanology of Venera and VeGa landing sites and the geochemistry of Venus', J.S. Kargel, G. Komatsu, V.R Baker and R.G. Strom. *Icarus*, vol. 103, p. 253, 1993.
74 *The Space Shuttle: roles, missions and accomplishments*, D.M. Harland. Wiley Praxis, p. 321, 1998.
75 'Magellan mission summary', R.S. Saunders and G.E Pettengill. *Science*, vol. 187, p. 247, 1995.
76 *Our neighbour worlds*, V.A. Firsoff. p. 199, 1952.
77 'The planet next door', A.T. Bazilevski. *Sky & Telescope*, p. 360, April 1989.
78 W.F. Libby and P. Corneil. In *Planetary atmospheres*, C.E. Sagan, T.C. Owen and H.J. Smith (Eds). IAU Symp. No.40 October 1969. Reidel Publishing, p. 55, 1971.
79 G.T. Sill. *Comm. Lunar and Planetary Laboratory*, vol. 171, p. 191, 1972.
80 A.T. Young. *Icarus*, vol. 18, p. 564, 1973.
81 For an account of the discovery that sulphuric acid is the main condensate in Venus's clouds, see *Planetary astronomy*, R.A. Schorn. Texas A&M University Press, p. 259, 1998.
82 Y.A. Surkov, F.F. Kirnozov and V.N. Glazov. *Letters to Astron. J.*, vol. 8, p. 700, 1982.
83 J.H. Hoffman, V.I. Oyama and U. von Zahn. *J. Geophys. Res.*, vol. 85, p. 7871, 1980.
84 *Venus*, D.M. Hunten, L. Colin, T.M. Donahue and V.I. Moroz (Eds). University of Arizona Press, 1984.
85 In fact, Pioneer Venus Orbiter surprised its designers by returning data until October 1992, at which time its orbit decayed and it fell into the planet's atmosphere. It had been designed to operate for one 243-day 'cycle' but continued operating for 20 cycles.
86 'Large-scale volcanic activity at Maat Mons: can this explain fluctuations in atmospheric chemistry observed by Pioneer Venus?' C.A. Robinson, G.D. Thornhill and E.A. Parfitt. *J. Geophys. Res.*, vol. 100, p. 11755, 1995.
87 'Impact craters and Venus resurfacing history', R.J. Phillips *et al. J. Geophys. Res.*, vol. 97, p. 15923, 1992.
88 'The global resurfacing of Venus', R.G. Strom, G.G. Schaber and D. Dawson. *J. Geophys. Res.* vol. 99, p. 10899, 1994.
89 'Explosive volcanism on Venus: transient volcanic explosions as a mechanism for localised pyroclastic dispersal', S.A. Fagents and L. Wilson. *J. Geophys. Res.*, vol. 100, p. 26327, 1995.
90 'Magma vesiculation and pyroclastic volcanism on Venus', J.B. Gavin, J.W. Head and L. Wilson. *Icarus*, vol. 52, p. 365, 1982.
91 'How climate evolved on the terrestrial planets', J.F. Kasting, O.B. Toon and J.B. Pollack. *Scient. Am.*, vol. 258, p. 90, February 1988.
92 'Atmospheres of the terrestrial planets', J.B. Pollack. In *The new Solar System*, J.K. Beatty and A. Chaikin (Eds). Cambridge University Press (third edition), p. 91, 1990.

93 'Magma reservoirs and neutral buoyancy zones on Venus: implications for the formation and evolution of volcanic landforms', J.W. Head and L. Wilson. *J. Geophys. Res.*, vol. 97, p. 3877, 1992.
94 'Magnetism and thermal evolution of the terrestrial planets', D.J. Stevenson, T. Spohn and G. Schubert. *Icarus*, vol. 54, p. 466, 1983.
95 'Geology of the Venus equatorial region from Pioneer Venus radar imaging', S.A. Senske. *Earth Moon Planets*, vol. 50/51, p. 305, 1990.
96 Although the names introduced by the radar astronomers were accepted, the International Astronomical Union decided to name features on Venus after female mythological and historical characters. So Maxwell Montes, the tallest point on the planet, named for the renowned 19th-century scientist James Clerk Maxwell, is the solitary monument to the male of the species.
97 'Styles of deformation in Ishtar Terra and their implications', W.M. Kaula *et al. J. Geophys. Res.*, vol. 97, p. 16085, 1992.
98 'Structural evolution of Maxwell Montes: implications for Venusian mountain belt formation', M. Keep and V.L. Hansen. *J. Geophys. Res.*, vol. 99, p. 26015, 1994.
99 'Gravitational spreading of high terrain in Ishtar Terra, Venus', S.E. Smrekar and S.C Solomon. *J. Geophys. Res.*, vol. 97, p. 16121, 1992.
100 'Long-term survival of the topography of Ishtar Terra, Venus', A.M. Freed and H.J. Melosh. *Proc. Lunar Planet. Sci. Conf.*, p. 421, 1995.
101 'Caldera-related volcanism and collapse at Ishtar Terra, Venus', J.J. Willis and V.L. Hansen. *Am. Geophys. Union EOS Trans.*, vol. 76, p. 341, 1995.
102 'Volcanism in northwest Ishtar Terra, Venus', L.R. Gaddis and R. Greeley. *Icarus*, vol. 87, p. 327, 1990.
103 'Calderas on Venus', M.H. Bulmer, J.E. Guest and E.R. Stofan. *Proc. Lunar Planet. Sci. Conf.*, p. 177, 1992.
104 Colette and Sacajawea Paterae are comparable in size to Loki, Io's largest caldera.
105 *Atlas of Venus*, P. Cattermole and P. Moore. Cambridge University Press, p. 47, 1997.
106 'Western Ishtar Terra and Lakshmi Planum, Venus: models of formation and evolution', K.M. Roberts and J.W. Head. *Geophys. Res. Lett.*, vol. 17, p. 1341, 1990.
107 'Distribution of tessera terrains on Venus', D.L. Bindschadler *et al. Geophys. Res. Lett.*, vol. 17, p. 171, 1990.
108 'Tessera terrain, Venus: characterisation and models for origin and evolution', D.L. Bindschadler and J.W. Head. *J. Geophys. Res.*, vol. 96, p. 5889, 1991.
109 'Tectonic and magmatic evolution on Venus', R.J. Phillips and V.L. Hansen. *Ann. Rev. Earth Planet. Sci.*, vol. 22, p. 597, 1994.
110 'Wrinkle ridges, stress domains, and kinematics of Venusian plains', G.E. McGill. *Geophys. Res. Lett.*, vol. 20, p. 2407, 1993.
111 'Evidence for divergent plate boundary characteristics and crustal spreading of Venus', J.W. Head and L.S. Crumpler. *Science*, vol. 238, p. 1380, 1987.
112 'Tectonic evolution of the terrestrial planets', J.W. Head and S.C. Solomon. *Science*, vol. 213, p. 62, 1981.
113 'Venus: limited extension and volcanism along lines of lithospheric weakness', G.G. Schaber. *Geophys. Res. Lett.*, vol. 9, p. 499, 1982.
114 'Geology and distribution of impact craters on Venus: what are they telling us?', G.G. Schaber, R.G. Strom, H.J. Moore, L.A. Soderblom, R.L. Kirk, D.J. Chadwick, D.D. Dawson, L.R. Gaddis, J.M. Boyce and J. Russell. *J. Geophys. Res.*, vol. 97, p. 13257, 1992.
115 'Sapas Mons, Venus: evolution of a large shield volcanoe', S.T. Keddie and J.W. Head. *Earth Moon Planets*, vol. 65, p. 129, 1994.

116 'Small domes on Venus: probable analogues of Icelandic lava shields', J.B. Gavin and R.S. Williams. *Geophys. Res. Lett.*, vol. 17, p. 1381, 1990.
117 'Volcanoes and centres of volcanism on Venus', L.S. Crumpler, J.C. Aubele, D.A. Senske, S.T. Keddie, K.P. Magee and J.W. Head. In *Venus II: geology, geophysics, atmosphere and solar wind environment*, S.W. Bougher, D.M. Hunten and R.J. Phillips (Eds). University of Arizona Press, p. 697, 1997.
118 'Relation of major volcanic centre concentration on Venus to global tectonic patterns', L.S. Crumpler, J.W. Head and J.C. Aubele. *Science*, vol. 261, p. 591, 1993.
119 'Cold spots and hot spots: global tectonics and matle dynamics on Venus', D.L. Bindschadler, G. Schubert and W.M. Kaula. *J. Geophys. Res.*, vol. 97, p. 13495, 1992.
120 'Ridge belts: evidence for regional and local-scale deformation on the surface of Venus', M.T. Zuber. *Geophys. Res. Lett.*, vol. 17, p. 1369, 1990.
121 'Anatomy of a Venusian hot spot: geology, gravity and mantle dynamics of Eistla Regio', R.E. Grimm and R.J. Phillips. *J. Geophys. Res.*, vol. 97, p. 16035, 1992.
122 'Regional topographic rises on Venus: geology of western Eistla Regio and comparison to Beta Ragio and Atla Regio', D.A. Senske, G.G. Schaber and E.R. Stofan. *J. Geophys. Res.*, vol. 97, p. 13395, 1992.
123 'Venus gravity anomalies and their correlations with topography', W.L Sjogren *et al. J. Geophys. Res.*, vol. 88, p. 1119, 1983.
124 'Lithospheric thickness and mantle/lithosphere density contrast beneath Beta Regio, Venus', W.B. Moore and G. Schubert. *Geophys. Res. Lett.*, vol. 22, p. 429, 1995.
125 'A mantle plume model for the equatoral highlands of Venus', W.S. Kiefer and B.H. Hager. *J. Geophys. Res.* vol. 96, p. 20947, 1991.
126 'Venus volcanism and rift formation in Beta Regio', D.B. Campbell, J.W. Head, J.K. Harmon and A.A. Hine. *Science*, vol. 225, p. 167, 1984.
127 'Venus volcanism: classification of volcanic features and structures, associations and global distributin from Magellan data', J.W. Head, L.S. Crumpler and C.J. Aubele. *J. Geophys. Res.*, vol. 97, p. 13153, 1992.
128 'Two global concentrations of volcanism on Venus: geologic association and implications for global pattern of upwelling and downwelling', L.S. Crumpler and J.C. Aubele. *Proc. Lunar Planet. Sci. Conf.*, p. 275, 1992.
129 'Height and altitude distribution of large volcanoes on Venus', S.T. Keddie and J.W. Head. *Planet. Space Sci.*, vol. 42, p. 455, 1994.
130 'Evidence for active hot spots on Venus from analysis of Magellan gravity data'. S.E. Smrekar. *Icarus*, vol. 112, p. 2, 1994.
131 'A global survey of possible subduction sites on Venus', G. Schubert and D.T. Sandwell. *Icarus*, vol. 117, p. 173, 1995.
132 'Tectonics of Artemis Chasma: a Venusian "plate" boundary', D.D. Brown and R.E. Grimm. *Icarus*, vol. 117, p. 219, 1995.
133 'Global distribution and characteristics of coronae and related features on Venus: implications for origin and relation to mantle processes', E.R. Stofan *et al. J. Geophys. Res.*, vol. 97, p. 13347, 1992.
134 'Quantitative tests for plate tectonics on Venus', W.M. Kaula and R.J. Phillips. *Geophys. Res. Lett.*, vol. 8, p. 1187, 1981.
135 'Orogenic belts on Venus', L.S. Crumpler, J.W. Head and D.B. Campbell. *Geology*, vol. 14, p. 1031, 1986.
136 'Formation of mountain belts on Venus: evidence for large-scale convergence, underthrusting and crustal imbrication in Freyja Montes, Ishtar Terra', J.W. Head. *Geology*, vol. 18, p. 99, 1990.

137 'The structure of Lakhsmi Planum: an indication of horizontal asthenospheric flow of Venus', A.A. Pronin. *Geotectonics*, vol. 20, p. 271, 1986.
138 'Mantle downwelling and crustal convergence: a model for Ishtar Terra, Venus', W.S. Kiefer and B.H. Hager. *J. Geophys. Res.*, vol. 96, p. 20967, 1991.
139 'Formation of Ishtar Terra, Venus: surface and gravity constraints', V.L. Hansen and R.J. Phillips. *Geology*, vol. 23, p. 292, 1995.
140 'The implications of basalt in the formation and evolution of mountains on Venus', M.G. Jull and J. Arkani-Hamed. *Phys. Earth Planet. Interiors*, vol. 89, p. 163, 1995.
141 'The geology and geomorphology of the Venus surface as revealed by the radar images obtained by Veneras 15 and 16', V.L. Barsukov *et al. J. Geophys. Res.*, vol. 91, p. 378, 1986.
142 'Radially fractured domes: a comparison of Venus and Earth', D.M. Janes and S.W. Squyres. *Geophys. Res. Lett.*, vol. 20, p. 2961, 1993.
143 'Coronae on Venus and Mars: implications for similar structures on Earth', T.R. Watters and D.M. Janes. *Geology*, vol. 23, p. 200, 1995.
144 'Magellan observations of Venusian coronae: geology, topography and distribution', G. Schubert *et al. Am. Geophys. Union EOS Trans.*, vol. 72, p. 175, 1991.
145 'The morphology and evolution of coronae on Venus', S.W. Squyres *et al. J. Geophys. Res.*, vol. 97, p. 13611, 1992.
146 'Caronae on Venus: morphology and origin', E.R. Stofan, V.E. Hamilton, D.M. Janes and S.E. Smrekar, In *Venus II: geology, geophysics, atmosphere and solar wind environment*, S.W. Bougher, D.M. Hunten and R.J. Phillips (Eds). University of Arizona Press, p. 931, 1997.
147 '[Venus:] Tectonic overview and synthesis', V.L. Hansen, J.L. Willis and W.B. Banerdt. In *Venus II: geology, geophysics, atmosphere and solar wind environment*, S.W. Bougher, D.M. Hunten and R.J. Phillips (Eds). University of Arizona Press, p. 797, 1997.
148 'Global distribution and characterisation of coronae and related features on Venus: implications for origin and relation to mantle processes'. E.R. Stofan *et al. J. Geophys. Res.*, vol. 97, p. 13347, 1992.
149 'Large topographic rises on Venus: implications for mantle upwelling', E.R. Stofan, S.E. Smrekar, D.L. Bindschadler and D.A. Senske. *J. Geophys. Res.*, vol. 100, p. 317, 1995.
150 'Volcanism without plumes: melt-driven instabilities, buoyant residuum and global implications', P.J. Tackley and D.J. Stevenson. *Am. Geophys. Union EOS Trans.*, vol. 74, p. 188, 1993.
151 'Effects of planetary thermal structure on the ascent of cooling magma on Venus', S.E.H. Sakimoto and M.T. Zuber. *J. Volcan. Geothermal Res.*, vol. 64, p. 54, 1995.
152 'Venus: limited extension and volcanism along zones of lithospheric weakness', G.G. Schaber. *Geophys. Res. Lett.*, vol., p. 499, 1982.
153 'Flexural ridges, trenches and outer rises around Venus coronae', D.T. Sandwell and G. Schubert. *J. Geophys. Res.*, vol. 97, p. 15923, 1992.
154 'Features on Venus generated by plate boundary processes', D. McKenzie *et al. J. Geophys. Res.*, vol. 97, p. 13533, 1992.
155 'Asymmetric Venusian rifts: arguments against subduction', V.L. Hansen. *Am. Geophys. Union EOS Trans.*, vol. 74, p. 377, 1993.
156 'Venus volcanism: initial analysis from Magellan data', J.W. Head *et al. Science*, vol. 252, p. 276, 1991.
157 'Channels and valleys on Venus: preliminary analysis of Magellan data', V.R. Baker. *J. Geophys. Res.*, vol. 97, p. 13421, 1992.
158 'Channels and valley morphology: an updated classification', V.C. Gulick, V.R. Baker and G. Komatsu. *Proc. Lunar Planet. Sci. Conf.*, p. 465, 1992.

Notes 181

159 'Turbulent lava flow and the formation of sinuous rilles [on the Moon]', G. Hulme. *Modern Geol.*, vol. 4, p. 107, 1973.
160 'A review of lava flow processes related to the formation of lunar sinuous rilles', G. Hulme. *Geophys. Surveys*, vol. 5, p. 245, 1982.
161 'Venusian channels and valleys: distribution and volcanological implications', G. Komatsu, V.R. Baker and V.C. Gulick. *Icarus*, vol. 102, p. 1, 1993.
162 'Channels and valleys [on Venus]', V.R. Baker, G. Komatsu, V.C. Gulick and T.J. Parker. In *Venus II: geology, geophysics, atmosphere and solar wind environment*, S.W. Bougher, D.M. Hunten and R.J. Phillips (Eds). University of Arizona Press, p. 757, 1997.
163 'Formation of Venusian canali: considerations of lava types and their thermal behaviours', T.K.P Gregg and R. Greeley. *J. Geophys. Res.*, vol. 98, p. 10873, 1993.
164 'Venusian channels and valleys: distribution and volcanological implications', G. Komatsu, V.R. Baker, V.R. Gulick and T.J. Parker. *Icarus*, vol. 102, p. 1, 1993.
165 'Stratigraphic studies in the Baltis Vallis region, Venus: implications for areal extent and timing of volcanic resurfacing events', A.T. Basilevsky and J.W. Head. *Geophys. Res. Lett.* 1996.
166 'Meander properties of Venusian channels', G. Komatsu and V.R. Baker. *Geology*, vol. 22, p. 67, 1994.
167 'Canali-type channels on Venus: some genetic considerations', G. Komatsu, J.S. Kargel and V.R. Baker. *Geophys. Res. Lett.*, vol. 19, p. 1415, 1992.
168 'Formation of canali: considerations of lava types and their thermal behaviours', T.K.P. Gregg and R. Greeley. *J. Geophys. Res.*, vol. 98, p. 10873, 1993.
169 'The rivers of Venus', J.S. Kargel. *Sky & Telescope*, p. 32, August 1997.
170 'Carbonatite–sulphate volcanism on Venus', J.S. Kargel, B. Fegley, A. Treiman and R.L. Kirk. *Icarus*, vol. 112, p. 219, 1994.
171 *Principles of physical geology*, A. Holmes. Nelson (second edition), p. 1067, 1965.
172 'Venus: a contrast in evolution to Earth', W.M. Kaula. *Science*, vol. 24, p. 1191, 1990.
173 For a thorough review of crustal thickness issues see 'Mantle convection and thermal evolution of Venus', G. Schubert, V.S. Solomatov, P.J. Tackley and D.L. Turcotte. In *Venus II: geology, geophysics, atmosphere and solar wind environment*, S.W. Bougher, D.M. Hunten and R.J. Phillips (Eds). University of Arizona Press, p. 1245, 1997.
174 'Long-term survival of the topography of Ishtar Terra, Venus', A.M. Freed and H.J. Melosh. *Proc. Lunar Planet. Sci. Conf.*, p. 421, 1995.
175 'Viscous relaxation of impact crater relief on Venus: constraints on crustal thickness and thermal gradient', R.E. Grimm and S.C. Solomon. *J. Geophys. Res.*, vol. 93, p. 11911, 1988.
176 'The deep structure of the Venusian plateau highlands', R.E. Grimm. *Icarus*, vol. 112, p. 89, 1994.
177 'Estimating lithospheric properties at Atla Regio, Venus', R.J. Phillips. *Icarus*, vol. 112, p. 147, 1994.
178 'Isostatic compensation of equatorial highlands on Venus', A.B. Kucinskas and D.L. Turcotte. *Icarus*, vol. 112, p. 104, 1994.
179 'Mantle convection and crustal evolution on Venus', W.M. Kaula. *Geophys. Res. Lett.*, vol. 17, p. 1401, 1990.
180 'On the tectonics of Venus', J. Arkani-Hamed. *Phys. Earth Planet. Interiors*, vol. 99, p. 2019, 1993.
181 'Stagnant lid convection on Venus', V.S. Solomatov and L.N. Maresi. *J. Geophys. Res.*, vol. 101, p. 4737, 1996.
182 'Resurfacing history of Venus', R.R. Herrick. *Geology*, vol. 22, p. 703, 1994.

183 'Global decoupling of crust and mantle: implications for topography, geoid, and mantle viscosity on Venus', W.R. Buck. *Geophys. Res. Lett.*, vol. 19, p. 2111, 1992.
184 'Convection-driven tectoncs on Venus', R.J. Phillips. *J. Geophys. Res.*, vol. 95, p. 1301, 1990.
185 'How does Venus lose heat?', D.L. Turcotte. *J. Geophys. Res.*, vol. 100, p. 16931, 1995.

6

Ancient Mercury

EARLY OBSERVATIONS

As viewed from the Earth, Mercury never strays more than 28 degrees from the Sun in the sky. In 1610 Galileo Galilei viewed Mercury through the telescope with which he had discovered that Jupiter possessed moons and Venus underwent lunar-like phases, but nothing was revealed. In 1627 Johann Kepler applied his empirically derived laws of planetary motion to Mercury and predicted that the planet would pass over the solar disk in 1631. This transit was duly observed by Pierre Gassendi in France. The fact that Mercury displayed lunar-like phases was noted by the Italian astronomer Giovanni Zupus in 1639.[1,2]

Mercury was a difficult object to study because it was visible only for short periods and in twilight, so only a few observers persevered. It was 1800 before albedo variations suggestive of surface features were reported, but the renowned telescopic observer William Herschel saw nothing. German astronomer J.H. Schroter reported mountains rising 20 kilometres, and after observing these for a period he inferred that the planet rotated in precisely 24 hours.

After extensive observations in the 1880s, Giovanni Schiaparelli concluded that Mercury's rotation was synchronised with its 88-day orbital motion around the Sun, making it a world of extremes in which one hemisphere always faced the Sun and the other was always in perpetual darkness.[3] In fact, the libration effects from the 0.2 orbital eccentricity would have made the Sun oscillate above and below the horizon along the terminator line. The energy of insolation decreases by a square-law with increasing distance from the Sun, so on average the energy density irradiating Mercury is almost an order of magnitude greater than that reaching the Earth. Because the orbit is eccentric the energy density will vary by a factor of two. At perihelion, the surface near the subsolar point is hot enough to sweat lead, zinc and tin from the rocks.[4] The surface must be composed only of refractories. In 1897 Percival Lowell, who agreed with Schiaparelli that the planet's rotation was synchronised with its orbital motion, charted linear features on Mercury, which he inferred must be vast cracks in the crust opened by the extreme heat.[5] In the 1920s, E.M. Antoniadi of the Meudon Observatory near Paris published a map in a book that served as the standard reference until a spacecraft was dispatched to take a close look at the planet.[6]

In 1962, however, W.E. Howard led a team from the University of Michigan that used a radio telescope to measure the temperature of the surface. To their surprise they found that the dark hemisphere was considerably hotter than it would be if it never faced the Sun. Evidently, Mercury's rotation was *not* synchronised. In 1965 Rolf Dyce and Gordon Pettengill used the Arecibo radio telescope in Puerto Rico as a radar, and established that the planet turns once on its axis in 59 days.[7] Noting that this was close to two-thirds of 88, Guiseppe Columbo suggested that the planet's rotation was *precisely* this fraction. It was promptly realised that the planet would make three rotations in two solar orbits. Because Mercury is difficult to study, and is well positioned only near elongation, the same hemisphere had been on display each time it had been studied by Schiaparelli, Antoniadi and the others. For once, Occam's Razor that the correct solution was probably the simplest solution had been misleading. It is not known *why* the planet has adopted the two-thirds resonance. Once it had been established that Mercury did not maintain the same hemisphere towards the Sun, the International Astronomical Union arbitrarily defined the zero of longitude to pass through the subsolar point at the first perihelion passage of the 1950 epoch.

In 1865 Johann Zollner, who had developed the first effective astronomical photometer a few years earlier, found that the variation in Mercury's brightness with its orbital phase closely resembled the lunar pattern,[8] implying that the planet was airless. However, in 1871 H.C. Vogel's spectroscopic study suggested that the planet had an atmosphere as dense as the Earth's[9] and in 1889 Schiaparelli, having noted 'white obscurations', reported that Mercury possessed an 'appreciable atmosphere'.[10] Antoniadi confirmed these obscurations, but argued that they were clouds of fine dust. In 1953, however, Audouin Dollfus rightly contradicted them by arguing that the surface was so hot that even the 4.24 kilometres per second escape velocity would not be able to retain any gases which might be leaking from the crust.[11]

MARINER 10

As the first spacecraft set off for Mercury, we knew a great deal about the planet's orbital motion – the subtleties of which had confirmed Albert Einstein's theory of relativity[12] – but very little of it as a planetary body.[13,14] The thermal characteristics of the surface measured by radio telescopes implied that (similar to the Moon) it is covered by a regolith of dark fine-grained material, suggesting that impact processes had been dominant in the formation of its crust.[15] In fact, as early as 1901 T.J.J. See in Washington had reported the presence of craters on Mercury.[16,17] Mariner 4's discovery that Mars was heavily cratered had come as a shock, but as Mariner 10 closed in on Mercury few doubted that it, too, would be scarred.[18]

The fact that parts of Mercury were as heavily cratered as the Moon provided insight into the nature of the early Solar System. Although craters had been expected, it had been assumed that Mercury would not be as battered as the Moon because it is further from the asteroid belt, which had been assumed to be the source of the impactors. In reality, this early bombardment was due to the newly accreted planets sweeping up the rest of the protoplanetary objects from the solar nebula. Whereas the planets would have accreted in 100 million years, another half a billion years would have elapsed before they had attracted the rest of the material in their neighbourhoods. Contrary to expectations, therefore, all the inner planets were similarly bombarded by a population of impactors which is now extinct. Studies of lunar samples revealed that this bombardment tailed off some 3.8 billion years ago. Impacts continue from comets and objects perturbed from the asteroid belt, but at a far lower rate.

Mariner 10 was not only the first spacecraft to be assigned to Mercury, it was also the first multiple-planet mission because it used a Venus fly-by to reach its objective. Furthermore, the highly elliptical orbit facilitated a series of encounters every 176 days, so Mariner 10 was able to achieve far more than Mariner 4 had at Mars during its brief fly-by. In contrast to the earlier imaging systems which had stored a series of images on tape and later replay, Mariner 10 was able, in effect, to provide real-time television.[19] A tape recorder was only required to store the imagery secured while it passed behind the planet as viewed from Earth, at which time the line of sight transmission was blocked.

The first approach produced a 760-kilometre fly-by on 29 March 1974. The far-encounter imaging sequence was initiated at a range of 5 million kilometres and provided a view of the planet comparable to the best telescopic observations. The images taken at the closest point of approach had a resolution of 100 metres. Although Mariner 10 provided a 5,000-fold increase in resolution compared to telescopic studies, the result was a view of the planet as clear as that of the Moon seen through a terrestrial telescope,[20] and because so much of what we now know of the Moon derives from spacecraft observations it is evident that Mercury's surface still retains many secrets – not least because every time the spacecraft returned the planet had completed two orbits, and as the same hemisphere was illuminated, the other half was *never* presented. Furthermore, because most of the terrain was under high-angle illumination (which was ideal for albedo studies) the subtle topography could be mapped only towards the terminator, so the resulting maps do not provide a uniform viewpoint.

Soon after its third encounter, Mariner 10 exhausted its attitude-control propellant and had to cease operations because it could no longer hold its antenna towards the Earth. Nevertheless, as the 'baseline' mission had called for only one fly-by of Mercury, the others were a significant bonus.

MASSIVE CORE

Although at 4,880 kilometres in diameter Mercury is the smallest of the inner planets, it is only slightly larger than the Moon.[21] Nevertheless, its bulk density (5.45

186 Ancient Mercury

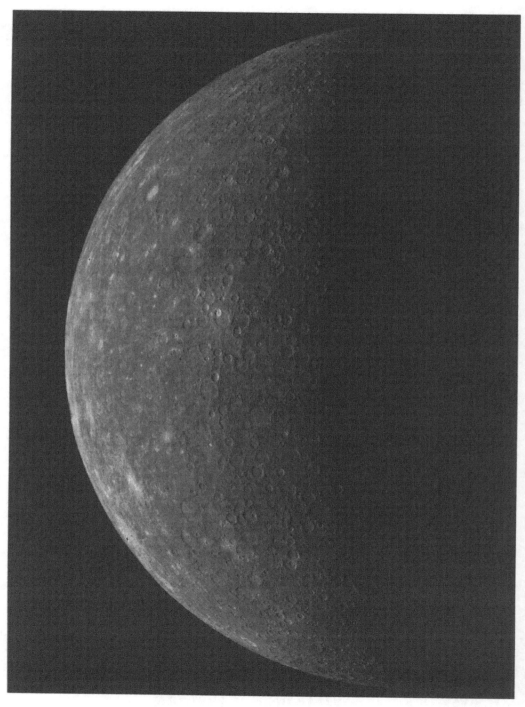

A mosaic of imagery of Mercury by Mariner 10 as it approached the planet. The superficial resemblance to the Moon is deceptive.

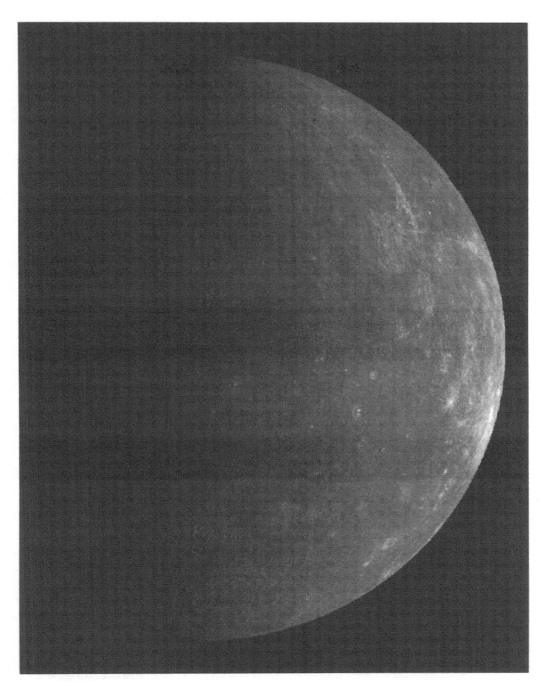

A mosaic of Mariner 10 imagery of Mercury as the spacecraft departed, showing the multiple-ringed Caloris basin on the terminator.

g/cm^3) compares with that of Earth (5.51 g/cm^3), suggesting that Mercury has a large iron core. In fact, bulk density figures are biased because the pressure inside a planet 'densifies' rock and squeezes silicates into more compact phases. If this bias could be eliminated, the Earth's bulk density would be 4.0 g/cm^3. However, Mercury's densification is insignificant because the planet is smaller and the internal pressure is much weaker, but as the adjusted bulk density is still a high 5.3 g/cm^3, it must be predominantly made of iron-nickel. It has been calculated that the core accounts for 75 per cent of the planet's radius, 50 per cent of its volume and 70 per cent of its mass. The englobing mantle of lower density silicate rock is just 640 kilometres thick. If the mantle could be stripped off, the exposed core would be physically larger than the Earth's Moon.[22]

It is widely assumed that the planets condensed from the solar nebula at more or less their current distances from the Sun. In Mercury's orbit, the nebula would have been 1,440 K,[23,24] so volatiles such as sodium, potassium, chlorine and water would not have condensed; only the refractories would have condensed. The innermost planet should therefore mark the high end of a decreasing progression of bulk density with distance from the Sun.[25] While there is such a trend, Mercury's bulk density is 'anomalously' high;[26,27] it 'ought' to be only 4 g/cm^3.

So, why had the planet accreted so much iron? At first, it was thought that Mercury had developed such a massive core because it had initially been so thoroughly molten that almost all of the dense iron–nickel grains that it had accreted from the solar nebula had been drawn down by the force of gravity. However, computer models of the accretion process established that many more protoplanets condensed out of the solar nebula than went on to form planets.[28] Although it was no surprise that the smaller protoplanets had been swept up and accreted by the larger ones, the sheer scale of these impacts had not been realised. It is now believed that Mercury's core is disproportionately large because much of the mantle was blasted off by at least one giant impact, and the denser material coalesced to form the core.[29,30,31]

PHYSIOGRAPHY

As soon as Mariner 10 began its far-encounter sequence, it was evident that the surface of Mercury bore little relationship to the maps made by telescopic observers. Once the principal physiographic surface features had been characterised,[32] the International Astronomical Union discarded Antoniadi's nomenclature and introduced its own scheme.

Heavily cratered terrain

As a result of Mariner 10, the impact statistics span the range from craters only 100 metres in diameter up to basins over 1,000 kilometres across, and the results indicate that craters on Mercury differ from those on the Moon in terms of how they create rays, ejecta blankets, and secondaries.[33] Because the albedo of the Mercurian ray material is greater than that of the lunar rays, the Mercurian rays appear more prominent.[34] The continuous ejecta immediately beyond the rims of craters is more closely confined on Mercury, and the secondaries are closer, deeper and more sharply defined than those on the Moon.

Evidently, the fact that Mercury's gravitational pull is greater than that of the Moon meant that the ejecta flying on a ballistic trajectory from an impact of a given energy did not reach as far.[35] Furthermore, the continuous ejecta blanket is thicker than that around a comparable lunar crater because the ejecta is piled up near the rim.[36] Because the secondaries are clustered closer to the primary than on the Moon they did not cause so much damage, and the pre-existing terrain is preserved in the form of intercrater plains.[37]

The depth-to-diameter ratio of Mercurian craters matches that of lunar craters,[38] so it would appear that impacts on airless worlds create craters with specific characteristics.[39] Some of the large craters, however, have uncharacteristically 'shallow' floors that seem to have undergone isostatic adjustment, and the associated lithospheric fractures later provided routes for magma to emerge and produce smooth volcanic plains.[40]

At diameters of approximately 140 kilometres, the crater morphology shows concentric rings and there is a transition to basins. At first it was thought that basins were rare on Mercury,[41] but a detailed analysis identified many that had been partially buried by plains units.[42] Some basins are more significantly deformed by isostatic adjustment than their lunar counterparts. Indeed, the floors of some basins have risen to the point that they are no longer depressions. In these cases, the still-preserved secondaries trace the faint outline of the rim.[43]

In some areas, craters upwards of about 25 kilometres in diameter are so common that they overlap – their ejecta blankets are indistinct and their secondaries are not readily identified. The floors of many of these craters contain smooth plains whose sparse cratering implies that their material was emplaced much later. The similarity between this heavily cratered terrain and the lunar highlands has prompted the suggestion that this Mercurian terrain dates to the tail-end of the bombardment.[44]

The Caloris basin

The youngest and best preserved of the Mercurian multiple-ringed impact basins is Caloris, so-called due to its location at one of the 'hot poles'. Although Mariner 10 observed only half of the structure because it was on the terminator, the visible hemisphere was dominated by its 'sculpture', indicating that it formed towards the close of the heavy bombardment. As a result, Caloris has played the same role in deciphering how the planet's surface developed as did the Imbrium basin when stratigraphic analysis in the 1960s provided the first insight into the early history of the Moon.

The rim is marked by the Caloris Montes, which comprise rectilinear massifs up to 50 kilometres across and several kilometres high.[45] As with the Apennines (the southeastern rim of Imbrium) these blocks were instantaneously pushed up on deep faults by the shock of the impact and were mantled by the 'base surge' of material excavated from deep in the cavity. With the exception of a gap to the southeast, the Caloris Montes comprise a continuous chain. A thick blanket of ejecta lies on the

190 Ancient Mercury

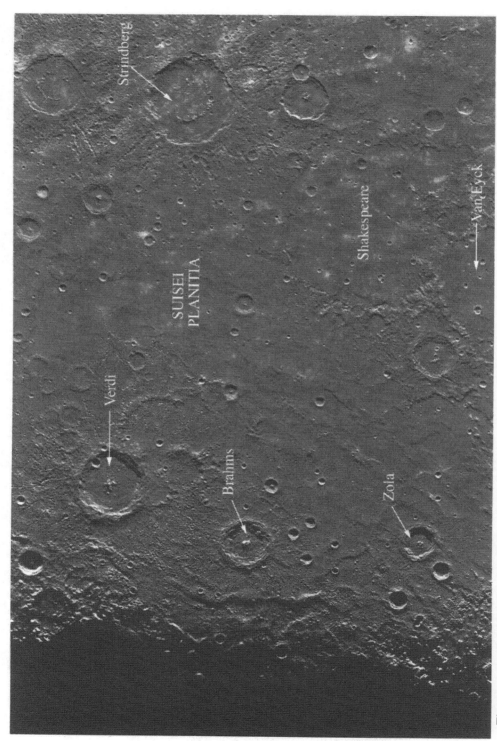

The area northeast of Caloris (which is off-frame bottom left) is dominated by basin ejecta and Suisei Planitia, one of the peripheral low-lying smooth plains.

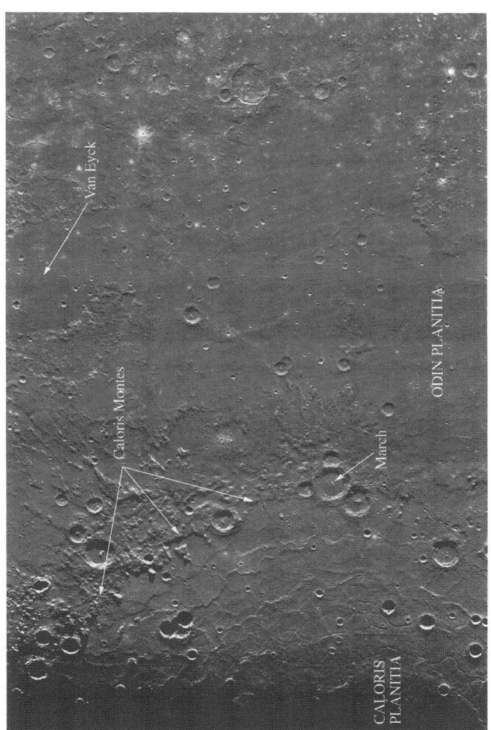

A mosaic showing the transition in terrains from the intensely ridged and grooved central plain of the Caloris basin, over the mountainous rim, and out through the ejecta to Odin Planitia, one of the peripheral low-lying smooth plain.

The area southeast of Caloris is dominated by Odin Planitia (top) and Tir Planitia (bottom), two of the peripheral low-lying smooth plains.

surface for a hundred kilometres or so beyond the rim and, in places, there is evidence of outer ring faults. Beyond this inner zone, in an annulus at about 1,000 kilometres (one full basin diameter), the terrain is dotted with secondary impacts and is strongly lineated by sculpture in the shape of radial ridges and grooves. In effect, therefore, this Van Eyck Formation is the counterpart of Imbrium's Fra Mauro Formation.

Smooth plains

Although only half of the planet was mapped, it is clear that the distribution of smooth plains is not random: they are widely distributed, filling the floors of the ancient basins and many of the large craters, and there is a significant concentration in and around the periphery of the Caloris basin. Radar studies have shown that the smooth plains around Caloris extend beyond the terminator which truncated Mariner 10's view, and that they reside in depressed areas.[46] It has been suggested that these plains are depressed by subsidence induced by the emplacement of the plains material.[47] The strip between 20 and 60 degrees longitude is conspicuously lacking in smooth plains. This focus on Caloris gives Mercury an asymmetric appearance that is reminiscent of the lunar dichotomy in which most of the maria are located on one side of the globe.

The well-defined contacts where the smooth plains embay older terrains indicate that they are stratigraphically younger. The fact that they are peppered only by small craters with sharp rims and bright halos shows that they are relatively recent in the Mercurian timescale. Further, the uniform cratering density indicates that all these plains were formed more or less simultaneously.

The smooth plains inside the rim of the Caloris basin are similar to those around its periphery,[48] but they are ridged and grooved by isostatic forces in the floor. The rounded ridges – which rise several hundred metres, are a dozen kilometres wide, and run for several hundred kilometres – tend to be larger and more numerous towards the basin's centre.[49] The grooves are on a similar scale, but have flat floors and steep walls. A clue to their origin lies in the fact that grooves have transected ridges, but no ridges have disrupted grooves. The ridges were probably formed in the same manner as the wrinkle ridges on the lunar maria, by compressional stresses resulting from the settling of the lava that erupted from deep fractures in the basin's floor. The grabens would have been opened by the extensional forces induced when the floor rose isostatically in response to the excavation of so much of the crust by the impact.

It has been suggested that the smooth plains around Caloris are fluidised ejecta from the impact which formed the basin.[50,51] If this is so, then they would be the counterpart of the Cayley Formation that was laid down by the impact that made the Moon's Imbrium basin. Until the Apollo 16 astronauts sampled this light-toned plain in 1972 it was thought to be the definitive example of highland volcanism, so its impact origin came as a shock and prompted a thorough reassessment of putative lunar volcanism. Such an analogy with the Mercurian plains is, however, weak. Cratering implies that a significant interval elapsed between the formation of the basin and the emplacement of the smooth plains around its periphery, so these plains cannot be ejecta, and in all likelihood are lava extruded from deep ring fractures.

Interestingly, the smooth plains are more densely cratered than the lunar maria and are therefore older; indeed, their formation may have ceased even before the basin in-fill began on the Moon. As the spectral characteristics of the smooth plains imply a homogeneous composition, the magma must have drawn upon a common source and in this respect the smooth plains differ from the lunar maria (whose varied spectral characteristics indicate either that they drew upon different sources or that the magma chemistry evolved over time). Mercury's large core and thin mantle imply that primary mafic magma was erupted. This chemistry is consistent with the low-viscosity lava required to form such smooth plains. The lunar surface's albedo range derives from the differing compositions of the anorthositic highlands and the mafic maria basalts. Although Mercury's smooth plains are distinctly darker, the uniformity of the rest of its surface suggests that there is no chemical distinction between the various terrains – evidently, the planet did not develop an aluminous crust.

Volcanism has clearly played a major role in forming the surface of Mercury, but this was of the plains-forming variety, and there is little evidence of edifices such as shields, domes or cones.

Intercrater plains

In about 30 per cent of the hemisphere mapped by Mariner 10, the density of large craters is significantly lower than in the heavily cratered terrain and the areas between form gently rolling intercrater plains. Did the craters strike the pre-existing plain (and if so, do the plains represent the primitive crust) or did the plains material erupt later and flow between the craters?

The fact that the plains have numerous small craters which are clearly secondaries to the larger ones shows that the primary impacts struck the plain.[52] But there are cratered terrains that are embayed by the plains material.[53] In some areas 'ghost' craters are visible, indicating that some of the plains are thin. The fact that ancient multiple-ringed basins were seen to underly some of the intercrater plains indicates that they are *not* the primitive crust, but magma that was extruded during the bombardment.[54] Some of the intercrater plains formed early in the bombardment, some towards its conclusion, and some afterwards.[55,56] This timescale suggested that although the intercrater plains form a distinct morphology, they are not a single stratigraphic unit;[57] they are the result of sustained volcanism.[58] In many respects the intercrater plains resemble the cratered terrain on Mars.

The 'weird' terrain

Directly antipodal of the Caloris basin, there is an expanse of hilly and lineated terrain that covers some 250 square kilometres. The intercrater plains are fractured into hills and hollows, and the rims of the large craters are shattered; however, the plains interior to the craters are smooth, indicating that they were formed some considerable time afterwards.

It is believed that this terrain (which is unique on the hemisphere mapped by Mariner 10) formed instantaneously when seismic shock waves from the Caloris impact were transmitted through the massive iron core and magnified as they

Physiography 195

converged on the far side of the planet.[59] In effect, the lineations were formed when the crust was ripped apart, and the hills mark large blocks which were displaced. This crustal disruption has also produced troughs which range from 5 to 15 kilometres in width, up to 150 kilometres in length, and up to half a kilometre in depth. In some cases, their walls are scalloped and material has slumped onto the valley floor, and in other cases, the valleys run between chains of upthrusted blocks. As there are similar patches of terrain on the Moon antipodal to the major basins,[60,61] this disruptive process would seem to be a factor in giant impacts. The disrupted terrain on Mercury, however, is more expansive and more prominent.

Lineations

A clue to Mercury's history is expressed on the surface in the form of linearities, such as ridges and scarps.

The generally linear ridges run for up to 300 kilometres and have soft profiles rising from 100 metres to 1 kilometre. Some ridges transition into scarps, which display a number of forms. The arcuate scarps are cliffs up to 1 kilometre high which run as far as 500 kilometres around arcs with a radius of curvature of several hundred kilometres. They transect large craters and the intercrater plains, but in some case are modified by impacts, indicating that the process that created the lineaments was contemporaneous with the bombardment. However, the lineaments do not transect the smooth plains. A study of the circumference-shortening where a scarp transects a large crater concluded that the scarps were formed by lateral compression of the crust. Analysis of a 75-kilometre crater cut by Vostok Rupes has been modelled by 6 kilometres of displacement by an overthrusting fault dipping at 11 degrees.[62] It has been proposed that this compression was in response to global shrinkage as the core cooled and the minerals assumed a denser phase.[63] However, the observed shrinkage is only a fraction of the 40 kilometres predicted. In fact, the observed 4 kilometres of shrinkage closely matched the value expected if only the mantle had cooled.[64] Such an endogenic process would be expected to generate a spherically symmetrical deformation pattern, but the orientation of the scarps is not random.

An exogenic alternative proposed that the scarps formed in response to the forces imparted as the planet's rotation was slowed by solar tidal forces, creating the spin-orbit resonance.[65] In this case, compressional stresses would be oriented east–west, linear shear oriented northwest–southeast, and the thrusting stresses oriented north–south. Unfortunately, the scarps cannot be analysed statistically because such relief features can only be observed under a low angle of illumination, and as much of the surface was highly illuminated there is observational bias in the sample.

The lobate scarps are similar to the arcuate scarps in that they curve, but the terrain within the lobe is generally higher than the surrounding terrain. It has been proposed that they are lava flows, but the scarps imply viscous flows up to 1 kilometre thick. The elevated areas are more likely to be 'slabs' of crust uplifted by global shrinkage.[66] Some of the smooth plains within large craters are marred by short scarps up to 400 metres tall, and these may be lava flows extruded from fractures opened by the isostatic adjustment of the crater floors.[67]

It can be inferred that the core formed very early in Mercury's history – and certainly before the most ancient exposed crust was created, otherwise this would show evidence of the global expansion caused by core formation.[68,69] That is, in the aftermath of accretion, heat from short-lived heavy radioactives would have induced melting and chemical differentiation, and as iron 'fell' towards the centre to form the core the release of gravitational energy would have further raised the temperature of the interior, inducing global expansion and increasing the planetary radius, possibly by as much as 30 kilometres. There is no evidence of this in the most ancient terrain. Subsequent to this brief period of expansion, Mercury's evolution was dominated by secular cooling and contraction. The scarps put a constraint on the time over which the volcanic plains were formed because the global shrinkage would have sealed the conduits through the lithosphere.

MAGNETIC FIELD

A major Mariner 10 discovery was that Mercury has a magnetic field. The magnetometer detected a rotation in the field vector towards the planet and an increase in the field strength.[70] Although only 1 per cent of the strength of the Earth's magnetic field, it was a 20-fold increase over the ambient interplanetary field. The magnetic axis is offset 10 degrees from the planet's rotational axis. The spacecraft's plasma detector noted a well-defined 'bow shock' in the solar wind 1,500 kilometres from the surface. The solar wind flows around the planet, but no plasma is trapped by the magnetosphere.[71] In a solar storm, the solar wind will impinge upon the surface. The third fly-by was deliberately low over the night hemisphere specifically to measure the magnetic field, and the results confirmed that the field is a planetary dipole. The discovery that Mercury has a magnetic field was a considerable surprise because it had been thought that only a rapidly rotating iron core could generate such a field.[72] Slow rotators such as the Moon, Venus and Mercury were not expected to create magnetic fields. Interestingly, the presence of the magnetic field raised the possibility that the core is still warm. In fact (upon reflection) it is certainly fluid. The planet is too small for its central pressure to have induced the solidification phase change. Furthermore, if the core contains even a small amount of sulphur this will have lowered the melting point sufficiently to inhibit the process of solidification. In the absence of the release of latent energy by a solidifying 'inner' core to drive convection in the surrounding fluid, convection is possible only near the core–mantle boundary. If so, then this may be the source of the weak dipole field.

As an alternative to a dipole resulting from magnetohydrodynamic effects within the core, it has been suggested that the field is the relic of an ancient magnetic field frozen into the surface rocks – remanent magnetism. Yet another alternative hypothesis is that Mercury's field is being induced in the iron core by the magnetic field that is carried by the plasma of the solar wind in which case its strength will vary with solar activity, but we will not acquire the data to determine this until a spacecraft can be placed into orbit around the planet to conduct a long-term study.

EXOSPHERE

Mariner 10's ultraviolet spectrometer found that Mercury has a helium exosphere, but at a pressure of 10^{-15} bar this is essentially a vacuum. It may be the result of outgassing of helium produced by the decay of radioactive elements.[73] The negative detection of argon (which is the decay product of radioactive potassium) is consistent with the absence of potassium, because the solar nebula would have been too hot for this volatile element to condense when the planet formed.

SINCE MARINER 10

Little new has been learned about Mercury since Mariner 10. The Hubble Space Telescope cannot observe the planet even when it is at elongation, because the telescope's operating rules prohibit it being aimed so close to the Sun. Because the planet rotates so slowly, it is possible to use radar interferometry to construct images showing topography, but the resolution is low. Nevertheless, the results confirmed that the cratered surface is rough.[74] Mariner 10's infrared radiometer indicated that the regolith is not uniformly distributed. In fact, there are large areas that radiate heat so rapidly that they must be bare rock – either fields of boulders or outcrops of bedrock.[75] The 'missing' hemisphere (the one that was in darkness during each of Mariner 10's fly-bys) was recently imaged using an advanced camera on the 60-inch telescope on Mount Wilson with a surface resolution comparable to that achieved by Mariner 10 during its far-encounter sequence.[76]

Although Mercury has received comparatively little attention by spacecraft, its sun-baked surface is a unique environment. Unfortunately, when Mariner 10 was developed, the remote-sensing technology did not exist to make a mineralogical survey of the crust by near-infrared imaging. Nevertheless, a recent analysis of the data revealed a different chemical composition from the other inner planets, variations in composition across the surface, and lava flows and pyroclastic deposits from explosive volcanic eruptions.[77] We should be able to pursue this finding with spacecraft that are now under development.[78,79]

NOTES

1. Although Galileo Galilei had discovered that Venus displayed phases, his telescope had been too crude to see them on Mercury.
2. E. M. Antoniadi's book says that Polish astronomer Johannes Hevelius was first to note the phases, but apart from saying "in the 17th century" he doesn't give a date! However, Hevelius made his map of the Moon in 1647, so he probably observed Mercury at around this same time. In *Planetary astronomy*, R.A. Schorn. Texas A&M University Press, p. 29, 1998, it says that the phases were noted by Francesco Fontana in the late 1630s but his results were not published until 1646.
3. 'Sulla rotazione di Mercurio', G.V. Schiaparelli. *Astron. Nach.* No.2944, 1889.

4 The temperature at the subsolar point at perihelion peaks at 700 K; it falls to 100 K in darkness.
5 *Planetary astronomy*, R.A. Schorn. Texas A&M University Press, p. 83, 1998.
6 *The planet Mercury*, E.M. Antoniadi. Gauthier-Villars, 1934; English translation by P. Moore. Reid Co., 1974.
7 'A radar determination of the rotation of the planet Mercury', G.H. Pettengill and R.B. Dyce. *Nature*, vol. 206, p. 1240, 1965.
8 *Photometrische Untersuchungen*, Z.K.F. Zollner. Leipzig, 1865.
9 *Untersuchungen über die Spektra der Planeten*, p. 90.
10 *Astron. Nach.* No.2944, 1889.
11 'The atmosphere of Mercury', S.J. Rasool, S.H. Gross and W.E. McCowern. *Space Sci. Rev.*, vol. 5, p. 565, 1966.
12 The argument of Mercury's perhelion point is migrating. In effect, the orbit itself is rotating around the Sun. Although the rate of advancement is twice that predicted by Newtonian physics, it was precisely the rate which Einstein's theory required.
13 For an excellent account of knowledge of Mercury in the 1950s see *Guide to the Planets*, P. Moore. Eyre & Spottiswoode, 1954.
14 *The planet Mercury*, W. Sandner. Faber & Faber, 1963.
15 T. Gehrels. NASA SP-267, p. 95, 1971.
16 'Mercury's craters from Earth', A. Young. *Icarus*, vol. 34, p. 208, 1978.
17 'Historical sightings of the craters of Mercury', R.M. Baum. *Strolling Astronomer*, vol. 28, p. 17, 1979.
18 The first crater to be individually resolved on Mercury during the far-encounter sequence was named in honour of the late Gerard P. Kuiper, a key member of the TV imaging team, and the founder of the Lunar and Planetary Laboratory at the University of Arizona, who died in December 1973. Kuiper's crater is about 60 kilometres in diameter.
19 'The Mariner 10 pictures of Mercury: an overview', B.C. Murray. *J. Geophys. Res.*, vol. 80, p. 2342, 1975.
20 A. Kenneth and K.A. Goethel. NASA TM X-3511, p. 47, 1977
21 The Moon's mass is 0.012 that of the Earth, and Mercury's is 0.056, so the planet is four and a half times as massive as the Moon.
22 'Mercury's heart of iron', C.R. Chapman. *Astronomy*, p. 22, November 1988.
23 'Mercury: internal structure and thermal evolution', R.W. Sigfried and S.C. Solomon. *Icarus*, vol. 23, p. 192, 1974.
24 'The temperature gradient of the solar nebula', J.S. Lewis. *Science*, vol. 186, p. 440, 1974.
25 'Metal/silicate fractionation in the Solar System', J.S. Lewis. *Earth Planet. Sci. Lett.*, vol. 15, p. 286, 1972.
26 J.S. Lewis. *Icarus*, vol. 16, p. 241, 1972.
27 'The strange density of Mercury', A.G.W. Cameron, B. Fegley, W.Benz and W.L. Slattery. In *Mercury*. University of Arizona Press, p. 692, 1988.
28 'Accumulation of the terrestrial planets and implications concerning lunar origin', G.W. Wetherill. In *Origin of the Moon*, W.K. Hartmann, R.J. Phillips and G.J. Taylor (Eds). Lunar and Planetary Institute, 1986.
29 'Accumulation of Mercury from planetesimals', G.W. Wetherill. In *Mercury*, University of Arizona Press, p. 670, 1988.
30 'Mineralogy of the planets', J.V. Smith. *Mineral. Mag.*, vol. 43, p. 1, 1979.
31 'Occurrence of giant impacts during the growth of the terrestrial planets', G.W. Wetherill. *Science*, vol. 228, p. 877, 1985.

32 'Preliminary geologic terrain map of Mercury', N.J. Trask and J.E. Guest. *J. Geophys. Res.*, vol. 80, p. 2461, 1975.
33 'Some comparisons of impact craters on Mercury and the Moon', D.E. Gault, J.E. Guest, J.B. Murray, D. Dzurisin and M.C. Malin. *J. Geophys. Res.*, vol. 80, p. 2444, 1975.
34 In fact, whereas the albedo of the lunar maria is 0.07, the Mercurian smooth plains are 0.12, so the material which forms the smooth plains is more reflective than that of the lunar highlands (0.10), and the albedo of the Mercurian highlands is 0.16, which is about the same as the ray material on the Moon. The albedo of the rays on Mercury is 0.4, so they stand out prominently. This probably indicates a difference in crustal composition between the two planets.
35 Mercury's surface gravity is $0.38g$, which is more than twice the Moon's $0.16g$, so the secondaries around a lunar crater are distributed over an area six times larger than those around a Mercurian crater. However, the large core means that Mercury's surface gravity is comparable to that of Mars, so ejecta ballistics on these two worlds can be directly compared, and the effects of Mars's atmosphere contrasted.
36 'Moon–Mercury: relative preservation states of secondary craters', D.H. Scott. *Phys. Earth Planet. Sci.*, vol. 15, p. 173, 1977.
37 'Some comparisons of impact craters on Mercury and the Moon', D.E. Gault, J.E. Guest, J.B. Murray, D. Dzurisin and M.C. Malin. *J. Geophys. Res.*, vol. 80, p. 2444, 1975.
38 'Landform degradation on Mercruy, the Moon and Mars: evidence from crater depth/diameter relationships', M.C. Malin and D. Dzurisin. *J. Geophys. Res.*, vol. 82, p. 376, 1977.
39 As will be made evident, craters on Mars, which has an atmosphere, are significantly different.
40 'Endogenic modification of impact craters on Mercury', P.H. Schultz. *Phys. Earth Planet. Interiors*, vol. 15, p. 207, 1977.
41 'Large impact basins on Mercury and relative crater production rates', H. Frey and B.L. Lowry. *Proc. Lunar Planet. Sci. Conf.*, p. 2669, 1979.
42 'New identification of ancient multiple-ring basins on Mercury and implications for geologic evolution', P.D. Spudis and M.E Strobell. *Proc. Lunar Planet. Sci. Conf.*, p. 814, 1984.
43 'Moon–Mercury: large impact structures, isostasy, and average crustal viscosity', G.G. Schaber, J.M. Boyce and N.J. Trask. *Phys. Earth Planet. Interiors*, vol. 15, p. 189, 1977.
44 'Mercury', D.E. Gault, J.A. Burns, P. Cassen and R.G. Strom. Ann. Rev. *Astron. Astrophys.*, vol. 15, p. 97, 1977.
45 'Stratigraphy of the Caloris basin', J.F. McCauley, J.E. Guest, G.G. Schaber, N.J. Trask and R. Greeley. *Icarus*, vol. 47, p. 184, 1981.
46 'Radar observations of Mercury', J.K. Harmon and D.B. Campbell. In *Mercury*, University of Arizona Press, p. 101, 1988.
47 'Radar altimetry of Mercury: a preliminary analysis', J.K. Harmon, D.B. Campbell, D.L. Bindschadler, J.W. Head and I.I. Shapiro. *J. Geophys. Res.*, vol. 91, p. 385, 1986.
48 The largest smooth low-lying plains on Caloris's eastern periphery are called Suisei, Odin and Tir Planitia.
49 'Surface history of Mercury: implications for terrestrial planets', B.C. Murray, R.G. Strom, N.J. Trask and D.E. Gault. *J. Geophys. Res.*, vol. 80, p. 2508, 1975.
50 'Mercurian volcanism questioned', D.E. Wilhelms. *Icarus*, vol. 28, p. 551, 1976.
51 'Comparative studies of lunar, martian and mercurian craters and plains', V.R. Overbeck, W.L. Quaide, R.E. Arvidson and H.R. Aggrawal. *J. Geophys. Res.*, vol. 82, p. 1681, 1977.

52 'Preliminary geologic terrain map of Mercury', N.J. Trask and J.E. Guest. *J. Geophys. Res.*, vol. 80, p. 2461, 1975.
53 'Observations of intercrater plains on Mercury', M.C. Malin. *Geophys. Res. Lett.*, vol. 3, p. 581, 1976.
54 'New identification of ancient multiple-ring basins on Mercury and implications for geologic evolution', P.D. Spudis and M.E. Strobell. *Proc. Lunar Planet. Sci. Conf.*, p. 814, 1984; NASA TM-86246, p. 87, 1984.
55 'Plains formation on Mercury: tectonic implications', P.G. Thomas. *Moon and Planets*, vol. 22, p. 261, 1980.
56 'Global volcanism and tectonism on Mercury: comparison with the Moon', P.G. Thomas, P. Masson and L. Fleitout. *Earth. Planet Sci. Lett.*, vol. 58, p. 95, 1982.
57 'Surface history of Mercury: a review'. J.E. Guest and W.P. O'Donnell. *Vistas in Astronomy*, vol. 20, p. 273, 1977.
58 'Mercury: a post-Mariner 10 assessment', R.G. Strom. *Space Sci. Rev.*, vol. 24, p. 3, 1979.
59 'Seismic effects from major basin formation on the Moon and Mercury', P.H. Schultz and D.E. Gault. *The Moon*, vol. 12, p. 159, 1975.
60 'Lunar nearside magnetic anomalies', L.L. Hood, P.J. Coleman and D.E. Wilhelms. *Proc. Lunar Planet. Sci. Conf.*, p. 2235, 1979.
61 'The lunar swirls: distribution and possible origin', L.L. Hood and C.R. Williams. *Proc. Lunar Planet. Sci. Conf.*, p. 99, 1989.
62 *Planetary landscapes*, R. Greeley. Allen & Unwin, p. 124, 1985.
63 'Some aspects of core formation in Mercury', S.C. Solomon. *Icarus*, vol. 28, p. 509, 1976.
64 'The tectonics of Mercury', H.J. Melosh and W.B. McKinnon. In *Mercury*. University of Arizona Press, p. 374, 1988.
65 'Global tectonics of a despun planet', H.J. Melosh. *Icarus*, vol. 31, p. 221, 1977.
66 'The tectonic and volcanic history of Mercury as inferred from studies of scarps, ridges, troughs and other lineaments', D. Dzurisin. *J. Geophys. Res.*, vol. 83, p. 4883, 1978.
67 'Tectonism and volcanism on Mercury', R.G. Strom, N.J. Trask and J.E. Guest. *J. Geophys. Res.*, vol. 80, p. 2478, 1975.
68 'The relationship between crustal tectonism and internal evolution in the Moon and Mercury', S.C. Solomon. *Phys. Earth Planet. Interior*, vol. 15, p. 135, 1977.
69 'Formation, history and energetics of cores in terrestrial planets', S.C. Solomon. *Phys. Earth Planet. Interior*, vol. 19, p. 168, 1979.
70 'Magnetic field of Mercury confined', N.F. Ness, K.W. Behannon, R.P. Lepping and Y.C. Whang. *Nature*, vol. 255, p. 204, 1975.
71 'Observations at Mercury encounter by the plasma science experiment on Mariner 10', K.W. Ogilvie *et al. Science*, vol. 185, p. 145, 1974.
72 At this point, neither Venus nor Mars had shown any evidence of possessing even a weak magnetic field.
73 'Mercury's atmosphere from Mariner 10: preliminary results', A.L. Broadfoot, S. Kumar, M.J.S Belton and M.B. McElroy. *Science*, vol. 185, p. 185, 1974.
74 'Surface features on Mercury', S. Zohar and R.M. Goldstein. *Astron. J.*, vol. 79, p. 85, 1974.
75 'Preliminary infrared radiometry of the night side of Mercury from Mariner 10', S.C. Chase, E.D. Miner, D. Morrison, G. Munch, G. Neugebauer and M. Schroeder. *Science*, vol. 185, p. 142, 1974.
76 'A digital high-definition imaging system for spectral studies of extended planetary atmospheres. 1. Initial results in white light show features on the hemisphere of Mercury unimaged by Mariner 10', J. Baumgardner, M. Mendillo and J.K. Wilson. *Astron. J.*, vol. 119, p. 2458, 2000.

77 'Recalibrated Mariner 10 color mosaics: implications for mercurian volcanism', M.S. Robinson and P.G. Lucey. *Science*, vol. 275, p. 197, 1997.
78 'BepiColumbo: a multidisciplinary mission to a hot planet', R. Grard, M. Novara and G. Scoon. *ESA Bull.* No.103, p. 11, August 2000.
79 The Messenger spacecraft which the Applied Physics Laboratory at Johns Hopkins University is building for NASA is scheduled for launch in 2004, with insertion into a highly elliptical orbit around Mercury in 2009.

7

Precambrian Earth*

THE BASEMENT

As the fossil hunters of the 19th century worked their way 'down' through exposed strata they defined geological 'periods' and assigned them names derived from the sites at which they were initially studied.

Sandwiched between the intricately stratified sediments of the Cambrian and the basement of granites, there was a thick layer of coarse sediment in which there did not appear to be any fossils. When the miners of the Harz Mountains first encountered this, they dubbed it 'grauwacke', meaning 'grey grit', which became 'greywacke' when anglicised. It differed from a sandstone by having a richer variety of constituents with the quartz, feldspar and mica mineral grains supplemented by poorly weathered ferromagnesian minerals and clasts of eroded rocks, and also by the fact that the fine-grained matrix of flakey alteration products of weathering (such as clay) constituted a significant fraction of the rock.

In the late 18th century, the Saxon mining engineer, A.G. Werner, believed that the early Earth's 'nucleus' was covered by an ocean that was rich in dissolved minerals and with rocky debris in suspension. He believed that granite was a 'primitive' rock that precipitated first; the coarse greywacke was a 'transitionary' rock that formed as the suspended debris settled; and the mature sediments of the Cambrian formed once a significant area of dry land was exposed to fluvial erosion.

Assuming that the origin of life occurred early in the Earth's history, geologists tended to dismiss the "interminable greywacke" and the granitic basement as "the Precambrian" and focused their efforts upon the fossiliferous strata which they assumed constituted the majority of the geological timescale.[1]

With the advent of radiometric dating of rocks and the appreciation of the enormity of geological time that this facilitated, it was realised that the Pangean supercontinent's break up in the last 200 million years accounted for only 5 per cent

---

* This chapter is necessarily fairly technical. The reader's attention is drawn to the glossary on page 419.

of the Earth's history. The orogeny associated with Pangea's assembly pushed this 'horizon' back only another few hundred million years. To investigate the first 90 per cent of the geological timescale, it would be necessary to locate the most ancient parts of the continents. If plate tectonics had been at work for several 'cycles' of continental assembly and break-up, this search could not simply involve studying rocks in the centres of plates, because the most ancient rocks may well be on the peripheries of today's continents.

On the southern side of the Baltic Sea and the Gulf of Finland, the Precambrian rocks are masked by a thin veneer of almost unfolded sediments, but to the north the schists, gneisses and granites are largely in outcrop. At first, it was thought that these crystalline rocks represented the original crust and therefore predated the development of life, but in the mid-1800s fossils were found in Alpine schists. The fossils they contained revealed that the schists were metamorphosed 200-million-year-old sediments, so schists are not necessarily Precambrian. Further studies established that schist could be 'feldspathised' into gneiss. Nevertheless, the outcrops in the northern Baltic have been radiometrically dated at 2.4 to 3.2 billion years old; with the oldest on the Kola Peninsula.[2]

The Earth has a bimodal distribution of surface elevation. Although mountains are tall and ocean trenches are deep, these are minor features in terms of surface area. The mean elevation of the continents is 850 metres above sea level and the abyssal plains of the ocean floor are some 6 kilometres below sea level, so there is a 7 kilometre difference between the planet's two dominant surface units. On land, a 'continental shield'[3] is an expanse of low and essentially level terrain comprising highly deformed sequences of metamorphic rocks that are extensively intruded by granite. Both types of rock formed in the high-temperature and high-pressure environment several kilometres below the surface, but through the ages this 'roof' has been eroded away to expose a cross-section of the structure. The shields imply that prior to somehow becoming 'stabilised', the rocks of the early Precambrian were formed by volcanism, erosion and immature sedimentation, folding, and repeated periods of metamorphism.

CANADIAN SHIELD

At 2.6 million square kilometres, the Canadian shield is the world's largest. It runs west from the northern Appalachian Range across Canada to the Rockies, and north from the shore of Lake Superior across Hudson Bay and Baffin Island. Much of it lies within a few hundred metres of sea level and displays very little surface relief. Large areas of the shield have been overlain by up to a kilometre of mature sandstone, shale and limestone whose presence means that the periphery of the shield was once inundated by a shallow sea. The Great Plains of the interior of the United States sit on the shield's southern margin. Although they appear totally flat on the local scale, they are 'warped' into broad shallow depressions and domes. The state of preservation of this 'stable platform' indicates that the basement has not been subjected to deformational stress for a considerable time.

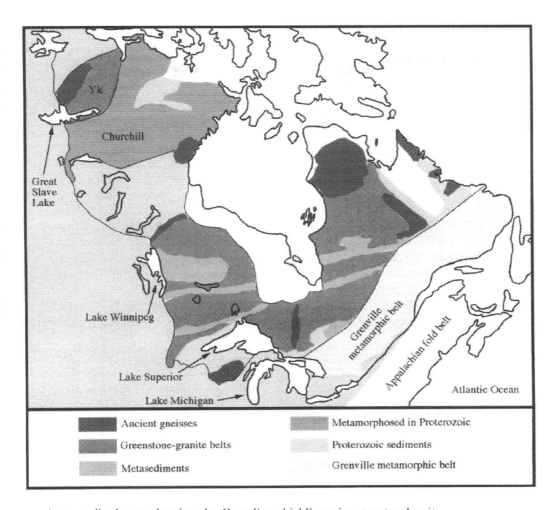

A generalised map showing the Canadian shield's major structural units.

Most of the 2.5 to 2.7 billion years old Keewatin Group of rocks to the south of Hudson Bay are metamorphosed volcanics with intrusive granites and they form a broad belt running east west between Hudson Bay and Lake Superior. They are masked by younger sediments south of Lake Winnipeg, where they form the Prairies, but they reappear at the eastern fringe of the Rocky Mountains in Wyoming and Montana in the United States. To the north there is an ancient strike-slip fault several hundred kilometres in length which forms a 300-metre-high scarp and marks the boundary between the Provinces of Churchill and Yellowknife. Although the Churchill exposure is 1.8 billion years old, the Yellowknife rocks on the north side of the fault are a billion years older. It demonstrates that the shield is a composite of 'cratons' which formed individually in the Archean and were then swept together by lateral crustal movements in the Proterozoic.[4]

In the 1960s, the oldest known rocks on the Canadian shield were in the far northwest, in Slave Province. As on the Kola Peninsula – which may once have been

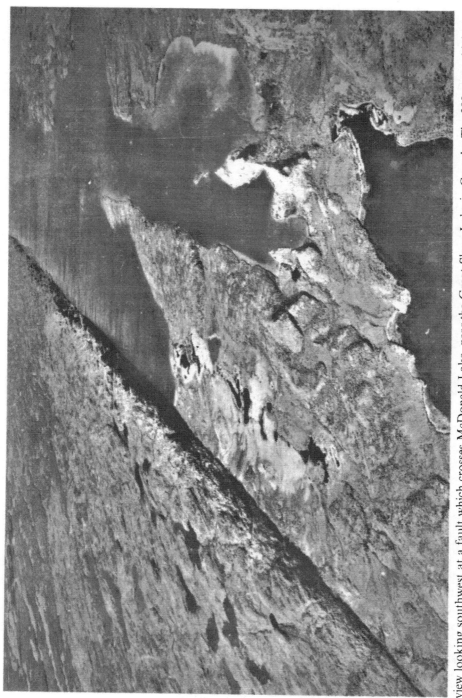

A view looking southwest at a fault which crosses McDonald Lake, near the Great Slave Lake in Canada. The 300-metre-high scarp runs for at least 500 kilometres. Churchill Province (on the left) has been dated at 1.8 billion years old, but Yellowknife (on the right) is fully a billion years older. (Aerial image A5120-105R/4-8-1935 1935. Her Majesty the Queen in Right of Canada, reproduced from the collection of the National Air Photo Library with permission of Natural Resources Canada.)

part of the Canadian shield – the oldest rocks were some 3.2 billion years old. Although this pushed the 'horizon' far into the Precambrian, it still left in excess of a billion years of geological time unaccounted for. Where were the older rocks?

BARBERTON MOUNTAINLAND

In the 1940s, in southern Rhodesia, now Zimbabwe, A.M. Macgregor found a cluster of two dozen granitoid domes 20 to 200 kilometres wide in a 'sea' of greenish rock. This is the Shamvaian–Bulawaysan–Sebakwian formation – the Shamvaian is 2.6 billion years old; the Bulawaysan is 2.8 to 3.0 billion years old; and the Sebakwian is 3.52 billion years old.[5,6,7] The Limpopo Fold Belt immediately to the south of this Zimbabwean craton near the Sand River contains some early Archean gneisses.[8,9] A ribbon-shaped outcrop of deformed rock contains 3.73-billion-year-old granulitic-to-amphibolitic metasediments caught up in highly aluminous quartz-rich metasediments which are 3.78 billion years old.[10,11] Massive dykes transected this entire complex 3.57 and 3.1 billion years ago.

Further south, in the Transvaal, many strata that had been assumed to be late Precambrian were found to be much older. Although it was high in the sequence, the basalt of the Bushveld Complex was 2 billion years old. Lower down, the Dominion Reef system is 2.9 billion years old.[12] The Basement Complex was obviously older still. Between Johannesburg and Pretoria, the 'Old Granite' of this Basement is 3.2 billion years old. An outcrop in southeast Transvaal and adjacent Swaziland that had been thought to be a continuation of this formation turned out to be distinct, and even older.[13]

In the 1880s, 'gold rush' miners founded the town of Barberton in the hilly terrain of the northern Transvaal. They referred to the greenish rocks that predominated as 'greenstone'. In fact, greenstone is mafic volcanic rock transformed into greenschist by low-pressure regional metamorphism. The pillow-form of the lavas indicates that they were extruded by submarine eruptions. Compared to the lavas that erupt on the ocean floor today, some of the greenstones are of a magnesium-rich chemistry which has been named komatiite, after the Komati River in the Barberton where it was first encountered.[14] This high-temperature orthopyroxene magma was evidently drawn directly from the mantle. Two decades later, in surveying what had become the Barberton Mountainland, A. Hall, a mining geologist, inferred from their context that the greenstone was extremely old, which Maarten de Wit, a Dutch geologist, confirmed upon remapping the area in the 1980s.

Although the Kaapvaal craton spans 600,000 square kilometres, most (85 per cent) of it is buried by Proterozoic deposits and most of the Archean exposure comprises granitoid domes and granitoid gneisses; less than 10 per cent of it is greenstone. Nevertheless, the considerable stratigraphic structure of the 20-kilometre-thick greenstone sequence, which has been tipped onto its side,[15] provided our first real insight into the Archean environment.[16,17,18]

There is a progressive evolution of rock types and composition through the sequence. The Onverwacht Group forms the majority of the sequence (in fact, the

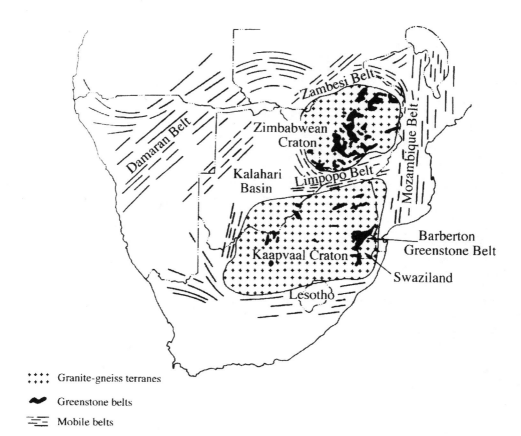

- ∴∴∴ Granite-gneiss terranes
- ～ Greenstone belts
- ≡ Mobile belts

An outline map showing the distribution of the Archean greenstone belts on the granite-gneiss masses of the Zimbabwean and Kaapvaal cratons of southern Africa. The Limpopo Fold Belt was formed when the cratons were swept together some time in the Proterozoic.

lower 15 kilometres of the 20-kilometre stack) and it transitions from predominantly ultramafic to felsic lavas, and also to an increasing ratio of pyroclastics and other volcanogenic sediments. A 6-metre-thick layer of chert halfway up the Onverwacht Group has been called the Middle Marker.[19] The ultramafic lavas (that is, the komatiites) are confined to the older strata. The volcanics above the marker display a repeating pattern of mafic lavas, felsic lavas, pyroclastic deposits, and a thin layer of chert, each recording a cycle of volcanic activity interspersed with sedimentation in a marine environment. The pillow-form of the lavas throughout shows that as the stack accumulated it settled sufficiently rapidly to remain submerged. The early sediments in the upper part of the sequence are volcanogenic sediments and immature sandstones (that is, greywacke) that were deposited in deep water by turbidity currents, but the later sediments show a cycle of conglomerates, mature sandstones, shales, chert and banded iron with characteristics implying deposition in shallow water by run-off from onshore weathering. There was a transition in the

(b.y.)	(km)	thickness	description
> 3	20	2.4	greywacke, shale, quartzite conglomerate and banded iron
		1.0	conglomerate and volcanogenic sediments
		1.0	greywacke, shale and chert
		2.8	mafic lavas, felsic volcanogenic sediments and chert
	12		
		4.8	felsic lavas and volcanogenic sediments
3.36			mainly mafic lavas
3.38			chert
3.53		3.5	mafic and ultramafic lavas
	4		
		2.0	mafic and ultramafic lavas, felsic tuff and chert
	2		
		2.0	ultramafic (komatiite) and mafic pillow lavas
	0		

An idealised Archean stratigraphic column in the Barberton Mountainland. Nowhere is the entire sequence preserved intact, but by correlating exposures it has been possible to reconstruct the stack.

volcanism during this period, with the ultramafic lavas yielding to felsic (andesites and calc-alkaline) volcanics. The clastic volcanogenic sediments suggested immature weathering close to the sea, which implied a volcanic island setting.

The rocks that surround the greenstone sequence form several distinct groups. The oldest gneisses (which predominate in Swaziland to the south) are nowhere in direct contact with the greenstones.[20] At 3.5 billion years old these gneisses are about the same age as the ultramafics at the base of the Onverwacht Group. As they display a higher 'grade' of metamorphism than the greenstone, they were processed much deeper within the crust. Some of the granitoids abut the Onverwacht Group, but their contacts are tectonic rather that intrusive, indicating that they were probably swept together by lateral crustal displacements. However, some granitoids did rise diapirically through the greenstone strata, locally deforming them.[21] Other assemblages of granitoids were emplaced in the late Archean, through the transition to the Proterozoic and on to 2.2 billion years ago, at which time activity ceased.

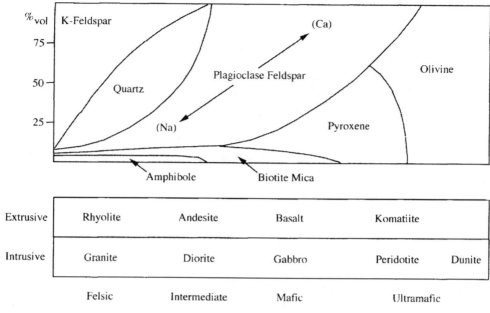

The chemical compositions of various magmas.

Further, there was a chemical evolution from sodic to potassic granite. In fact, the Archean gneisses are distinctive for being 'tonalitic' (tonalite is primarily sodic plagioclase feldspar and quartz[22]).

The nature of the surface on which the greenstone sequence accumulated is hotly debated. It has been suggested that the greenstones were erupted onto a granitic basement, and that the gneisses are relics of this.[23] However, the precursor to the metamorphism which produced the gneisses seems to have included mafic and ultramafic rock, and it has been suggested that not only did the greenstones accumulate on an oceanic crust but that the ultramafics in the base of the sequence are relics of this.[24] Supporting this case is the fact that rocks resembling those at the base of the greenstone sequence are enclaved in the gneisses. In this latter hypothesis, the weight of the extruded lavas induced isostatic settling, and the granitoids resulted from partial melting as the base of the sinking structure was driven deep. If so, then this occurred rapidly because the oldest gneisses appeared within 100 million years of the oldest greenstones in the stack. Without an unambiguous field relationship between them, however, it is difficult to be sure which came first. Irrespective of the nature of the crust, the pillow-form of the lavas and the character of the sediments imply that the greenstone sequence accumulated in a submarine environment that persisted for half a billion years. Later in the Archean, lateral displacements swept together the Zimbabwean and Kaapvaal cratons, trapping the intervening terrains as the Limpopo Fold Belt to form a stable platform that has managed to survive several billion years of tectonism remarkably well.[25]

GREENLAND

The Greenland craton has been fragmented by the opening of the Atlantic, with segments embedded in Scotland, Norway and Labrador.[26] Most of the craton lies beneath the inland ice sheet, so our knowledge of it is based on surveys of southwestern Greenland and the coast of Labrador across the Davis Strait, and stratigraphic analysis was the key to interpreting its history.

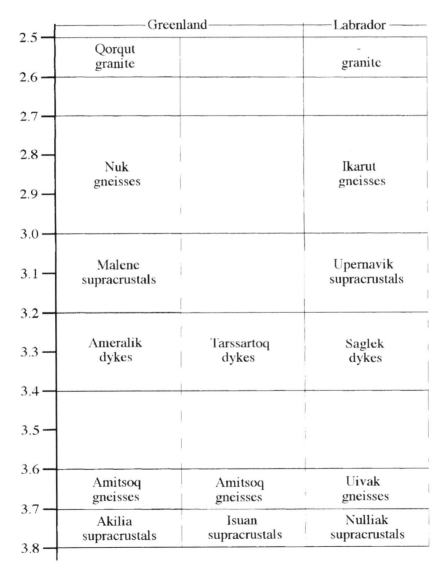

This chart illustrates the correlations between the early Archean sequences of the Greenland and Labrador fragments of the Greenland craton.

In 1967, New Zealander Vic McGregor was surveying the Ameralik fjord to the south of Godthaab, the capital, also known as Nuk, and he found an extremely deformed band of dark metamorphic rock within a mass of light-toned gneiss that was known to be very old. Oxford University geochronologist Stephen Moorbath determined that the dark rocks were 3.8 billion years old.[27,28,29,30]

The area south of Godthaab is predominantly granitoid gneisses that are composed mainly of tonalite and granodiorite. The enclaves of dark rock that McGregor found in these Amitsoq gneisses form ribbons stretching for several hundred metres. These metamorphosed rocks – collectively known as the Akilia supracrustal sequence – are relics of mafic lavas, mafic and ultramafic intrusives and clastic sediments.[31] The gneisses and supracrustals are transected by the mafic Ameralik dykes, which range in thickness from a few centimetres to several metres, but the younger rocks such as the Malene supracrustals – which consist of mafic volcanics, ash tuffs and volcanogenic sediments plus some chert, limestone and shale – are not. However, the Nuk gneisses intrude the Amitsoq gneisses, Amiralik dykes and Malene supracrustals and are therefore younger. They resemble the Amitsoq gneisses but can be distinguished by the absence of Ameralik dykes – which therefore serve as a useful stratigraphic marker. Radiometric dating indicated that the Amitsoq and Nuk gneisses spanned the Archean, and that they represented the build-up of the craton by pulses of igneous activity interspersed with periods of extensive weathering and sedimentation. Later in the Archean, the original granitoids and supracrustals were metamorphosed. The fact that the granitoid dykes which accompanied the appearance of the Qorqut granite plutons in the early Proterozoic are undeformed implies that the craton had already become stabilised against internal tectonic stresses. The ancient supracrustals included in the Uivak gneisses on the Labrador coast at Saglek and Hebron are similar to the Akilia assemblage. The similarity in the sequences is striking, because the Uivak gneisses are transected by the mafic Saglek dykes (just as are the Amitsoq gneisses by the Amiralik dykes) and the later Upernavik supracrustals are undisturbed (as are the Malene).[32,33,34,35]

In 1971, McGregor revisited Greenland with Moorbath. At a mine in the Isua hills on the edge of the ice sheet 150 kilometres from Ameralik there is a large arc of supracrustals included in the Amitsoq gneisses. This is enclaved (as is the Akilia sequence) but the Isuan supracrustals are less metamorphosed and so they have retained their characteristics. The sequence includes mafic and ultramafic lavas and intrusives, fine-grained clastic sediments, conglomerates, cherts and banded iron. Their state of preservation enabled the environment in which they formed to be inferred. The pillow lavas implied extrusion onto the sea floor. The conglomerates suggested a pebble beach. The boulders and clasts of felsic volcanic rocks embedded in the fine-grained limestone matrix implied coarse sedimentation of volcanogenic debris in water. The mineral concentrations implied hydrothermal activity. The banded iron was formed by precipitation in shallow water. As in the case of the Barberton, the sequence suggests a volcanic island setting.[36,37,38,39,40,41,42] The oldest Isuan supracrustals are 3.8 billion years old, and therefore date to the start of the Archean.[43] As at Godthaab, the Labrador Archean sequence ends with the younger granitoids, but in the Isua area the late-Archean rocks are absent and the sequence ends with the equivalent of the Amiralik dykes.

Greenland's is not the only craton to have been split by recent rifting. The opening of the South Atlantic left fragments of 3.77-billion-year-old gneisses in Venezuela in South America and in Liberia and Sierra Leone in Africa.[44,45]

SUPERIOR PROVINCE

The Canadian shield is exposed in the USA in Minnesota around Lake Superior.[46] To the south, it forms the basement of the Great Plains. In fact, the Midcontinental Rift (characterised as a 'failed rift') that runs southwest from the western tip of the lake has actually split that part of the Superior craton, and a piece of the Minnesota outcrop has been displaced 100 kilometres to the east into northern Wisconsin and adjacent Michigan.

The southern part of the US section of the craton – both east and west of the rift – consists of highly deformed and metamorphosed gneisses which were emplaced as granitoids some 3.5 billion years ago. The rest of the Minnesota section forms the southern extremity of an extensive sequence of greenstones which were formed 2.8 to 2.6 billion years ago during the transition between the Archean and the Proterozoic.[47] The greenstone comprises mafic and felsic lavas and intrusives, volcanogenic sediments and banded iron. Although deformed and intruded by granitoids the sequence is only mildly metamorphosed. There is a narrow strip of greenstone on the northern fringe of the section of the craton that was displaced eastwards by the rift. The highly deformed tectonic zone that forms the 30-kilometre-wide contact between this strip of greenstone and the gneiss complex to the south indicates that the two were swept together at some time in the Proterozoic, probably as the shield accreted.[48] West of the rift, the oldest rocks are in the Minnesota River Valley. To the east the oldest rocks are at Watersmeet, in the broad transition zone just north of the Wisconsin–Michigan border. In both cases, amphibolites are included in the ancient gneiss complexes, the chemical composition of which indicates that they are earlier mafic and ultramafic lavas that were metamorphosed deep below the surface. The gneisses at Morton in Minnesota were initially thought to be 2.6 billion years old, but they were found to be 3.66 billion years old, making them contemporaneous with the Greenland and Labrador gneisses. Granitoids were intruded between 3.0 and 2.6 billion years ago.[49,50] At Granite Falls (100 kilometres northwest of Morton) the gneisses are derived from granitoids that were emplaced 3.6 and 3.1 billion years ago.[51] As at Morton, the older gneisses have amphibolite enclaves but there is also a gneiss unit derived from mafic lavas and intrusives and associated sediments. Both of these sites underwent intense metamorphism and deformation by granitoid emplacement 1.8 billion years ago, in the mid-Proterozoic. To the east of the Midcontinental Rift, in the transition zone between the greenstones and the older gneisses, this late metamorphism has degraded most of the radiometric dating signal, but zircons indicate that the oldest gneisses are about 3.6 billion years old.[52]

The early Archean mafic supracrustals are fairly well preserved at Isua, more metamorphosed at Ameralik, and transformed into amphibolite on the southern margin of the Superior Province. In no case was the depositional sequence as extensive as that at Barberton. However, the best was yet to come, because in the 1970s mining companies started to survey Australia's rocky western desert.

WESTERN AUSTRALIA

Western Australia is dominated by two Archean cratons. The Pilbara runs eastwards along the northwest coast for several hundred kilometres, forming a rectangular block that includes the Kimberley Plateau. In fact, it continues south for a hundred kilometres as the basement for younger sediments beyond which is the Yilgarn which (at 600,000 square kilometres) is more expansive and is still largely unexplored.[53] Both are granitoid-greenstone complexes with thick sequences of mildly metamorphosed volcanic and sedimentary rocks into which several pulses of granitoids have been emplaced. However, while the Yilgarn (like Greenland and Superior Province) has granitoid gneisses with enclaves of ancient supracrustals, the Pilbara does not. It is likely that the Pilbara and Yilgarn cratons formed separately and were swept together in the Proterozoic, forcing up the fold belt between them.

The Pilbara consists of domical granitoids surrounded by deformed basin-like greenstone belts. At first, it was thought that the greenstones had formed in elongated basins between the domes, but detailed mapping indicated that the greenstones were deposited horizontally across the entire craton and deformed by the subsequent appearance of the domes.[54,55] However, the way in which the granitoids formed is still disputed.[56] In one theory, the greenstone sequence was deposited on a mafic crust that was later intruded by granitoid magma. In another theory, the greenstone sequence accumulated on a felsic basement, parts of which rose diapirically to form the granitoid domes.[57] Supracrustals enclaved in the Shaw Batholith at the eastern end of the craton are clearly derived from the nearby greenstones, and they show signs of solid-state deformation, which argues in favour of the granitoids having been emplaced in a plastic rather than a molten state.[58] Overall, the domes account for about 60 per cent of the Pilbara's surface exposure, so their emplacement considerably deformed the initially tabular greenstone strata. Zircons in the Shaw Baltholith showed that it was emplaced 3.47 billion years ago.[59] Most of the rest of the domes were emplaced between 3.3 and 3.0 billion years ago.[60] A few granitoid plutons invaded the craton and crystallised during the transition to the Proterozoic, 2.7 to 2.6 billion years ago, but little has happened since then and the greenstones are well preserved.

The 2-kilometre-thick North Star Basalt that lies at the base of the greenstone sequence is primarily composed of basalt flows but includes lavas ranging from ultramafic to felsic types and is 3.56 billion years old.[61] The 5-kilometre-thick layer of felsic volcanics that is somewhat above it in the sequence is 3.45 billion years old.[62] A tremendous amount of volcanic activity during this 100-million-year interval piled up almost 10 kilometres of lavas and volcanogenic sediments before the granitoids began to appear.

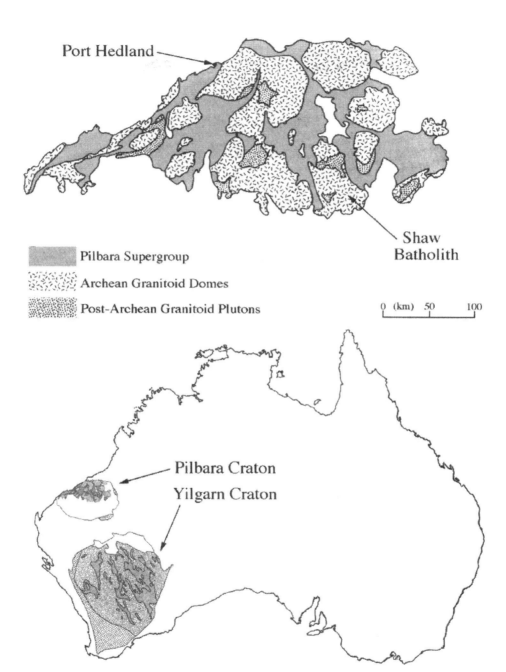

A map showing the location of the Pilbara craton and its isolation from the Yilgarn craton by a fold belt which suggests that the two cratons are distinct and were swept together some time in the Proterozoic.

(b.y.)	(km)	thickness (km)	description
3.0	28	1.5	mafic to intermediate volcanics and metasediments
		5.0	sandstone, shale and tuff
	23		
		3.0	sandstone, conglomerate and tuff
	20	1.0	basalt
		1.0	banded iron and chert
	17	1.0	pillow lavas
		1.5	greywacke and other metasediments
		1.0	rhyolite
		2.0	pillow lavas and komatiites
3.34		1.0	felsic volcanics and chert
		1.5	pillow lavas and komatiite
	9	1.0	chert
		5.0	felsic volcanics
3.55	4		
		2.0	pillow lavas, komatiite, chert and banded iron
		1.9	
3.6	0		

An idealised view of the Archean stratigraphic column on the Pilbara craton. Nowhere is the entire sequence preserved intact, but by correlating exposures across the craton it has been possible to reconstruct the stack.

The full stratigraphic sequence of the greenstone is nowhere complete, but if all its various pieces could be reconstructed they would form a stack, 30 kilometres thick, that has been called the Pilbara Supergroup. Like most Archean greenstones, it is predominantly volcanic near the bottom of the stack, with clastic sedimentary rocks increasing upward. The presence of pillow lavas, chert, banded iron and a variety of clastic sediments throughout the stack implies that it spent much of its time under *shallow* water. Indeed, sandwiched between the thick layers of pillow lava are sediments containing stromatolites, which form only in shallow water. Clearly, the sea floor was subsiding in pace with vast extrusions which were on a par with anything seen in a modern continental flood basalt eruption. Layers of silica-rich dacite pyroclastics are evidence of explosive eruptions, but this activity was not continuous. Deposits of chert and immature sediments from erosion of islands mark quiescent intervals and the banded iron formation that once covered the entire craton to a depth of 1 kilometre represents a *lot* of iron precipitation in shallow water during a lengthy period which was completely free of volcanic activity.

Disruptions of the depositional environment by tectonic movements, uplift and denudation are represented by unconformable erosional surfaces. In many instances vertical displacement resulted in the emergence at the surface of granitoids that became small continental-style units. In the early-1990s, while mapping the central Strelley area, Roger Buick of the University of Western Australia discovered an unconformity that ran for 75 kilometres.[63,64] The 3.52-billion-year-old Coonterunah Group had been laid down in water, then raised to form dry land, tilted, eroded and buried by the 3.5-billion-year-old rhyolite of the Warrawoona Group. This was a discovery almost as significant as the classic unconformity in Scotland that was interpreted by James Hutton two centuries earlier. By establishing that continental-style crust had existed 3.6 billion years ago, this unconformity lent further support to the theory that the basement of this greenstone sequence was felsic, rather than mafic. As the world's thickest, best exposed, and least dismembered Archean outcrop, the Pilbara is a magnificent study of the early Earth.[65] It records the deposition of greenstones from 3.5 to 3.0 billion years ago, on a crust of tonalitic granite that is at least 3.65 billion years old.

ARCHEAN PROCESSES

Precambrian rock is found on every continent, but Archean rock contributes no more than 12 per cent to the exposed crust. Of course, if extensions inferred to exist beneath Proterozoic sediments are included, then this fraction increases dramatically.[66] Much Precambrian rock has been repeatedly subjected to metamorphism, in some cases with sufficient intensity to induce such severe chemical alteration that the progenitor rock type is obscure.

Although the gneisses constitute 85 per cent of the Archean exposures, most research has focused on the greenstones because they contain valuable minable minerals. The processes that produced them were evidently active throughout the

218 **Precambrian Earth**

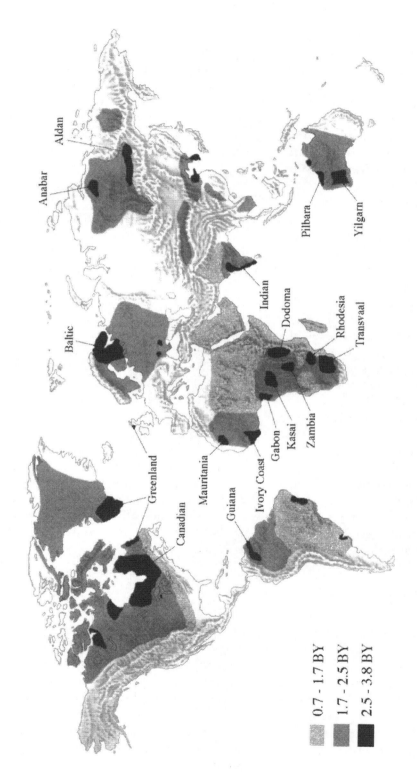

A map showing the distribution of Archean cratons

Archean and well into the Proterozoic. In fact, most of the surviving outcrops were created early in the Proterozoic.

The fact that the oldest surviving rock structures on the planet were sedimentary came as a surprise. If the oldest rocks had turned out to be crystalline, geologists would have concluded that this was the primitive crust. The base of every greenstone stack is ultramafic lava that was extruded under water. The clasts in the oldest greenstones were derived from felsic volcanics, implying that granitoids were being produced early on and being eroded ashore, but only in a few cases is there evidence to suggest that the local basement was felsic.

Greenstones typically form belts whose major axis is 800 kilometres in length, suggesting a large syncline. The onset of the Archean 3.8 billion years ago coincided with the conclusion of the heavy bombardment, and the depressions in which the greenstones accumulated were perhaps large impact craters. Regardless of whether the crust was mafic or felsic, the sudden removal of a large amount of crust by an impact would have 'decompressed' the mantle and induced it to melt, and the deep faults in the crater's floor would have provided such rising magma with access to the surface. This primary magma would have had ultramafic composition; it would have flowed out onto the surface in massive quantities and filled in the crater which, with the crust already weakened, would have promptly settled isostatically producing a downwarp of the crust.[67] The ultramafic flows would therefore be the terrestrial equivalent of the lunar maria which (as it happened) were being extruded at about the same time. The rise of granite plutons would have compressed and deformed these lava-filled basins and mildly metamorposed them. Although the Earth would have been as cratered as the Moon's, uniformitarianists eschew any association (because impacts smack of catastrophism) favouring instead an endogenic stimulus for the initial ultramafic flows, but thereafter they invoke a similar sequence of accumulation, metamorphism and erosion to expose a cross-section of the intensely folded syncline.[68]

Since the greenstones were formed in a volcanic island setting, one non-impact theory proposed that they accumulated in early 'marginal basins'. If a slab of dense lithosphere slipped beneath a block of buoyant lithosphere, the partial melting above the subducting slab would have produced a line of volcanic islands. Thinning in the back-arc zone would have induced extrusions, which in turn would have induced isostatic settling. If the buoyant block was above sea level, the basin would have accumulated both continental run off and volcanogenic sediments from the islands. Once this sedimentation broached the surface, the basin would have been integrated into the continental mass. Later, it would have been domed and deformed by diapiric granites, and eventually eroded.[69] The Pilbara unconformities indicate that there was *some* buoyant crust in the early Archean. Seismic evidence has established that there is a subducted slab beneath a 2.7-billion-year-old suture in Superior Province.[70] There is evidence that subduction of one piece of mafic lithosphere by another was prevalent, so back-arc spreading in the late Archean is plausible.[71]

The *scale* of the early granitoid intrusions was comparable to the batholiths which intrude mountain ranges behind subduction zones today. Although the crust was much thinner in the Archean, the pressures and temperatures required to induce

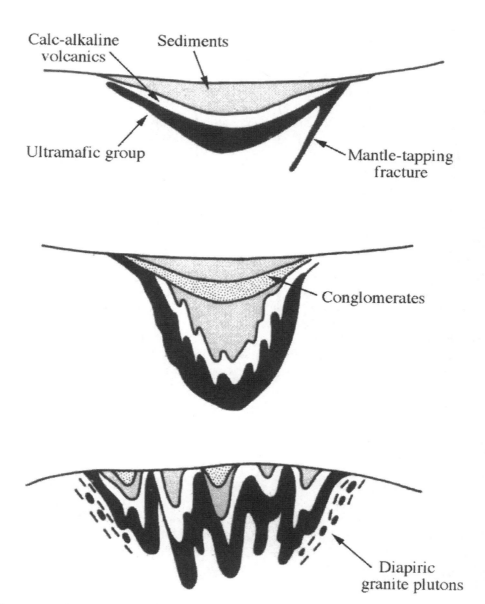

One model of how an Archean greenstone-granite structure might have formed. First, a deep fracture in the crust enabled ultramafic magma to rise from the mantle and 'flood' out onto the surface. Calc-alkaline volcanics followed. The presence of this layer of dense rock stimulated isostatic settling. Sediments interlayered with conglomerate filled in the resulting basin. As the basin sank the crust thickened and the deeply buried original crust was metamorphosed. Rising granitic plutons then deformed the basin, and subsequent erosion exposed the structure with its 'belt' character. (From *The Evolving Continents*, B.F. Windley, 1984. Copyright John Wiley & Sons Ltd. Reproduced with permission.)

Archean processes 221

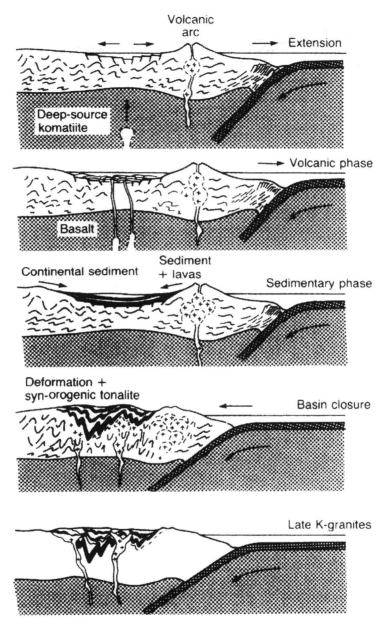

A model of how an Archean greenstone-granite structure might have formed in the context of a subduction zone, using the Rochas Verdes complex in south Chile as a model. (Based on 'Marginal basin Rochas Verdes Complex from south Chile: a model for Archean greenstone belt formation', J. Tarney, I. Dalziel and M.J. de Wit. In *The Early History of the Earth*, B.F. Windley (Ed.) 1976. Copyright John Wiley & Sons Ltd. Reproduced with permission.)

high-grade metamorphism suggest that the basins in which the greenstone sequences accumulated warped the crust so badly that its keel projected to a depth comparable to the base of today's continents.[72] If the metamorphism of the tonalite into granulite did not occur at such a great depth, it must have occurred at a shallower depth in a physical environment with an unusually high pressure. The distinctive thermal and pressure characteristics associated with subduction induce different degrees of metamorphism. In the compressional zone on the 'front' of the overriding plate, the cold dense plate imparts a low thermal gradient of 10 °C per kilometre. This high-pressure but low-temperature zone runs to a depth of about 30 kilometres and transforms trapped sea floor sediments into blueschist. However, the 'dewatering' of the descending plate as it dives below this depth induces partial melting in the rock above and creates a very high geothermal gradient of 25 to 50 °C per kilometre, and the high-grade metamorphism in this high-temperature and low-pressure environment transforms the rocks above into greenschist, amphibolite and granulite.

When the Japanese island arc was being geologically mapped, it was found that these two very different environments had produced *paired* metamorphic belts spaced 150 kilometres or so apart.[73,74] Similar paired belts were later recognised behind Andean-style subduction zones. In fact, the vast tonalite batholiths in the Andean-style plate margins are geochemically similar to the Archean gneisses.[75] Despite erosion, this tell-tale juxtaposition of rocks types can identify sites of subduction in ancient times. If several small buoyant blocks were separated by dense crust and subduction was active in the Archean, a succession of collisions may have created a fairly large proto-continental block comprising alternating greenstone belts

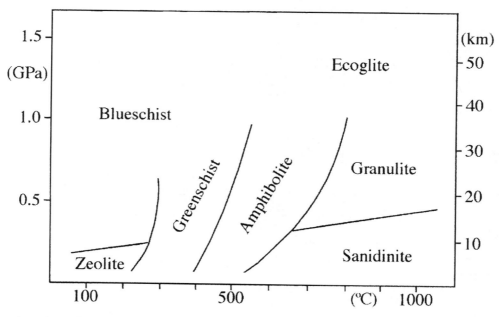

A variety of metamorphic rocks can be produced within the Earth's crust, depending upon the extant temperatures and pressures.

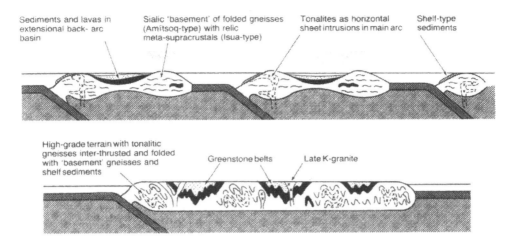

A model of how cratons may have accreted to form protocontinents in the Archean by subducting the dense crust. Widespread small continental plates with back-arc basins formed and the Andean-type margins were invaded by batholiths of tonalites. Later aggregation produced an extensive continental plate consisting of greenstone belts and granulite gneiss belts. (From *The Evolving Continents*, B.F. Windley, 1984. Copyright John Wiley & Sons Ltd. Reproduced with permission.)

and gneisses with folds of sediment sandwiched in between.[76] The Rocas Verdes complex of Patagonia in southern Chile may well be a modern analogue for how greenstone belts were formed.[77] Its ophiolites have been interpreted as evidence of a marginal basin in the Cretaceous.[78] The particular geometry of this back-arc region led to oceanic-style spreading in the south, and to intrusion into the continental crust in the north where the basin narrowed and the activity is 'intermediate' between oceanic and continental.

The processes that created the greenstone belts are evidently *not* at work today, however. 'Sedimentary basins' may be the best modern analogue, but these later basins were not filled by the ultramafic lavas that characterise the ancient greenstone belts, they simply accumulated erosional products. In 1859, James Hall, the State Geologist of New York, was mapping the structure of the northern Appalachian Range when he realised that its folds recorded a stack of shallow-water Proterozoic sandstones, shales and limestones 12 kilometres thick. What was remarkable was that each layer was ten to twenty times as thick as the equivalent unfolded strata in the interior lowlands to the west. The bedrock must have subsided as the marine sediments accumulated. Evidently, a long period of downwarping preceded the formation of the mountain range. This discovery gave rise to the concept of a geosyncline. At first, it was thought that the basement must have settled isostatically in response to the developing overburden of sediment, but J.D. Dana, one of Hall's contemparies, proved that lightweight sediments could have induced only a fraction of the subsidence required to account for the Appalachian sequence, and the shallow sea would eventually have silted up. Some other process had caused the basement to subside; the sedimentation was a consequence rather than the cause. The thickest coal seams formed in the Carboniferous by a similar process. Even with the insight of plate tectonics, the formation of intraplate basins remains a matter for debate.

The distribution of continents influences sea level. A mid-ocean ridge displaces an enormous volume of water. In fact, today's ridge system contributes 300 metres to the current sea level and the extent of the ridge has a major effect on low-lying continents. As a supercontinent forms, the sea level falls as ocean floor is pinched out, but when a supercontinent breaks up and the fragments move apart, the spreading ridges that they leave in their wake increase the sea level and turn the low-lying areas of the stable platforms into shallow seas. Sedimentation in an inland sea stretching from the Gulf of Mexico to Canada produced the Great Plains, but the sedimentation resulting from these regional inundations is limited by the extent to which the sea rises. The vast thicknesses in the sedimentary basins must have accumulated in some other manner. One widely accepted model posited that this subsidence occurred in response to lithospheric stretching.[79] A study of the forces acting upon plates concluded that the continental lithosphere suffers regional tensile stresses in only two situations. One case is associated with continental rifting in response to a mantle plume impinging upon the plate. The second situation develops when a continental plate overrides an oceanic plate and the trench is *close* offshore. If a trench lies far offshore, the back-arc spreading will create a marginal basin, but if the trench is close offshore it can 'suck' upon the continental plate and induce tensile stress *beyond* the Andean-style mountain range.

In the case of the Appalachian geosyncline, the tectonic stresses that created the sedimentary basin occurred long before the mountain range was created (which was during the integration of the Pangean supercontinent). In fact, many of the sedimentary basins of North America, Europe, Africa and South America were formed during a 50-million-year period in the early Cambrian. The fact that they have similar sedimentation histories and dates of regional unconformities due to erosion suggests that they formed in close proximity, as the late Precambrian supercontinent broke up,[80] which suggests a formation mechanism involving continental rifting. Many of these sedimentary basins were subsequently folded into mountain ranges during the creation of Pangea.

There are older sedimentary basins: starting 1.6 billion years ago and continuing for 800 million years much of what is now northern Idaho and western Montana was a basin in which a 20-kilometre-thick sequence of marine sediment accumulated.[81] Because it was first studied in the Belt Mountains in Montana, it is known as the Belt Basin. Puzzlingly, there is evidence that much of the sediment came from the west, even though there are no mountains of that age to the west that could have been eroded. In fact, the Belt rocks terminate abruptly close to Idaho's western border. To the north, they swing into northeastern Washington, and continue on into British Columbia in Canada. Evidently, 800 million years ago the continent was split by a rift that transected the basin and left the continental margin near Idaho's western border. The location of the other part of the continent is not known. Half a billion years later, as the North Atlantic opened and began the western movement of North America, a trench formed off Idaho and what remained of the sediment basin was folded to form the Rockies, which were then intruded by the Idaho batholith.

Archean processes 225

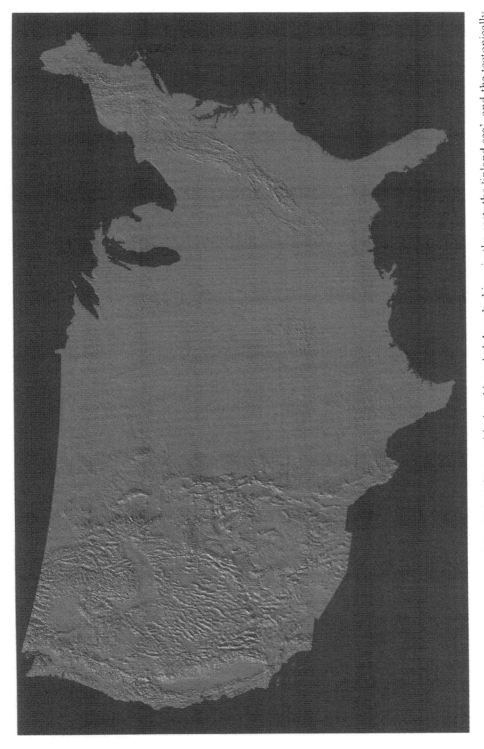

A topographic representation of the United States with the old eroded Appalachians in the east, the 'inland sea', and the tectonically active western margin. (Courtesy of USGS.)

Observe that even a sedimentary basin that formed in the late Proterozoic did not develop a greenstone character.

THE HADEAN

The Isuan supracrustals did not retain the record as the oldest surviving rocks on Earth for long.[82] In the mid-1980s, zircons from a gneiss retrieved from Mount Sones in Enderby Land, Antarctica, indicated that a granitoid (subsequently metamorphosed into a gneiss) crystallised 3.93 billion years ago.[83] Most of Slave Province on the northwestern extremity of the Canadian shield is late Archean supracrustals intruded by gneisses, but the gneisses in the small Acasta exposure are much older. In the late 1980s, zircons revealed that the progenitor granodiorites of these gneisses crystallised 3.96 billion years ago,[84] and a decade later, zircons in metamorphosed tonalites nearby were dated at slightly over 4 billion years old;[85] but even then, the Earth was already half a billion years old.

There is no evidence of Hadean rock in its massif form, but inferences as to its nature can be drawn from the clasts in the immature sediment in the oldest greenstones. So, was the earliest crust a scum on a magma ocean by flotation of chemically fractionated lightweight minerals, as occurred on the Moon? Or was the crust a direct expression of the ultramafic mantle, as the ocean floor is today?

There is a sequence of metamorphosed sedimentary rock surrounded by 'banded' gneiss near Mount Narryer on the northwestern section of the Yilgarn craton.[86,87] Zircons showed that this gneiss is 3.69 billion years old.[88] The sediments include quartzite (that is, metamorphosed sandstone) composed of weathered mineral grains. In the early 1980s the Australian National University set out to study the Hadean by way of zircons extracted from this ancient quartzite. Most of the zircons were found to be 3.7 to 3.5 billion years old but several were 4.15 billion years old.[89] One zircon in a conglomerate in the Jack Hills metasedimentary belt 60 kilometres northeast of Mount Narryer was later dated at 4.28 billion years old.[90]

Although the composition of the early crust was a mystery, there was an indication that it formed very early on – within 100 million years of the planet accreting, in fact.[91,92] In 1993 Harvard geochemists Stein Jacobsen and Charles Harper analysed isotopic ratios in the rocks retrieved from the Greenland craton. In effect, they were able to measure the degree to which the mantle had thermally differentiated when the rocks crystallised 3.8 billion years ago. The results indicated that the mantle had already undergone significant partial melting. This meant that the oldest known rocks were not the first to 'emerge' from the mantle. By extrapolating the isotopic trend back in time, Jacobsen and Harper found that the mantle and the crust started to separate very early. The Earth was battered for half a billion years by the bombardment, so it had been expected that its surface would have been a magma ocean for much of this time. If a crust did form in only 100 million years, then it would have been brecciated to a considerable depth by the ongoing bombardment.[93]

In 2001 Simon Wilde of Curtin University's School of Applied Geology, based in

Perth, Western Australia, announced finding a 4.4-billion-year-old zircon in the Yilgarn's Jack Hills metasediments.[94] Zircon is an accessory in felsic rock, so the earliest crust evidently included granite. The presence of the zircons in sediments meant that the host had been eroded. A wet environment could be inferred from the fact that exposure to water increases a rock's ratio of the uncommon isotope oxygen-18 to the more common isotope oxygen-16. This can occur in one of three ways: when water chemically exchanges with minerals in rock; when crystals precipitate from solution in ground water; or when mineral veins are formed underground in hydrothermal systems. Evidently, the zircons worked free as this granite was eroded and were flushed by water into sandy stream sediment that was consolidated 3 billion years ago.[95] A few zircons grains would seem to be our only clue to the Earth's primitive crust.[96]

It had been evident from the sedimentary origin of the ancient cratonic rocks that there was surface water at the start of the Archean, but this new discovery meant that the Earth cooled so rapidly that it was able to support surface water within 100 million years[97]; it wasn't a magma ocean after all.

THE ONSET OF PLATE TECTONICS

The crucial physical factor determining the Earth's evolution is its heat flow. For the first few hundred million years the accretional energy would have created a heat flow an order of magnitude greater than the current value. Short half-life radioactive isotopes would have been decaying profusely early on. At the close of the bombardment, 3.8 billion years ago, the heat flow would have been something like 2.5 times its current value.[98] Today's heat flow derives from the heat that is leaking from the core, and long-lived radioisotopes.[99] Mantle convection allows a planet to regulate its internal temperature. The creation and subduction of mafic crust is a very efficient way of shedding that heat. Despite the high initial heat flow, the temperature of the lower crust could not have exceeded 800 °C, otherwise it would have prompted extensive melting.[100,101] Nevertheless, this early lithosphere would have been thin and pliable. It is likely that mafic crust was being created at a rate that was an order of magnitude greater than that at even the most active of today's spreading ridges.[102] In 1994, Francis Albarede of the Ecole Normale Supérieure of Lyon in France and Janne Blichert-Toft of Copenhagen's Geological Museum undertook an isotopic study of several hundred samples of late-Archean basalts, gabbros and komatiites collected from all around the world, and found that they were all derived from the same primary magma, indicating that the convection was still sufficiently vigorous to prevent 'pockets' from forming and evolving local characteristic chemistries.[103] In fact, they calculated that for the mantle to maintain such a uniform composition it must have been churning at a rate ten times faster than it does today.[104,105] In the Archean the crustal blocks would have been smaller than those participating in today's global plate tectonics. A finely detailed computer model of the mantle, developed by Hans-Peter Bunge and John Baumgardner of the Los Alamos National Laboratory in New Mexico and Mark Richards of the

University of California at Berkeley, showed in 1996 that when the mantle was uniformly viscous and was capped by only a thin crust the convection cells, although vigorous, would have been narrow. But as the heat flow diminished and the lower mantle thickened these cells would have broadened out and rapidly achieved a scale comparable to that today.[106,107] Early in the Earth's history, therefore, the large number of hot spots would have burned right through the thin crust to make a world dominated by volcanic islands. There would have been rapid recycling of the mafic crust, but this would have taken place on a local scale, with the crust firstly being domed by a rising mantle plume and then 'spread' radially with the crust moving downslope towards a circular subduction zone (as may be happening to the coronae on Venus today). Crust above the descending part of a convection cell would have been downwarped and literally sucked down. However, it is unlikely that *large* lithospheric plates could have formed until the cratons were 'underplated' by a mafic basement as the uppermost mantle rigidised. Some people consider that a granitic crust formed early and was extensively recycled through the mantle.[108,109] Others argue that it was accumulated progressively in a monotonic fashion.[110,111]

In 1995 Andrew Calvert of Montreal's Ecole Polytechnique studied seismic data from two adjacent units of Superior Province on the Canadian shield, and found evidence that a piece of ocean floor had been subducted 2.7 billion years ago when this cratonic suture was formed.[112] Horizontal crustal displacements were therefore taking place as early as the transition between the Archean and the Proterozoic, and may have been active as long as a billion years earlier. Where the Archean cratons are now clustered together (such as in southern Africa) they are separated by the relics of ancient intensely folded mountain belts. Interestingly, just as we find sections of ocean floor trapped in continental margins today in the form of ophiolites, there are isolated fragments of Archean greenstone belts embedded in these Proterozoic mountain belts (such as on the Sand River in the Limpopo Fold Belt). A study published in 2001 reported the discovery in the North China craton of a well-preserved section of a 2.5-billion-year-old ophiolite sequence, showing both that ocean floor was being formed at a spreading ridge and that the process of cratonic accretion was taking place along a subduction zone.[113]

The Archean cratons were 'swept up' during the early Proterozoic to form large continents.[114] As the scale of the buoyant masses grew, so too did the scale of the slabs undergoing recycling. Eventually, the spreading ridges and subduction trenches linked up to adopt their current global character.

NOTES

1 *Growth of a prehistoric timescale*, W.B.N. Berry. Freeman Co., 1968.
2 'Age measurements on rocks from the Finnish Precambrian', G.W. Wetherill, O. Kuovo, G.R. Tilton and P.W. Gast. *J. Geology*, vol. 70, p. 74, 1962.
3 A continental shield is not to be confused with a volcanic shield.
4 'Archean cratons', J. A. Percival. In *Searching for diamonds in Canada*, D.G. Richardson,

R.N.W. DiLabio and K.A. Richardson (Eds). Geol. Surv. Canada, Open File 3228, p. 161, 1996.
5. 'Age relationships between greenstone belt and 'granites' in the Rhodesian Archean craton', C.J.S. Hawkesworth, S. Moorbath, R.K. O'Nions and J.F. Wilson. *Earth Planet. Sci. Lett.*, vol. 25, p. 251, 1975.
6. '3500 myr old granite in southern Africa', M.H. Hickman. *Nature*, vol. 251, p. 295, 1974.
7. 'Early Archean ages for the Sebakwian Group at Selukwe, Rhodesia', S. Moorbath, J.F. Wilson and P. Cotterill. *Nature*, vol. 264, p. 536, 1976.
8. *Crustal evolution of Southern Africa: 3.8 billion years of Earth History*, A.J. Tankard, M.P.A. Jackson, K.A. Ericksson, D.K. Hobday, D.R. Hunter and W.E.L. Minster. Springer-Verlag, 1982.
9. 'Rb–Sr and U–Th–Pb isotopic studies of the Sand River gneisses, central zone, Limpopo mobile belt', J.M. Barton, B. Ryan and R.E.P. Fripp. In *The Limpopo Belt*, W.J. van Biljon and J.E. Legg (Eds). Geol. Soc. S. Africa Special Pub. no.8, p. 9, 1983.
10. 'The pattern of Archean crustal evolution in southern Africa as deduced from the evolution of the Limpopo mobile belt and the Barberton granite-greenstone terrain', J.M. Barton. In *Archean geology*, J.E Glover and D.I. Groves (Eds). Geol. Soc. Australia Special Pub. no.7, p. 21, 1981.
11. 'The relationship between Rb–Sr and U–Th–Pb whole-rock and zircon systems in the >3790 m.y. old Sand River gneisses, Limpopo mobile belt, southern Africa', J.M. Barton, B. Ryan and R.E.P. Fripp. In *Short Papers of the Fourth International Conference on Geochronology, Cosmochronology, Isotope Geology*, R.E. Zartman (Ed.). USGS Open File Report 78–701, p. 27, 1978.
12. 'Evidence for the extreme age of certain minerals from the Dominion Reef conglomerates and the underlying granite in the western Transvaal', L.O. Nicolaysen. *Geochimica Cosmochimica Acta*, vol. 26, p. 15, 1962.
13. 'Rb–Sr age measurements on various Swaziland granites', H.L. Allsopp, H.R. Roberts and G.D.L. Schreiner. *J. Geophys. Res.*, vol. 67, p. 5307, 1962.
14. *The geology and geochemistry of the lower ultramafic unit of the Onverwacht Group and a proposed new class of igneous rock [komatiite]*, M.J. Viljoen and R.P. Viljoen. Geol. Soc. S. Africa Special Pub. no.2, p. 55, 1969.
15. The greenstone sequence of the Barberton Mountainland and Swaziland is called the Swaziland Supergroup.
16. *An introduction to the geology of the Barberton granite-greenstone terrain*, M.J. Viljoen and R.P. Viljoen. Geol. Soc. S. Africa Special Pub. no.2, p. 9, 1969.
17. 'The granitic-gneiss greenstone shield', C.R. Anhaeusser and J.F. Wilson. In *Precambrian of the Southern Hemisphere*, D.R. Hunter (Ed.). Elsevier, p. 423, 1981.
18. 'Notes on the provisional geological map of the Barberton Greenstone Belt and surrounding granitic terrane, eastern Transvaal and Swaziland', C.R. Anhaeusser, L.J. Robb and M.J. Viljoen. In *Contributions to the geology of the Barberton Mountainland*, C.R. Anhaeusser (Ed.). Geol. Soc. S. Africa Special Publication no.9, p. 221, 1983.
19. 'Ancient age of the Middle Marker horizon, Onverwacht Group, Swaziland sequence, South Africa', P.M. Hurley, W.H. Pinson, B. Nagy and T.M. Teska. *Earth Planet. Sci. Lett.*, vol. 14, p. 360, 1972.
20. 'The ancient gneiss complex in Swaziland', D.R. Hunter. *Trans. Geol. Soc. S. Africa*, vol. 73, p. 107, 1970.
21. 'Geochronological and Sr-isotopic studies of certain units in the Barberton granite-greenstone terrane, South Africa', J.M. Barton, L.J. Robb, C.R. Anhaeusser and D.A.

van Nierop. In *Contribution to the geology of the Barberton Mountainland*, C.R. Anhaeusser (Ed.). Geol. Soc. S. Africa Special Pub. no.9, p. 63, 1983.
22 Was the Earth boiling off its sodium?
23 'Crustal development in the Kaapvaal craton', D.R. Hunter. *Precambrian Res.*, vol. 1, p. 259, 1974.
24 'The geologic evolution of the primitive Earth: evidence from the Barberton Mountainland', C.R. Anhaeusser. In *Evolution of the Earth's crust*, D.H. Tarling. Academic Press, p. 71, 1978.
25 'The pattern of Archean crustal evolution in southern Africa as deduced from the evolution of the Limpopo mobile belt and the Barberton granite-greenstone terrain', J.M. Barton. In *Archean geology*, J.E Glover and D.I. Groves (Eds). Geol. Soc. Australia Special Pub. no.7, p. 21, 1981.
26 'The Archean craton of the North Atlantic region', D. Bridgewater, J. Watson and B.F. Windley. *Phil. Trans. Roy. Soc. Series A*, vol. 273, p. 493, 1973.
27 'Mineral age patterns in ca. 3700 m.y. old rocks from West Greenland', R.J. Pankhurst, S. Moorbath, D.C. Rex and G. Turner. *Earth Planet. Sci. Lett.*, vol. 20, p. 157, 1973.
28 'The early Precambrian gneisses of the Godthaab district, West Greenland', V.R. McGregor. *Phil. Trans. Roy. Soc. Series A*, vol. 273, p. 343, 1973.
29 'Further Rb–Sr age determinations on the very early Precambrian rocks of the Godthaab district', S. Moorbath, R.K. O'Nions, R.J. Pankhurst, N.H. Gale and V.R. McGregor. *Nature Phys. Sci.*, vol. 240, p. 78, 1972.
30 'Rb–Sr ages of early Archean supracrustal rocks and Amitsoq gneisses at Isua', S. Moorbath, J.H. Allaart, D. Bridgewater and V.R. McGregor. *Nature*, vol. 270, p. 43, 1977.
31 'Petrogenesis and geochemistry of metabasaltic and meta sedimentary enclaves in the Amitsoq gneisses', V.R. McGregor and B. Mason. *Am. Mineral.*, vol. 62, p. 887, 1977.
32 *Field character of the early Precambrian rocks from Saglek*, D. Bridgewater, K.D. Collerson, R.W. Hurst and C.W. Jesseau. Geol. Surv. Canada Paper no.75-1a, p. 287, 1975.
33 'The major petrological and geochemical character of the 3600 m.y. Uivak gneisses from Labrador', D. Bridgewater and K.D. Collerson. *Contrib. Mineralogy Petrology*, vol. 54, p. 43, 1976.
34 'Crustal development of the Archean gneiss complex of eastern Labrador', K.C. Collerson, C.W. Jesseau and D. Bridgewater. In *The early history of the Earth*, B.F. Windley (Ed.), p. 237, Wiley 1976.
35 'Geochronology and evolution of the late Archean gneisses in northern Labrador: an example of reworked sialic crust', K.D. Collerson, A. Kerr and W. Compston. In *Archean geology*, J.E. Glover and D.I. Groves (Eds). Geol. Soc. Australia Special Pub. no.7, p. 205, 1981.
36 'Early Archean age for the Isua iron formation', S. Moorbath. R.K. O'Nions and R.J. Pankhurst. *Nature*, vol. 245, p. 138, 1973.
37 'The evolution of early Precambrian crustal rocks at Isua: geochemical and isotopic evidence', S. Moorbath, R.K. O'Nions and R.J. Pankhurst. *Earth Planet. Sci. Lett.*, vol. 27, p. 229, 1975.
38 'The pre-3760 m.y. old supracrustal rocks of the Isua area, and the associated occurrence of quartz-banded ironstone', J.H Allaart. In *The early history of the Earth*, B.F. Windley (Ed.). Wiley, p. 177, 1976.
39 'Late event in the geological evolution of the Godthaab district', R.J. Pankhurst, S. Moorbath and V.R. McGregor. *Nature Phys. Sci.*, vol. 243, p. 24, 1973.

40 'Further Rb–Sr age and isotopic evidence for the nature of the late Archean plutonic event in West Greenland', S. Moorbath and R.J. Pankhurst. *Nature*, vol. 262, p. 124, 1976.
41 'Early (3700 Ma) Archean rocks of the Isua supracrustal belt and adjacent gneisses', A.P. Nutman, D. Bridgewater, E. Dimroth, R.C.O. Gill and M. Rosing. Geol. Surv. Greenland Report no.12, p. 5, 1983,
42 'Stratigraphic and geochemical evidence for the depositional environment of the early Archean Isua supracrustal belt, West Greenland', A.P. Nutman, J.H. Allaart, D. Bridgewater, E. Dimroth and M. Rosing. *Precambrian Research*, vol. 25, p. 365, 1984.
43 'The zircon geochronology of the Akilia association and Isua supracrustal belt', H. Baadsgaard, A.P. Nutman, D. Bridgewater, M. Rosing, V.R. McGregor and J.H. Allaart. *Earth Planet. Sci. Lett.*, vol. 68, p. 221, 1984.
44 'Progress report on early Archean rocks in Liberia, Sierra Leone and Guayana, and their general stratigraphic setting', P.M. Hurley, H.W. Fairbairn and H.E. Gaudette. In *The early history of the Earth*, B.F. Windley (Ed.). Wiley, p. 511, 1976.
45 'U–Pb geochronology of the Archean Imataca Series, Venezuela Guayana Shield' C.W. Montgomery. *Contrib. Mineralogy Petrology*, vol. 69, p. 167, 1979.
46 'Archean rocks in the southern part of the Canadian Shield: a review', P.K. Sims and Z.E. Peterman. In *Archean geology*, J.E. Glover and D.I. Groves (Eds). Geol. Soc. Australia Special Pub. no.7, p. 85, 1981.
47 *Archean greenstone belts*, K.C. Condie. Elsevier, 1981.
48 'Boundary between Archean greenstone and gneiss terranes in northern Wisconsin and Michigan', P.K. Sims. In *Selected studies of Archean gneisses and lower Proterozoic rocks, southern Canadian Shield*, G.B. Morey and G.N. Hanson (Eds). Geol. Soc. Am. Special Paper no.182, p. 113, 1980.
49 'Origin of the Morton gneiss, southwestern Minnesota: geochronology', S.S. Goldich and J.L. Wooden. In *Selected studies of Archean gneisses and lower Proterozoic rocks, southern Canadian Shield*, G.B. Morey and G.N. Hanson (Eds). Geol. Soc. Am. Special Paper no.182, p. 77, 1980.
50 *Insight into the geological history of ancient gneisses in USA from microanalysis of zircon*, P.D. Kinney, I.S. Williams and W. Compston. Ann. Rep. Australian National University, p. 98, 1984.
51 'Archean rocks of the Granite Falls area, southwestern Minnesota', S.S. Goldich, C.E. Hedge, T.W. Stern, J.L. Wooden, J.B. Bedkin and R.M. North. In *Selected studies of Archean gneisses and lower Proterozoic rocks, southern Canadian Shield*, G.B. Morey and G.N. Hanson (Eds). Geol. Soc. Am. Special Paper no.182, p. 19, 1980.
52 'Early Archean Sm–Nd model agents for tonalitic gneiss, [Watersmeet] northern Michigan', M.T. McCulloch and G.J. Wasserburg. In *Selected studies of Archean gneisses and lower Proterozoic rocks, southern Canadian Shield*, G.B. Morey and G.N. Hanson (Eds). Geol. Soc. Am. Special Paper no.182, p. 135, 1980.
53 'Structural framework of the Australian Precambrian', R.W.R. Rutland. In *Precambrian of the Southern Hemisphere*, D.R. Hunter. Elsevier, p. 1, 1981.
54 *Definitions of new and revised stratigraphic units of the eastern Pilbara region*, S.L. Lipple. Ann. Rep. Geol. Surv. Western Australia: 1974, p. 58, 1975.
55 *Precambrian structural geology of part of the Pilbara region*, A.H. Hickman. Ann. Rep. Geol. Surv. Western Australia: 1974, p. 68, 1975.
56 'Crustal evolution of the Pilbara block, Western Australia', A.H. Hickman. In *Archean geology*, J.E. Glover and D.I. Groves (Eds). Geol. Soc. Australia Special Pub. no.7, p. 57, 1981.

57 'Horizontal tectonic interaction of an Archean gneiss belt and greenstones, Pilbara Block, Western Australia', M.J. Bickle, L.F. Bettenay, C.A. Boulter, D.I. Groves and P. Morant. *Geology*, vol. 8, p. 525, 1980.

58 *Archean tectonics of the Shaw Batholith, Pilbara Block, Western Australia: structural and metamorphic tests of the batholith concept*, M.J. Bickle, P. Morant, L.F. Bettenay, C.A. Boulter, T.S. Blake and D.I. Groves. Geol. Assoc. Canada Special Paper no.28, p. 325, 1985.

59 *Zircon U–Pb ages from the Shaw Batholith, Pilbara Block, determines by ion microprobe*, I.S. Williams, R.W. Page, J.J. Foster, W. Compston, K.D. Collerson and M.T. McCulloch. The Australian Nat. Univ. Res. School of Earth Sci. Ann. Rep. for 1982, p. 199, 1983.

60 'Geochronological investigation of granite batholiths of the Archean granite-greenstone terrain of the Pilbara Block, Western Australia', R.T. Pidgeon. *Proc. Archean Geochemistry Conf. U. of Toronto*, p. 360, 1978.

61 'Sm-Nd dating of the North Star Basalt, Warrawoona Group, Pilbara Block, Western Australia', P.J. Hamilton, N.M. Evensen, R.K. O'Nions, A.Y. Glikson and A.H. Hickman. In *Archean geology*, J.E. Glover and D.I. Groves (Eds). Geol. Soc. Australia Special Pub. no.7, p. 187, 1981.

62 '3450 m.y. old volcanics in the Archean layered greenstone succession of the Pilbara Block, Western Australia', R.T. Pidgeon. *Earth Planet. Sci. Lett.*, vol. 37, p. 421, 1978.

63 'The First Landscape', B. Wright. *Earth Mag.*, p. 16, December 1995.

64 'Record of emergent continental crust 3.5 billion years ago in the Pilbara craton of Australia', R. Buick, J.R. Thornett, N.J. McNaughton, J.B. Smith, M.E. Barley and M. Savage. *Nature*, vol. 375, p. 574, 1995.

65 *Atlas of north Pilbara: geology and geophysics*, R.S. Blewett, P. Wellman, M. Ratajkoski and D.L. Huston. Australian Geol. Surv. Org. AGSO 2000/4, p. 5, 2000.

66 *Archean greenstone belts*, L.C. Condie. Elsevier, 1981.

67 'Archean greenstone belts may include terrestrial equivalents of lunar maria?', D.H. Green. *Earth Planet. Sci. Lett.*, vol. 15, p. 263, 1972.

68 'Precambrian tectonic style: a liberal uniformitarian interpretation', R.B. Hargraves. In *Precambrian plate tectonics*, A. Kroner (Ed.). Elsevier, p. 21, 1981.

69 'Marginal basin Rocas Verdes complex from south Chile: a model for Archean greenstone belt formation', J. Tarney, I. Dalziel and M.J. de Wit. In *The early history of the Earth*, B.F. Windley (Ed.), Wiley, p. 131, 1976.

70 'Archean subduction inferred from seismic images of a mantle suture in the Superior Province', A. Calvert, E.W. Sawyer, W.J. Davis and J.N. Ludden. *Nature*, vol. 375, p. 670, 1995.

71 'Marginal basins throughout geological time', J. Tarney and B.F. Windley. *Phil. Trans. Roy. Soc. Series A*, vol. 301, p. 207, 1981.

72 'Aspects of the chronology of ancient rocks related to continental evolution', S. Moorbath. In *The continental crust and its mineral deposits*, D.W. Strangway (Ed.), Geol. Assoc. Canada, p. 89, 1980.

73 'Metamorphism and related magmatism in plate tectonics', A. Miyashiro. *Am. J. Sci.*, vol. 272, p. 629, 1972.

74 'Paired and unpaired metamorphic belts', A. Miyashiro. *Tectonophysics*, vol. 17, p. 241, 1973.

75 'Precambrian rocks in the light of the plate tectonic concept', B.F. Windley. In *Precambrian plate tectonics*, A. Kroner (Ed.). Elsevier, p. 1, 1981.

76 *The evolving continents*, B.F. Windley, Wiley Ltd (second edition), 1984.

77 'Marginal basin Rocas Verdes complex from south Chile: a model for Archean greenstone belt formation', J. Tarney, I. Dalziel and M.J. de Wit. In *The early history of the Earth*, B.F. Windley (Ed.), Wiley, p. 131, 1976.
78 'Variations in the degree of crustal extension during formation of a back-arc basin', M.J. de Wit and C.R. Stern. *Tectonophysics*, vol. 72, p. 229, 1981.
79 'Some remarks on the development of sedimentary basins', D.P. McKenzie. *Earth Planet. Sci. Lett.*, vol. 40, p. 25, 1978.
80 'Origin of cratonic basins', G. de V. Klein and A.T. Hsui. *Geology*, vol. 15, p. 1094, 1987.
81 *Northwest exposures: a geologic story if the Northwest*, D. Alt and D.W. Hyndman. Mountain Press, p. 13, 1995.
82 *The Archean: search for the beginning*, G.J.H. McCall (Ed.). Dowden, Hutchinson & Ross Co., 1977.
83 'Four zircon ages from one rock: the history of a 3930 Ma-old granulite from Mount Sones, Enderby Island, Antactica', L.P. Black, I.S. Williams and W. Compston. *Contrib. Mineralogy Petrology*, vol. 94, p. 427, 1986.
84 '3.96 Ga gneisses from the Slave Province, NW Territories, Canada', S.A. Bowring, I.S. Williams and W. Compston. *Geology*, vol. 17, p. 971, 1989.
85 'Priscoan (4.00–4.03) orthogneisses from [Acasta] northwestern Canada,' S.A. Bowring and I.S. Williams. *Contrib. Mineralolgy Petrology*, vol. 134, p. 3, 1999.
86 'Growth of Archean crust within the western gneiss terrain, Yilgarn block, Western Australia', M.T. McCulloch, K.D. Collerson and W. Compston. *J. Geol. Soc. Australia*, vol. 30, p. 155, 1983.
87 'The age and metamorphic history of the Mount Narryer banded gneiss', W. Compston, I.S. Williams, D.O. Froude and J.J. Foster. *The Australian National University Research School of Earth Sciences Ann. Rep. for 1982*, p. 196, 1983.
88 'Early Archean gneisses from the Yilgarn Block, Western Australia', J.R. de Laeter, I.R. Fletcher, K.J.R. Rosman, I.R. Williams, R.D. Gee and W.G. Libby. *Nature*, vol. 292, p. 322, 1981.
89 'Ion microprobe identification of 4,100–4,200 Myr-old terrestrial zircons', D.O. Froude, T.R. Ireland, P.D. Kinney, I.S. Williams and W. Compston. *Nature*, vol. 304, p. 616, 1983.
90 'Jack Hills, evidence of more very old detrital zircons in Western Australia', W. Compston and R.T. Pidgeon. *Nature*, vol. 321, p. 766, 1986.
91 'Evidence from coupled 147Sm–143Nd and 146Sm–142Nd systematics for very early (4.5-Gyr) differentiation of the Earth's mantle', C.L. Harper and S.B. Jacobsen. *Nature*, vol. 360, p. 728, 1992.
92 'Accretion and early differentiation of the Earth based on extinct radionuclides', S.B. Jacobsen and C.L. Harper. In AGU Geophysical Monograph no.95 *Earth Processes: Reading the Isotopic Code*, A. Basu and S. Hart (Eds), p. 47, 1996.
93 'Earth's violent birth', J. Alper. *Earth Mag.*, p. 56, December 1994.
94 'Evidence from detrital zircons for the existence of continental crust and oceans in Earth 4.4 Gyr ago', S.A. Wilde, J.W. Valley, W.H. Peck and C.M. Graham. *Nature*, vol. 409, p. 175, 2001.
95 'In the beginning', A.N. Halliday. *Nature*, vol. 409, p. 144, 2001.
96 'The earth's oldest known crust: a geochronological and geochemical study of 3900–4200 Ma old detrital zircons from Mt. Narryer and Jack Hills, Western Australia,' R. Maas, P.D. Kinny, I.S. Williams, D.O. Froude and W. Compston. *Geochimica Cosmochimica Acta*, vol. 56, p. 1281, 1992.
97 'Oxygen isotope evidence from ancient zircons for liquid water at the Earth's surface

4,300 Myr ago', S.J. Mojzsis, T.M. Harrison and R.T. Pidgeon. *Nature*, vol. 409, p. 178, 2001.
98 *Precambrian plate tectonics*, A. Kroner. Elsevier, 1981.
99 'The flow of heat from the Earth's interior', N.H. Pollack and D.S. Chapman. *Scient. Am.*, vol. 232, p. 60, 1963.
100 The temperature of the lower crust today is typically 500 °C.
101 'Precambrian rocks in the light of the plate tectonics concept', B.F. Windler. In *Precambrian plate tectonics*, A. Kroner (Ed.). Elsevier, p. 1, 1981.
102 'Heat loss from the Earth: a constraint on Archean tectonics from the relation between geothermal gradients and the rate of plate production', M.J. Bickle. *Earth Planet. Sci. Lett.*, vol. 40, p. 301, 1978.
103 See news item on, p. 16 of *Earth Mag.*, September 1994.
104 In this respect, the early Earth would have resembled Io today, which seems to have a vigorously convecting mantle of uniform composition capped by a thin mafic crust peppered with vast volcanic calderas.
105 For a comprehensive review of current understanding of the convection within the mantle, see 'Deep-mantle high-viscosity flow and thermochemical structure inferred from seismic and geodynamic data', A.M. Forte and J.X. Mitrovica. *Nature*, vol. 410, p. 1049, 2001.
106 'Matching tops and bottoms', I. Wickelgren. *Earth Mag.*, p. 22, June 1997.
107 'Imaging 3-D spherical convection models: what can seismic tomography tell us about mantle dynamics?', C. Megnin, H.P. Bunge, B. Romanowicz and M. Richards. *Geophys. Res. Lett.*, vol. 24, p. 1299, 1997.
108 'Punctuated evolution of tectonic style', R.B. Hargraves. *Nature*, vol. 276, p. 459, 1978.
109 'Crust formation and destruction', W.S. Fyfe. In *The continental crust and its mineral deposits*, D.W. Strangway (Ed.), Geol. Assoc. Canada, p. 89, 1980.
110 'Evolution of the precambrian crust from strontium isotope evidence', S. Moorbath. *Nature*, vol. 254, p. 395, 1975.
111 'Early precambrian tonalite-trondhjemite sialic nuclei', A.Y. Glikson. *Earth Science Rev.*, vol. 15, p. 1, 1979.
112 'Archean subduction inferred from seismic images of a mantle suture in the Superior Province', A. Calvert, E.W. Sawyer, W.J. Davis and J.N. Ludden. *Nature*, vol. 375, p. 670, 1995.
113 'The Archean Dongwanzi Ophiolite Complex, North China Craton: 2.505-billion-year-old oceanic crust and mantle' T.M. Kusky, J.-H. Li and R.D. Tucker. *Science*, vol. 292, p. 1142, 2001.
114 'The oldest rocks and the growth of continents', S. Moorbath. *Scient. Am.*, vol. 236, p. 92, 1977.

8

Giant Jupiter

MR GALILEO

In the realm of naked-eye astronomy, astronomers concentrated on tracking the motions of the planets across the sky, but on 7 January 1610 Galileo Galilei in Padua aimed his telescope at Jupiter and resolved the planet as a disk for the first time. He also saw small star-like points in a line close by the planet which changed their relative positions from night to night, leading him inevitably to the conclusion that these subsidiary objects were moving *around* Jupiter.[1,2,3]

This was contrary to the accepted view, in Claudius Ptolemy's *Almagest*, that everything in the sky moved around the Earth. Galileo favoured the heliocentric view proposed by Nicolaus Copernicus that only the Moon circled the Earth and that everything else circled the Sun,[4] but the Vatican had rejected this view. Although the fact that Jupiter had a system of satellites did not actually prove Copernicus's theory, it was strong evidence in favour of it, and by lending his support Galileo found himself in serious strife with the Church.[5,6]

It is possible that Galileo was not actually the first person to see what he referred to as "these marvellous things which have lain hidden for all ages past", as Simon Marius, his long-standing German rival, had received a Dutch-built telescope and may have seen them several weeks earlier.[7] However, Marius did not appreciate their nature until Galileo's announcement,[8] and Galileo is therefore credited as he was first to publish the observations.[9,10] Galileo called the moons the 'Medicean Stars' in honour of the influential local merchant Cosimo de Medici. Johann Kepler referred to them as 'the Galilean satellites' and in 1614 Marius awarded them the individual names (in order of increasing distance from the planet) of Io, Europa, Ganymede and Callisto, drawing the names from mythology.[11]

In 1675, Danish astronomer Olaus Roemer was pondering the reason for the inaccuracies in the tables that he had devised to predict the times that Jupiter's moons entered and emerged from the planet's shadow. In a moment of inspiration he realised that the times were accurate but the light did not reach us immediately.

Contrary to contemporary wisdom, light travelled at a finite speed.[12] The discrepancies in the observed timings were due to the changing distances between the two planets as they independently moved around the Sun. This absolute measure served to calibrate the relative scale provided by Kepler's laws of orbital motion.[13]

GAS GIANT

What of the planet itself? Jupiter's radius is about 10 times that of the Earth. When astronomers interpreted the periods of Jupiter's satellites employing the laws of gravity formulated by Isaac Newton in 1687, they found that although the planet is 143,000 kilometres in diameter, and is 318 times as massive as the Earth, its density of 1.33 g/cm^3 required it to be predominantly composed of hydrogen, making it a 'gas giant'. In fact, as more planets were identified farther from the Sun, it was realised that Jupiter is not only the largest planet, it is more massive than all the other planets combined.

It was also apparent that Jupiter's disk does not show a solid surface, but an atmospheric circulation system comprising latitudinal belts and zones, decorated by a multitude of 'spots', 'ovals', 'barges' and 'plumes', some of which are long-lived. The belts were first glimpsed by Francesco Fontana in Naples in 1630. Jupiter's axis of rotation is almost perpendicular to its orbital plane, so there is no 'seasonal' influence on its weather over its 12-year-long orbit. Nevertheless, the weather system is far from monotonous. As individual features in Jupiter's upper atmosphere were tracked moving across the disk it was realised that the planet rotates in only 10 hours. This makes its equator bulge, and the rotational rate on the equator is faster than that at higher latitudes (this is called differential rotation). Furthermore, each atmospheric feature has its own rate so they progressively drift both in their longitude and with respect to each other.[14]

What is a gas giant?

In 1860 G.P. Bond determined that Jupiter radiates twice as much energy into space as it receives from the Sun.[15] It must therefore be in the process of contracting and is transforming gravitational potential into heat. It was concluded from this that the planet's interior was a hot gas.[16] A decade later, R.A. Proctor posited, "Jupiter is still a glowing mass, fluid probably throughout, still bubbling and seething with the intensity of the primaeval fires, sending up continuous enormous masses of cloud, to be gathered into bands under the influence of the swift rotation of the giant planet." He was thinking of Jupiter as a 'failed' star. In the 1920s Harold Jeffreys proved that the visible surface could not be hot. He argued instead for a rocky core englobed by a mantle of ice and solid carbon dioxide, surrounded by a very deep but tenuous gaseous envelope.[17,18]

A key advance was the spectroscopic identification of ammonia and methane in the Jovian atmosphere, as well as the expected hydrogen and helium. Other compounds were minor constituents which, together, contributed only 1 per cent. In 1932 Rupert Wildt refined Jeffreys' idea by proposing that the rocky core was surrounded first by a thick layer of water ice and, in turn, an ocean of condensed gases.[19] The next real advance was the independent suggestion in 1951 by W.R. Ramsey in England and by W. DeMarcus in America that the core was not rock but metallic hydrogen, and that this was surrounded first by an ocean of liquid hydrogen and then by the hydrogen-rich gaseous envelope. Such a metallic core would readily conduct electrical currents, and these currents would in turn generate a magnetic field. The likelihood of Jupiter having a strong magnetic field was a prediction-in-waiting. In 1955, in America, B.F. Burke and K.L. Franklin found that Jupiter emitted radio waves, which was a chance discovery as they were not studying the planet; in tracking down a 'noise' they realised that the source came from the giant planet. These emissions suggested that Jupiter possessed a strong magnetic field. As observations of the radio emission accrued, it became evident that there was a periodicity correlated with Jupiter's axial rotation because the magnetic field in the core was dragging the entire magnetosphere around with it; this periodicity of 9 hours 55 minutes therefore represented the rate of rotation of the core. It was eventually realised that the 'burst' emissions correlated with the motion of the innermost Galilean moon, Io, which – in some way – was interacting with the Jovian magnetic field. The discovery in 1958 by Explorer 1, America's first satellite, that the Earth is circled by intense belts of charged particles prompted speculation that similar belts might encircle Jupiter. As the magnetic axis is inclined at 9.5 degrees to Jupiter's rotational axis, the moon, which orbits in the planet's equatorial plane, is swept by an alternating magnetic field and this stimulates bursts of radio emission.

JUPITER'S MOONS

Once the resonances in the orbital motions of the Galilean satellites were understood, most astronomers promptly lost interest in them, but in 1892 E.E. Barnard decided to use the newly commissioned 36-inch refractor at Lick Observatory near San José to find out whether Jupiter had any small moons, and within minutes of turning to the planet he found Amalthea circling well inside Io's orbit.[20] At that time, and for the first half of the 20th century, Jupiter's moons were not referred to by the names that Marius assigned, but by the numerical order of their distance from the planet. Although referring to Barnard's moon as 'Satellite V' ran against this scheme, everyone knew what was meant. However, as further satellites were discovered orbiting at a variety of distances, this scheme became impracticable. The International Astronomical Union dropped the numerical scheme and named each object.

Although to Galileo the moons of Jupiter were mere star-like points, subsequent telescopic observers were able to resolve them as disks. Both William Herschel and J.H. Schroter noted that the disks showed albedo variations. Furthermore, Herschel pointed out that as the moons moved around their orbits they showed periodic variations in brightness. This was confirmed by J.H. Madler. In the early years of the 20th century, P. Guthnick accumulated photometric observations which established that the rotations of the satellites were synchronised with their orbital periods. Hence, just as in the case of the Moon whose rotation has been 'captured', the Jovian

Giant Jupiter

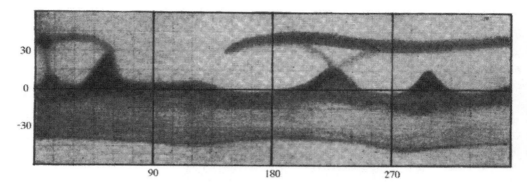

In 1951 E.J. Reese compiled a map of Ganymede based on telescopic observations by W.R. Dawes, E.E. Barnard, P. Lowell and E.M. Antionadi that was considered to be definitive.

moons maintain one face towards their primary. The first successful attempt to measure the diameters of the Jovian moons was by William Herschel, which he did by timing how long it took for each moon to disappear behind the limb of the planet during an occultation. It was a difficult observation, because he had to try to allow for effects as the light from the moon passed through the outer part of the planet's atmosphere. Upon realising that Ganymede would occult a bright star on 13 August 1911, F. Ristenpart, director of Santiago Observatory in Chile, organised a network of observers all across the southern hemisphere, and the results showed that Ganymede is considerably larger than our Moon. The diameters of the others were estimated by assuming that they all had similar albedos and that their sizes were proportional to their apparent brightnesses. It was a crude method, but it was sufficient to show that Amalthea was only about 150 kilometres across.

In 1849, W.R. Dawes made a concerted study of Ganymede, and concluded that its most prominent surface feature was a bright 'polar spot'. Further observations were made by E.E. Barnard, Percival Lowell and E.M. Antionadi, and then in 1951 E.J. Reese made a map that integrated their best observations. A few years later B.F. Lyot published maps of all four Galilean satellites based on his observations at the Pic du Midi Observatory in the Pyrenees.[21] His Ganymede map was in general agreement with Reese's map on the distribution of albedo features, but the maps published by A. Dollfus in 1961 were generally regarded as being the most accurate.[22]

The photometric observations which enabled Guthnick to establish that the rotations of all of the Galilean satellites were synchronised also provided the basis for a study of the reflectivity of their surfaces. In order to derive the 'phase coefficient', he measured the change in brightness for each 1 degree of arc change in the viewing (phase) angle relative to the angle of incidence of the sunlight, and from this he concluded that Callisto had a fairly smooth surface, that Ganymede was rougher than the Moon and that Io and Europa were rougher still.[23] There was evidently a correlation with albedo, because Io and Europa were highly reflective, Ganymede was moderately so, and Callisto's reflectivity was low. Although Callisto

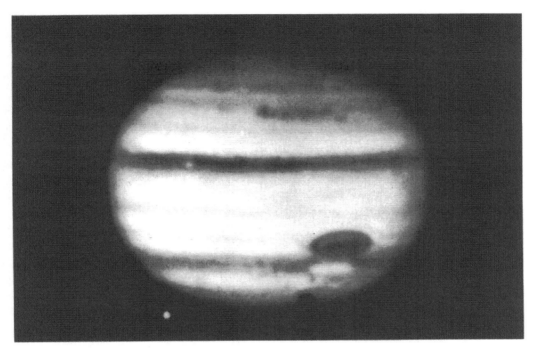

A pre-'Space Age' photograph of Jupiter taken by the 200-inch Hale telescope of the Palomar Observatory.

was the darkest of the four, it was noted that its reflectivity was twice that of the Moon (averaged over the entire lunar disk). Guthnick interpreted temporal irregularities in his light curves as evidence that the satellites possessed atmospheres that were occasionally laden with clouds, and, by drawing an analogy with Venus, it seemed reasonable to conclude that Io and Europa were possessed of shallow but dense atmospheres, Ganymede had a less substantial atmosphere, and Callisto's atmosphere was negligible. It was generally believed that the surfaces of all the moons would be covered with a layer of frozen gases, the favoured candidates being carbon dioxide and methane. The basis for measuring temperature is kinetic energy. The planet Mercury is almost the same size as Ganymede, but it is considerably closer to the Sun. On Mercury, molecules of these gases would literally shake themselves apart and the component atoms would have sufficient energy to exceed the planet's escape velocity. In Jovian space, however, the gases would be sluggish because the energy in sunlight (which falls off with the inverse square of distance, and Jupiter is 14 times further from the Sun than is Mercury) is some 200 times less powerful. Therefore, the likelihood of Jupiter's moons having atmospheres seemed reasonable. In the 1940s, G.P. Kuiper used the 82-inch reflector of the McDonald Observatory on Mount Locke in Texas, and detected no spectroscopic evidence to indicate an atmosphere on any of the moons. However, an infrared absorption band in their reflection spectra suggested that Europa was covered by a substantial layer of water-snow and that Ganymede possessed a patchy cover of snow that was possibly confined to the 'polar spot'. Neither Io nor Callisto showed any such feature.

Observations of the motions of the moons around Jupiter enabled their bulk densities to be estimated. At 3.5 g/cm^3, Io's density is comparable to the Earth's Moon (which is clearly a rocky body). The other Galilean satellites have lower densities, which means that they contain significant fractions of water. It was noticed that Io brightens somewhat upon emerging from Jupiter's shadow; evidently, a reflective material settles onto the surface in the darkness and then sublimates when illumination is restored. Further, multi-colour photometry revealed that Io's spectrum is distinctly reddish – in fact, it is the reddest body in the Solar System, even redder than Mars, which is often called the Red Planet.

There was, however, only so much that could be discovered about Jupiter and its system of satellites by viewing them from afar.[24] With the dawning of the 'Space Age', the role of Solar System study passed from telescopic observers to robotic exploring machines, but it was far from certain that the route to Jupiter was clear.

ASTEROIDS

Upon contemplating the broad gap between the orbits of Mars and Jupiter, Johann Kepler speculated that there might be an undiscovered planet in this zone.[25] J.D. Titius of Wittenberg noted in 1766 that there was a consistent progression in the sizes of the orbits of the planets.[26] The relative distances from the Sun of the six planets that he knew matched the values derived by adding 4 to the series 0, 3, 6, 12, 24, 48 and 96, except that the entry 24 was mysteriously vacant. In 1772 J.E. Bode of Berlin, reviving Kepler's speculation, proposed that there was indeed an as yet undiscovered planet corresponding to this gap in what became known as the Titius–Bode 'law'.

On 21 September 1800, F.X. von Zach convened a meeting of a few astronomical friends with an interest in searching for this 'missing' planet at J.H. Schoter's observatory in Lilienthal, near Hanover. It was agreed that the object would most probably lie near the ecliptic, so letters would be sent to colleagues urging that over the coming year they methodically search a series of zones of sky in an effort to locate it. The group nicknamed itself the 'celestial police'.

On 1 January 1801, however, Giuseppe Piazzi of Palermo in Sicily, who was not actually a member of the group, and was nine years into the task of drawing up a star

Table 8.1 The Titius–Bode 'law'

Planet	Series	TB law	Actual
Mercury	0	4	3.9
Venus	3	7	7.2
Earth	6	10	10.0
Mars	12	16	15.2
–	24	28	–
Jupiter	48	52	52.0
Saturn	96	100	95.4

catalogue, saw a star-like object that was moving slowly across the sky from night to night. After following its westward motion for several weeks he watched it come to a halt and then reverse its path, just as do the outer planets as a result of the changing line of sight as the Earth overtakes an object further out. Unfortunately, he then lost the object in the twilight as it slipped behind the Sun.

However, the mathematician C.F. Gauss managed to process the meagre data to calculate the object's orbit (thereby proving that it was located in the gap between Mars and Jupiter) and after it emerged from superior conjunction it was recovered by Zach on 31 December. In fact, the next day (precisely a year after it had been discovered) it was also recovered by Heinrich Olbers in Bremen. Since at 27.7 AU the object's mean distance from the Sun was in excellent agreement with the Titius–Bode prediction, the mystery was considered to have been resolved and the object was named Ceres in honour of the patron goddess of Sicily.

On 28 March 1802, however, Olbers, having reverted to searching for comets, was rather surprised to discover a *second* object in the same part of the sky and following a similar orbit. William Herschel estimated that Ceres was no more than 260 kilometres across, so it was not much of a planet. Since they appeared at the highest magnification only as 'star-like' points he suggested that they be designated 'asteroids'.[27] Piazzi countered that 'planetoids' was a better term because they were clearly unrelated to stars. In an attempt to explain why there was more than one object, Olbers suggested that the planet that originally orbited at that distance from the Sun had been shattered by a collision with a comet. After K.L. Harding, one of Schroter's assistants, discovered the third one on 2 September 1804, Olbers found the fourth (his second) on 29 March 1807. After a few years without another sighting, the 'celestial police' disbanded itself. In 1830, however, K.L. Hencke started his own systematic search, and was rewarded with his first asteroid in 1845 and his second two years later. This opened the flood gate and by 1850 there were ten asteroids (Table 8.2). At first, they were eagerly sought, but with the advent of photography the pace of discoveries accelerated, and they were derided as being no more than the 'vermin' of the sky.[28]

Table 8.2 The first ten asteroids

No.	Asteroid	Year	Discoverer	Diameter (km)
1	Ceres	1801	G. Piazzi	715
2	Pallas	1802	W. Olbers	486
3	Juno	1804	K.L. Harding	192
4	Vesta	1807	W. Olbers	382
5	Astraea	1845	K.L. Henke	–
6	Hebe	1847	K.L. Henke	–
7	Iris	1847	J.R. Hind	–
8	Flora	1847	J.R. Hind	–
9	Metis	1848	Graham	–
10	Hygeia	1849	de Gasparis	–

Note: The diameters are those derived by E.E. Barnard in 1894.

It was evident from the start that the asteroids were 'minor' planets at best. In 1853, U.J.J. Leverrier showed by consideration of possible perturbations that the total mass of the 'asteroid belt' was no more than that of Mars. E.E. Barnard estimated the diameters of the brightest members in 1894 and confirmed that they were very small. In fact, we now know that their total mass is less than 0.1 per cent the mass of the Earth.

A mathematical study conducted by Daniel Kirkwood in 1866 (when the total had risen to 90) revealed the existence of statistically significant 'zones of avoidance', and he realised that these 'gaps' occurred at mean distances from the Sun corresponding to orbits with periods in resonance with Jupiter's, from which he concluded that the planet's immense gravitational field was perturbing the asteroids from their orbits.

When NASA started to consider a mission to Jupiter, there were doubts as to whether the spacecraft would survive the passage through the asteroid belt. The known large objects were not a serious concern, because they were few in number and spaced far apart. The worry was the small rocks and specks of micrometeoroid debris that seemed likely to litter the entire zone.

PIONEERING DEEP SPACE

After an abortive early attempt by the US Air Force to send a series of probes towards the Moon, NASA's Ames Research Center in Mountain View, California, picked up the 'Pioneer' name and assigned it to probes that were dispatched into solar orbit to report on the state of the solar wind. Some of these spinning 'particles and fields' spacecraft were stationed slightly within the Earth's orbit and some were just beyond it. After these successes, Ames decided to dispatch an Advanced Pioneer on a fly-by of Jupiter to study the electromagnetic environment in the outer Solar System. Two identical spacecraft were built and launched as Pioneers 10 and 11, in March 1972 and April 1973, respectively. By launching them a year apart, it would be possible to postpone the final commitment to Pioneer 11's fly-by trajectory until after Ames saw how its predecessor fared in the Jovian magnetosphere.

Pioneer 10 lived up to its name by becoming the first spacecraft to fly through the asteroid belt. On 26 November 1973, rather earlier than expected, its instruments first detected the planet's magnetosphere which, because the spacecraft was still 8 million kilometres distant, was evidently considerably larger than had been thought. Although the spacecraft did not have a television camera, it did have a photometer which scanned Jupiter as the spacecraft rotated axially, so it was possible to assemble pictures by 'stacking' these strips side by side. It was a time-consuming process, and the spacecraft was only able to take two dozen images during the final day of its approach. When processed at Ames, these images revealed features in the Jovian atmosphere in much greater resolution than had ever been achieved by even the largest terrestrial telescope, and showed the weather system to be unexpectedly intricate.

The photometer also returned images of the Galilean satellites. A radiation glitch prevented the first spacecraft from scanning Io, but it was imaged by the second

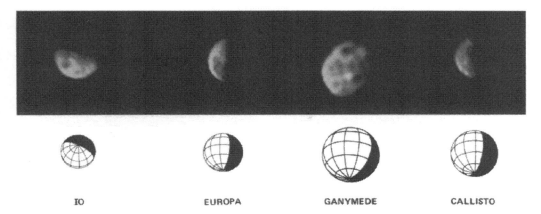

Although computer-enhanced, these Pioneer line-scan views of the Galilean moons of Jupiter show little detail. Nevertheless, they were very welcome because they provided broad hints at the surface morphology.

from the perspective of peering over its north pole.[29] Pioneer 10 secured a single image of Europa but it was not close enough to reveal much detail on the extremely reflective surface. Both Ganymede and Callisto were scanned several times. Ganymede's north polar region was confirmed to be very bright, and a number of large dark regions were evident, but the imaging technique was too primitive to resolve features smaller than about 400 kilometres across. Pioneer 10's trajectory had been chosen so that it would fly behind Io, as viewed from Earth, to enable its radio signal to 'sound' for any atmosphere: it identified a tenuous ionosphere at an altitude of 120 kilometres.

By surviving the radiation at its closest approach of 130,000 kilometres above the planet's cloud tops, Pioneer 10 cleared the way for its successor to go deeper into the magnetosphere and use Jupiter's gravity to deflect its trajectory towards Saturn.[30] By passing under Jupiter's south pole, Pioneer 11 was able to avoid the intense equatorial radiation belt on the way in and be accelerated through it on the far side, so even though it approached three times closer to the planet than its predecessor, its total radiation dose was no worse. After its Jovian encounter on 3 December 1974, Pioneer 11 set off on the long trip to Saturn, which it reached September 1979. Both Pioneers are now heading out of the Solar System, and their instruments continue to report on the solar wind as they seek the 'heliopause' where the Sun's magnetic field yields to the interstellar medium.

John Naugle, NASA's associate administrator for space science, had emphasised that the Pioneers were "precursor missions" whose function was to "lay the groundwork for the outer planet exploration programme". They had certainly opened the door, but the spacecraft which followed up were not drawn from the same stable.

Voyager 2 rides a Titan-III/Centaur rocket into space on 20 August 1977.

VOYAGERS AT JUPITER

As planetary imaging was the speciality of the Jet Propulsion Laboratory in Pasadena, which is run jointly by NASA and the California Institute of Technology, it was assigned the task of exploring the outer Solar System.

In the early 1970s, JPL proposed the development of a sophisticated 'Mark 2' Mariner spacecraft for what it promoted as a 'Grand Tour' of *all* the outer planets in succession. This mission emerged from a study by Gary Flandro in 1965 in which he explored the utility of trajectories incorporating gravitational slingshots. Mariner 4 had just made the first fly-by of Mars, so the prospect of a 12-year-long mission to explore the outer planets was a visionary piece of long-term planning.[31,32] The 'Grand Tour' title had actually been coined by G.A. Crocco, who presented an analysis of multiple-planet trajectories to the International Astronomical Union's Congress in Rome in 1956, a year before the 'Space Age' dawned. However, in the cash-strapped 1970s, with the development of the Space Shuttle dominating NASA's budget, funding was awarded only to adapt the proven Mariner design to operate far from the Sun and to send a pair of these spacecraft, renamed Voyagers, on a rather less ambitious mission that was officially restricted to following up the Pioneers at Jupiter and Saturn but, if all went well, could be extended.[33,34]

Both Voyager spacecraft were launched in the same window in 1977 so that they would encounter Jupiter a few months apart in 1979. In addition to particles and fields instruments, they carried a comprehensive suite of instruments to provide ultraviolet spectroscopy, infrared spectrometry and radiometry, photopolarimetry and high-resolution imaging. Furthermore, a series of images taken over a 10-hour period during the Jupiter fly-by were later sequenced to produce a movie depicting the planet rotating on its axis, and to document the highly dynamic atmosphere in unprecedented detail.

The Galilean satellites were the 'stars', however. Voyager 1's passage through the Jovian system provided an opportunity to view the small inner moonlet Amalthea on the way in, and to image Io, Ganymede and Callisto on the way out. Its fly-by was well within the orbit of Io. In fact, the spacecraft passed within 20,000 kilometres of Io's south pole, and in so doing it documented much of the side of Io that permanently faces the planet. Although Voyager 1's trajectory did not offer an opportunity to study Europa in detail, Voyager 2 made a close pass on the way in, and this time it was Io that was inconveniently on the other side of the planet. Ganymede and Callisto received the best overall surface coverage because Voyager 2 viewed the hemispheres which had been in darkness when its predecessor passed through, but only portions of Europa and Io could be mapped in detail. Nevertheless, at long last, the "marvellous things" that Galileo had discovered were revealed as miniature worlds with their own highly distinctive geological histories.[35]

Voyager 1 took this photo of Jupiter when still 32 million kilometres from the planet. It was a revelation. In addition to eddies and whorls in the cloud structure around the Great Red Spot, the smaller spots caught up in the latitudinal atmospheric circulation system were seen clearly for the first time.

Io

As Voyager 1 approached the Jovian system in early 1979, very little was known about Io with certainty. It orbits only 350,000 kilometres above Jupiter's cloud tops. This is approximately the distance between the Moon and the Earth, but whereas Io is comparable in size to the Moon, mighty Jupiter is 318 times more massive than the Earth. Because Jupiter's gravitational field must draw in material and accelerate it, the expectation was that Io would be a heavily cratered object, and so when the early low-resolution Voyager 1 imagery showed vaguely circular albedo features these were interpreted as craters. However, as the spacecraft closed in it became apparent that these dark features were not impact craters. Astonishingly, a thorough survey revealed that Io has *no* impact craters. Since no body can completely escape suffering impacts, their absence indicated that some process was continuously resurfacing Io and had 'removed' the craters. Io was evidently volcanically active. Nevertheless, no one seriously expected to catch a volcano in the process of erupting. As it departed, the spacecraft secured an image showing Io's position relative to the stars in order to confirm its trajectory for Saturn, and saw it as a crescent illuminated by 'Jupiter shine'. When Voyager navigation engineer Linda Morabito enhanced this image to look for stars close to the limb she spotted a faint plume projecting 280 kilometres into space. It was an active volcano! There was also an anomalous glow on the dawn terminator. A second volcano! Although this vent site was in darkness, its plume of dusty gas projected so high that it reflected the first rays of the Sun.[36,37] In fact, there were nine active vents with plumes rising up to 300 kilometres,[38] and eight were still active when Voyager 2 flew by four months later. The infrared spectrometers identified 'hot spots' on the surface, so there were obviously many active vents that were not producing plumes.[39]

Just before Voyager 1 ventured into the Jovian system, a paper by Stanton Peale, Patrick Cassen, and Ray Reynolds reported the results of modelling the tidal stresses acting upon Io.[40] Europa's orbital period is twice that of Io, and Ganymede's is twice that of Europa, and these orbital resonances make Io's orbit slightly elliptical – this eccentricity is only 0.0041 but it has a significant effect. The overall gravitational force acting on Io varies cyclically. All the Galilean satellites rotate synchronously and thus maintain one hemisphere facing Jupiter. The 'tidal bulge' on Io rises and relaxes in response to the varying gravitational attraction, and this mechanical stress is converted into heat. The calculations suggested that Io gains fully two to three *orders of magnitude* more heat from tidal stress than it is likely to derive from the decay of radioactive elements. Peale's team not only concluded that Io's interior had to be thermally differentiated, they tentatively predicted that the surface might be volcanically active. However, they were as surprised as everyone else to find that Io is the *most* volcanically active object in the Solar System. By radiating energy to the vacuum of space, the intense volcanic activity serves to cool Io from the relentless tidal heating. In fact, terrestrial infrared observations undertaken in 1981 established that Io's heat flow is 30 times greater than that of the Earth.[41]

It was speculated that Io's silicate lithosphere had been transformed into the floor of a global ocean of molten sulphur several kilometres deep, and that the visible surface is actually a layer of frozen sulphur a few hundred metres thick capping this ocean.[42,43] Volcanism, it was argued, occurs when a localised build-up of sulphur dioxide gas within this sulphurous ocean opens a vent in the crust. In this model, a plume is powered by explosive venting and, in some cases, there is an accompanying extrusion of fluid sulphur. Once the gas pressure has been relieved, the vent seals itself. The sulphurous ocean (it was said) was the natural result of the extreme thermal differentiation. There was also a precedent for a magma ocean: as the Earth's Moon accreted from the debris ejected by an impact with the Earth, the infall kept its outer few hundred kilometres molten, and chemical differentiation induced lightweight aluminium-based silicates to rise to the surface and crystallise as a crust of anorthosite, which was later flooded by extrusions of dark mafic lavas that welled up through fractures.[44] Although the details were different in Io's case, the result was strikingly similar. Of course, this was disputed by other researchers, but at last there was real data to work with.

As far as we can tell, Io is desiccated. Any water it may have possessed (and being so far from the Sun, it must have originally accreted a fair proportion of ice) has long since been lost to space. The water molecules would have been dissociated by solar ultraviolet, the hydrogen leaked away, and the oxygen was partly recovered and chemically bound into the crustal material. In fact, early on, when Io still possessed hydrated minerals, volcanic plumes would have been driven by water steam, and

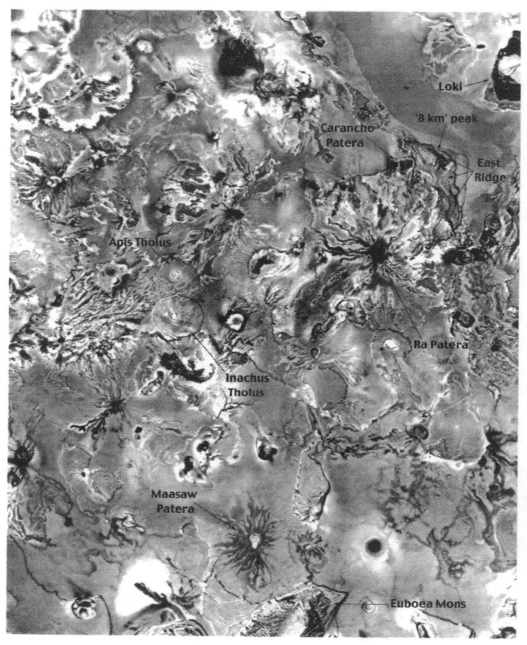

This airbrushed map based on a Voyager mosaic highlights Io's intensely volcanic landscape southwest of Loki, the largest caldera (top right). Many of the vents – most notably Ra Patera – show radial lava flows, but other calderas appear 'clean'. The low shields and overlapping flows of the Snake River Volcanic Plain in Idaho may be a close terrestrial analogue for this type of terrain, particularly around the 'domes' of Apis and Inachus Tholus. (Courtesy of USGS.)

geyser activity would have been global in extent, continuous and ongoing for millions of years. When the supply of water was finally exhausted, this activity would have ceased and, denied its means of cooling, the crust would have heated up until the temperature was sufficient to cause another volatile compound to boil, with the result that Io has now lost its water, nitrogen, carbon dioxide and neon.[45]

At the present time, Io is boiling off its sulphur dioxide. As sulphur dioxide in a volcanic plume condenses and falls back, it 'paints' a frosty halo on the surface around the vent. Other volatiles will also condense out, creating a chemically diverse pyroclastic blanket, and sulphur atoms will link up to form molecules of varying numbers of atoms – that is, 'allotropes' of sulphur. Temperature is a measure of kinetic energy, and as an allotrope of a specific length will be shaken apart above a specific temperature, sulphur chemistry is temperature dependent. At 'room temperature' in a terrestrial environment, sulphur is a pale yellow solid, and at 400 K it melts and becomes orangey. As the temperature is further increased, both the viscosity and the colour of the liquid change: at 435 K it becomes clear pink, and at 465 K it becomes viscous and turns red. A transition in the molecular structure occurs at 500 K and it develops a black tarry constituency. It begins to lose its viscosity at 600 K and is a dark runny fluid at 650 K. In fact, it reaches its minimum viscosity just before it vaporises. The boiling point of sulphur is dependent on pressure: it will boil at 715 K at 1 bar, but it will also do so at 450 K at 0.001 bar. Impurities will raise the boiling point at any pressure. And, of course, just as the colour and the viscosity of sulphur evolve as it is heated, they do so as it cools. Evidently, Io's 'pizza-like' surface represents subtle sulphur chemistry. As a hot dark sulphur flow cools, the colour changes, and it becomes viscous and slows – which correlates with the fact that the material in the calderas and close to the vents is dark. The variety of hues further out from the calderas correspond to different chemical compositions, ages and temperatures. A study by Alfred McEwen and Laurence Soderblom concluded that the plumes are driven by explosive gas venting, and that sulphurous 'lava' extrusions are less violent events.[46] This combination of processes seemed to be adequate to explain the observed structure of Ra Patera, a large 'shield' volcano on the summit of the tidal bulge facing Jupiter.

With such clear evidence of sulphurous volcanism, geologists began to doubt whether Io underwent terrestrial-style basaltic volcanism, because the temperatures required to melt silicate rock are significantly higher than will boil sulphur.[47,48] However, as photogeologists mapped the moon in fine detail they were astonished by the steep mountain slopes and caldera walls.[49] The base of Haemus Mons in the south polar region spans almost 200 kilometres. The peak rises about 10 kilometres above the surrounding plain. How was this constructed? It does not seem to be a volcanic edifice. It displays the appearance of being a fractured crustal block that has been forced up by tectonic forces – a massif. If the observable surface is just the frozen cap on a sulphurous ocean, are the peaks deep-rooted structures poking through it? Or are the sulphurous flows only a veneer on a silicate lithosphere which has undergone fracturing and uplift? Furthermore, the large vents have steep cliff-like walls plunging several kilometres to caldera floors. If these volcanic complexes were sulphur constructs, the caldera walls would have slumped. The temperatures of the Pele and Loki calderas (the vents spotted by Morabito) were measured by the Voyager infrared spectrometers as several hundred degrees warmer than their environs, but the spatial resolution of the instrument was low, the instrument readings were averaged over wide areas, and the surrounding terrain was freezing. Although the derived temperatures were consistent with sulphurous chemistry, it was also possible that there were rapidly cooling molten silicate flows.[50] In 1983, NASA's Infrared Telescope Facility on Hawaii began regular monitoring of Io for signs of activity. In August 1986 an 'outburst' was noted.[51] Analysis suggested that an area several tens of kilometres across had suddenly become heated to 900 K. A reanalysis of the data subsequently revised this to a smaller area at around 1,550 K.[52] This was sufficiently hot to prove that silicate volcanism is active, at least on an intermittent basis, and it lent strong support to the argument that the sulphurous surface is just a thin veneer upon the 'real' silicate lithosphere.

The art of photogeology involves four stages: (1) identify the various geological units in terms of their surface characteristics; (2) chart the distribution of each unit; (3) determine the relative sequencing; and (4) formulate a model to explain their origins. While mapping can be done objectively, interpreting the data is a subjective act (witness the debate over silicate versus sulphurous volcanism). Impact craters traditionally provide a means of determining the relative sequence of surficial deposits, but craters on Io are notable for their complete absence. If Io's volcanoes release 10 billion tonnes of material per year, and if this is averaged over the whole moon, then it would produce a 100-metre-thick blanket in just 1 million years, which is more than sufficient to smother the occasional impact crater.[53] Nevertheless, analysis indicated three primary physiographic units.[54,55]

The term 'plains material' was introduced as a generic name for several types of flat terrain that is collectively the most widespread. The 'inter-vent plains' are intermediate albedo and are smooth, but in some cases they are crossed by low scarps. They are interpreted as plume fall-out interbedded locally with flows and fumerolic materials. The 'layered plains' are broad flat surfaces crossed by grabens and prominent scarps, with localised undercutting, slumping and etching due to venting of sulphur dioxide from shallow depth.[56] Once a fracture formed, liquid sulphur dioxide would be able to rise through the crust. As it approached the surface, and the pressure decreased, it would sublimate to gas while still underground, expand, and then vent explosively, eroding the surface expression of the fracture. The vented sulphur dioxide would promptly condense and fall nearby as white frost. An undercut scarp would eventually slump. An explosive release of gas might even create a collapse structure; a maar. The multi-coloured surficial patterns were therefore interpreted as being anhydrous mixtures of sulphur allotropes, sulphur dioxide frost and sulphurous salts of sodium and potassium, and the depositional and erosional patterns indicated a complex geological history. The 'vent-related materials' derive directly from calderas. The floors of the calderas are dark, implying that they contain lakes of either silicate lava or molten sulphur. The sheer walls suggest the structural integrity of a silicate lithosphere. It is possible that vents initially opened by silicate volcanism are subsequently exploited by sulphurous volcanism. The 'mountain material' forms individual edifices standing above the level of the plains. Euboea Mons in the south polar zone is one of the tallest peaks, rising 13 kilometres.[57] Although the absolute ages of the mountains are unknown, the superposition relationships indicate that they are the oldest exposed features.

A comprehensive analysis by Susan Kieffer concluded that Io's plumes are very similar to terrestrial water geysers.[58] She modelled the sulphur dioxide in a plume as being derived from a reservoir at a depth of several kilometres. If it came into contact with silicate lava at 1,400 K, the sulphur dioxide would boil under extreme pressure; its force would then open pre-existing fractures in the brittle silicate crust, rise adiabatically, and be released explosively upon approaching the surface. Despite the supposition that plumes are driven by gas pressure, there was no evidence of terrestrial explosive Plinian-style eruptions. On the contrary, the process of lava extrusion appeared to have been effusive. Several types of extrusive activity were evident, including 'shields', 'discoids' and large central features with multiple ('digitate') flows. The distribution of vents does not suggest that there are any large-scale convection cells operating in the interior; nor are there any patterns of activity to indicate that the lithosphere is mobile in a plate tectonic sense. Io's means of dealing with its tremendous heat flow has simply been to form a large number of open volcanic sores over the 'hot spots'.

In December 1973, when Pioneer 10 was occulted by Io, the refraction of the radio signal at the limb-crossing indicated that the moon had an ionosphere. Astronomers had noticed that Io tended to brighten slightly immediately after emerging from Jupiter's shadow and this was taken to mean that a gas was settling onto the surface during the chilly darkness and resublimating upon the return of sunlight. Telescopic spectroscopy had noted a 'halo' around Io, and the yellow line-emission indicated the presence of neutral sodium. Jupiter turns on its axis in just under 10 hours, dragging the inner portion of its magnetosphere around with it but Io orbits the planet in two days, so the magnetic field sweeps past the moon. Once the gas blasted into space by the volcanic plumes is ionised, it is 'snatched' by the magnetic field and drawn right around Io's orbit, forming a torus of plasma that has been likened to "the beating heart of the Jovian magnetosphere". Voyager 1 observed an aurora extending for 30,000 kilometres in an arc around the planet's north polar zone, and its plasma wave instrument detected 'whistling' radio emissions. It had been thought that the solar wind would penetrate the magnetosphere's 'polar cusps', as occurs in the case of the Earth's magnetosphere, but the Jovian aurora was found to be due to Io's presence. Because its plumes pump out material, the moon is electrically linked to the planet by a pair of 'flux tubes' that run down the magnetic field lines to the polar zones. As the Galileo spacecraft penetrated the Jovian system on 7 December 1995, it passed 892 kilometres in front of Io's leading limb and its particles and fields instruments studied its magnetospheric 'wake'. The energetic particle detector reported bidirectional electron flows aligned with the planet's magnetic field lines. This was the first direct observation of the flux tubes. The charged particle flow constitutes an electrical current of several million amps and has an overall energy deposition into Jupiter's atmosphere of a trillion watts (so it is the most powerful 'direct current' in the Solar System) and this is the energy that stimulates visible emissions at the auroral 'footprints' where the flux tubes impinged upon the atmosphere in the polar zones.

While surveying the inner region of the Jovian magnetosphere and Io's plasma torus, Galileo's magnetometer made a significant discovery concerning the moon itself. Instead of increasing continuously as the spacecraft approached Jupiter, the field strength fell some 30 per cent as it passed Io, as if the moon orbits the planet inside a 'bubble' within the Jovian magnetosphere. Although this strongly implied that Io possesses an intrinsic magnetic field, the evidence was not conclusive because the signal may have been due to an interaction between the flux tubes and the torus. However, the plasma wave spectrometer detected a very dense cloud of ionised oxygen, sulphur and sulphur dioxide from Io's plumes. The fact that these ions had not been swept up by the Jovian magnetosphere lent support to the case for Io having a magnetosphere of its own. This situation could only be clarified by returning Galileo to Io for a much closer fly-by. The intense radiation meant that the only encounter with Io on the spacecraft's nominal schedule was the one on the initial run in to Jupiter. However, if Galileo managed to survive its two-year primary mission, it was to be sent back to study Io.

Monitoring the doppler on Galileo's radio signal throughout the initial Io fly-by measured the moon's gravitational field so that the moment of inertia, which is a key dynamical parameter, could be computed.[59] The low value implied that Io is not a homogeneous sphere. It has a metallic core, most likely composed of iron and iron sulphide, englobed by a partially molten silicate mantle. The data indicated that if the core was iron mixed with iron sulphide it would comprise 52 per cent of Io's radius, but if it was pure iron it would be 36 per cent of the radius; in either case it was a 'giant' core.

As Galileo performed its primary mission over the next two years, it observed Io from a distance, monitoring long-term trends in volcanic activity and mapping the composition of the surface. Its solid-state sensors were able to measure the temperature of Io's surface while the moon was in Jupiter's shadow, therefore the thermal emission from the 'hot spots' could be studied in the absence of reflected sunlight. Upon completing its nominal mission Galileo was still in excellent condition so its mission was extended, and in October 1999 it made the eagerly-awaited return to Io with an equatorial fly-by at an altitude of just 600 kilometres. The camera suffered somewhat from the radiation but survived, and the spacecraft was sent back to make passes over the moon's poles.[60] The results revolutionised our understanding of this volcanic moon.

At some 200 kilometres in diameter, Loki is Io's largest caldera and, like many other calderas, it is D-shaped. The sheer wall plunges several kilometres. There is a plateau-like island within the caldera whose white sulphur dioxide crust is in marked contrast to the dark lava lake. A linear feature to the northeast may be a related fissure; in fact, whereas the plume seen by Voyager 1 was rising from one end of this feature, the plume seen by Voyager 2 was rising from the other end. Galileo found Loki little changed since 1979, which was surprising, because it was one of the most persistently active of the 'outburst' sites. Diana Blaney had monitor Io in the decade leading up to Galileo's arrival using the Infrared Telescope Facility and established that Loki undergoes periods of intense volcanic activity, interspersed by periods of relative quiescence. Occurring about once per year, these 'brightenings' typically last several months. The lava lake is evidently only thinly encrusted, and upwellings frequently flood large areas of the caldera's floor. The Infrared Telescope Facility noted a significant brightening within Loki's caldera on 12 March 1997, and when Galileo next observed Io this eruption was at its peak. The subsequent evolution of the thermal profile indicated that a fresh lava flow had covered a large area of the caldera floor and was rapidly cooling.[61] The caldera brightened again in May 1998, and was observed by Galileo the following month. The Infrared Telescope Facility saw Loki glowing faintly on 9 August 1999, but the next day it had 'bloomed' by a factor of 10. The Wyoming Infrared Observatory followed the brightening's progress through September. When Galileo made its first close fly-by of Io on 11 October it flew directly over Loki and saw it glowing in the darkness. Its infrared spectrometer's thermal scan across Loki marked a real milestone in the mission, because it was the first time that it had been able to map the thermal structure of a hot spot at high spatial resolution.[62] It was able to map the *inside* of the caldera with resolutions varying from about 25 kilometres per pixel for general context, down to 500 metres per pixel for the detailed study. It secured a thermal map with a very high signal-to-noise ratio in which local temperature variations could be discerned with confidence. The investigation focused on the southeastern quadrant of the elevated and remarkably flat 'island'. The centre was warmer than the periphery, indicating that the heat flow from the centre was greater. Significantly, the warmer part was marked by albedo variations. Despite being within an extremely active caldera, the white sulphur dioxide deposit was frozen, suggesting that the island is a long-term feature of the caldera.

The infrared spectrometer continued to record data as the scan platform swung southeast towards Pele, sampling a zig-zag pattern across the floor of Loki's caldera, and over its rim onto the frozen surface beyond. The heat flowing from the dark caldera floor was about ten times greater than from the island, which was in turn ten times greater than the surrounding plain.[63] The integrated photopolarimeter and radiometer's study of the temperatures in and around the caldera found that most of the dark horseshoe was at a remarkably uniform 250 K. Although 'subzero', enormous amounts of volcanic heat are required to maintain such a large area at that temperature. At 125 K, the surrounding terrain was typical of the sulphur dioxide frost on an inter-vent plain. The site of the recent upwelling in the southwestern part of the caldera was still 400 K. Even though the outburst was so energetic, the vast caldera had accommodated it with very little physical disruption. Remarkably, Loki accounts for 25 per cent of Io's overall energy flow.[64,65] As the most 'powerful' volcano in the Solar System, Loki releases more heat than all the Earth's active volcanoes combined. In fact, its power is 100,000 times greater than a resurgent caldera such as Yellowstone in Wyoming, the most violent type of terrestrial eruption. However, in Loki's case, the heat is released on a continuous basis rather than by a series of catastrophic eruptions.

As the first plume ever to be observed on Io, Pele was of historical significance. When Galileo made its fly-by in October 1999, it was able to peer into the caldera and observe the lava glowing brightly. An image taken using a near-infrared filter that selected only material hotter than 1,000 K detected only a thin curving line some 50 metres across and 10 kilometres long. Temperatures of 1,300 K had previously been measured in Pele. Such fresh lava would cool and become invisible to this filter within a few minutes, so Galileo evidently saw a part of the lava lake that was at most *several minutes old*. Superimposing the thermal data onto the best visual image showed that the hot lava was on the periphery of the caldera, which implied that the lava lake had a thin scum that was repeatedly being disrupted where it abutted the cliff-like walls.

In contrast to volcanoes which erupt in pulses and flood large areas with lava which cools over time, Pele's intensely hot vent is remarkably stable, suggesting that this extremely active lava lake is constantly exposing fresh lava. As Galileo made its first close fly-by, the Hubble Space Telescope detected molecular sulphur vapour in Pele's plume.[66] Once the data had been coordinated, a model based on plume chemistry suggested that Pele's lava is 1,450 K, which was in excellent agreement with Galileo's measurements of 1,200 to 1,470 K.[67] Pele's plume seems to be the result of the lava lake exsolving volatiles, rather than by the explosive venting of sulphur dioxide from fractures. The sulphurous compounds in the plume have produced a prominent red halo 1,000 kilometres in diameter. Io's plumes can 'paint' such a broad feature because the moon has a weak gravitational field and no atmosphere to impart air resistance.

At 450 kilometres across, Ra Patera is the largest of Io's small number of shield volcanoes. It is a very shallow edifice located in a cluster of vents near the sub-Jovian point. Colourful lava flows run out as far as 300 kilometres from its dark caldera. The profusion of low shields and overlapping lava flows of the Snake River Volcanic Plain in Idaho may be a close terrestrial analogue for this part of Io.[68,69]

Ra Patera had not been active when the Voyagers inspected Io, but in July 1995 when the Hubble Space Telescope turned to Io it noted the appearance of a yellowish-white patch some 350 kilometres across where, a year earlier, there had been a small white spot.[70] However, the telescope's resolution was insufficient to map this new feature in detail. When Galileo arrived, it observed a 100-kilometre-tall plume from Ra Patera rising over the limb. The plume's blue hue implied that sulphur dioxide was condensing out as the gas expanded and rapidly cooled, and was falling as snow. A comparison of Voyager and Galileo imagery revealed that an area of 40,000 square kilometres had been resurfaced. The yellow surface flows visible in the Voyager imagery indicated that Ra Patera extruded sulphurous lava.[71,72,73,74] The new flow extending off to the southeast of the caldera was dark,[75] so if it was sulphurous it was still warm; however, Galileo did not detect any hot spots near the caldera. The eruption was studied by processing Voyager imagery to create a topographic map of the area.[76] Although broad, Ra's summit rises only 1 kilometre. To flow so far on gradients as shallow as 0.1 degree, the lava must have been of extremely low viscosity. Stereoscopic analysis indicated that there is a plateau some 500 metres high located about 50 kilometres east of Ra, which progressively rises as it runs north for 600 kilometres, terminating on the 8-kilometre-tall peak to the east of Carancho Patera. The recent lava flow followed the plateau's southern periphery.

On 5 July 1997, the Hubble Space Telescope discovered a plume, 200 kilometres

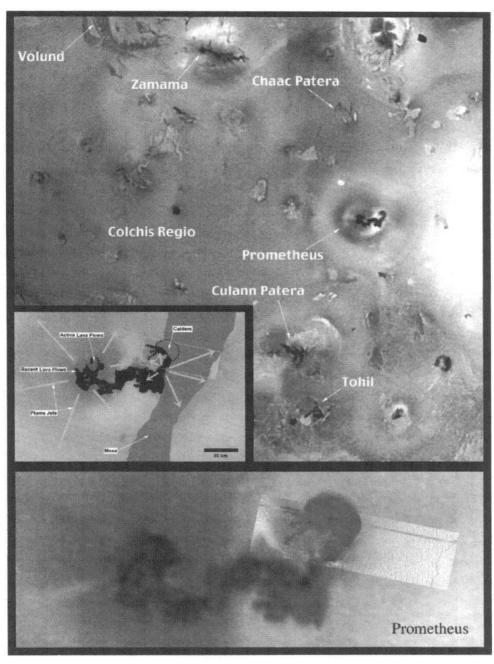

Colchis Regio on Io's anti-Jovian hemisphere is a smooth inter-vent plain. Note how different it look to the Ra Patera area on the sub-Jovian side. Galileo discovered that the D-shaped feature at the eastern end of the Prometheus volcano is a caldera and that this has extruded a lava flow over 100 kilometres to the west. In the close up, a high-resolution image of the caldera refines an earlier view in which the spacecraft was peering straight down through the plume.

tall, rising from Pillan Patera – a previously quiescent caldera on the northeastern fringe of the plume halo around Pele. A week earlier, Galileo had recorded an intense high-temperature hot spot within the caldera but the spacecraft's imagery was not transmitted to Earth until later. In 1979 Pillan had appeared as a simple caldera free of surrounding deposits and, upon its first look, Galileo had found it unchanged, but albedo variations within the caldera had soon indicated increasing activity. After the Hubble's detection of the plume, terrestrial infrared telescopes had identified a hot spot in July.[77,78] When Galileo returned, the plume was still present but there was now a circle of dark pyroclastics about 400 kilometres across that had disfigured Pele's reddish halo. Most of Io's plume deposits were white, yellow or red due to sulphur compounds. However, the new deposit appeared to be silicate-rich ash. In fact, Pillan is similar to Babbar Patera which had disfigured the far side of Pele's halo with dark material at some time in the past.

In addition to spewing out a great deal of pyroclastics, Pillan's eruption had also extruded a dark flow extending 75 kilometres northeast from the vent. A pair of hot spots observed when Io was in eclipse corresponded to the vent and to the tip of the flow. The temperatures of these hot spots exceeded 1,700 K.[79] In fact, because the thermal emissions saturated the spacecraft's sensors, the hot spots were probably 2,000 K. Silicate volcanism had long been inferred on Io, but this was the first example to be documented by before-and-after imagery. Not only was Pillan conclusive proof of silicate volcanism, the high temperatures almost certainly yield silicate compositions that are unusual on Earth. As the hottest terrestrial lavas are about 1,500 K, Io's lavas are much hotter than anything that has erupted on Earth in the last few billion years. The nearest terrestrial equivalent is probably the ultramafic orthopyroxene-rich komatiite lavas which were extruded during the early Archean and form the base of the greenstone sequences of the Barberton in South Africa or the Pilbara in Western Australia. During its October 1999 fly-by, Galileo recorded the fringe of Pillan's new lava flow. As the Sun had just risen, the shadows emphasised the topographic relief of the complex mix of smooth and rough terrains, and the sheer variety of types of lava flow features was surprising. On the local scale, there were clusters of pits and domes, some of which were house-sized. A lava channel 140 metres wide and 3 kilometres long had fed the flow which was significantly below the level of the adjacent plain. Although the flow may have pooled in a pre-existing hollow, it was more likely that the super-hot lava had eroded the sulphurous plain and made its own cavity, in so doing limiting the extent of its excursion.

Although not as energetic as Loki, Io's Prometheus is remarkable because it would appear to have been continuously active ever since its discovery – hence its nickname 'Old Faithful of the Outer Solar System'. When Galileo inspected Prometheus in detail, it found that the northeastern end of the long dark feature was actually a D-shaped caldera whose floor was coated by plume fall-out and so was cold. There was a warm spot on the caldera's southern rim, but the hottest site was at the western tip of the lava flow. At 1,100 K, this had to be molten silicate. The fact that the plume rose from the flow front was significant. It was concluded (1) that the deep magma reservoir was beneath the caldera but it has broached the surface 15

The plume from Prometheus was found to be produced when the western tip of the dark lava flow encroached upon, and sublimated to gas, the light-toned frozen plain of sulphur dioxide pyroclastic.

kilometres south of the rim, and (2) that the lava was being transported 100 kilometres westward through tubes to the front, at which point it disgorged onto the surface and vaporised sulphur-dioxide frost, creating the plume. In 1979, the plume had risen from the hot spot by the caldera. Kilauea in Hawaii is a reasonable terrestrial analogue for Prometheus; some of its lava flows through tubes to fronts which vent into the ocean and raise plumes of steam. However, Kilauea's caldera is much smaller and its lava tubes are only 10 kilometres long. Significantly, this was a second case in which the plume was not due to explosive venting by a deep fracture, as had been inferred from the Voyager imagery; Io was rather more diverse than had been thought.

Several hundred kilometres east of Prometheus, Emakong Patera is essentially surrounded by yellow-white lava flows, prompting speculation that its caldera has spilled moderately hot sulphurous lava rather than very hot silicate lava. There are dark sinuous channels in the light flows which so resemble the lava channels that meander down the shallow slopes of terrestrial volcanoes that this is almost certainly what they are.[80] However, at about 100 kilometres long and half a kilometre wide, Io's channels are rather larger. Evidently, the dark material is black sulphur that spills from the channels and yellows as it cools. Emakong is one of the best examples of the sulphurous volcanism that was once believed to be predominant on Io.

Tvashtar Catena on the northern plains comprises a chain of some of Io's largest calderas. Although there was no record of activity, Galileo inspected it and

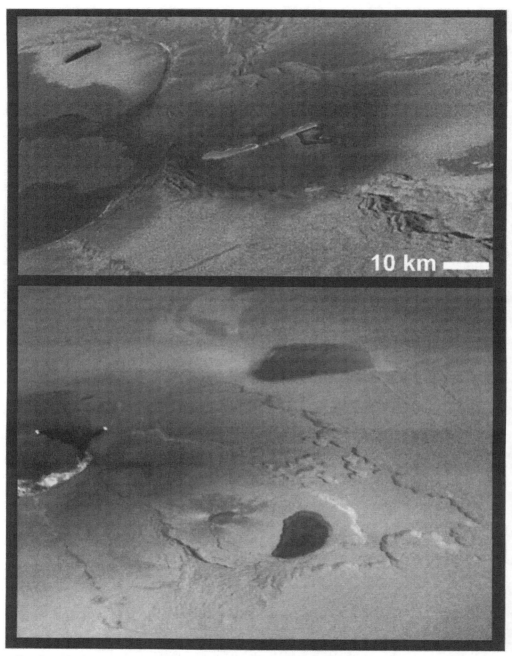

By sheer luck, during its close fly-by in November 1999, the Galileo spacecraft caught a 'fire fountain' spewing lava over a kilometre into the sky in one of the calderas of Tvashtar Catena. Upon its return in February 2000, Galileo found that the site of the eruption had switched to a larger fissure to the west which, judging from the dark surface in the earlier view, had clearly been previously active. The two bright spots on the later image are where molten lava was exposed at the flow fronts.

discovered a fire fountain in one of the calderas which was sending a curtain of lava a kilometre into the sky! Similar activity in terrestrial volcanoes rarely rises more than a few hundred metres. The 'adaptive optics' on the 10-metre Keck Telescope in Hawaii recorded this eruption at a range of infrared wavelengths, and integrating the ground-based and spacecraft data indicated that the lava was 1,800 K, so it was a silicate eruption. In December 2000, Galileo caught a second, and even larger eruption in progress close by. Furthermore, there was now a red pyroclastic halo on a scale comparable to Pele's.[81] The Cassini spacecraft, which was passing through the Jovian system at that time, observed a large, Pele-type plume over Tvashtar.[82] The prospect of fire fountains on Io had been proposed by Laszlo Keszthelyi of the Lunar and Planetary Laboratory in Arizona to explain some observations of Loki, but it had not been possible to verify this. Catching the fire fountain in action was a 1-in-500 chance observation. The detection was all the more remarkable because it was not as if the spacecraft had 'noticed' it and fired off a picture as a target of opportunity. If Galileo had not been instructed to take a picture, the nature of this eruption would have remained unidentified. What else had we been missing?

Io's mountains do not form ranges, they stand in isolation. This indicates that they are not the result of large-scale tectonism in the style of the Earth's fold belts. The fact that they rise directly off the plain, without foothills, makes them all the more impressive. When it had been thought that Io's volcanism was primarily sulphurous, it had been obvious that the mountains could not be volcanic edifices since sulphur did not have the structural strength to create peaks on such a scale. Galileo had shown that much of Io's volcanism involved silicates, but this, however, did not shed much light on the mountains.

Even as Galileo was exploring the Jovian system, the origin of Io's mountains remained a topic of research. Paul Schenk of the Lunar and Planetary Institute in Houston and Mark Bulmer of the Center for Earth and Planetary Science at the National Air and Space Museum analysed the shape of Euboea Mons near the south pole to assess a possible landslide.[83] Galileo had yet to improve upon Voyager's coverage of this feature, so they made a topographical map using a stereo pair of images from 1979. A prominent curving ridge crest about 10 kilometres high divides the steep southern flank from the much smoother and shallower northern slope which leads to a thick and heavily ridged deposit possessing a scalloped outer fringe. It seemed that the entire north face had slumped and that the northern deposit was an accumulation of debris from a major avalanche. If Euboea Mons was ancient, then its steep southern flank suggested structural strength. On the other hand, the collapsed northern flank indicated structural failure. The mountain's polygonal shape lent support to the proposition that it is a faulted and uplifted block of crustal material that was displaced on a shallow over-thrusting fault that dips to the north. This neatly avoided the contradiction of a single edifice displaying both structural strength and structural collapse by positing that the massif is basically rigid; the failure was triggered by the act of uplift and the debris is the supracrustal material which had been deposited on the block. The fact that the northern flank is smooth implied that it was once a horizontal surface. The 6-degree slope presumably matches that of the dipping fault. The southern flank is a heavily eroded, but still 10-

kilometre-tall vertical exposure that presents a magnificent opportunity for a future intrepid Ionian field geologist. The altitude of these peaks demonstrates that the silicate lithosphere is at least 10 kilometres thick.

So what caused the fault and the associated uplift? Io is so active that it is resurfaced at an average rate of 1 centimetre per year. Schenk and Bulmer reasoned that older buried deposits will be forced to subside as new volcanic material is added. As it sinks, it will undergo lateral compression which will induce thrust faulting. This idea is appealing because it draws together several lines of evidence. The mountains are massifs and even though they appear to be the most ancient features on Io they may still have been upthrusted recently. This suggested that Io's mountains would tend to be polygonal and combine steep slopes with shallower slopes showing signs of mass wastage Galileo studied the mountains it found topography ranging from angular peaks to low mounds surrounded by gently sloping debris aprons, and the more angular appeared to be younger than the more rounded. The ridges parallel to the margins suggested progressive slumping under the force of the moon's weak gravity.[84] Clearly, the mountains are in various states of deterioration, with tilted blocks in early, middle, and late stages of collapse. The structures evidently start out tall and steep and progressively collapse into flattened mounds. Are the faulted blocks eased up gently or by sudden uplift in stupendous Io-quakes? The sharp features of the youngest peaks suggest that they were forced up rapidly.

Significantly, Schenk and Bulmer saw no evidence of lava flows, vents, calderas, or any other volcanic features on Euboea Mons. This indicated that the deep fracture which prompted the uplift did not provide a route for lava to reach the surface to modify the massif. Evidently, although Io's volcanism has given rise to vast calderas, it has *not* created edifices similar to Earth's stratovolcanoes. Io's calderas show several unique characteristics.[85] In many cases they abut mountains, an excellent example being the massive fractured block alongside Pele's caldera. Juxtapositions such as this indicate a structural relationship between the mountains and the calderas.[86,87,88] According to one recent study, the calderas and mountains are both manifestations of Io's seething interior, in that the mountain-forming faults are induced by slight changes in the rate of lava build-up on Io's surface and the heating of the cooler crust at some depth[89] – that is, heat slowly making its way out from the interior by conduction becomes 'trapped' in the rapidly subsiding strata. If volcanic activity ceases in a given area and no new lava flows are deposited, then the crust will stabilise. Once the heat is able to rise, it will cause the crust to expand. The combination of compressive and thermal stresses should fracture the crust and push up and tilt large blocks. However, the juxtaposition of so many mountains and calderas prompted the suggestion that the subcrustal heating resulting from subsidence near active sites is itself sufficient to break up the crust.[90] Another theory, agreeing that the mountains involved compressional stress brought on by the high resurfacing rate, explored the possibility that rising diapirs impinge upon the underside of the crust and play a part in the upthrusting process.[91,92]

As a planetary interior melts, the amount of iron and magnesium in the melt increases with the *degree* of melting. Depending upon the pressure, a lava temperature of 1,800 K suggests up to 30 per cent melting. This prompted Laszlo

Keszthelyi and Alfred McEwen, with Jeffrey Taylor of Hawaii's Institute of Geophysics and Planetology, to re-examine the discarded early concept that Io's interior is a partially molten 'mush' of crystals in magma.[93,94] In 1979, when Stanton Peale, Patrick Cassen and Ray Reynolds had reasoned that Io might be volcanically active, they had envisaged an ocean of magma beneath a thin volcanic crust. However, there were problems with this. One argument was that a magma ocean would soon lose heat due to rapid convection and would solidify. Another problem was that mountains would be eroded at their bases faster than they could be built up. The idea was discarded as inappropriate when it was concluded that Io's lava was primarily sulphurous. Although it was widely accepted that Io's interior was mostly solid, M. Ross and Gerald Schubert proposed in 1985 that it was partially molten and mushy. Analysis of the reflection spectra shows that the lavas at all of Io's active volcanoes are similar in composition, in containing a high proportion of iron and magnesium, almost certainly in the form of orthopyroxene-rich komatiite. Galileo's long-term volcano-monitoring established that the geographical distribution of volcanic activity is uniform, which indicates that the source of this mafic magma is global. Keszthelyi therefore suggested that Io's interior is a sticky slurry of molten rock and suspended crystals beneath a lithosphere about 100 kilometres thick. A mushy mantle would not convect so vigorously and would not cool so rapidly, but mixing would maintain the high proportion of mafic minerals in the zone feeding the volcanism. A mushy mantle would tend to eat away at the lower lithosphere and induce partial melting, which would help to break the crust into large blocks which are then tilted and upthrusted to form the massifs.

A study of the composition of the materials around Io's vents suggested a model in which the crust is a thick stack of ultramafic lava, sulphurous lava, and various types of pyroclastic, with a reservoir of liquid sulphur sandwiched between this volcanically created crust and the underlying ultramafic lithosphere.[95,96,97] The composition of any given activity depends upon the depth of its source, with the silicate lavas coming from the well-mixed deep mantle and the sulphurous lavas being derived from individual shallow reservoirs.

After the Voyagers revealed that Io was extremely volcanic, the debate centred on whether this was silicate or sulphurous magma. We now know not only that *both* types of activity are involved, but also that the silicate volcanism extrudes super-hot ultramafic magmas that are no longer extruded onto the Earth's surface. In the case of the Earth, on which plate tectonics is active, the interior releases its heat by a process of recycling in which the mantle extrudes mafic magma at a spreading rift, and this cools as it is transported horizontally to a trench, whereupon, as it is subducted and reassimilated into the mantle, it prompts partial melting in the overriding plate and progressively builds up silicic crust. On Io, however, the crust undergoes continuous vertical recycling on a *random* basis. As material piles up around a vent its weight forces the thin crust to sink, which in turn causes the subcrust to melt and be reassimilated. Instead of being the anticipated frozen impact-scarred relic, Io serves as a 'working model' of the processes by which the hot early Earth might have released its heat, prior to the onset of global plate tectonics. It is an extraordinary parallel.

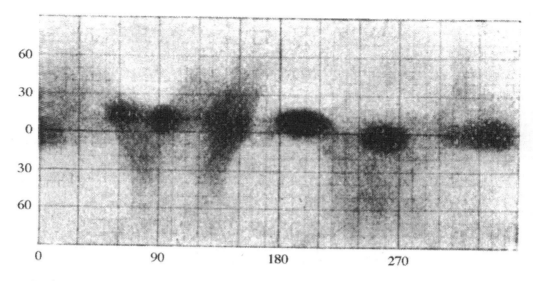

In the early 1950s B.F. Lyot of the Pic du Midi Observatory in the Pyrenees released a map of Europa. Because it was predominantly white, it was speculated that icy caps were so large that only a narrow strip of dark rocky terrain was exposed in the equatorial zone.

EUROPA

At about 1,500 kilometres radius Europa is the smallest of the Galilean satellites, but it is nevertheless comparable in size to the Earth's Moon. Prior to the 'Space Age', there was little reason to expect Europa to be significantly different from Io. Both were expected to be battered by ancient impacts and, as a result of their remoteness from the Sun, laced with frozen gases. However, Europa had an intriguingly high albedo of 0.64, which made it one of the most reflective bodies in the Solar System.

When Voyager 1 flew by Europa at a range of 732,000 kilometres in March 1979, it found Europa to be very different to Io. Firstly, Io was reddish, but Europa was predominantly white. It had been speculated on the basis of maps drawn by telescopic observers that icy caps may have left only a narrow strip of dark rocky terrain along the equator. In fact, Europa was revealed to be *completely* enshrouded in ice. A few months later, Voyager 2 closed to within 204,000 kilometres of Europa and showed that its surface is darkly 'mottled' and criss-crossed by linear features which are thousands of kilometres long but only a few kilometres wide. One scientist, watching the raw imagery come in, is reputed to have mused that G.V. Schiaparelli and Percival Lowell, both of whom had reported seeing linearities on Mars, would have *loved* Europa.

Although the best imagery had a resolution of 2 kilometres per pixel, only a small fraction of Europa's surface was covered at anything like that resolution, so the mapping project was severely constrained. Photogeologists prefer to base their initial

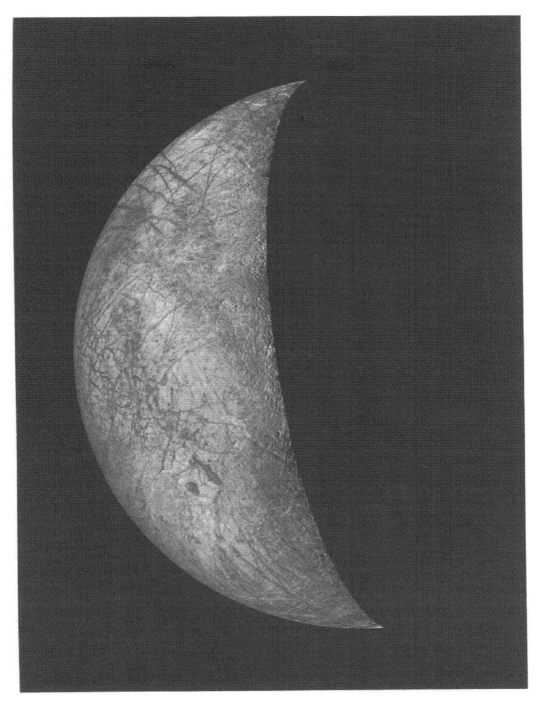

Voyager 2 showed Europa to have long dark linea, mottled terrain, irregular dark maculae and fractured plains. The sharply defined terminator indicated that there is very little surface relief.

analyses of planetary surfaces on purely observational aspects, such as the albedo, colour and texture, in order to distinguish between objective mapping and subjective theorising as to which processes formed those features. A number of observations were immediately obvious. The exceptionally sharp terminator line from pole to pole indicated that the surface is essentially devoid of topographic relief. In fact, the elevation range spans only a few hundred metres – in other words, it is 'billiards ball' smooth. One physiographic analysis based on albedo, colour, and textural variations identified two principal surface units referred to as 'plains' and 'mottled terrain' (the mottling was divided into 'brown' and 'grey'), with a number of isolated irregular spots that were classified as maculae (meaning literally 'fuzzy spot').[98] The surprising absence of craters indicated that some active process had to be 'removing' them. Four types of plains were noted: the 'undifferentiated plains' are smooth, tend to be gradational with adjacent terrains, and are transected by numerous linear features; 'bright plains' are located preferentially towards high latitudes, and are criss-crossed by a variety of types of lineation; 'dark plains' are similar but are darker; 'fractured plains' appear to have been shattered and they bear curvilinear grey streaks and numerous brown spots. The brown mottling is moderately textured. It forms distinctive contacts with adjacent terrains, suggesting that it is younger. In contrast, the grey mottling is smooth and has diffuse boundaries. So what is the texturing on the brown terrain? It seems to be hummocky. Is this related to the hue? What is the brown material? Is it endogenic or exogenic in origin? The brightest mottled terrain is on the leading hemisphere and the darkest is on the trailing hemisphere. As the Jovian magnetosphere rotates with the planet, the trailing hemisphere of the slowly orbiting moon is bombarded with charged particles, and the leading hemisphere is shielded by being within the magnetospheric 'wake'. Is the difference in albedo between the hemispheres due to this irradiation? If the ice is darkened by exposure to charged particles, then this would imply that any light plains close to dark plains must be younger. As to the linearities, the fact that some of them extend almost halfway around the moon and trend northwest in the northern hemisphere and southwest in the southern hemisphere hinted that they are the result of gravitational stresses induced by orbital eccentricities. The shapes of the 'dark wedges' which transected other types of terrain hinted at crustal spreading. Apparently, the bright plains formed when a global ocean froze over. This was disfigured by the dark and fractured plains and by the grey mottling and then – some time later – fluid oozed out as brown mottling. The various lineations were added still later. However, deriving the relative ordering is only half of the solution. Fewer than a dozen or so craters up to 25 kilometres across were evident. The paucity of impact craters made estimating the exposure ages of the various terrains tricky, but suggested that the entire surface is *extremely* young. Furthermore, a few palimpsests 100 kilometres across with concentric fractures and low surface relief look as if they may have been caused by large impacts, and their subdued form implies that the ice was thin when the impacts occurred. The prospect that the surface froze over 'recently' raised the amazing possibility that there might be an ocean beneath the ice, and one of the Galileo mission's primary objectives was to find out.

On 27 June 1996, after its first perijove, Galileo imaged Europa from 155,000 kilometres. Although this was only 20 per cent closer than the best Voyager fly-by, Galileo's camera was better. The highest resolution of its regional mosaic was only 1.6 kilometres, but this was enough to reveal a major surprise in the intensely fractured terrain southwest of the anti-Jovian point. Dark linear, curved, and wedged-shaped bands had fragmented the icy crust into blocks up to 30 kilometres across, and as these had drawn apart and rotated, slush had oozed up to seal the gap. It was strikingly reminiscent of 'pack ice' on the Earth's polar seas, but on a gargantuan scale. The fact that the ice had been broken into blocks which had been able to move strongly supported the case for the ice being a thin shell perhaps only a few kilometres thick – capping an ocean.

A Europan ocean would freeze from the top down. Unless there was a source of internal heat to maintain its base in a liquid state, the ocean would freeze solid in several million years. Although some heat could be derived from the decay of radioactive elements deep within the silicate interior, Europa is more likely to be kept warm by gravitational tidal action. Its orbital period is twice that of Io and half that of Ganymede, so it is in a double resonance. Europa is farther than Io from Jupiter, however, so it will endure less stress than Io. Europa has not boiled off its volatiles, and so is enshrouded with ice. It may be volcanically active, but this will occur in the silicate lithosphere which forms the floor of the ocean. Hydrothermal activity on the seabed could provide an energy source for any primitive form of life that may have developed, so Europa was put on the short-list of places where life may have developed.[99]

On its first close fly-by, at a range of 34,000 kilometres, Galileo revealed the dark wedges in unprecedented detail. The symmetric ridges within the wedges confirmed that these are sites of spreading. The central feature marks the spreading axis and the succession of parallel ridges were formed as the crust was pulled apart and fluid oozed from the axial fracture and froze; in this way the ages of the ridges increase away from the axis. Several months later, Galileo flew within 700 kilometres of Europa – fully 300 times closer than the Voyager 2 fly-by. The doppler on the spacecraft's signal yielded a measure of the moon's moment of inertia.[100] This indicated a layered structure with a dense core englobed by a rocky mantle and a shell of water some 100 to 200 kilometres thick, but the data could not determine how much of the water in this shell was ice. However, the spacecraft's magnetometer noted a magnetic signature which prompted a suggestion that the Jovian magnetosphere induces electric currents in 'salty' water, which in turn generate the observed magnetic field. If this were true, then most of the shell of water would still be in the liquid phase.

The impact that excavated the 26-kilometre-wide crater Pwyll had deposited bright rays of ice for a thousand kilometres across the trailing hemisphere, and the far end of one of the rays crossed a 100-kilometre-wide patch of mottled terrain, called Conamara, north of the equator. Galileo revealed this to have been broken into polygonal 'rafts', individually 3 to 6 kilometres in size. An analysis of the manner in which the rafts were rotated and tilted concluded they had floated in a turbulent fluid which froze and locked them in position. As icebergs, 90 per cent of their bulk lies below the waterline.[101] To some researchers, this was the 'smoking

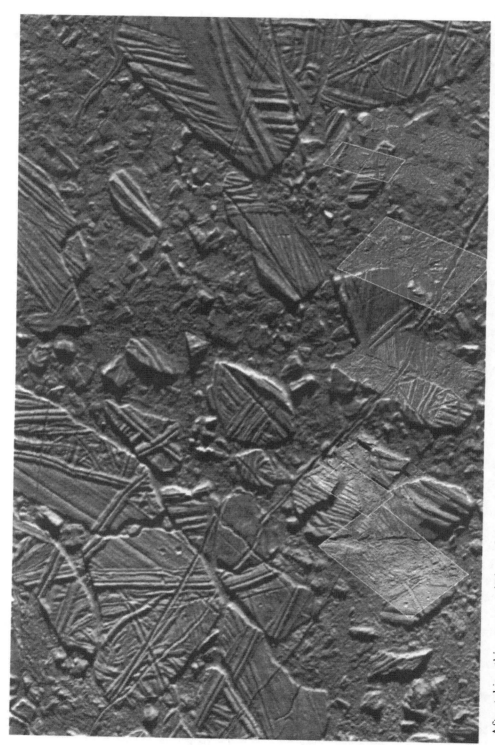

After taking this contextual shot of the iceberg 'rafts' of Conamara on Europa, Galileo returned to take a series of high-resolution images running east-to-west across the chaotic terrain.

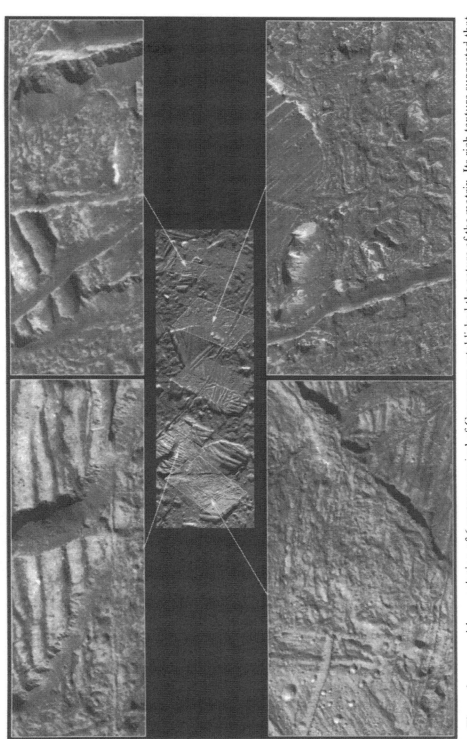

These close-ups with a resolution of 6 metres per pixel of Conamara established the nature of the matrix. Its rich texture suggested that a jumble of variously-sized blocks of ice had swirled in a turbulent fluid and then been frozen in position. Some of the rafts are fairly flat but others are heavily ridged, reflecting the features extant on the disrupted plain.

gun' that proved that Europa had once had liquid water exposed at the surface – at this site, at least. One study of this and other chaos zones found that the thickness of the ice increases away from the equator to mid-northern latitudes, varying from 2 to 6 kilometres.[102] On an ocean up to 200 kilometres deep, such a thin and brittle frozen shell would be completely decoupled from the underlying silicate lithosphere. Some rafts carry segments of ridges. An effort to reconstruct the disrupted ridges found that more than half of the original surface is missing.[103] This could only have been completely melted. It looked as if some internal heat source, perhaps a volcano on the ocean floor, had created a plume of warm water which had melted through the ice and jostled the rafts before freezing again.

Of course, the age of the crust is contentious. A study by Gene Shoemaker of an impactor population dominated by comets concluded that the surface froze 10 million years ago.[104] Clark Chapman of the Southwest Research Institute in Boulder, Colorado, interpreted the paucity of small craters to mean that the surface may be even younger. A million years ago is 'yesterday' in terms of geological processes, and it would be almost certain that such processes are still active. Michael Carr, on the other hand, has argued that Europa's surface could be a billion years old. Although this could be considered to be fairly 'recent' in terms of the age of the Solar System, any endogenic processes that were active then would very likely have expired by now. Ironically, like Chapman, Carr based his conclusion on cratering. He noted that the largest impacts on nearby Ganymede (which can reasonably be attributed to the tail-end of the bombardment 3.8 billion years ago) have not been disfigured by smaller impacts to the same extent as the lunar basins, which implied that the rate for smaller impactors was lower in the Jovian environment than in the inner Solar System, so the paucity of craters on Europa did not necessarily imply *extreme* youth. Once Galileo had accumulated cratering statistics for Ganymede, a study that assumed an early asteroidal impactor population and scaled this rate for Europa's position closer in to Jupiter concluded that Europa's surface froze 3 *billion* years ago.[105] It is notable that the debate concerning the age of Europa's surface spans a range of *three orders of magnitude*.

A 200-kilometre fly-by by Galileo yielded a strip of images with a resolution of 6 metres per pixel on an east–west line running across the Conamara chaos to confirm that the icebergs had floated in a turbulent fluid. The improved resolution also demonstrated that, in addition to painting western Conamara with a ray of light-toned fines, the Pwyll impact had peppered the area with a large number of secondary craters ranging from 30 to 450 metres in diameter. The fact that the ejecta is superimposed on the chaos meant that the melting event could not be any younger than Pwyll, whose prominent rays indicated that it was 'recent'.

Astypalaea Linea in Europa's southern polar region is a strike-slip fault where vast blocks of the icy shell have slipped past one another horizontally, free of vertical displacement.[106] On Earth, such faults are an integral feature of plate tectonics. In the case of the San Andreas, the coast of southern California has moved 300 kilometres north. Astypalaea Linea is comparable in size, and Galileo built up a mosaic of a section of it. Correlating features to either side showed that it has undergone about 50 kilometres of motion. Along much of its length the actual fracture is marked by a ridge. In places, the structure forms a 'staircase', the far sides of which can be matched to recreate the terrain that has been displaced by the fault. As the fault moved, its zig-zagging line made rhomboidal openings and warm ice welled up to create areas of new ice. A similar crustal extension has been observed in deep troughs on strike-slip faults in Death Valley in California and in the Dead Sea in the Middle East, except that there the upwelling is buried by sediments.

Although both Earth and Europa possess long transform faults, the processes that create them are different. In the terrestrial case, the lithospheric plates are driven by convection in the silicate mantle. This cannot be the case on Europa. The icy surface is only a thin brittle shell on a very deep ocean, so it will be particularly susceptible to cracking under gravitational stresses. One theory is that convection currents in the putative subsurface ocean could have dragged the sheets of icy crust, driving the fault in much the same way as mantle currents drag the Earth's plates. However, the boundaries of the icy plates should show signs of twisting and shearing where they rub past each other, just as the main fracture of the San Andreas has given rise to a swarm of secondary faults. The absence of such shearing supported an alternative hypothesis, in which the fault's motion is due to diurnal gravitational tides early in the moon's history. In the 'morning', tidal tension would have opened a pre-existing fault, and the build up of stress made it move in one direction. In the 'afternoon', when the extensional force relaxed, the fault would have locked. Because the fault never returned to its original position, the diurnal cycles would have produced a steady accumulation of lengthwise offset motions; a process referred to as 'walking'. The fact that Astypalaea is marred by many fractures shows that it has not been active recently. It may well be an ancient structure which ceased moving either when the tidal forces diminished, or when the icy crust thickened sufficiently to inhibit the tidal forces from making it 'walk'.

Voyager imagery of the south polar area showed chains of scalloped lines linked arc-to-arc at their cusps, extending over the icy plains for hundreds of kilometres. When Galileo arrived, the origin of these 'cycloids' (or 'flexi' as the International Astronomical Union called them) was still a mystery. When the process that created them was finally identified, it was further evidence for there once having been a deep ocean beneath the icy crust. Evidently, the cuspate form is the result of pressure on the underside of the thin icy shell induced by diurnal tides in the enclosed ocean.[107] While Europa's rotation is synchronised so that it turns on its axis once per orbit of Jupiter, resonances with Io and Ganymede make its orbit slightly elliptical. When Europa is closer to Jupiter, the increased gravitational attraction draws the water in the ocean towards the sub-Jovian point; then as the moon moves out, the water ebbs. The Moon raises tides in the Earth's oceans with an amplitude of 2 metres. Although Europa is comparable in size to the Moon, Jupiter is 318 times more massive than the Earth and raises tides in Europa's ocean with an amplitude of 30 metres. Because the ocean is capped by ice, the flow of water causes the icy shell to flex in a cyclical manner. If this tidal force exceeds the tensile strength of the ice, a crack propagates relatively slowly across the ever-changing stress field, following a curving path. Detailed modelling revealed that each 'morning', when the propagation resumes, it does so along a new arc. The distinctive scalloped appearance results from the fact

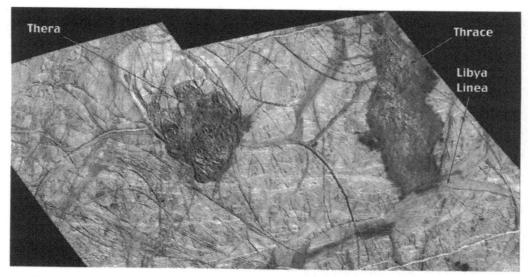

It would appear that endogenic processes disrupted Europa's icy plain and formed the irregular dark maculae of Thrace and Thera as zones of chaos, and then stained them with dark brown fluid.

that successive arcs have shared cusps. While this model accurately accounted for the shape of the cycloids, it could do so only if the tidal bulge was decoupled from the interior, which meant that when these features were formed a deep ocean must have separated the thin icy shell from the underlying silicate lithosphere, and since the cracks probably extended only a few kilometres into the ice, the shell may have been as little as 15 kilometres thick.

North of Astypalaea Linea are the irregular dark splotches of Thrace and Thera Maculae. They are suspected of being similar to the chaotic terrain in which icebergs were trapped in a fluid matrix, only in this case with a dark matrix. At about 80 kilometres across, Thera, the smaller of the two, is slightly lower than the level of the surrounding plain. The curved fractures along its periphery suggested that its creation may have involved collapse. Thrace, however, stands at or slightly above the surrounding terrain. A study of its southern margin revealed that it has been flooded by an upwelling of a brown fluid.[108] The central area consists of many broken and disrupted blocks, and is indeed a chaos zone, but its blocks have not been moved around. This is in contrast to Conamara, which has been extensively melted and is 60 per cent matrix. Around the periphery of Thrace, the topography of the surrounding plain can be traced through a transition zone poking through the dark flooding material. In the Voyager imagery, the macula appeared to terminate abruptly where it abuts Libya Linea to the south, but Galileo showed that the grey band is stained with the dark macula material. Several theories have been offered to explain the formation of the maculae, ranging from a total melt-through of the icy crust by the ocean beneath to a partial melting and disruption of the surface by upwelling warm ice. Their lobate character had suggested that they might be the result of extensive flooding following the ascent and eruption of large quantities of aqueous magma.[109]

However, Galileo's high-resolution views contradicted this. The process appeared to have been more gentle. Perhaps the crustal melting was the result of an infusion of brine into the ice from below?

The geological evidence in favour of there being an immensely deep ocean beneath a thin shell of ice was mounting,[110] but the surface studies were subject to a variety of debatable (and hotly debated) assumptions – witness the ambiguity concerning the age of the surface. Only Galileo's magnetometer could investigate the state of the putative ocean *as it is today*. A close fly-by was arranged specifically to test the working hypothesis that the magnetic field derives from electric currents induced in salty water by the Jovian magnetosphere. As Jupiter's magnetic axis is inclined with respect to its spin axis, the ambient field at Europa's location reverses its polarity periodically. The moon's magnetic poles ought therefore to be near the equator, and they ought to migrate. The fly-by geometry was arranged to determine whether this was the case, and Galileo found that the north pole was changing alignment precisely as predicted – every five and a half hours. Other than observing a cryovolcanic flow underway, this was about as definitive as the circumstantial case was likely to be.

Being the silicate lithosphere, the ocean floor is Europa's *real* surface. If the moon was a little closer to Jupiter, the gravitational stresses from its orbital resonances would surely have induced such intense volcanism that its volatiles boiled off, in which case Europa could easily have ended up like Io. It is remarkable that these two moons, so similar in size and orbiting so close to one another, should have ended up so different – literally a case of 'fire' and 'ice'.

GANYMEDE

Ganymede is not only the largest of the Jovian moons, it is the largest satellite in the Solar System. In fact, being somewhat larger than the planet Mercury, it is really a small planet that happens to be orbiting around Jupiter. However, because the temperature of the solar nebula decreased with distance from the newly formed Sun, the rocky bodies that condensed far out incorporated a high proportion of water, creating 'ice moons'. The bulk density of 2.0 g/cm^3 implies that Ganymede is a roughly equal mix of rock and ice. The 100 K surface temperature means that the ice at the surface must be as hard as rock.

The Voyager missions mapped 80 per cent of Ganymede with a resolution of 5 kilometres or better per pixel, and two major physiographic terrains were defined, one dark and the other light in tone.[111] The 'dark cratered terrain' consists of several large circular features, and a large number of small polygonal blocks. The largest, appropriately named Galileo Regio, is an oval some 2,800 to 3,200 kilometres in diameter. In fact, it is so large that it spans about one-third of the anti-Jovian hemisphere. Although it had actually been seen by telescopic observers, the early maps otherwise bore little resemblance to the moon's surface. This type of terrain is etched by parallel curvilinear 'furrows' which run for hundreds of kilometres and have been interpreted as grabens due to crustal extension.

The lighter 'grooved terrain' is characterised by parallel alternating ridges and

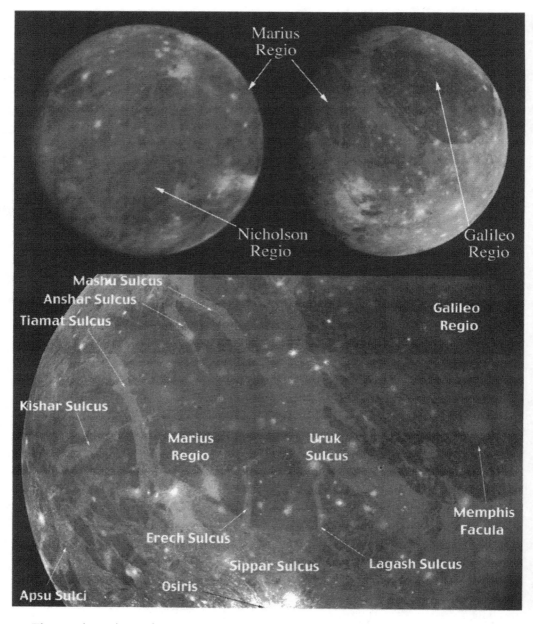

The two long-shots of Ganymede show the hemispheres dominated by the large dark ovals of Nicholson Regio and Galileo Regio. The fragmentation of Marius Regio (located in between) by narrower light-toned sulci is suggestive of crustal spreading.

grooves running for hundreds of kilometres with the troughs and crests separated by 5 to 10 kilometres horizontally and several hundred metres vertically. The ridge/groove pairs are 'bundled' to form lineations. In some places, these cut across one another at various angles, forming a 'reticulate terrain' which documents a complex record of tectonic activity. This terrain often includes small hills and so presents a 'hummocky' texture. There are patches of remarkably smooth material within the grooved terrain, and the grooves warp around these 'smooth plains'. Since there is no albedo change over the boundaries, the smooth plains may be either grooved terrain that has been melted from beneath or submerged by fluid extrusions – cryovolcanism. As the grooved terrain transects the dark cratered terrain, it is younger. Furthermore, matching edge-relationships between dark units separated by a strip of light terrain imply that the dark terrain has been fractured and drawn apart by the formation of the lighter-toned material. This is shown starkly by the symmetry with which Uruk Sulcus divides Marius Regio from Galileo Regio and by the manner in which Marius Regio has been fractured into small polygonal blocks. Such 'spreading' suggests nascent plate tectonics in the mainly icy crust. Tiamat Sulcus, one of the light-toned strips that fragment Marius Regio, has been transected by a strike-slip fault which documents horizontal motions by large blocks of icy crust. The sulci is wider to the south of the fault line than to the north of it, and differences in the grooving across the fault indicated that the two sides underwent crustal extension at different rates.[112] Most of Ganymede's north polar region is sulci terrain. Its high albedo vindicated the early telescopic mappers who had charted a 'polar spot'.

There are several multiple-ringed basins, the largest of which is Gilgamesh. It has a 150-kilometre-wide central depression which forms a smooth plain surrounded by disrupted terrain and hummocky ejecta. The outermost ring, which is delineated a scarp, is 275 kilometres out. Overall, the structure is rather subdued. Its impact origin is evident from the secondary craters and other characteristic 'sculpture' ranging out to a radius of 1,000 kilometres. Superposition relationships indicate that Gilgamesh formed subsequent to the adjacent sulci, which dates the tectonic activity that made the sulci to the tail-end of the bombardment, some 3.8 billion years ago. Unlike on the Moon and Mars, craters on Ganymede become progressively 'flatter' with increasing size. Actually, this had been predicted in 1973, on the basis that an icy crust would be plastic and would flow to relax vertical relief.[113] The rim of a crater would gradually sink and the floor would rise to form a dome. It was isostasy at work. This process would operate most rapidly on the largest craters with the most deeply excavated cavities and the most piled-up ejecta, and it has transformed large impact craters into palimpsests ranging up to 400 kilometres across that are still surrounded by their well-formed secondaries. They are preferentially (although not exclusively) found on the older, more densely cratered, dark terrain. Taken together, the palimpsests account for a quarter of the mapped surface, and thus they represent a major part of the early cratering history. The larger impacts probably date to the tail-end of the bombardment. The morphology of the palimpsests that exceed 50 kilometres in diameter implies that the lithosphere was only about 10 kilometres thick at that time.

The darker terrain was deemed to be ancient on the basis that it was more heavily cratered, and the light sulci were deemed to be 'clean' ice extruded as a result of tectonic activity. It was therefore concluded that Galileo Regio, Marius Regio, Nicholson Regio, Barnard Regio, and Perrine Regio were the oldest terrains.[114] Although the surface is predominantly ice, it has been speculated that it grows 'dirty' with age. Certainly the brightest areas are the 'ray craters', such as Osiris, where fresh white ice has been excavated. However, if the light terrains are ancient (as indicated by the superposition of Gilgamesh on the sulci), the 'darkening' must have occurred very early in the moon's history. The furrows of Galileo Regio have a radius of curvature that is focused in the vicinity of the anti-Jovian point, suggesting that they may have been induced by forces extant while the moon's rotation was being synchronised with its orbit. However, this may be simply a coincidence. Alternatively, Ganymede's tectonism might be the result of crustal expansion. A study proposed that both the furrows crossing the dark cratered terrain and the flat-floored grooves of the sulci could be accounted for if the ice in the interior underwent a phase change and expanded as the moon cooled, cracking the lithosphere.[115] This model suggested that the lithosphere was about 10 kilometres thick when the furrows formed, but only half as thick when the sulci formed. Since the sulci formed after the dark cratered terrain, this crustal thinning is consistent with 'warm ice' upwelling through the fractures and, where this broached the surface, cryovolcanic activity. When the Galileo spacecraft returned a high-resolution view of Uruk Sulcus showing dozens of long parallel ridges and valleys – almost as if it had been graded by a giant rake – this lent support to the hypothesis that the sulci resulted from lithospheric expansion. Albedo variations correlate with topography, in that bright icy material is exposed on the crests of the ridges and the dark material has 'pooled' in the lower-lying areas. Stereoscopic analysis showed that the sulci typically lie as much as 1,000 metres below the adjacent dark terrain, with the smoothest sulci being the deepest,[116] supporting the proposal that low-viscosity cryogenic fluid had accumulated in the low-lying areas. Galileo Regio, to the north, also provided several surprises, with at least three terrain types in terms of both albedo and morphology.[117] Comparing craters roughly a kilometre across, the number density is about an order of magnitude greater on Galileo Regio than on Uruk Sulcus.[118] When the statistics for small craters were merged with those for larger craters derived from the Voyager imagery, the curve bore a striking similarity to that of the Moon. If the youngest multiple-ring structure, Gilgamesh, is assigned an age of 3.8 billion years in order to anchor the curve, the ancient terrains date back to the period 4.1 to 4.3 billion years. So, even though Galileo Regio is ancient, it has been reworked by repeated episodes of shearing, rifting, and cryovolcanic activity. Only one-third of the early surface persists, in the form of the dark terrains. The remainder of the moon has been resurfaced by bright terrain,[119,120] and if the oldest dark terrain dates to about 4.2 billion years and the youngest of the light-toned sulci to 3.7 billion years, the tectonism occurred over a period of half a billion years.[121]

A major Galileo surprise was that Ganymede has an intrinsic magnetic field. Although it is only one-thousandth of the strength of the Earth's field, it is sufficient to form a 'bubble' within the Jovian magnetosphere. The 'standard model' for an intrinsic dipole magnetic field involved an iron core, but the Voyager data suggested an icy lithosphere less than 100 kilometres thick that capped a convecting mantle of 'soft ice' 400 to 800 kilometres deep, which in turn englobed a large silicate body that was not expected to contain a significant iron core. However, Galileo's close fly-bys established that the moon is more 'evolved' than had been expected: there is an 800-kilometre-thick layer of warm ice beneath the warped and faulted icy crust, an equally thick mantle of rock, and a dense core which, depending on whether it is pure iron or a mixture of iron and iron sulphide, could constitute as little as 1.4 per cent, or as much as 30 per cent, of the mass. The differentiation indicated that the moon had undergone significant internal heating. Since Ganymede is effectively a small planet, the accretional and likely radiogenic heating would certainly have been sufficient to invoke partial melting. This may have been further stimulated by a pulse of heat from the gravitational stress as the moon's rotation was synchronised with its orbital motion. The extant orbital resonance with Europa probably serves to keep Ganymede's interior warm. The magnetometer showed unambiguously that – as in the case of Europa – the magnetic field is derived from electric currents flowing in a salty fluid beneath the surface. If there is any ongoing cryovolcanism, it does not appear to modify the surface to any significant extent, and any activity is probably limited to small water geysers venting under pressure.

Ganymede's remarkable geological history derives from its proximity to Jupiter and the orbital resonances with Europa and Io. With a differentiated structure, Ganymede is remarkably similar to the Earth, and is more of a 'terrestrial planet' than a frozen remnant from the formation of the Solar System.

CALLISTO

Although 4,800 kilometres in diameter, Callisto's bulk density is just 1.83 g/cm^3, indicating that it is a roughly half ice and half rock mix. Its surface is the least reflective of Jupiter's primary moons, so the ice evidently has a significant fraction of impurities. Indeed, it may be dark for the same reason that Ganymede's oldest terrain is dark – whatever that might have been – and, if so, then the absence of bright sulci-terrain implies that Callisto has not been subjected to the tectonism that characterises its larger neighbour.

The closest Voyager approach was 124,000 kilometres, and some 80 per cent of Callisto's surface was mapped at a resolution of 5 kilometres per pixel or better, revealing a monotonously cratered landscape. The fact that Callisto was peppered with craters to the limiting resolution of the imagery implied that it is an ancient surface.

The largest, and most striking topographic feature is the Valhalla multiple-ringed basin. It comprises a smooth central plain some 600 kilometres in diameter and a series of rings, the largest of which has a radius of almost 2,000 kilometres. The rings are 20 to 100 kilometres apart with the spacing increasing with increasing radius, a relationship that places a constraint on any model for the formation of the structure. The impact which excavated the central plain may well have come close to punching right through the icy lithosphere. The population density of craters on this central plain is low, implying that the basin formed towards the end of the period of bombardment and that there has been a low impact rate since that time. The number density of craters increases with radial distance through the ring system and merges with the prevailing rate beyond. The existence of *older* craters within the rings indicates that the rings are 'frozen' relics of the shock wave that propagated out from the impact site through the crust.

Although Valhalla bears a striking resemblance to the Moon's Orientale basin, the fact that Callisto's crust is icy means that the morphologies are different in detail. Orientale's rings are composed of fractured and upthrusted crustal blocks but Valhalla's rings are ridges, troughs and outward-facing scarps. The central plain, which marks the extent of the excavation, is not rimmed by an inward-facing scarp front, as is strikingly so for lunar basins. Three distinct morphological zones are found within Valhalla.[122] There is an inner zone comprising the central plain and the first few rings. A dark fluid appears to have welled up though cracks in the badly fractured cavity floor to form the central plain.[123,124] The rings in this zone consist of bands of intermediate-albedo terrain that are sinuous and scalloped and might actually be ridges. In the transition zone, the rings are discontinuous. In the outer zone the rings take several forms: to the south and east they are narrow sinuous troughs with light-toned floors, while to the north and west they are outward-facing scarps with dark heavily cratered backslopes and light-toned material running along the base of the scarp and flooding into nearby small craters. The albedo variations which highlight the ring system therefore derive from mass wastage on steep slopes that has exposed a bright subsurface material.

The Voyagers did not detect anything to suggest that Callisto's surface has been shaped by endogenic processes so, in contrast to its inner neighbours, and apart from continuing cratering, it has evidently been geologically inert. This state undoubtedly reflects both the moon's greater distance from Jupiter and the fact that it is not tormented by any orbital resonances. In fact, as the Galileo spacecraft approached for its first inspection, Callisto's ancient appearance was its most distinctive feature. Galileo flew almost 100 times closer than Voyager, and when the early 30 metres per pixel imagery of a chain of small craters on Valhalla's northern fringe was presented to a packed von Kármán Auditorium at JPL, even the 'old hands' of the planetary exploration community were astonished. There was a startling *paucity* of small craters. It had been expected that there would be an increasing profusion of ever-smaller craters, as on other ancient cratered surfaces. In fact, the local topography appeared to have been blanketed by dark material which formed a seemingly smooth surface. What was this material? How had it been applied? Had the low-lying terrain been levelled by a 'splash' of fluidised ejecta from a major impact? Would such ejecta have come from one of the multiple-ringed basins? As the area imaged was in the fringe of the Valhalla ring structure, had the dark material been ejected by that impact? Was it a non-ice material? The Asgard basin is similar but smaller.[125] Galileo's infrared spectrometer showed that although its central plain was primarily 'clean' ice, the surface in the peripheral ring system contained a significant fraction of non-ice material. The 'raised floor' of the 55-kilometre crater, Doh, on the central plain suggested that the impactor struck an already weakened shallow crust, and broke through to a 'slushy' zone.

One way to estimate the impact rate for the Galilean satellites is to assume that conditions in Jovian space have been similar to those of the inner Solar System and that the size-to-frequency curve derived from counting craters on the Earth's Moon can be applied. This presupposes that the large multiple-ring structures were excavated by asteroidal bodies impacting during the bombardment which ended around 3.8 billion years ago, and that the population of impactors is now dominated by smaller bodies – with the smallest being the most numerous. However, applying the cratering rate calibrated in the inner Solar System to the Jovian environment is complicated by the fact that Jupiter's tremendously far-reaching gravity draws in asteroids and comets, increasing the flux of potential impactors and 'focusing' their effect on the satellites orbiting closer to the planet, so the statistics have to be 'scaled' to compensate for the positions of the satellites relative to their primary. A further complication is that while the population of large asteroids diminished rapidly, the population of comets did not. To derive an age estimate for a surface based on an impactor population that is dominated by comets, it is necessary to determine the current population, assume it to be more or less constant, and then (scaling for Jupiter's proximity) run the rate backwards in time to calculate how long it would have taken to crater each satellite to the observed degree. It sounds simple, but correlating the model with observation is not straightforward, and there are widely differing opinions.

Gene Shoemaker had previously concluded that the cratering rate calibrated in the inner Solar System could *not* be applied to Jovian space where, he believed, the ongoing flux of comets predominates.[126] He conducted a search for comets in order to refine the statistics, and reached the conclusion that Europa's smooth surface is extremely young. If this rate is scaled to the outer moons, it implies implausibly young ages for the 'ancient' terrains on Ganymede and Callisto and suggests that the multiple-ring structures were produced a billion years after such massive impacts had ceased in the inner Solar System. Shoemaker's statistics were based on Voyager imagery, however, which could not resolve craters smaller than a few kilometres in diameter, so the curve was frustratingly truncated. The Galileo imagery provided a 200-fold increase in resolution. Although Galileo was able to study only limited areas in fine detail, this facilitated the counting of craters down to 100 metres in diameter, and the statistics could be superimposed on the curves derived from Voyager to form a record covering craters up to 100 kilometres in diameter. For both Ganymede and Callisto, these curves are consistent with an ancient bombardment by asteroids.[127] In fact, Callisto is typically somewhat more heavily cratered than Ganymede's dark terrain, and is therefore even older. If Ganymede's well-preserved Gilgamesh multiple-ring structure is assumed to date to the tail-end of the bombardment 3.8 billion years ago, and the flux rate is scaled for Callisto, this yields an age for Callisto's most ancient terrain of 4.3 billion years, and makes it the oldest planetary surface so far observed anywhere in the Solar System.

Although the Voyager encounters had been fairly remote, the doppler data secured by the Deep Space Network had enabled Callisto's moment of inertia to be estimated. This had suggested that the moon was differentiated with a rock and ice lithosphere some 200 to 300 kilometres thick, a 1,000-kilometre-thick mantle of convecting 'soft' ice, and a silicate core. Galileo's first fly-by was expected to refine the parameters of this model. However, doppler tracking of its early fly-bys found no evidence of a central mass concentration. It found that the outermost 300 kilometres is predominantly ice, and that the interior is a *homogeneous* mix of 40 per cent water ice and 60 per cent silicates, iron, and iron sulphide in ratios reflecting solar abundances because the moon formed from the solar nebula.[128] However, the model was revised after later fly-bys. The interior was found to be neither differentiated nor completely uniform. It seems that the interior materials have *partially* separated but are still largely mixed together, with the percentage of rock increasing towards the centre. Compared to its inner neighbours which have been intensely heated by gravitational tides, Callisto is 'half-baked', which raises the possibility that (the ancient appearance of its surface notwithstanding) it might suffer cryovolcanic activity. The real surprise, however, came from the particles and fields instruments. The early results indicated no evidence of a magnetic field, but following the discovery that Europa's weak magnetic field was due to Jupiter's magnetosphere inducing electrical currents to flow within a layer of salty fluid at shallow depth beneath an icy shell, the Callisto data was reanalysed and a very weak 'signature' with similar characteristics was revealed. The currents probably flow in a transition zone of slushy ice global in extent, 100 to 200 kilometres below the surface and something like 10 kilometres deep. Because Callisto is so far from Jupiter and it is free of orbital resonances. However, it undoubtedly accumulated its share of radioactive elements during accretion, and in its undifferentiated state these will be fairly uniformly distributed and the heat that prevents the slush solidifying is probably radiogenic. Over time, Callisto will have progressively frozen from the outside in, so it will still be warm only far below the surface. The electrical currents which generate the secondary magnetic field require the slush to be no more salty than Earth's oceans, and the salt will act as a natural antifreeze. Nevertheless, the outermost Galilean satellite formed a thick lithosphere very early in its history and this preserves the impacts from the tail-end of the accretional process.[129]

THE 'FREE LUNCH'

Because the majority of the energy of sending a payload into space is devoted to climbing out of the Earth's 'gravitational well', low orbit has been called 'halfway to anywhere in the Solar System'. Jupiter is the key to the outer Solar System. Once a spacecraft has been set on course for Jupiter, and as long as the positions of the planets are conducive to using slingshots, it requires no more fuel to reach the planets beyond. Remarkably, although flying from one planet to the next in sequence greatly increases the distance that has to be travelled to reach one of the outer planets compared to the 'direct' route using a Hohmann transfer orbit, the accelerations from the encounters *en route* actually shorten the travel time. In providing access to the outer Solar System, Jupiter has served up the ellusive 'free lunch'.[130]

Although JPL was determined to preserve the option of undertaking the 'Grand Tour', by flying close to Jupiter in order to investigate Io, the boost that Voyager 1 received meant that it would reach Saturn too soon for a 'gravity assist' to destinations beyond. Its planetary exploration would therefore conclude at Saturn. The two spacecraft had been launched a few weeks apart, and they had arrived at Jupiter a few months apart, but by keeping its distance from Jupiter, Voyager 2 flew a slower trajectory to Saturn timed to arrive when conditions were right for this slingshot. However, it would be released to try for the 'Grand Tour' only if its predecessor's Saturn encounter was successful.

NOTES

1. 'Galileo's telescopic observations', G.V. Coyne. In *The three Galileos: the man, the spacecraft and the telescope*, C. Barbieri, J.H. Rahe, T.V. Johnson and A.M. Sohus (Eds). Kluwer Academic, p. 1, 1997.
2. 'The discovery by Galileo of Jupiter's moons', E. Bellone. In 'The three Galileos: the man, the spacecraft and the telescope', C. Barbieri, J.H. Rahe, T.V. Johnson and A.M. Sohus (Eds). Kluwer Academic, p. 7, 1997.
3. Galileo's first telescope magnified 9 times and he could not see very much using it. His improved instrument magnified 30 times.
4. Copernicus developed his heliocentric theory in 1512, and wrote it up in a book in 1530, but this was not published until his death in 1543.
5. In fact, Galileo was found guilty of heresy by the Vatican for advocating Copernicus's theory, and he was put under house arrest until his death in 1642. It was not until 1992 that the Church formally conceded that he had been correct.
6. 'How Galileo changed the rules of science', O. Gingerich. *Sky & Telescope*, March 1993.
7. J.H. Johnson. *J. Brit. Astron. Assoc.*, vol. 41, p. 164, 1931.
8. P. Pagnini. *J. Brit. Astron. Assoc.*, vol. 41, p. 415, 1931.
9. Galileo announced his discovery in March 1610 in a periodical entitled *The Starry Messenger*.
10. According to a Chinese manuscript, an observer named Gan De saw a faint object close to Jupiter in 364 BC; *Patrick Moore's history of astronomy*, P. Moore. Macdonald Co. (fifth edition), p. 52, 1977.
11. As Marius explained his rationale for naming the moons: "Jupiter is much blamed by the poets on account of his irregular loves. Three maidens are especially mentioned as having been clandestinely courted by Jupiter with success: Io, the daughter of the River Inachus; Callisto of Lycaon; Europa of Agenor; and Ganymede, the son of King Tros ...".
12. Light travels at about 300,000 kilometres per second.
13. *Satellites of the Solar System*, W. Sandner. The Scientific Book Club, p. 42, 1965.
14. *The planet Jupiter*, B.M. Peek. Faber & Faber, 1958.
15. *Mem. Am. Ac.*, vol. 8, p. 221.
16. D.H. Menzel. *Ap. J.*, vol. 58, p. 65, 1923.
17. H. Jeffreys. *Mon. Not. Roy. Astron. Soc.*, vol. 83, p. 350, 1923.
18. H. Jeffreys. *Mon. Not. Roy. Astron. Soc.*, vol. 84, p. 534, 1924.

19 R. Wildt. *Ap. J.*, vol. 87, p. 508, 1938.
20 E.E. Barnard. *Astron. Astrophys.*, vol. 11, October 1892.
21 Bernard Lyot's maps were published in *L'Astronomie* upon his death in 1952.
22 *Jupiter*, G. Hunt and P. Moore. Mitchell Beazley, 1981.
23 *Satellites of the Solar System*, W. Sandner. The Scientific Book Club, p. 64, 1965.
24 For a comprehensive review of knowledge of Jupiter prior to the 'Space Age', see *The planet Jupiter*, B.M. Peek. Faber & Faber, 1958.
25 *A popular history of astronomy*, A.M. Clerke. Black Co., p. 93, 1885.
26 Titius presented his simple progression as an annotation while translating Charles Bonnet's *Contemplation de la nature* of 1764 for its German edition in 1766. Later (in 1783) Titius said that Christian Wolff had given him the idea of the numerical sequence. For a reconstruction of the origin of the idea, see 'The early history of the Titius–Bode Law', S.L. Jaki. *Am. J. Phys.*, vol. 40, p. 1014, 1972.
27 *Phil. Trans.*, vol. 92, part 2, p. 228.
28 For a comprehensive review of the history of research concering asteroids, see *Asteroids: their nature and utilisation*, C.T. Kowal. Wiley–Praxis (second edition) 1996.
29 *Pioneer odyssey*, R. Fimmel, W. Swindell and E. Burgess. NASA SP-396, 1977.
30 'Pathfinder to the rings – the Pioneer Saturn trajectory decision', M. Wolverton. *QUEST*, 2000.
31 For an account by G.A. Flandro of how the 'Grand Tour' was conceived see *Planet's beyond: discovering the outer Solar System*, M. Littmann. Wiley, p. 95, 1988.
32 'Fast reconnaissance missions to the outer Solar System utilising energy derived from the gravitational field of Jupiter', G.A. Flandro. *Astronautica Acta*, vol. 12, p. 329, 1966.
33 *Journey into space – the first thirty years of space exploration*, B.C. Murray. Norton Co., 1989.
34 *Far travelers: the exploring machines*, O.W. Nicks. NASA SP-480, p. 241, 1985.
35 Voyager 1's Jovian results were published in *Science*, vol. 204, pp. 945–1008, 1979; *Nature*, vol. 280, pp. 725–806, 1979; and *Geophys. Res. Lett.*, vol. 7, pp. 1–68, 1980; Voyager 2's results appeared in *Science*, vol. 206, pp. 925–996, 1979; and a 'round up' of both Jovian encounters appeared in *J. Geophys. Res.*, vol. 85, pp. 8123–8841.
36 The limb plume, initially referred to simply as 'Plume 1', was subsequently named Pele by the International Astronomical Union, and the one on the terminator (Plume 2) was named Loki.
37 'Discovery of currently active extraterrestrial volcanism', L.A. Morabito, S.P. Synnott, P.N. Kupferman and S.A. Collins. *Science*, vol. 204, p. 972, 1979.
38 'Volcanic eruption plumes on Io', R.G. Strom, R.J. Terrile, H. Masursky and C. Hansen. *Nature*, vol. 280, p. 733, 1979.
39 'Volcanic features on Io', M.H. Carr, H. Masursky, R.G. Strom and R.J. Terrile. *Nature*, vol. 280, p. 729, 1979.
40 'Melting of Io by tidal dissipation', S.J. Peale, P. Cassen and R.T. Reynolds. *Science*, vol. 203, p. 892, 1979.
41 'Heat flow from Io', D.L. Matson, G.A. Ransford and T.V. Johnson. *J. Geophys. Res.*, vol. 86, p. 1664, 1981.
42 'The Jupiter system through the eyes of Voyager 1', B.A. Smith *et al*. *Science*, vol. 204, p. 951, 1979.
43 'The Galilean satellites and Jupiter – Voyager 2 imaging science results', B.A. Smith *et al*. *Science*, vol. 206, p. 927, 1979.
44 *Lunar sourcebook: a user's guide to the Moon*, G.H. Heiken, D.T. Vaniman and B.M. French (Eds). Lunar and Planetary Institute and Cambridge University Press, 1991.

45 'Volcanic eruptions on Io: implications for surface evolution and mass loss', T.V. Johnson and L.A. Soderblom. In *Satellites of Jupiter*, D. Morrison (Ed.). University of Arizona Press, p. 634, 1982.
46 'Two classes of volcanic plume on Io', A.S. McEwen and L.A Soderblom. *Icarus*, vol. 55, p. 191, 1983.
47 'Sulphur flows on Io', C.E. Sagan. *Nature*, vol. 280, p. 750, 1979.
48 'Physics and chemistry of sulphur lakes on Io', J.I. Lunine and D.J. Stevenson. *Icarus*, vol. 64, p. 345, 1985.
49 'Volcanic features on Io', M.H. Carr, H. Masursky, R.G. Strom and R.J. Terrile. *Nature*, vol. 280, p. 729, 1979.
50 'Silicate volcanism on Io', M.H. Carr. *J. Geophys. Res.*, vol. 91, p. 3521, 1986.
51 'Io: evidence for silicate volcanism in 1986', T.V. Johnson, G.J. Veeder, D.L. Matson, R.H. Brown, R.M. Nelson and D. Morrison. *Science*, vol. 242, p. 1280, 1988.
52 'Io's heat flow from infrared radiometry: 1983–1993', G.J. Veeder, D.L. Matson, T.V. Johnson, D.L. Blaney and J.D. Goguen. *J. Geophys. Res.*, vol. 99, p. 17095, 1994.
53 'Volcanic resurfacing rates and implications for volatiles on Io', T.V. Johnson, A.F. Cook, C.E. Sagan and L.A. Soderblom. *Nature*, vol. 280, p. 746, 1979.
54 'The surface of Io: geological units, morphology and tectonics', G.G. Schaber. *Icarus*, vol. 43, p. 302, 1980.
55 'The geology of Io', G.G. Schaber. In *Satellites of Jupiter*, D. Morrison (Ed.). University of Arizona Press, p. 556, 1982.
56 'Erosional scarps on Io', J.F. McCauley, B.A. Smith and L.A Soderblom. *Nature*, vol. 280, p. 736, 1979.
57 The tallest mountain currently measured is Boosaule Mons, which is 16 kilometres high. 'Formation models of Ionian mountains', E.P. Turtle, W.L. Jaeger, L.P. Keszthelyi, A.S. McEwen, *Proc. Lunar Planet. Sci. Conf.*, p. 1960, 2000.
58 'Dynamics and thermodynamics of volcanic eruptions: implications for the plumes on Io', S.W. Kieffer. In *Satellites of Jupiter*, D. Morrison (Ed.). University of Arizona Press, p. 647, 1982.
59 'Initial Galileo gravity results and the internal structure of Io', J.D. Anderson, W.L. Sjogren and G. Schubert. *Science*, vol. 272, p. 709, 1996.
60 For a comprehensive account of activities in Jovian orbit, see *Jupiter odyssey: the story of NASA's Galileo mission*, D.M. Harland. Springer–Praxis, 2000.
61 'Eruption evolution of major volcanoes on Io: Galileo takes a close look', A.G. Davies, L.P. Keszthelyi, R. Lopes-Gautier, W.D. Smythe, L. Kamp and R.W. Carlson. *Proc. Lunar Planet. Sci. Conf.*, p. 1754, 2000.
62 'A close-up view of Io in the infrared: NIMS results from the Galileo fly-bys', R. Lopes-Gautier, W.D. Smythe, R.W. Carlson, A.G. Davies, S. Doute, P.E. Geissler, L.W. Kamp, S.W. Kieffer, F.E. Leader, A.S. McEwen, R. Mehlman and L.A. Soderblom. *Proc. Lunar Planet. Sci. Conf.*, p. 1767, 2000.
63 'The thermal structure of Loki seen in Galileo's NIMS data from the I24 orbit', W.D. Smythe, R. Lopes-Gautier, L. Kamp, A.G. Davies, R.W. Carlson. *Proc. Lunar Planet. Sci. Conf.*, p. 2049, 2000
64 'Volcanic eruptions on Io: heat flow, resurfacing and lava composition', D.L. Blaney, T.V. Johnson, D.L. Matson and G.J. Veeder. *Icarus*, vol. 113, p. 220, 1995.
65 'Thermal emission from lava flows on Io', R.R. Howell. *Icarus*, vol. 127, p. 394, 1997.
66 M. McGrath. American Geophysical Union Fall Meeting 1999 (presented by F. Bagenal).
67 'Eruption conditions of Pele volcano on Io inferred from chemistry of its volcanic plume', M.Yu. Zoltov and B. Fegley. *Proc. Lunar Planet. Sci. Conf.*, p. 2098, 2000.

68 'Road log from American Falls to Split Butte', R. Greeley and J.S. King. In *Volcanism of the eastern Snake River Plains, Idaho: a comparative planetary geology guidebook*, NASA CR 15554621, p. 295, 1977.
69 'The Snake River Plain, Idaho: representative of a new category of volcanism', R. Greeley. *J. Geophys. Res.* 1982.
70 'Volcanic resurfacing of Io: post-repair HST imaging', J.R. Spencer, A.S. McEwen, M. McGrath, P. Sartoretti, D.B. Nash, K.S. Noll and D. Gilmore. *Icarus*, vol. 127, p. 221, 1997.
71 'Sulphur flows of Ra Patera', D. Pieri, S.M. Baloga, R.M. Nelson and C.E. Sagan. *Icarus*, vol. 60, p. 865, 1984.
72 'Dynamic geophysics of Io', A.S. McEwen, J.I. Lunine and M.H. Carr. In *Time-variable phenomena in the Jovian system*, NASA SP-494, 1989.
73 'Observations of industrial sulphur flows: implications for Io', R. Greeley *et al. Icarus*, vol. 84, p. 374, 1990.
74 'Phase transformations and the spectral reflectance of solid sulphur: can metastable sulphur allotropes exist on Io?', J Moses and D. Nash. *Icarus*, vol. 89, p. 277, 1991.
75 'Galileo's first images of Jupiter and the Galilean satellites', M.J.S. Belton *et al. Science*, vol. 274, p. 377, 1996.
76 'Geology and topography of Ra Patera in the Voyager era: prelude to eruption', P.M. Schenk, A.S. McEwen, A.G. Davies, T. Davenport, K. Jones and B. Fessler. *Geophys. Res. Lett.*, vol. 24, p. 2467, 1997.
77 'Active volcanism on Io as seen by Galileo SSI', A.S. McEwen, L.P. Keszthelyi, P.E. Geissler, D.P. Simonelli, T.V. Johnson, K.P. Klaasen, H.H. Breneman, T. Jones, J. Kaufman, K.P. Magee, D. Senske, M.J.S. Belton and G. Schubert. *Icarus*, vol. 135, p. 181, 1998.
78 'The 1997 mutual event occultations of Io', R.R. Howell, J.R. Spencer and J.A. Stansberry. In *Io during the Galileo era*. Lowell Observatory, Flagstaff.
79 'High-temperature silicate volcanism on Jupiter's moon Io', A.S. McEwen, L.P. Keszthelyi, J.R. Spencer, G. Schubert, D.L. Matson, R. Lopes-Gautier, K.P. Klaasen, T.V. Johnson, J.W Head, P.E. Geissler, S. Gagents, A.G. Davies, M.H. Carr, H.H. Breneman and M.J.S. Belton. *Science*, vol. 281, p. 87, 1998.
80 'Lava channels on Io: latest Galileo imaging results', D.A. Williams and R. Greeley. *Proc. Lunar Planet. Sci. Conf.*, p. 1723, 2000.
81 J.R. Spencer. *Newsletter of the International Jupiter Watch Satellite Discipline*, No.3, 2001.
82 News item in *Tvashtar Sun*, No.2, 2001.
83 'Origin of mountains on Io by thrust faulting and large-scale mass movements', P.M. Schenk and M. H. Bulmer. *Science*, vol. 279, p. 1514, 1998.
84 J.M. Moore *et al. Proc. Lunar Planet. Sci. Conf.*, p. 1531, 2000.
85 J Radebaugh *et al. Proc. Lunar Planet. Sci. Conf.*, p. 1983, 2000.
86 E.P. Turtle *et al. Proc. Lunar Planet. Sci. Conf.*, p. 1948, 2000.
87 'Formation models of Ionian mountains', E.P. Turtle, W.L. Jaeger, L.P. Keszthelyi and A.S McEwen. *Proc. Lunar Planet. Sci. Conf.*, p. 1960, 2000.
88 'Observations of Ionian mountains', E.P. Turtle, L.P. Keszthelyi, A.S. McEwen, M. Milazzo and D.P. Simonelli. *Proc. Lunar Planet. Sci. Conf.*, p. 1948, 2000.
89 'Chaos on Io: A model for formation of mountain blocks by crustal heating, melting, and tilting', W.B. McKinnon, P.M. Schenk and A.J. Dombard. *Geology*, vol. 29, p. 103, 2001.
90 'Orogenic tectonism on Io', W.L. Jaeger, E.P. Turtle, L.P. Keszthelyi and A.S. McEwen. *Proc. Lunar Planet. Sci. Conf.*, p. 2045, 2001.

91 'Formation models of Ionian mountains', E.P. Turtle, W.L. Jaeger, L.P. Keszthelyi and A.S. McEwen. *Proc. Lunar Planet. Sci. Conf.*, p. 2104, 2001.
92 'Formation modes of Ionian mountains', J. Perry. *Tvashtar Sun*, No.2, 2001.
93 'Revisiting the hypothesis of a mushy global magma ocean on Io', L.P. Keszthelyi, A.S. McEwen and G.J. Taylor. *Icarus*, vol. 141, p. 415, 1999.
94 'Does Io have a mushy magma ocean?', L.P. Keszthelyi, A.S. McEwen and G.J. Taylor. *Proc. Lunar Planet. Sci. Conf.*, p. 1224, 1999.
95 'Io spectra from Hubble: volatile distribution, composition and processes,' J.S. Kargel, J.R. Spencer, L.A. Soderblom, T. Becker and G. Bennett. *Proc. Lunar Planet. Sci. Conf.*, p. 1988, 2000.
96 Komatiite basalt has not been extruded onto the Earth's surface since the Archean, which drew to a close about 2.7 billion years ago.
97 'Komatiites from the Commondale Greenstone Belt, South Africa: a potential analog to Ionian ultramafics?', D.A. Williams, A.H. Wilson and R. Greeley. *Proc. Lunar Planet. Sci. Conf.*, p. 1353, 1999.
98 'The geology of Europa', B.K. Lucchitta and L.A. Soderblom. In *Satellites of Jupiter*, D. Morrison (Ed.). University of Arizona Press, p. 521, 1982.
99 This is in addition to the Earth (which is known to have life), Mars (where there may be fossilised relics of ancient, extinct life), and Saturn's moon Titan (where prebiotic chemistry may be well advanced). Fifteen years earlier, with typical prescience, Arthur C. Clarke had written a life-bearing ice-capped Europan ocean into his novel *2010: Odyssey Two*.
100 'Europa's differentiated internal structure', J.D. Anderson, E.L. Lau, W.L. Sjogren, G. Schubert and W.B. Moore. *Science*, vol. 276, p. 1236, 1997.
101 'A sea ice analog for the surface of Europa', R.T. Pappalardo and M.D. Coon. *Proc. Lunar Planet. Sci. Conf.*, p. 997, 1996.
102 'Order from chaos: determining regional ice lithosphere thickness variations on Europa using isostatic modelling of chaos regions', S. Kadel and R. Greeley. *Proc. Lunar Planet. Sci. Conf.*, p. 2091, 2000.
103 'Chaos terrain on Europa: characterisation from Galileo E12 very-high-resolution images of Conamara', J.W. Head, R.T. Pappalardo, N.A. Spaun, L.M. Prockter and G.C. Collins. *Proc. Lunar Planet. Sci. Conf.*, p. 1285, 1999.
104 'The age of Europa's surface', E.M. Shoemaker. In *Europa Ocean Conference*, San Juan Capistrano Res. Inst. Calif., p. 65, 1996.
105 'Bombardment history of the Jovian system', G. Neukum. In *The three Galileos: the man, the spacecraft and the telescope*, C. Barbieri, J.H. Rahe, T.V. Johnson and A.M. Sohus (Eds). Kluwer Academic, p. 201, 1997.
106 'Strike-slip faults on Europa: global shear patterns driven by tidal stress', G. Hoppa, B.R. Tufts, R. Greenberg and P.E. Geissler. *Icarus*, vol. 141, p. 287, 1999.
107 'Formation of cycloidal features on Europa', G. Hoppa, R. Tufts, R. Greenberg and P. Geissler. *Science*, vol. 285, p. 1899, 1999.
108 'Thrace Macula, Europa: characteristics of the southern margin and relations to background plains and Libya Linea', B. Kortz, J.W. Head and R.T. Pappalardo. *Proc. Lunar Planet. Sci. Conf.*, p. 2052, 2000.
109 'Eruption of lava flows on Europa', L. Wilson, J.W. Head and R.T. Pappalardo. *J. Geophys. Res.*, vol. 102, p. 9263, 1997.
110 'Does Europa have a subsurface ocean? Evaluation of the geological evidence' R.T. Pappalardo *et al*. *J. Geophys. Res.*, vol. 104, p. 24015, 1999.
111 'The geology of Ganymede', E.M. Shoemaker, B.K. Lucchitta, J.B. Plescia, S.W. Squyres

and D.E. Wilhelms. In *Satellites of Jupiter*, D. Morrison (Ed.). University of Arizona Press, p. 435, 1982.
112 D. Senske, J.W. Head, R.T. Pappalardo, G. Collins, R. Greeley, K. Magee, G. Neukum and C. Chapman. *Proc. Lunar Planet. Sci. Conf.* 1997.
113 'Topography on satellite surfaces and the shape of asteroids', T.V, Johnson and T.R. McGetchin. *Icarus*, vol. 18, p. 612 1973.
114 The regio were named for Galileo Galilei, Simon Marius, Seth Nicholson, Edward Barnard, and Charles Perrine.
115 'Constraints on the expansion of Ganymede and the thickness of the lithosphere', M.P. Golombek. *J. Geophys. Res. Supp.*, vol. 87, p. 77, 1982.
116 'Flooding of smooth terrain on Ganymede by low-viscosity aqueous lavas: direct evidence from VGR-GLL stereo synergism', P.M. Schenk, D. Gwynn, W.M. McKinnon and J.M. Moore. *Proc. Lunar Planet. Sci. Conf.*, p. 2037, 2000.
117 L. Prockter, J.W. Head, D.A. Senske, G. Neukum, R. Wagner, U. Wolf and R. Greeley. *Proc. Lunar Planet. Sci. Conf.* 1997.
118 'Bombardment history of the Jovian system', G. Neukum. In *The three Galileos: the man, the spacecraft and the telescope*, C. Barbieri, J.H. Rahe, T.V. Johnson and A.M. Sohus (Eds). Kluwer Academic, p. 201, 1997.
119 'Estimate of areal coverage of bright terrain on Ganymede', P.M. Schenk, S. Sobieszczyk. *Bull. Astron. Soc. Am.*, vol. 31, p. 1182, 1999.
120 'A global database of grooves and dark terrain on Ganymede enabling quantitative assessment of terrain features', G, Collins, J.W. Head and R.T. Pappalardo. *Proc. Lunar Planet. Sci. Conf.*, p. 1034, 2000.
121 C.M. Weitz, J.W. Head, R.T. Pappalardo, G. Neukum, B. Giese, J. Oberst, A. Cook, B. Schreiner, R. Greeley, P. Helfenstein and C. Chapman. *Proc. Lunar Planet. Sci. Conf.* 1997.
122 'Origin of the Valhalla ring structure: alternative models', W. Hale, J.W. Head and E.M. Parmentier. In *Conference on multi-ring basins*, Lunar and Planetary Institute, Houston, 1980.
123 K.C. Bender, K.S. Homan, R. Greeley, C. Chapman, J. Moore, C. Pilcher, W.J. Merline, J.W Head, M.J.S. Belton and T.V. Johnson. *Proc. Lunar Planet. Sci. Conf.*, p. 89, 1997.
124 C. Chapman, W.J. Merline, B. Bierhaus, J. Keller, S. Brooks, A.S. McEwen, R. Tufts, J. Moore, M.H. Carr, R. Greeley, K.C. Bender, R. Sullivan, J.W. Head, R.T. Pappalardo, M.J.S Belton, G. Neukum, R. Wagner and C. Pilcher. *Proc. Lunar Planet. Sci. Conf.*, p. 217, 1997.
125 'Model assessment and refinement of multi-ring structures on Callisto from Galileo SSI data analysis', J.E. Klemaszewski and R. Greeley. *Proc. Lunar Planet. Sci. Conf.*, p. 2064, 2000.
126 'Cratering timescales for the Galilean satellites', E.M. Shoemaker and R.F. Wolfe. In *Satellites of Jupiter*, D. Morrison (Ed.). University of Arizona Press, p. 277, 1982.
127 'Bombardment history of the Jovian system', G. Neukum. In *The three Galileos: the man, the spacecraft and the telescope*, C. Barbieri, J.H. Rahe, T.V. Johnson and A.M. Sohus (Eds). Kluwer Academic, p. 201, 1997.
128 'Gravitational evidence for an undifferentiated Callisto', J.D. Anderson, E.L. Lau, W.L. Sjogren, G. Schubert and W.B. Moore. *Nature* 1997.
129 'The great Solar System revision: the past 25 years have changed our view of the Solar System forever', W.K. Hartmann. *Astronomy*, p. 40, August 1998.
130 In fact, the 'free lunch' of a gravitational slingshot is gained at the expense of the planet that accelerates the spacecraft, because the planet loses precisely the energy that is transferred to the spacecraft. One radical group, the Pasadena Society for the Preservation of Jupiter's Orbit, criticised NASA for flying the Voyager missions because doing so altered the orbits of the planets.

9

Saturn's retinue

RINGED PLANET

In July 1610, Galileo turned his telescope upon Saturn. Although it displayed the expected disk, there was something strange about it. At first, he thought that it possessed a pair of huge moons, one on either side, but when they maintained their positions he realised that they were not satellites. Baffled, he referred to the planet as 'Saturnus triformis'. When he looked again in 1612 and discovered that the appendages were no longer present he decided that his earlier telescope must have been flawed. Because there were no satellite motions to monitor, he made no further observations of the planet. In 1614, however, C. Schreiner noticed that the oddities had reappeared and drew them as 'handles' on either side of the planet, as indeed Galileo had four years earlier. When G.B. Riccioli examined the planet in 1640 he drew a trio of distinct objects arranged in line. However, from 1647 to 1650 he saw them as 'handles' of various shapes and in 1651 he published a book summarising all that was known concerning the planet.[1] The mystery was solved by Christiaan Huygens. In 1655, he saw the appendages as thin 'arms', and when these disappeared for several months the following year he realised that he was viewing "a thin, flat ring not connected with the planet at any place, and inclined at an angle to the ecliptic". Nevertheless, because this was so unusual, he waited until 1659 before announcing his conclusion.[2] Soon after beginning his study of Saturn, Huygens realised that the planet had a satellite orbiting far beyond the ring. This was subsequently named Titan[3]. In 1671, G.D. Cassini spotted Iapetus somewhat farther out, and the following year discovered Rhea closer in; and on a memorable night in March 1684 he identified both Tethys and Dione orbiting even closer to the planet.

In 1665, William Ball was astonished to see a thin dark line marring the ring, and in 1675 G.D. Cassini interpreted this to mean that the ring was not a monolithic disk: the line divided an inner ring from an outer ring. W. von Struve promptly designated the rings 'A' (outer) and 'B' (inner). Ball's precedent notwithstanding, the thin dark gap between the rings has come to be known as Cassini's Division. There the situation rested until 1837, when J. Encke detected a thin gap in the 'A' ring. This duly became known as Encke's Division. The following year, J.G. Galle saw a very faint ring inside the inner edge of the 'B' ring.[4] However, its existence was considered suspect until W.C. Bond in America spotted it on 15 November 1850. While W.R. Dawes was playing host to William Lassell a fortnight later in England, and the day before a report of Bond's discovery was published in a London paper, they turned to Saturn and noted "something like a crepe veil covering a part of the sky within the inner ring."[5] This was designated the 'C' ring, in keeping with the nomenclature. Given that it is so translucent that the planet can be seen through it, the 'C' ring is also known as the 'Crepe' ring. It has no inner edge, it simply fades away about halfway towards the planet's surface. It was belatedly realised that this faint ring had been discerned by Campani in 1664,[6] Jean Picard in 1673[7] and John Hadley in 1720[8] crossing Saturn's disk. In 1907, G. Fournier detected indications of a faint ring outside the 'A' ring, and this was confirmed the following year by M.E. Schaer. When E.E. Barnard was unable to see it using the Yerkes Observatory's 40-inch refractor, the reports of this outer ring were taken to have been in error[9], but when R.M. Baum reported seeing it again in 1952 it was speculated that the structure of the ring system might be variable. Indeed, Percival Lowell had reported in 1915 that Encke's Division had 'doubled' and in 1954 T.R. Cave saw it close up and there was only an abrupt change in brightness from the 'A' ring to the 'B' ring where the dark gap had been.

At first, it had been assumed that the rings formed a solid disk, but Simon Laplace proved this to be mathematically impossible. B. Peirce and G.P. Bond independently suggested that the ring material might be a fluid, but in 1850 Eduard Roche found mathematically that a fluid body could not exist as a coherent structure within about 2.4 times a planet's radius if the two objects were of similar overall densities and composition. This prompted the suggestion that the rings were the debris from a disrupted satellite. In 1856 J.C. Maxwell proposed that they comprised untold numbers of fragments orbiting the planet in accordance with Kepler's laws. That is, the fragments in the inner edge of the ring system were travelling more rapidly than those beyond. The rings could not be rigid because the stresses of this differential rotation would rip it apart. In 1888 H. von Seelinger undertook photometric observations to try to prove that this was the case, and he concluded that only a ring composed of small particles would show the observed surface brightness of the rings under varying degrees of solar illumination, but it was not until J.E. Keeler of the Allegheny Observatory in America published spectroscopic observations in 1895 that the case was considered proved.[10] Keeler had exploited the doppler effect to measure the speed of rotation of the ring material at various distances from the planet, and instead of the straight-line relationship of a rigid structure, the 'rotation curve' that he measured was just as predicted by Kepler's laws of orbital motion. In fact, to explain the translucency of the fainter rings it had been speculated that they might be composed of fragments broken from the primary ring structure. Once it was realised that the rings comprised individually orbiting fragments, it seemed likely that the variations in the appearance of the ring system were due to gravitational perturbance altering the distribution of the material. The fine structure of Saturn's ring system became a major research topic for many astronomers. Mathematical investigations established that the periods of the gaps in the rings are in simple proportion to the orbital periods of some of the large moons which orbit just beyond the ring system. Daniel Kirkwood, fresh from his mathematical analysis of the 'gaps' in the asteroid belt, pointed out in 1867 that a fragment of material in Cassini's Division would have an orbital period that resonated with no fewer than *four* of the main moons, and so would soon be drawn out of that orbit[11]. In effect, therefore, the nearby moons 'sweep' clear lanes in the ring system and keep them free of fragments. Although Roche had established that a loosely consolidated satellite could not survive if it strayed within about 2.4 times Saturn's radius, Harold Jeffreys showed in 1947 that because Saturn's density is so low, a dense rocky object only a few hundred kilometres across would survive intact within Roche's limit, so if the rings were the debris from a moon that had been broken apart by gravitational tides, there were likely to be sizeable objects drifing among the finer debris.

After Cassini's string of discoveries, there was a hiatus of over a century before any new moons were discovered. Then in July and August 1789, William Herschel discovered Mimas and Enceladus orbiting close together between Tethys and the ring system. On 16 September 1848, the father-and-son team of W.C. and G.P. Bond of the Harvard Observatory spotted a faint 'star' situated in the plane of the rings. Two days later William Lassell in England noted it too, and three days later they simultaneously recognised that it was not a background star but was moving with the planet. It was named Hyperion. W.H. Pickering set out in 1888 for the Harvard Observatory's telescope at Arequipa in Peru to make a photographic search for more satellites. A decade later, in August 1898, he found Phoebe. It was orbiting far beyond Titan. Furthermore, it was orbiting in the 'wrong' direction, in a retrograde fashion. Pickering reported another satellite in 1905, which was named Themis, but his report was premature because the faint speck of light was later found to be a background star.

A pre-'Space Age' photograph of Saturn taken by the 200-inch Hale telescope of the Palomar Observatory.

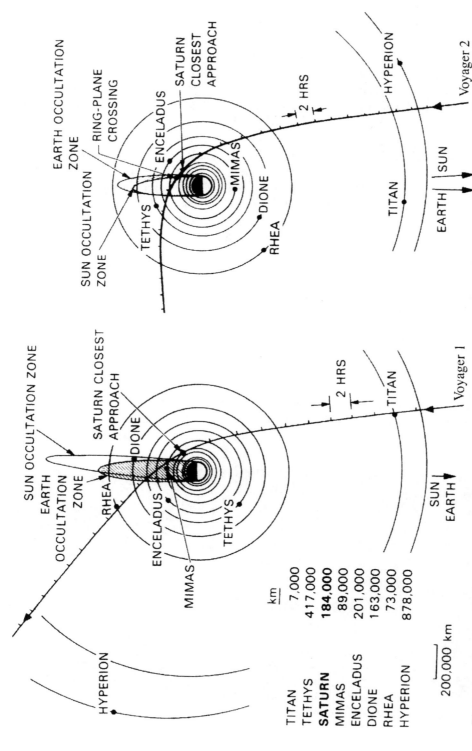

The trajectories through the Saturnian system of Voyager 1 in November 1980 and Voyager 2 in August 1981. The leading spacecraft's primary mission was the Titan fly-by on the way in, and achieving this released its successor to fly a trajectory for a gravitational slingshot for the 'Grand Tour' of the outer Solar System.

Even the approach imagery of Saturn and its ring system by the Voyagers were spectacularly better than the earlier telescopic photographs.

As for Saturn itself, although its disk was frustratingly bland there were sufficient features of a transitory nature to enable William Herschel in 1794 to estimate its rotational period as 10 hours and 16 minutes. In 1876, Asaph Hall of the Naval Observatory in Washington observed a bright equatorial spot that lasted for 60 revolutions and from this he was able to calculate a rotational period of 10 hours 14 minutes 24 seconds.[12] However, he warned that if the planet's atmosphere underwent differential rotation (as did Jupiter's) then this period would apply only to that particular spot. During their fly-bys, the Voyager spacecraft observed various spots that resembled small counterparts of Jupiter's Great Red Spot. They also revealed that much of the structure of the deeper atmosphere is masked by a hydrocarbon haze, which is why the disk is so bland telescopically. As in the case of Jupiter, Saturn is contracting and is therefore converting gravitational energy to heat. However, Saturn's interior differs from Jupiter's in one crucial respect: being much less massive, it is not converting so much gravitational energy into heat, and the reduced pressure does not heat its interior so intensely. In this 'cooler' environment, helium is able to condense and settle gravitationally.[13] As it falls towards the core, the helium frictionally heats the dense mantle of hydrogen. It is believed that 30 per cent of Saturn's internal energy derives from this 'raining out' of helium. Jupiter's interior on the other hand, is too hot for this to occur.[14] Since Saturn is smaller than Jupiter it releases less heat in real terms, but because it is twice as far from the Sun and receives only 25 per cent as much energy from the Sun as does Jupiter, it still manages to emit more energy than it receives.[15]

A SYSTEM OF MOONS

Saturn's retinue is more extensive than Jupiter's, with a wider variety of small moons, and it was soon realised that, based on their distance from the planet, there are several 'groups'.

- Mimas, Enceladus, Tethys, Dione, Rhea and
- Titan and Hyperion
- Iapetus
- Phoebe

To gain a sense of perspective, note that Dione is about as far from the centre of Saturn as the Moon is from the centre of the Earth. However, Saturn is much larger than the Earth. Looked at another way, the ring system and the first five large moons orbit within the volume of space that our Moon has all to itself. Furthermore, while the Moon takes a month to orbit the Earth, Mimas circles its primary in just 22.6 hours. In fact, even Titan, out beyond this inner group, takes only 16 days to complete one orbit. H.L. d'Arrest in Berlin identified resonances in the periods of the orbits of this inner group, with Mimas paired with Tethys, and Enceladus with Dione. That is, Mimas takes twice as long as Tethys to complete an orbit and Enceladus twice as long as Dione. The outer satellites are similarly interrelated. The period of Hyperion's orbit is five times that of Rhea. For every three of Hyperion's orbits, Titan completes four. Titan is also paired with Iapetus. In the time that Iapetus takes to complete one orbit, Titan does so five times. By Kepler's laws of orbital motion, the period of an orbit is related to the radius of its orbit, so it was apparent that gravitational interations between the various satellites over the years had refined their orbits until these resonances were achieved. Pairs of moons repeat mutual positional relationships, although the perturbations from the others make these patterns evolve over long periods. H. von Struve utilised the gravitational interactions to calculate the masses of the various moons with a high degree of accuracy.

Titan is the only member of Saturn's retinue that is large enough to present a disk, but efforts to measure its diameter showed considerable variation. Nevertheless, it was soon evident that Saturn's primary satellite is nearly as large as Jupiter's Ganymede. F.G.W. von Struve estimated the sizes of the others by assuming that their albedos were the same as Titan's, then employed the photometric data recorded by P. Guthnick and E.C. Pickering as an indication of their true size, and with this he was able to calculate their bulk densities. These suggested that the inner group of satellites were predominantly water ice. Furthermore, the densities of the innermost trio were less than unity, the density of water, which implied that their material was loosely consolidated.[16] Soviet astronomer V.I. Tscherednitschenko proposed a mixture of water-snow and frozen methane. The trend was extended when Harold Jeffreys calculated that the surface density of the ring material was 0.31 g/cm^3, and it was concluded that the rings comprise chunks of water ice of sizes ranging up to several metres across. The greater densities for Titan and Hyperion implied that these moons were mixtures of rock and ice and hence were similar to the Galilean satellites of Jupiter. Orbiting outside Titan, and with an intermediate density, Iapetus appeared to be in a class of its own. One striking observation was that whereas in the Jovian system the densities of the moons decrease with distance from the primary, the densities of the Saturnian moons seemed to trend the other way. However, the weakness in F.G.W. von Struve's case was his assumption that all the

Table 9.1 Saturn's moons

	Diameter (km)		Mass (kg)		Density (g/cm^3)		Albedo
	Pre	Post	Pre	Post	Pre	Post	
Mimas	640	398	3.5×10^{19}	3.7×10^{19}	0.4	1.44	0.7
Enceladus	800	498	1.4×10^{20}	7.4×10^{19}	0.4	1.16	0.9
Tethys	1,280	1,046	6.2×10^{20}	6.2×10^{20}	0.6	1.21	0.8
Dione	1,200	1,120	1.1×10^{21}	1.05×10^{21}	1.3	1.43	0.5
Rhea	1,760	1,528	2.3×10^{21}	2.3×10^{21}	1.0	1.33	0.6
Titan	4,160	5,150	1.4×10^{23}	1.36×10^{23}	3.3	1.88	0.2
Hyperion	500	–	1.1×10^{20}	1.1×10^{20}	4.0	–	0.3
Iapetus	1,584	1,436	5.7×10^{21}	1.9×10^{21}	2.6	1.16	–

Notes
1. Pre-Voyager data from W. Sandner, using H. von Struve's mass estimates and F.G.W von Struve's density estimates.
2. Post-Voyager data from Moore and Hunt.
3. Iapetus's albedo varies depending upon its orbital position. One side of the moon is 0.5 and the other side is 0.05.
4. Hyperion is irregularly shaped, measuring 360 × 280 × 236 kilometres, and its density has not been estimated.
5. Although given as 0.9, Enceladus's albedo is marginally less than absolute.

satellites had similar albedos and the Voyager fly-bys established that this was false. Titan was found to be the least reflective of the retinue, so the diameters of the others were over estimated, and their densities under estimated. However, comparing the pre/post-Voyager data is illuminating. Firstly, H. von Struve's mass estimates stood up quite well, which shows the analytical power of Newton's law of Universal Gravitation. Enceladus turned out to have a remarkable albedo; it is the most reflective body in the Solar System, far exceeding even brilliant cloud-enshrouded Venus. The density of 'dark' Titan is significantly lower than predicted and is similar to Jupiter's Callisto. Iapetus's 'intermediate' density was found to be illusory. Clearly, no matter how much a planetary system is studied telescopically, there is no substitute for sending a spacecraft to take a close look.

Pioneer 11 had not observed Saturn's satellites during its fly-by in September 1979, so as Voyager 1 ventured into the system in November 1980 the mood in the spaceflight operations centre at JPL was expectant.[17] Titan was Voyager 1's primary objective. If it failed, Voyager 2, trailing a year behind, would back it up, but if the encounter was successful, Voyager 2 would be manoeuvred onto a trajectory that would set up the slingshot for the 'Grand Tour', taking in Uranus and Neptune. After Titan, Voyager 1 took a close look at Mimas, Dione and Rhea. Fortunately, the trajectory for Voyager 2's Saturn slingshot was complementary and it offered opportunities to study Iapetus, Hyperion, Enceladus and Tethys. However, the scan platform which carried all the aimed instruments jammed towards the end of the sequence and the best Enceladus and Tethys opportunities were lost.[18] Unavoidably, the imaging coverage was mixed, with some of the moons receiving greater areal coverage and others being recorded at higher resolution. The best imagery had a resolution of several kilometres per pixel. Until the Cassini spacecraft settles into Saturnian orbit in 2004 to conduct an in-depth study, conclusions derived from detailed comparisons of the surfaces of the satellites as documented by the Voyagers are necessarily tentative[19,20,21,22].

The expectation was that the large satellites would have undergone thermal differentiation, but the smaller moons would be homogeneous.[23] However, the chilly state of the solar nebula from which Saturn's moons condensed suggested that the ice would contain a significant ratio of ammonia, and because even a modest amount of radiogenically generated heat might have prompted melting, surface eruptions could not be ruled out simply because the objects were small.

MIMAS

As measured from Saturn's centre, the outer edge of the main ring system is 2.3 planetary radii, the slender 'G' ring lies at 2.9, and Mimas is just beyond it at 3.1, which is comfortably outside the Roche radius so it is in no danger of being tidally disrupted. At 390 kilometres in diameter, it is the smallest of the historically known satellites. Its rotation is synchronous with its orbit, so it maintains one hemisphere facing the planet. Voyager 1 was able to view most of the side that faces away from the planet and much of the southern part of the near side with a maximum resolution of about 1 kilometre per pixel.

All things considered, Mimas lived up to the stereotype of an ancient battered icy moon, in that it is heavily cratered. The cratering, however, is not uniform. Most of the craters form deep cavities, and the larger craters tend to have central peaks. Their ejecta blankets are indistinct, suggesting that on a small body with weak gravity the ejecta was distributed far and wide. The most prominent crater, appropriately named Herschel, is on the leading hemisphere. At about 130 kilometres across, it spans nearly one-third of Mimas's diameter, and it is remarkable that the moon survived

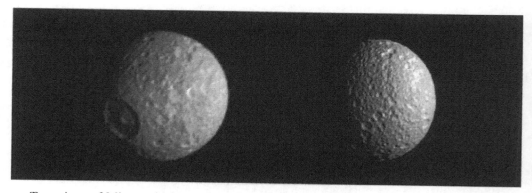

Two views of Mimas, the innermost of Saturn's main satellites. The south polar region on the right shows prominent grooves in addition to craters. The other vantage point is dominated by the crater Herschel.

the impact. Its floor is depressed about 10 kilometres and its very well defined central peak rises about 6 kilometres from the floor. When the raw imagery was flashed up on the JPL TV monitors, one wag whimsically exclaimed that Voyager had found a 'Death Star', as depicted by George Lucas in the 'Star Wars' movie. The moon's cratered surface has been transected by a number of prominent chasms up to 10 kilometres across, several kilometres in depth and 100 kilometres in length whose origin may be related to the Herschel impact.[24] One study of the 'gravitational focusing' by Saturn suggested that as the innermost moon, Mimas should be the most heavily cratered, with a preference for impacts to occur on its leading hemisphere.[25]

ENCELADUS

Voyager 1 did not approach Enceladus very closely, but it established that the moon has a *very* highly reflective surface. Voyager 2 provided imagery of the northern part of the trailing hemisphere at a resolution of 1 kilometre per pixel. As Enceladus is similar to Mimas in size and position it had been thought that they would be physically similar, but Enceladus has been extensively resurfaced by cryovolcanism. Physiographic analysis has identified cratered, smooth, and ridged plains. The cratered plains have an abundance of craters in the 10 to 20 kilometres size range. However, in some areas the craters are sharply defined and in other areas they are softened, possibly due to viscous flow. A study of the crater forms concluded that the moon's lithosphere is a mixture of ammonia ice and water ice.[26,27]

Over large areas the craters have been replaced by plains whose sparsity of impacts implies either a low cratering rate or the surface's relative youth. In some places, these smooth plains are criss-crossed by a rectilinear set of fractures which are believed to be grabens formed when subsurface ice froze, expanded, and cracked the brittle outer shell. There are also ridges rising a kilometre or so, with smooth plains lying between. These ridged plains, which predominate on the trailing hemisphere, could be fluids that oozed from fractures and solidified in place. In fact the extent of the resurfacing and tectonic activity on this small moon was surprising. One analysis concluded that Enceladus would have rapidly lost its accretional heat (due to its high ratio of surface area to volume) and it is far too small to have undergone significant radiogenic heating. It must therefore have endured an exogenic mode of heating at a later stage otherwise it would resemble Mimas.[28] Indeed, Enceladus appears to have undergone *several* phases of resurfacing over an extended period. The most plausible cause of heating is tidal stress. Enceladus has an orbital resonance with Dione, which orbits farther out. If the ice contains a significant fraction of either methane clathrate or ammonium hydrate then melting would have been both more readily achieved and more extensive.

Interestingly, the densest section of the tenuous 'E' ring is coincident with Enceladus's orbit. It is thought that the moon's surface is so reflective because it is coated with a layer of fine ice crystals, and it is likely that the moon is the source of the very small ice particles that form the ring material. The sparse cratering argues

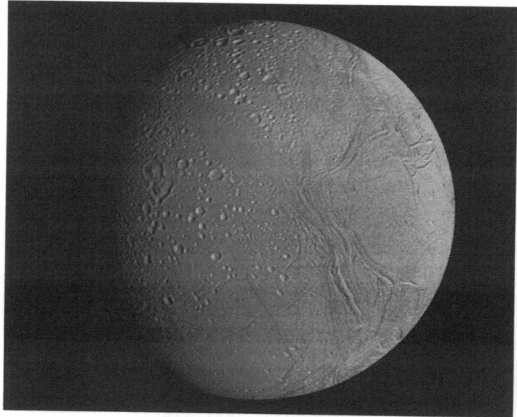

Much of the original cratered surface of Enceladus has been transformed by cryovolcanism into smooth plains. Long fractures run across these plains.

against this material having been ejected by impacts, so perhaps water is being vented into space by cryogenic plumes.[29]

TETHYS

Tethys received comparatively comprehensive imaging coverage, with only a strip on the far side's southern hemisphere being missed. It is more than twice the diameter of Enceladus, and it too displays evidence of tectonic deformation and cryovolcanic resurfacing.

A physiographic study identified several terrain types.[30] The oldest surface unit is the hilly cratered terrain that is characterised by rugged topography. It is densely cratered, but most of the larger craters are degraded. As in the case of Mimas, Tethys has a vast impact crater on its leading hemisphere. However, while Odysseus is in similar proportion to Mimas's Herschel, the fact that Tethys is larger means that Mimas itself would fit comfortably within Odysseus's rim. Although it is large, Odysseus is nevertheless very shallow and its floor actually follows the curvature of

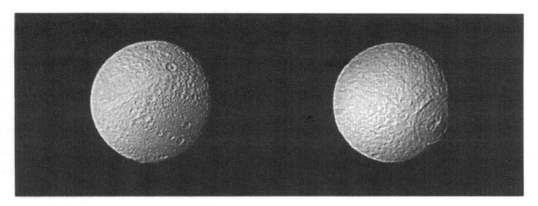

The left view of Tethys shows the canyon system of Ithaca Chasma. The other vantage point is dominated by the crater Odysseus.

the moon. This may be the result of isostatic adjustment of the icy lithosphere soon after the crater was formed. The rim is softened by a succession of terraces, and the central peak complex is a ring-like structure. Although the antipodal area could reasonably be expected to have been torn up by the focusing of seismic shock waves, it is less rugged and forms a plain displaying faint lineations and albedo variations; nevertheless, is possible that the eruption of material that formed the plain was prompted by the impact.[31]

Tethys's most spectacular feature is Ithaca Chasma, a vast canyonland that winds its way almost all the way around the globe. This narrow branching terraced canyon system is at least 1,000 kilometres in length, 100 kilometres in width and up to 4 kilometres in depth.[32] Sections of its rim are raised as much as 500 metres above the adjacent terrain. While it may be related to the Odysseus impact, the canyon system does not actually encroach upon the crater. In fact, if the crater's central peak is considered to be a 'pole', the chasm tends to trace the great circle of the associated 'equator'. It has been suggested that the Odysseus impact occurred when the moon's interior was still a 'soft' ice and that the force of the impact temporarily deformed the satellite into an ellipsoidal shape, compressing the interior and inducing transient tensional stresses sufficient to crack the lithosphere almost all the way around.[33]

DIONE

Dione displays striking albedo contrasts, and a physiographic study has identified several associated terrain types.[34] Predominant is a rugged 'highland' terrain, with many craters up to 100 kilometres in diameter. In general, these craters are shallower than on Tethys. Most of the larger craters have terraced walls and central peaks and scarps up to 100 kilometres in length run across this terrain. A less rugged cratered terrain was defined as cratered plains, and terrain with very few craters was defined as smooth plains. The smooth plains contain troughs that are typically less than 100 kilometres in length but in some cases exceed 500 kilometres; in most cases there is a

Three views of Dione showing the remarkable wispy bright streaks on its trailing lower albedo hemisphere. The bright streaks appear to be related to troughs, in some cases forming parallel strands. On the other hemisphere (left image) the heavily cratered terrain is cut by fissures that are independent of the albedo variations.

'pit' at either end. The largest craters (up to 200 kilometres across) are on the trailing hemisphere. However, Dione appears to have escaped an impact on the scale to match Tethys's Odysseus. While the trailing hemisphere is generally darker, it also displays a pattern of broad wispy intensely bright streaks. Some of these are associated with troughs and other lineaments. In the middle of the radiating pattern of streaks there is an elliptical structure (named Amata) several hundred kilometres across with a dark central patch which might mark an impact. However, the streaks are *not* rays of bright ejecta. They appear to follow the line of surface lineaments. The fact that the streaks seem to be associated with lineaments suggests that they are deposits of clathrate pyroclastic that vented from fractures after being warmed at shallow depth by a pocket of radioactive elements.[35] As for the dark background, the fact that the planet's magnetosphere rotates rapidly with the planet means that the trailing hemispheres of the slowly orbiting moons are bathed with charged particles, and it has been suggested that the dark material is a modification of the surficial material by this radiation.[36]

Dione is comparable in size to Tethys, but as its bulk density is higher, which indicates that it contains a higher proportion of rock, radiogenic heating would have kept its interior warm for longer. It has been suggested that after the formation of the brittle lithosphere, internal heat prompted fluid ammonia–water ice to erupt from fractures to form the plains.[37] Indeed, a series of ridges which rise only a few hundred metres but extend up to 100 kilometres in length could indicate the fronts of viscous surface flows.

The fact that both Enceladus and Dione, whose orbits are in mutual resonance, underwent sufficient internal melting to prompt surface flows is evidence for tidal stresses having been the main heating agent.

RHEA

At 1,530 kilometres across, Rhea is second in size among Saturn's moons only to Titan. When Voyager 1 flew within 60,000 kilometres of Rhea it secured imagery of the north polar zone at a resolution of about 500 metres per pixel. As with Dione, Rhea's trailing hemisphere is anomalously dark and displays a pattern of bright streaks, but in this case the pattern is not as prominent.

Rhea's cratered terrain is dominated by large degraded craters, most of which have central peaks and, in some cases, bright patches within their walls which could be 'clean' ice that has been exposed by slumping.[38] Like Tethys, Rhea has suffered at least one massive impact, which left a double-ringed structure some 450 kilometres in diameter. The inner ring seems to be composed of ridges and the outer ring comprises two closely spaced concentric inward-facing scarps (the inner of which is more prominent than the outer, which is discontinuous).[39] There is evidence of a larger impact on the trailing hemisphere whose cavity has been almost completely resurfaced, but the inner rings of this structure are visible as subtle ridges poking through this in-fill.[40] The degraded condition of these larger craters indicates that they are very old and that they struck when the interior was still warm and the lithosphere was sufficiently pliable to enable the surface deformation to relax.

TITAN

The dominant member of Saturn's retinue is clearly Titan, which was the first to be noted. There were occasional reports of extremely faint markings on its distinctly orangey-hued disk. B.F. Lyot of the Pic du Midi Observatory made a lengthy study of it during which he drew a variety of large splotches and light and dark bands. In 1942, G.P. Kuiper launched a study of Titan using the 82-inch reflector of the McDonald Observatory on Mount Locke, in Texas.[41] He reported spectroscopic indications of a substantial atmosphere. A subsequent study identified absorption bands due to methane.

Voyager 1 performed its key Saturnian objective of a Titan fly-by flawlessly. By this time, the presence of an atmosphere had been established and it was known that this sported clouds, but it was hoped that the spacecraft would be able to see the surface through gaps in the cover. However, the cloud completely enshrouded the moon. Furthermore, there was no observable weather system. Limb views showed layering, however, which is believed to be a succession of photochemical hazes. The spacecraft found that the atmosphere is predominantly molecular nitrogen. Up to this time, the Earth had been thought to be the only object in the Solar System to have a nitrogen-based atmosphere.[42] Furthermore, Titan's atmosphere is considerably more substantial than expected. The surface pressure had been expected to be 0.1 bar, whereas it is really 1.5 bars. The fact that nitrogen is transparent but Titan was enshrouded meant that there was another consitituent. The spectroscopic data indicated the presence of assorted carbon–nitrogen compounds, including the basic hydrocarbons ethane, acetylene and ethylene, and it is these that create the orangey hue. The visible surface is at an altitude of 200 kilometres, but a number of distinct layers continue up to about 750 kilometres, so it is a very deep atmosphere.[43] For atmospheric specialists this exotic atmosphere was fascinating,[44] but the fact that Titan's surface was obscured meant that this was a very disappointing fly-by for the geologists.

In October 1994, however, a team led by Peter Smith of the University of Arizona's Lunar and Planetary Laboratory used the Hubble Space Telescope to observe Titan in a near-infrared band in which the haze is sufficiently transparent for the surface to be characterised in terms of reflectivity.[45] Only the equatorial and mid-latitudes could be studied in this way and the surface resolution was just 500 kilometres per pixel, but it was sufficient to make the first crude albedo map showing a prominent infrared-bright feature that is comparable in size to the terrestrial continent of Australia.

At 5,150 kilometres in diameter, Titan is larger than the planet Mercury. It is only slightly smaller than Jupiter's Ganymede but its lower bulk density of 1.9 g/cm^3 implies that it is half rock and half ice. Nevertheless, it is thought to have had

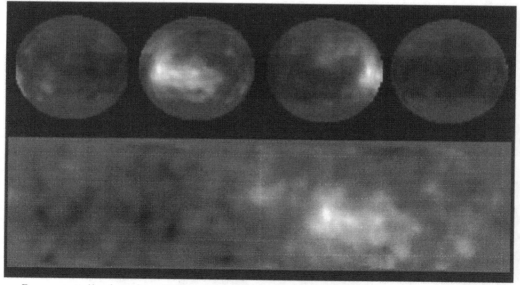

By repeatedly imaging Titan during its 16-day axial rotation at near-infrared wavelengths, the Hubble Space Telescope was able to sense the reflectivity of the moon's surface (masked at visible wavelengths by a hydrocarbon haze) and make an albedo map. One prominent bright area is comparable in size to Australia. The low-albedo areas may represent oceans of ethane. (Courtesy of the Space Telescope Science Institute and Peter Smith of the University of Arizona.)

sufficient radioactive elements for its interior to be differentiated. The resulting release of volatiles is thought to have generated an early atmosphere of methane (carbon drawn from the solar nebula) and ammonia (nitrogen).[46] Once the ammonia was dissociated by the ultraviolet component of insolation, the hydrogen escaped and the heavier nitrogen was retained. When methane in the upper atmosphere is dissociated, the free radicals recombine to create a variety of organic compounds. Titan's surface temperature was measured at 93 K, with a variation of only a few degrees from pole to pole, so the atmosphere is stagnant. The temperature is near the triple point of ethane and methane, so these precipitate out and fall as rain. In the mid-1990s, infrared observations detected variations suggestive of clouds in the lower atmosphere,[47,48] supporting the notion that there is a rain of hydrocarbons, which might even flow across the surface and drain into hydrocarbon seas. It has been calculated that the eccentricity of the moon's orbit would raise tides of several metres amplitude,[49] which, over time, would erode the shores of water ice. The exposed surface should also be physically and chemically eroded by the rain. As a result, Titan may be predominantly oceanic. We do not know if Titan has any large-scale surface relief. In the absence of ongoing tectonic activity to rebuild the topography, its surface may have been levelled long ago. We will not know for certain until the Cassini spacecraft produces a topographic map using radar to penetrate the clouds. The Huygens probe, which Cassini will deliver, will make a parachute descent to report on the physical conditions for up to half an hour after touching or perhaps splashing down,[50] and it is likely that the target will be at the interface between the dark and light albedo zones.

If Titan turns out to have lakes of hydrocarbons, the anoxic conditions in this thick gloop may have prompted the reactions necessary for the development of primitive life.[51] In fact, one day, Titan may 'bloom' because, in about 6 billion years when the Sun's core shrinks and its envelope inflates to form a 'red giant' that consumes the inner planets, advanced life may emerge from Titan's oceans as the temperature rises.[52]

HYPERION

Hyperion was found to be an irregular body measuring 350 by 235 by 200 kilometres with a passing resemblance to a very thick hamburger. In addition to craters 120 kilometres across, Hyperion has ridges 10 kilometres high, which is astonishing for such a small object. However, the arcuate form of these ridges is suggestive of rims of craters several hundred kilometres in diameter, implying that it is a small fragment of a much larger progenitor. It evidently retained its structural integrity, because it has not yielded to the gravitational urge to transform itself into a sphere. Its shape is all the more remarkable in light of the fact that its major axis is comparable in size to Mimas, whose sphericity derives from its origin as an independent entity.

Initially, Hyperion's rotation was a puzzle because the Voyager imagery did not identify its spin axis, but in 1983 it was realised that it has a 'chaotic' rotational state.[53] This was proved in 1988 by nearly continuous observations of the moon's light curve over a 3-month period.[54] As it travels its eccentric 21-day orbit, Hyperion is simultaneously and independently rotating on several axes and accelerating and decelerating randomly in response to the perturbations of Titan, with which it is in orbital resonance.[55]

IAPETUS

Iapetus was the most enigmatic of the Saturnian moons for the early telescopic astronomers. Soon after discovering it, G.D. Cassini realised that the moon's brightness was varying by fully two visual magnitudes in a systematic manner as it moved around its orbit. It was brightest at western elongation and faintest at eastern elongation. The fact that this pattern remained fixed implied that the moon's rotation was synchronised with its orbital period. For some reason, the leading hemisphere was considerably darker than its trailing hemisphere. At first, there was debate as to whether the surface was really darker or whether the moon was a peculiar (non-spherical) shape. William Herschel concluded that one hemisphere was partially obscured.[56] Iapetus's public profile was immeasurably boosted when Arthur C. Clarke, in his novelised form of Stanley Kubrick's movie '2001: A Space Odyssey', set a black monolithic 'star gate' in a white circle right in the middle of the dark hemisphere.[57] There was therefore a sense of *déjà vu* as Voyager approached.

Although Iapetus was not well placed for viewing by either Voyager, and the best imagery had a resolution of 10 kilometres per pixel, this was sufficient to indicate that the dark leading hemisphere was featureless. The boundary of this dark feature,

The best image of Iapetus showing Cassini Regio on the anomalously dark leading hemisphere.

which has been appropriately called Cassini Regio, is sharply defined. However, there are craters on the adjacent terrain up to 120 kilometres in diameter with dark floors, suggesting that the dark material is a mantling deposit. The material has an albedo of only 0.04, so it is as black as tar. In fact, it may be tar, because very dark material seen elsewhere in the outer Solar System is believed to be complex carbon compounds. On the other hand, the reddish spectral character of the material is similar to that of the carbonaceous material in some meteorites. These are the 'primitive' material left over from the formation of the Solar System. As little as 1 per cent carbonaceous impurity can reduce the reflectivity of normally bright water ice to 0.07. But how was it 'painted' onto the surface? The outermost satellite, Phoebe, is also very dark. Perhaps Iapetus's leading hemisphere is sweeping up material blasted off Phoebe by micrometeoroids.[58] Even if the dark material does not derive from Phoebe, it could be that because comets are the most likely intruders into the Saturnian system Iapetus took a head-on strike and its leading hemisphere has been splattered by black cometary residue.

Could the dark material be of endogenic origin? One proposal interpreted the dark-floored craters adjacent to Cassini Regio as suggesting that the dark material erupted as a carbonaceous icy slurry.[59] Although Iapetus is similar in size to Rhea, its lower density means that it is more ice than rock. Nevertheless, it is large enough to have accreted sufficient rock to have undergone modest radiogenic heating.[60] Convection beneath the frozen lithosphere may have prompted extrusion of cryogenic lava. But why only one hemisphere? Why the leading hemisphere? And why did it cover *all* of that hemisphere? Was there no high-standing terrain? Was such a surface flow accompanied by explosive venting whose dark fall-out painted the high ground? The absence of bright ray craters on Cassini Regio implies either that this deposit is too thick to have been penetrated by recent small impacts, or that the process is ongoing and the rays of bright ejecta are rapidly darkened by fall-out. Why would Iapetus be that active? The mystery of Iapetus's dark hemisphere is unlikely to be resolved until the Cassini spacecraft's remote-sensing suite determines the composition of the material.

PHOEBE

With the exception of Iapetus and Phoebe, Saturn's satellites all move in circular prograde orbits coplanar with both the planet's equator and the ring system, but Phoebe has an elliptical, inclined and retrograde orbit. Voyager 2's best imagery had a resolution of 20 kilometres per pixel, which was sufficient to show that this moon is spheroidal with a diameter of about 220 kilometres, rotates in about 9 hours, is cratered and, despite having a general albedo of 0.06, has a number of relatively 'bright' patches. It is almost certainly a captured asteroid or comet. Its dark surface (the darkest of any of Saturn's satellites) shows spectral properties indicative of carbonaceous material, so if it is an asteroid it is of the 'C' type, a 'primitive' relic of the solar nebula.

Because Phoebe's orbit is retrograde, dust motes blasted off its surface by micrometeoroid impacts will fall towards Saturn. As they spiral inwards in a retrograde manner they will meet Iapetus head-on with their combined velocities producing particularly energetic impacts which may produce unusual reactions in the ices on the surface. Hyperion, the next moon in towards the planet, does not seem to have been affected, but its chaotic rotation prevents it from maintaining one side facing 'forward'. Perhaps its entire surface has been uniformly darkened to some extent. Phoebe's orbit is inclined at 30 degrees to the system, so the dust will form a tenuous but deep annular disk. As Iapetus's orbit is inclined at 15 degrees, perhaps its gravity draws in most of the dust so that very little of it penetrates the inner system, because the moons within Titan's orbit display no darkening on their leading hemispheres that could be attributed to dust spiralling in from Phoebe. Significantly, Phoebe, Iapetus's dark side and Hyperion all have reddish spectral characteristics, so Phoebe may indeed have polluted them, primarily as a result of having been captured in a retrograde orbit.[61]

MOONLETS

As the 'Space Age' dawned, it was reasoned that stray asteroids were unlikely to be able to reach Saturn without being intercepted by mighty Jupiter, which was believed to be why Jupiter had several families of moonlets only a few dozen kilometres in diameter, whereas Saturn did not.[62] It was, however, acknowledged that Saturn was much farther from the Sun, and if such tiny objects were indeed orbiting it they would be very difficult to see.[63]

In 1979 Pioneer 11 made the first reconnaissance of the Saturnian system. Its photometer served as a crude imaging system. In addition to showing the planet's disk, this provided our first close look at the ring system; discovered a faint extremely narrow ring (known as the 'F' ring) beyond the main ring; and identified a moonlet some 200 kilometres in diameter (initially dubbed 'Pioneer Rock' by the imaging team, this was designated S-1-1979 and eventually named Epimetheus). Furthermore, the particles and fields data showed that the spacecraft had passed through the magnetic 'wake' of an object which, although initially listed as S-2-1979, turned out to be this same moonlet.[64] Another thin ring (the 'G' ring) was inferred from Pioneer 11's particles and fields data to lie farther out. All of these discoveries were confirmed by Voyager imagery, as was the 'D' ring that had been glimpsed by terrestrial observers within the 'C' ring, and it was shown to extend to about 3,200 kilometres of the planet's cloud tops as material spirals in.

The striking discovery of the 'F' ring supported a theory which posited that moonlets within the Roche radius would promote and stabilise very narrow rings.[65] When Voyager observed a pair of these 'shepherds' (named Pandora and Prometheus), one just inside and one just outside of the 'F' ring, this theory was confirmed. Kepler's laws of planetary motion control how the shepherds act. By being closer to the planet, the inner shepherd orbits a little faster than the material in the ring and its gravity accelerates the nearby ring material, causing it to rise, which sharply defines the inner edge of the ring. By being farther from the planet, the outer

The complexity of Saturn's ring system was a major surprise. It is composed of *thousands* of individual very narrow rings nested one within the other. The main rings form an annular disk 350,000 kilometres wide but only 1 kilometre thick. In fact, the outer edge of the 'A' ring is a mere 10 metres thick. The ring material is ice ranging from several microns to a several tens of metres across. Two additional rings were identified. The outer edge of the 'B' ring is actually elliptical, and as the material orbits the planet the ellipse maintains its short axis aligned with Mimas. Elliptical and discontinuous rings were found in the gaps in the main ring system at distances consistent with resonances with the larger satellites orbiting beyond.

shepherd orbits more slowly, decelerates ring material that rises too far, causing it to descend, thereby defining the outer edge of the ring. Since it cannot disperse radially, the tightly controlled ring will survive for a long time. If the shepherds are in resonance with larger outer moons, then the ring system's stability is further assured. Indeed, such a system might exist for billions of years. However, dynamicists were astonished to see that the distribution of material around the ring was clumpy, and in places it had a 'braided' appearance. Because the ring lies only just within the planet's Roche radius, the clumps may be accreting but are too loosely consolidated to resist the tidal stresses, and hence cannot integrate into a single object.

Saturn was found to have two moonlets following almost the same orbit. More accurately, they have a common *mean* orbit. At any time, one of the two is in a slightly lower orbit than the other, and by Kepler's laws of orbital motion travels slightly faster. As a result, it slowly catches up with its associate. As it draws close it is accelerated by the gravitational field of the moon in the higher orbit, causing the chaser to rise. At the same time, the leading moon is decelerated and drops down. Note, however, that while the two moons swap orbits during these encounters they do not actually *pass* one another because, by dropping down, the one that was ahead pulls away from its associate. They encounter one another every four years or so. They orbit several thousand kilometres beyond the rings and the difference between the orbital radii is just 50 kilometres. One of them was discovered by Audouin Dollfus at Pic du Midi in 1966, when the ring system was presented edge-on and named Janus. However, it is not sure which one he saw. In fact, Dollfus suspected that he might have been monitoring a pair of objects in similar orbits. Anyway, when Voyager confirmed that there were two, the other one was named Epimetheus; it is the satellite that Pioneer 11 encountered. They are almost certainly fragments of a common progenitor, with Janus having the larger mass.

Several cases of moonlets in co-orbital relationships with Saturn's major moons were also identified. Helene leads Dione and Tethys has Calypso and Telesto travelling in its leading and trailing Lagrange points, respectively.[66]

AND THEN THERE WAS ONE

By flying close to Jupiter Voyager 1 had been accelerated onto the 'fast track' to Saturn. If Voyager 1 had failed to make the Titan encounter, its back-up's aim would have been adjusted to try again for Titan and both vehicles would have terminated their primary missions at Saturn. By successfully achieving its objectives at Saturn, Voyager 1 released its running mate to attempt the 'Grand Tour'. To perform the slingshot, Voyager 2 would have to cross the plane of the ring system at 2.86 planetary radii, between the outer edge of the 'A' ring and Mimas's orbit. Pioneer 11 had flown this same route to investigate conditions in what appeared to be clear space, and had survived.[67] Voyager 2, which up to this point had been serving a supporting role, now took on the mission of exploring where no spacecraft had ventured before. Interestingly, the existence of the two planets that it was to tackle had not been suspected when Galileo Galilei first turned his telescope to the sky.

NOTES

1. *Almagestum Novum*, G.B. Riccioli. Bologna, 1651.
2. *Systema Saturnium*, C. Huygens. 1659.
3. In fact, Saturn's satellites were not named until John Herschel did so in 1850, at which time he assigned to all the known moons names from myths concerning Saturn. Titan was a child of Uranus and Gaia, and Saturn was the son of Titan. Iapetus and Hyperion were Saturn's brothers; Tethys, Dione, Rhea, Themis and Phoebe were Saturn's sisters; the offspring were collectively known as 'The Titans'. Enceladus, however, was the giant who lived atop the volcano Mount Etna on the island of Sicily.
4. *Astron. Nach.* No.756 (2 May) 1851.
5. *Mon. Not. Roy. Astron. Soc.*, vol. 11, p. 21.
6. *Mon. Not. Roy. Astron. Soc.*, vol. 13, p. 248.
7. *Mon. Not. Roy. Astron. Soc.*, vol. 15, p. 32.
8. *Phil. Trans.*, vol. 32, p. 385.
9. Over the centuries, there have been many reports of discoveries of satellites which proved to be false alarms upon further investigation, so the prospect that Fournier and Schaer were mistaken in seeing a faint outer ring was readily accepted.
10. 'A spectroscopic proof of the meteoritic constitution of Saturn's rings', J.E. Keeler. 1895.
11. Observatory, vol. 6, p. 335.
12. Am. J. Sci., vol. 14, p. 325.
13. This chemical separation process is called fractionation.
14. For a while after the 'quick look' data from the Galileo atmospheric entry probe concerning the abundance of helium in Jupiter was received in 1995, it was suspected that Jupiter might not be as hot as had been believed, and that helium fractionation might actually be occurring, but when the fully calibrated data was read back this was shown not to be the case.
15. 'A comparison of the interiors of Jupiter and Saturn', T. Guillot. *Planetary & Space Science*, vol. 47, p. 1183, 1999.
16. For example, although a rock, volcanic pummice is so full of cavities that its bulk density is only 0.3, hence it will float on water.
17. For a summary of the state of knowledge concerning Saturn's moons immediately prior to the arrival of the Voyagers, see S. Larson and J.W. Fountain. *Sky & Telescope*, vol. 60, p. 365, 1980.
18. It was the azimuthal motion of Voyager 2's scan platform that was impaired. The problem was the result of a coolant leak. Once beyond Saturn, the azimuthal action was able to be recovered, but only if the platform was slewed slowly, so the spacecraft had to undertake the Uranus and Neptune encounters with this limitation.
19. 'Encounter with Saturn: Voyager 1 imaging science results', B.A. Smith, L.A. Soderblom, R. Beebe, J. Boyce, G. Briggs, A. Bunker, S.A. Collins, C.J. Hansen, T.V. Johnson, J.L. Mitchell, R.J. Terrile, M.H. Carr, A.F. Cook, J. Cuzzi, J.B. Pollack, G.E. Danielson. A.P. Ingersoll, M.E. Davies, G.E. Hunt, H. Masursky, E.M. Shoemaker and D. Morrison. *Science*, vol. 212, p. 163, 1981.
20. 'A new look at the Saturn system: the Voyager 2 images', B.A. Smith, L.A. Soderblom, R. Batson, P. Bridges, J. Inge, H. Masursky, E.M. Shoemaker, R. Beebe, J. Boyce, G. Briggs, A Bunker, S.A. Collins, C.J. Hansen, T.V. Johnson, J.L. Mitchell, R.J. Terrile, A.F. Cook, J. Cuzzi, J.B. Pollack, G.E. Danielson, A.P. Ingersoll, M.E. Davies and G.E. Hunt. *Science*, vol. 215, p. 504, 1982.

21 *Saturn*, T. Gehrels and M.S. Matthews (Eds). University of Arizona Press, 1984.
22 *Satellites*, J.A. Burns and M.S. Matthews (Eds). University of Arizona Press, 1986.
23 'Satellites of the outer planets: their physical and chemical nature', J.S. Lewis. *Icarus*, vol. 15, p. 174, 1971.
24 A similar pattern of fractures on Mars's Phobos seems to be related to the impact that formed the large crater Stickney.
25 'A new look at the Saturn system: the Voyager 2 images', B.A. Smith, L.A. Soderblom, R. Batson, P. Bridges, J. Inge, H. Masursky, E.M. Shoemaker, R. Beebe, J. Boyce, G. Briggs, A Bunker, S.A. Collins, C.J. Hansen, T.V. Johnson, J.L. Mitchell, R.J. Terrile, A.F. Cook, J. Cuzzi, J.B. Pollack, G.E. Danielson, A.P. Ingersoll, M.E. Davies and G.E. Hunt. *Science*, vol. 215, p. 504, 1982.
26 'Viscosity of the lithosphere of Enceladus', Q.R. Passey. *Icarus*, vol. 53, p. 105, 1983.
27 'Volcanic and igneous processes in small icy satellites', D.J. Stevenson. *Nature*, vol. 298, p. 142, 1982.
28 'Evolution of Enceladus', S.W. Squyres, R.T. Reynolds, P.M. Cassen and S.J. Peale. *Icarus*, vol. 53, p. 319, 1983.
29 There are three forms of ice that are of concern for planetary surfaces: (1) amorphous, (2) cubic crystals and (3) hexagonal crystals. Cubic ice is made at low temperature (roughly 120 to 150 K), hexagonal ice is made at higher temperatures. Cubic ice will gradually transform to hexagonal ice, the lowest energy form. The shape of the reflectance spectrum is the same for cubic and hexagonal ice, so it is not possible to tell one from the other spectrally. However, VIMS on Cassini will be able to tell if Enceladus is covered with water ice, what the grain size is, and if it is amorphous or crystalline. A way to tell the difference between cubic and hexagonal ice is by the angular refraction pattern, looking for the 22 deg (hexagonal) or 46 deg (cubic) halo, (as was attempted Europa by the Galileo Spacecraft), but such observations require special geometry, and may not be easy to arrange.
30 'Tectonics and geological history of Tethys', J.M. Moore and J.L. Ahern. *Lunar Planet. Sci. Conf.*, vol. 13, p. 538, 1982.
31 'Rheology of ices: a key to the tectonics of the ice moons of Jupiter and Saturn', J.P. Poirier. *Nature*, vol. 299, p. 683, 1982.
32 'The geology of Tethys', J.M. Moore and J.L. Ahern. *J. Geophys. Res.*, vol. 88, p. 577, 1983.
33 'The moons of Saturn', L.A. Soderblom and T.V. Johnson. In *The planets*, B.C. Murray. W.H. Freeman, p. 95, 1983.
34 'The geology of Dione', J.B. Plescia. *Icarus*, vol. 56, p. 255, 1983.
35 'Volcanic and igneous processes in small icy satellites', D.J. Stevenson. *Nature*, vol. 298, p. 142, 1982.
36 'Interactions of planetary magnetospheres with icy satellite surfaces', A.F. Cheng, P.K. Haff, R.E. Johnson and L.J. Lanzerotti. In *Satellites*, J.A. Burns and M.S. Matthews. University of Arizona Press, p. 403, 1986.
37 'The tectonic and volcanic history of Dione', J.M. Moore. *Icarus*, vol. 59, p. 205, 1984.
38 'The geomorphologic features on Rhea', J.M. Moore and V.M. Horner. *Lunar Planet. Sci. Conf.*, vol. 15, p. 560, 1984.
39 'Geomorphology of Rhea: implications for geologic history and surface processes', J.M. Moore, V.M. Horner and R. Greeley. *J. Geophys. Res.*, vol. 90, p. 785, 1985.
40 *Planetary landscapes*, R. Greeley. Allen & Unwin, p. 235, 1987. Greeley cites a personal communication by R. Pike and P.D. Spudis, but (according to Spudis) this observation was never formally published.

41 '[Gerard P. Kuiper:] Forging a new Solar System', S.A. Stern. *Astronomy*, p. 40, March 1999.
42 Molecular nitrogen is very difficult to detect spectroscopically from Earth, because the signature is masked by the nitrogen in the Earth's atmosphere.
43 'Titan', T. Owen. In *The planets*, B.C. Murray (Ed.). W.H. Freeman, p. 84, 1983.
44 'On the origin of Titan's atmosphere', T.C. Owen. *Planetary & Space Science*, vol. 48, p. 747, 2000.
45 Peter Smith, Mark Lemmon, Ralph Lorenz, John Caldwell, Larry Sromovsky and Michael Allison. See STScI-PR94-55, December 1994.
46 'Abundances of the elements in the Solar System', A.G.W. Cameron. *Space Sci. Rev.*, vol. 15, p. 121, 1973.
47 C.A. Griffith *et al. Nature*, 8 October 1998.
48 'Titan's methane clouds'. *Sky & Telescope*, p. 23, March 1999.
49 'The tides in the seas of Titan', C.E. Sagan and S.F. Dermott. *Nature*, vol. 300, p. 731, 1982.
50 For a comprehensive review of the state of knowledge regarding Titan, and a preview to the Cassini mission, see *Titan: the Earth-like moon*, A. Coustenis and F. Taylor. World Scientific Series on Atmospheric, Oceanic and Planetary Physics, 1999.
51 'Destination Titan', S. Mirsky. *Astronomy*, p. 42, November 1997.
52 'Titan under a red giant Sun: a new kind of "habitable" world', R.D. Lorenz, J.I. Lunine and C.P. McKay. *Geophys. Res. Lett.*, vol. 24, p. 2905, 1997.
53 'Chaotic motion in the Solar System', C.D. Murray. In *Encyclopedia of the Solar System*, P.R. Weissman, L.A. McFadden and T.V. Johnson (Eds). Academic Press, p. 825, 1999.
54 The discovery of Iapetus's chaotic rotation was made by Jack Wisdom, Stanton Peale and Francois Mignard and the observations which confirmed it were made by James Klavetter, who published his results at the American Astronomical Society meeting.
55 As yet, Hyperion is the only planetary satellite known to behave in this random manner.
56 *Phil. Trans.*, vol. 82, p. 14.
57 The Iapetus aspect of the novel was elided when Stanley Kubrick made the movie version of this story.
58 See *Planetary landscapes*, R. Greeley, p. 243, who attributes the Phoebe hypothesis to Steve Soter of Cornell University who presented it at a conference in 1974.
59 'A new look at the Saturn system: the Voyager 2 images', B.A. Smith, L.A. Soderblom, R. Batson, P. Bridges, J. Inge, H. Masursky, E.M. Shoemaker, R. Beebe, J. Boyce, G. Briggs, A Bunker, S.A. Collins, C.J. Hansen, T.V. Johnson, J.L. Mitchell, R.J. Terrile, A.F. Cook, J. Cuzzi, J.B. Pollack, G.E. Danielson, A.P. Ingersoll, M.E. Davies and G.E. Hunt. *Science*, vol. 215, p. 504, 1982.
60 'Saturn's icy satellites: thermal and structural models', K. Ellsworth and G. Schubert. *Icarus*, vol. 54, p. 490, 1983.
61 'Small bodies and their origins', W.K. Hartmann. In *The new Solar System*, J.K. Beatty and A. Chaikin (Eds). Cambridge University Press (third edition), p. 251, 1990.
62 *Satellites of the Solar System*, W. Sandner. The Scientific Book Club, p. 95, 1965.
63 Nevertheless, as recent discoveries have demonstrated, Saturn has an astonishing number of tiny moons.
64 *Rings: discoveries from Galileo to Voyager*, J. Elliot and R. Kerr. MIT Press, p. 121, 1984.
65 'Towards a theory for the Uranian rings', P. Goldreich and S. Tremaine. *Nature*, vol. 277, p. 97, 1979.
66 *Voyages to Saturn*, D. Morrison. NASA SP-451, 1982.
67 *Rings: discoveries from Galileo to Voyager*, J. Elliot and R. Kerr. MIT Press, p. 119, 1984.

10

Planets beyond

'GEORGE'S STAR'

As telescopic astronomers began to catalogue the sky, the belief that the Solar System ended with Saturn was accepted without question, so the discovery of a new planet came as a considerable surprise.

William Herschel, a Hanovarian musician living in Bath, England, had a burning passion for astronomy. In 1781 he was charting the sky for double stars whose parallax might enable him to determine the distances to the stars as a step towards his ultimate objective of gauging the scale of the heavens. On 13 March, he happened upon a small sea-green disk that was not listed on his charts.[1] Assuming this to be a tail-less comet, he monitored its motion against the stars for several nights, then reported his discovery. Upon inspecting the object, Nevil Maskelyne, the Astronomer Royal at Greenwich Observatory, observed that the well-defined disk was "very different from any comet I ever read any description of, or saw" and he wrote to Herschel with the radical suggestion that it might actually be a previously unknown planet.

Observations during the summer enabled the orbit to be derived. In fact, it was calculated independently by A.J. Lexell in St Petersburg and P.S. Laplace in France and shown to be unlike any comet for which an orbit had been determined. It was twice as far from the Sun as Saturn and as it was in a nearly circular rather than an eccentric orbit, with a period of 84 years; it really was a planet.

Herschel decided to name the new planet 'Georgium Sidus' (George's Star) in honour of King George III, but this was not well received by the Europeans who had worked on its orbit. J.J. Lalande in Paris suggested 'Herschel's Planet', but this did not find favour either. J.E. Bode in Berlin proposed the name Uranus, because in mythology Uranus was the father of Saturn – just as Saturn fathered Jupiter, who in turn fathered Mars.

On a single night in January 1787, Herschel noted that Uranus had a pair of attendants.[2] Over the next ten years or so, he determined that these satellites orbit in a retrograde manner and that their orbital plane is inclined at 98 degrees to the ecliptic,[3] so it was clearly an unusual system. In October 1851, again on a single night, William Lassell observed two more satellites, orbiting nearer the planet.[4] Drawing upon English literature for his inspiration he named Herschel's pair of moons Titania and Oberon and his own Ariel and Umbriel. In 1931 W.H.M. Christie of the Mount Wilson Observatory made a photographic search, utilising the 60-inch reflector and very long exposures in order to reach faint objects, but he failed to discover anything. In 1948 Gerard Kuiper made another search using the McDonald Observatory's 82-inch reflector and spotted a moon orbiting very close to Uranus.[5] It was so faint that in the pictures that Christie took it was probably lost in the planet's glare. In fact, it is much closer to Uranus than any of the large Jovian or Saturnian moons are to their primaries. In the spirit of the established nomenclature, it was named Miranda.[6]

It had been presumed (by analogy with other systems) that Uranus's axial tilt would match the plane in which its satellites orbit. But then in 1870 J. Buffham reported faint markings on the disk whose apparent motions suggested a rotational period of 12 hours in a plane that was *not* coincident with the satellites[7]. G.V. Schiaparelli, however, noted that the disk was oblate, with the 'equatorial' bulge matching the satellites.[8] In 1883, C.A. Young at Princeton reported faint bands set at an angle to the satellite plane,[9] and in the first half of 1884 the Henry brothers in Paris reported a pair of grey parallel bands separated by a bright zone, all offset at 40 degrees from the orbital plane of the satellite system. A possible factor in this spate of reports of atmospheric features might have been that over the first 100 years following the planet's discovery it was moving towards perihelion, which not only brought it closer to the Earth, and hence made it somewhat easier to observe, but also closer to the Sun and the resultant heating may have stirred the outer atmosphere into activity.

The first spectrum of Uranus was by A. Secchi in 1869.[10] However, its half dozen broad absorption bands proved to be baffling. But V.M. Slipher of the Lowell Observatory in Flagstaff, Arizona, identified hydrogen in Uranus's atmosphere in 1902, and Rupert Wildt at Princeton detected methane in 1932.[11] The disk's pale greenish hue is due to the red end of the spectrum being absorbed by methane. By measuring the doppler effect across its disk, in 1912 Slipher estimated the planet's rotational period as 10.8 hours.[12] We now know that Uranus's axial tilt matches the satellites and is almost perpendicular to the ecliptic. Sometimes its northern, and sometimes its southern, hemisphere faces 'forward' as the planet moves along its orbital path and in between times it literally 'rolls' along on its side, which creates unusual 'seasons'. For large arcs of each orbit one or other hemisphere is subjected to prolonged illumination and the other is in darkness. The fact that the planet's equator is coincident with the plane in which its satellites orbit indicates that the process that induced the axial tilt was probably related to the formation of the satellite system.

Because none of the five satellites showed a disk to telescopic observers, their diameters were estimated from photometric observations. Initially, depending upon the assumptions regarding their albedoes, a wide variety of sizes were estimated.

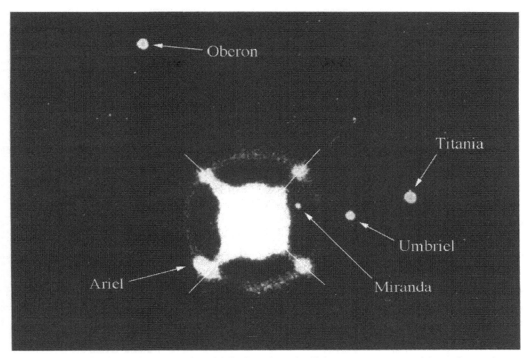

A telescopic view of Uranus in which the planet's disk was overexposed in order to be able to resolve the moons. The symmetrical quartet are telescope artefacts.

However, once the Voyagers provided data on Saturn's icy moons, the diameters of the Uranian moons were estimated by assuming that their surfaces were similar.[13]

Even in the world's largest telescopes, Uranus's satellites are mere points of light, so prior to Voyager 2's approach we could only draw inferences as to the nature of their surfaces from their spectroscopic and near-infrared characteristics. Studies in 1979 showed that their surfaces are predominantly water ice, just as with most of Saturn's moons. There were indications of a carbonaceous residue in the ice. Once the temperatures of their surfaces were measured in 1981, it became possible to estimate their heat flows in order to refine their sizes and densities. The results indicated that Ariel and Umbriel had bulk densities of around 1.3 g/cm^3 and Titania and Oberon were 2.6, from which it was inferred that Titania and Oberon had a higher proportion of rock than the smaller ice moons orbiting closer to the planet. This prompted speculation that even though they are not involved in orbital resonances, the larger moons might be thermally differentiated and (like Jupiter's Europa) have rocky cores englobed by oceans of liquid water capped by a few kilometres of ice.[14] Nevertheless, as Voyager 2 left Saturn on its long interplanetary cruise to Uranus, virtually nothing was known concerning the planet's moons.[15]

RINGS

On 10 March 1977, several teams of planetary astronomers observed as Uranus occulted a star in the hope that the timing would refine the planet's diameter and that the manner in which the starlight was refracted would yield insight into the physical state and chemical composition of the outer atmosphere. One team was observing in the infrared from aboard NASA's Kuiper Airborne Observatory. The instruments were switched on well in advance, so that everything would be running smoothly at the crucial moment. But almost immediately, the photometer signal decreased momentarily and then resumed its previous strength. In fact, it did this five times in rapid succession, then settled down. Following the occultation, another set of momentary dips were observed. When the data was processed it showed nine events timed symmetrically either side of the main occultation. The only reasonable answer was that the starlight had been attenuated by passing through a series of very thin rings.[16] Uranus had rings! Unlike Saturn's rings, however, which are spectacularly bright, Uranus's rings are dark, suggesting that they, like Jupiter's rings, are composed of rocky fragments rather than lumps of ice.[17,18] As such, the energy of sunlight that they absorb is re-emitted in the infrared, so they are readily observed in the near-infrared.[19]

FLY-BY

Owing to the angle at which the Uranian system is tilted, Voyager 2, travelling in the plane of the ecliptic, approached on a trajectory that was almost perpendicular to the plane in which the moons orbit. Instead of having several days in which to study each of the moons as the spacecraft worked its way through the system, it had only a few hours; it was almost a case of 'if it blinked, it would miss a moon'. Further, because the travel time for the radio signal was several hours, the spacecraft could not be controlled from Earth in real time, and had to be programmed with the sequence of activities and told to proceed. If it encountered a serious problem, it would simply have had to abandon the science, 'safe' itself, and await a signal from the Earth, and by the time that the fault was rectified – if indeed it could be rectified – the encounter would be over and Uranus would he fading fast.

On 24 January 1986, some four and a half years after leaving Saturn and eight and a half years into its mission, Voyager 2 passed through the Uranian system. Just over two centuries had passed since William Herschel discovered the planet. To utilise the planet's gravity for a slingshot on to Neptune, Voyager 2 flew right through the heart of the system, passing 81,600 kilometres above the planet's atmosphere. In fact, the trajectory passed very close by Miranda, the innermost of the known moons, so the smallest of the retinue received the closest attention. What might Herschel have thought of the Voyager imagery? Uranus's south pole was facing sunward, so Voyager 2 observed its sunlit hemisphere on the way in. On the way out it saw only the dark northern hemisphere. However, the departure angle provided an ideal opportunity to view the entire ring system lit up like a giant bull's-eye as it forward-

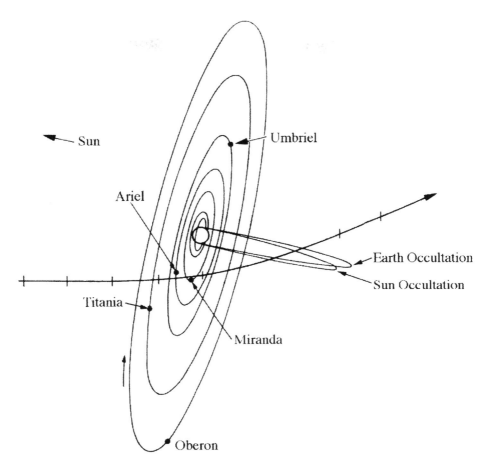

The axis of the Uranian system is so inclined to the ecliptic that Voyager 2's path was almost perpendicular to the system's equatorial plane, within which the satellites and rings reside. The fly-by was calculated to set up a departure trajectory for Neptune.

scattered sunlight.[20] On 28 January, with the encounter over, the scientists and engineers at JPL were preparing their triumphant 'wrap up' Press Conference when the media's attention was snatched away by the tragic loss of the Space Shuttle Challenger.[21]

THE GREEN GIANT

All the giant planets are derived from the solar nebula, so their chemical compositions are similar: they are all predominantly hydrogen with a secondary proportion of helium and trace amounts of other chemicals. However, it is these chemicals that produce the visual variety – which is most evident in the case of Jupiter.

Over the years, telescopic observers occasionally reported detail on Uranus's atmosphere, but Voyager 2 found it to be frustratingly bland. The atmospheric specialists had been fired up with anticipation as the interplanetary cruise drew to a close, then the spacecraft transmitted a series of pictures of a featureless disk. On the other hand, the sheer blandness was significant, because it implied that the deeper structures were completely masked by the hydrocarbon haze. Methane is dissociated by ultraviolet insolation, and the carbon that is released polymerises to form a haze of hydrocarbons. In Jupiter's case, the temperature of the upper atmosphere is not low enough for the methane to condense, so the haze is both thin and the patchy and the underlying structures are sufficiently prominent to be seen through it. In the case of Saturn, much of the underlying structure is masked by the haze. However, being further from the Sun, and hence much colder, Uranus is completely masked by this haze. The thin layer of haze that had been expected at an altitude of 200 kilometres above the 1-bar pressure level was found to span the altitude range from 50 to 180 kilometres.[22] The disk was not totally featureless, however. Computer processing of images through filters for wavelengths that were less readily absorbed by ethane and acetylene in the haze provided a glimpse of the deeper structure, including faint latitudinal banding resulting from lateral shear as the atmosphere rotated differentially. Although there was evidence of a layer of methane ice crystals a few kilometres thick some 8 kilometres below the 1-bar level,[23] the deeper ammonia and hydrogen sulphide cloud decks were out of sight.

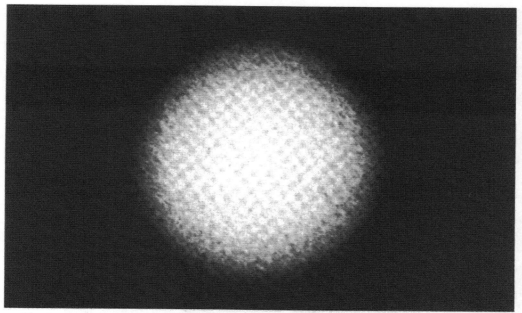

The best pre-Voyager photograph of Uranus was obtained by Project Stratoscope II utilising a balloon-borne telescope at an altitude of 80,000 feet above the Earth's surface. The resolution in this 'exceptionally clear' view was 2,000 kilometres per pixel, but the disk was featureless nevertheless. (Courtesy of Princeton University, NSF and NASA.)

It had been expected that because the south polar region had been in continuous sunlight it would be somewhat warmer than on the limb, where the Sun was no more than a few degrees above the horizon. However, Voyager 2 found that the 'surface' of the planet was a uniform temperature irrespective of whether it had been illuminated for years or in shadow for years.[24] Furthermore, the winds had been expected to radiate out from the subsolar point irrespective of the orientation of the planet, but this turned out not to be the case. Although the most intense winds were indeed near the subsolar point, the planet's rotation maintained a latitudinal weather system even when one of its poles was facing the Sun, which indicated that the planet is driven by internal heat.[25] Being far less massive, Uranus cannot generate as much energy from residual contraction. Overall, it radiates just 1.1 times as much energy as it receives from the Sun. In fact, it has been suggested that Uranus's interior might be 'cool', because the extreme axial tilt will have facilitated efficient cooling via the polar regions during their long periods of continuous darkness.[26]

When the International Ultraviolet Explorer satellite in Earth orbit turned its attention upon Uranus shortly before Voyager 2's arrival there, it detected an ultraviolet 'excess' which was interpreted as indicating auroral activity, which was considered to be proof of a magnetic field and a magnetosphere containing trapped radiation. In fact, Voyager 2 revealed the ultraviolet excess to be due to the illuminated hemisphere radiating in response to ultraviolet excitation. This has been dubbed 'dayglow'.[27] In addition, because Uranus is too small to hold onto hydrogen it is surrounded by a 'coma' of ultraviolet-radiating hydrogen. Neverthless, the particles and fields instruments confirmed that the planet has a magnetic field, and that the solar wind enters the polar cusps of the magnetosphere and excites auroral displays. The spacecraft was able to chart Uranus's magnetic field as it passed through the system, and revealed — to general astonishment — that not only was the field's axis inclined at about 60 degrees to the planet's rotational axis, but it was also being generated in a region that was *displaced* about 30 per cent of the way towards the surface.[28,29,30,31]

Whereas the Earth's magnetic field is the result of interactions at the boundary between the solid inner nickel iron core and the englobing fluid outer core, the magnetic fields of the giant planets are believed to be generated by flows of ionised fluids located deep within the mantles that surround their cores. Jupiter has the strongest magnetic field, which is inclined only a few degrees from the planet's rotational axis. Saturn, being somewhat smaller, has a weaker field closely aligned with its spin axis. Being even smaller, Uranus had been expected to have only a modest magnetic field, which it does, but its oblique axis and displaced generation centre posed a problem for the theorists. If the obliquity had been the only factor, it could reasonably have been argued that the planet was undergoing a reversal of polarity (as is known to happen occasionally with the Earth's magnetic field) and that we were simply lucky to catch it in the act of flipping over, but the displaced generation centre greatly complicated the picture. In fact, the spacecraft's passage through the Uranian system prompted a rethink concerning this planet's interior. It had been expected that Uranus, like Saturn, would simply be a subscaled version of Jupiter in which the core is englobed firstly by a mantle and then by a gaseous envelope, except that, in Uranus, the mantle was expected to be liquid water mixed with traces of methane, nitrogen and ammonia.[32] However, it had also been proposed that Uranus might actually have a two-layer structure in which its core was surrounded by a dense homogeneous gaseous mantle that extended to the surface.[33] Six of Uranus's rings are elliptical, and the manner in which these are perturbed provided insight into the interior, and confirmed the two-layer model. The magnetic field is most likely generated by flows of ionised liquid water deep within the mantle.[34] A detailed study showed that if the field was generated in the mantle rather than in the core, an offset centre of generation is inevitable and its location will migrate with response to turbulence in the flows. Given sufficient data and ingenuity, therefore, even the most peculiar physical situations can be explained.[35] The planet's bulk density implies that the rocky core is comparable in mass to the Earth, so Uranus is effectively a terrestrial planet which, due to its low temperature, is able to retain a voluminous envelope that would long since have been lost by a similar object in the inner Solar System.

With few (if any) atmospheric features whose motions could be timed, Uranus's rotational period was not known with any certainty. Even as late as a year before Voyager 2 approached the planet, telescopic studies were suggesting periods ranging from as short as 10 hours to as long as 24 hours, with the 'best' estimates falling in the range 15 to 17 hours. By monitoring the precession of the magnetic axis by the cyclic variation in the strength of the magnetic field, the spacecraft was able to establish that the planet's interior rotates in 17 hours and 15 minutes. However, the atmosphere undoubtedly rotates differentially.[36]

The bland state of Uranus's atmosphere in 1986 may have been anomalous.[37] The banding that had been reported by early telescopic observers, but was absent when Voyager 2 flew by, began to reappear in the 1990s.[38] The Hubble Space Telescope cannot rival the resolution of a passing spacecraft, but it can provide excellent views of the planet, and in July 1997 it observed a cluster of five high-altitude clouds strung out in line in the equatorial zone.[39] In contrast to a fly-by spacecraft, Hubble offers the advantage of being able to undertake *long-term* weather monitoring.[40]

OBERON

Oberon, the outermost of Uranus's satellites, was the least well resolved by Voyager 2. It lived up to the Callistoan model by being extensively cratered and having little regional variation. In addition to craters in the 100 kilometres size range there are faint arc-like features which hint at degraded multiple-ringed impact structures. Although the surface is dark, ray craters appear to have splashed out bright streaks of melted ice. Interestingly, although the ray craters are large, and hence are likely to be old, their rays remain bright. One ray crater (named Hamlet) near the south pole has a central peak surrounded by very dark patches which might be a carbonaceous eruption from fissures in the crater's floor, but there is no evidence of any regional endogenic activity.

TITANIA

At 1,610 kilometres in diameter, Titania is 50 kilometers larger than Oberon. Even so, it is less than half the diameter of our own Moon and has only 5 per cent of its mass. Nevertheless, Titania was a surprise. Although it must have been as exposed to the same early population of impactors as Oberon, there are only a few large craters. Something has erased the ancient structures and produced plains that have since been cratered by a random distribution of much smaller strikes. Evidently, 'soft' ice oozed from the floors of the early impact structures and flooded their depressions, and internal heat enabled the lithosphere to settle isostatically to smooth out the surface producing palimpsests. Although younger than Oberon's, Titania's surface is nevertheless very old. It is criss-crossed by grabens 20 to 50 kilometes in width and up to 1,600 kilometres in length, with steep walls which plunge several kilometres. A lighter-toned material is exposed in their walls, suggesting that mass wastage has exposed fresh ice. It is believed that as the 'warm' ice of the interior froze it expanded and cracked the crust. Why did Titania undergo extensive cryogenic resurfacing when neighbouring Oberon did not?

Doppler tracking of the spacecraft's radio signal enabled its trajectory to be monitored very accurately during the encounters with the satellites, and the perturbations provided insight into their internal structures.[41,42] It turned out that Uranus's satellites are somewhat denser than their Saturnian counterparts – only Titan had a higher density, and this was only slightly more so. Evidently, despite having formed far from the Sun, Uranus's satellites, listed below, have proportionately significantly more rock than Saturn's icy retinue.[43]

Miranda	1.3 g/cm^3
Ariel	1.7 g/cm^3
Umbriel	1.4 g/cm^3
Titania	1.6 g/cm^3
Oberon	1.5 g/cm^3

By virtue of being slightly larger and denser than Oberon, Titania could be expected to be more evolved. But to such an extent? Gene Shoemaker has suggested that the reason for their having such different histories is exogenic.[44] Perhaps Titania was disrupted by major collision early in its history and the debris, following very similar orbits, progressively reassembled itself via a series of low-energy 'gentle' collisions. If this occurred, then the energy from the re-accretion would have induced significant melting.[45] The temperatures of the surfaces of the moons were measured to be around 80 K (which is warmer than the planet's outer atmosphere because the moons have dark surfaces that absorb the meagre sunlight). It has been argued that methane, or perhaps carbon monoxide, may have been mixed with the water of the ice. Such clathrates would have melted and flowed more readily than pure water ice. A 'pulse' of heat might have produced a significant amount of cryogenic volcanism.

318   **Planets beyond**

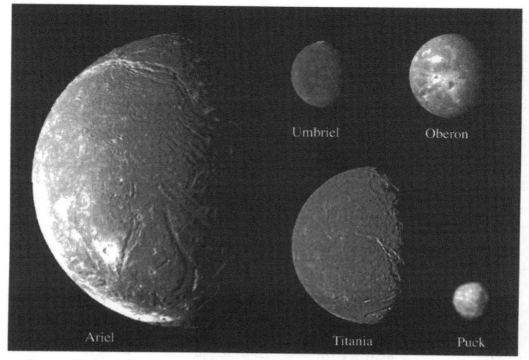

Voyager 2 snapped Uranus's outer moons from a distance as it sped through the system. Puck was discovered by the spacecraft early on in its approach, and a single image was programmed. Not to scale.

UMBRIEL AND ARIEL

Orbiting closer to the planet than the larger outer pair, Umbriel and Ariel are also similar in size but at 1,190 kilometres in diameter Umbriel is 30 kilometres larger. Unfortunately, it was on the far side of the planet when Voyager 2 passed through the Uranian system, so it was not well observed. At 0.19 albedo, it has the darkest surface of the retinue.[46] Apart from a bright ring of material on the floor of a crater named Wunda, there is remarkably little variation in albedo. As Wunda was viewed on the limb, it is in the equatorial zone, and its ring is broken by a dark linear streak running from the central peak to the southern rim. Its origin is a mystery. The fact that none of the craters has laid down lighter-toned ejecta indicates either that the dark material is a very thick deposit or that it is just a thin veneer formed long after the craters. One theory proposed that the charged particles that circulate in Uranus's magnetosphere might have irradiated methane ice and darkened it by driving out the hydrogen to form a cabonaceous residue.[47] However, if the magnetic field is the agent, why was Ariel not similarly darkened? In fact, with an albedo of 0.40, Ariel has the brightest surface in the system. Another theory suggested that Umbriel has been 'painted' by dark material from a satellite that broke up. Might the rings be the last remnant of this? But if Umbriel swept up dark material why was it not concentrated on the leading hemisphere? And, of course, how did Ariel manage to escape?

Umbriel shows no evidence of having suffered endogenic activity, while Ariel appears to have been extensively resurfaced. In fact, *most* of Ariel's surface has been masked by cryogenic flows which have erased the large craters. The ancient cratered terrain is extensively fractured by faults and grabens, some of which have sinuous valleys (possibly carved by fluid flows) on their floors. Some of the crater walls appear to have been breached, and their floors flooded by surface flows, many of which are young and seem to be intimately related to the faults. Some of the grabens have smooth floors suggesting that fluids erupted. In some cases, this material has flooded across the surrounding terrain. The oldest and broadest grabens are on the hemisphere that faces the planet, where the flow activity is most extensive. The extent of Ariel's endogenic activity is clear in the form of its fractures, which are even more pronounced than those on Titania. Indeed, the fractures seem to form a *global* network. One crack, which winds its way right around the moon, is 30 kilometres deep. An insight into the resurfacing process can be drawn from the fact that some flows have halted at crater walls, leaving scarps up to 1 kilometre high. This suggests an analogy with glacial activity, in which 'warm' ice flows across the surface. The absence of large craters suggests that the resurfacing occurred quite early on. The subsequent cratering of the plains indicates that such flows ended 3 billion years ago. However, the smooth sparsely cratered plains imply that endogenic activity continued until 'recent' times. Was this later surface modification prompted by clathrates first melting and erupting onto the surface? Where did the heat come from? The Uranian system occupies a volume of space that barely exceeds that of the Earth–Moon system. Just as the Earth's rotation is slowed by the oceanic tides which the Moon raises, and the Moon is slowly receding to preserve the angular momentum of the system as a whole, the Uranian moons raise tides in the planet's atmosphere and, as this slows the planet's rotation, the moons recede. Because the satellites would have moved out at different rates, the inner ones would have found themselves in a succession of resonances, and the resulting tidal stresses might have stimulated pulses of melting within Ariel.[48]

MIRANDA

Simulations have shown that there is a significant 'scaling' effect in the cratering rates for moons orbiting giant planets, such that a single population of impactors causes inner satellites to receive a heavier bombardment. The long reach of the planet's gravitational field draws in stray debris. The deeper it falls into the gravity well, the more it is accelerated. Calculations implied that the flux for Miranda would have been 14 times that of distant Oberon and as a result Miranda was expected to be saturated with craters.[49] However, because Uranus is far from the asteroids concentrated in the belt between Mars and Jupiter, the stray objects drawn in by Uranus's gravity are more likely to have been comets, for which there is a reverse scaling effect: that is, for every long-period comet to penetrate the inner Solar

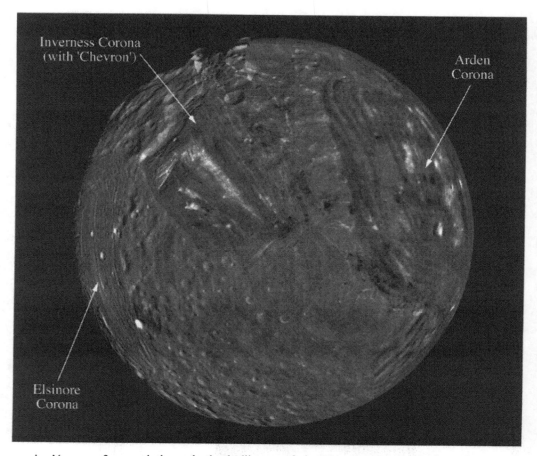

As Voyager 2 passed through the bull's-eye of the Uranian system, it flew close to Miranda, the smallest of the main satellites. The complexity of its surface was a considerable surprise. Arden Corona shows a broad pattern of concentric light and dark bands, and a bright jumbled zone at its centre. In contrast, the structure of Elsinore Corona is a fairly uniform albedo. Inverness Corona, which contains the bright 'Chevron', is located near the south pole (due to the unusual viewing angle this is centred) and appears to lie lower than its surroundings.

System, many more will penetrate only as close as Uranus, so the ongoing impactor population will be dominated by comets.

Orbiting outside the ring system, Miranda, at 485 kilometres in diameter, is the smallest of Uranus's primary moons. However, because Voyager 2 had to fly close to the planet to set up the slingshot that would send it on to Neptune this tiny moon received the spacecraft's closest attention. This proved fortuitous, because it enabled this moon's uniquely complex surface to be imaged with a resolution of better than 1 kilometre per pixel.

As with all of the Uranian moons, owing to the tilt of the system, it was Miranda's southern hemisphere that was illuminated. Most of its surface forms a heavily cratered rolling plain but there are at least three large peculiar ovoidal features with very distinct contacts. The structure of these features is quite unlike anything seen elsewhere in the Solar System. The smallest of the three is located near the south pole and is about 200 kilometres across. The others, one on either side of the moon, are at least 300 kilometres across (they were not fully visible because they extended beyond the limb). Considering that Miranda is less than 500 kilometres across, the scale of these features is remarkable. Interestingly, Miranda is ellipsoidal in shape, with its main axis aimed towards the planet.

The three ovoids, which initially were informally referred to as the 'trapezoidal', 'banded' and 'ridged' ovoids, were subsequently classified as 'coronae' and named Inverness, Arden and Elsinor respectively. Arden is on the leading hemisphere, Inverness is near the south pole (and hence in the centre of the view) and Elsinor is on the trailing hemisphere. Inverness, the smallest of the trio, is dominated by a light-toned feature that was dubbed the 'chevron'. The other two show a succession of ridges and grooves, whose concentric arrangement prompted their being referred to as 'race tracks'. They are physiographically distinct, however: Arden shows considerable albedo variation, but Elsinor has a more uniform albedo that is similar to that of the cratered terrain; Arden is ridged and grooved, but Elsinor is generally less rugged and its concentric structures appear to be ridges; and wide canyons radiate out across the older cratered terrain from Arden, but not from Elsinor. Arden and Inverness are bounded by trench-like depressions which are often flanked by parallel scarps which project far beyond the coronae. Inverness seems to lie generally lower than its surroundings. Both 'arms' of the chevron are paralleled and overlain by closely spaced ridges. The sharp albedo contact might indicate that the chevron is a flow of 'clean' ice. Otherwise, it is difficult to account for such a distinctive shape.

As the images of Miranda streamed in, planetary geologists did not know what to make of them. Miranda had been expected merely to be Uranus's version of Saturn's Mimas. The cratered terrain is clearly ancient, but many of the oldest craters have been softened, so the original surface must have been 'masked' by a deep blanket of material, perhaps fluidised ejecta, perhaps surface flows. Even more remarkable are the systems of fractures. Both the cratered plain and the ovoids are deeply incised by grooves and sinuous valleys, some of which may actually extend all the way around the moon. One of the grooves (named Verona Rupes) was recorded in detail as it passed over the terminator. The bright material exposed in this 20-kilometre-high sheer scarp implies that there is a thick layer of 'clean' ice beneath the dark surface. Topographic relief on such a scale on such a small moon is remarkable.[50,51] One scientist reflected that it appeared as if Miranda had been designed by a committee, with a little of everything thrown in to secure a planning certificate.[52] How should a surface as strange as this be interpreted? How could a body with only 1 per cent of the Moon's mass – in effect a respectably sized asteroid – have undergone such activity?

One theory suggested that Miranda had not started out as a homogeneous mix of rock and ice, but had been lumpy, and before the initial heat of accretion had worn off, the interior had been sufficiently mobile to enable the rock to start to settle and the ice to rise, but the moon then froze before the process of separation was

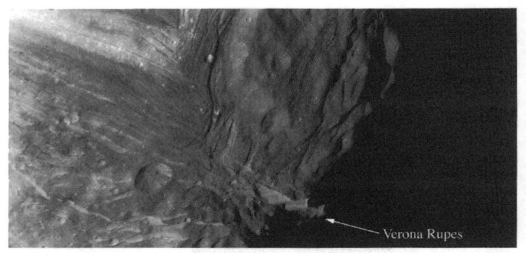

The height of the Verona Rupes scarp varies from 10 to 20 kilometres. It clearly indicates that Miranda has been subjected to extreme tectonic forces. Note the variation in albedo within the cliff's exposure.

complete.[53] According to this theory the ovoids are icy masses which broached the surface as the denser rocks settled.[54] If this process had run its course (it is argued) the ice could have flowed out over the adjacent terrain and resurfaced the moon. However, this theory cannot account for the fact that the three coronae appear to be of different ages. A refinement observed that Miranda's orbital eccentricity is the greatest of this system, and that the moon is the innermost of the primary satellites, and argued that the internal heat for the mobilisation was generated by the tidal stress from a short-term orbital resonance with Ariel.[55] This correlated with the fact that Ariel has also been heated. According to this theory, Arden was initially a vast impact basin whose ejecta had masked the peripheral terrain, and its cavity was flooded by upwelling cryogenic fluid. The other coronae, it seemed, were formed by subsequent episodes of tidal heating.

The surface within the ovoids appears to be *unrelated* to the adjacent rolling plains.[56] Indeed, there is a patch of jumbled terrain at the centre of Arden that is morphologically distinct. Had Miranda been broken into large fragments by a massive impact soon after undergoing partial differentiation, and then progressively reassembled by low-energy collisions? By this theory, the coronae mark where reaccreted rocky material came to rest near the new surface, and then settled isostatically as tidal heating induced internal melting. In fact, Miranda might have been shattered several times before assuming its present configuration.[57] Why should Miranda have suffered so? It is not only Uranus's innermost main moon, it orbits closer to its primary than any large moon of any other planet, and the gravitational focusing effect would therefore have been even greater. Miranda's unique characteristics may simply derive from its location. Despite the profusion of competing theories, Miranda's strange surface is likely to remain a mystery until another spacecraft ventures out to Uranus – and there are no such plans.

It is noteworthy that the Urianian satellites, as a group, have undergone greater geological activity than their similarly sized Saturnian counterparts.

RING EROSION

In the case of Saturn's broad ring system, the many narrow gaps are kept vacant by resonances with the large moons orbiting beyond the ring system. Uranus's rings, however, are narrow, and are separated by wide gaps. Resonances with outer moons cannot sweep wide gaps clear of material and concentrate it into narrow rings. Instead, it was argued, the narrow rings must be defined by pairs of 'shepherd' moonlets, one just inside the ring and the other just outside, which together inhibit the ring material from dispersing.[58,59] The Epsilon ring is the outermost, brightest and most substantial of the series. It is clumpy, varying in width from 30 to 100 kilometres; in fact, it appears to be composed of two 'strands' which display complex interactions. Voyager 2 identified its shepherds, both of which are 40 to 50 kilometres in diameter. Ophelia controls the outer edge and Cordelia controls the inner edge. Although each of the nine narrow rings was expected to have two shepherds, none was identified.[60,61] However, the imaging resolution of the rings was 20 kilometres per pixel, so any objects smaller than this would not have been detected. Two new rings were discovered, one of which is diffuse and extends at least 2,500 kilometres down towards the planet, so clearly this ring does not have an inner shepherd.

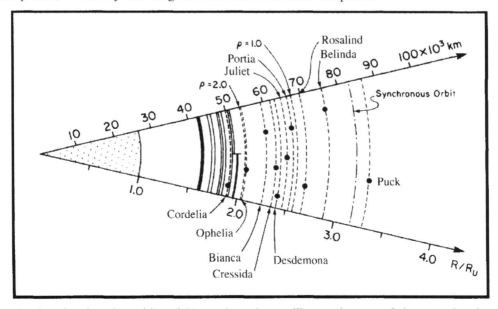

A plot showing the orbits of Uranus's main satellites and some of the moonlets in relation to the ring system. (Modified from fig-1 of 'The rings of Uranus', J.L. Elliot and P.D. Nicholson. In *Planetary Rings*, R. Greenberg and A. Brahic (eds). Copyright © 1984 The Arizona Board of Regents. Reprinted by permission of the University of Arizona Press.)

Once past Uranus, Voyager 2 turned to image the rings by forward-scattered sunlight, and this enabled the size distribution of the material to be determined. In fact, the Uranian rings are remarkably free of dust (the term 'dust' being defined as particulates a few microns across). It turned out that Uranus's rings are composed of 'particles' up to a metre in diameter. This was surprising, because the rings were believed to be material blasted off the moonlets by impacts, and thus to have a specific size distribution. However, the rings were clearly not in 'collisional equilibrium'. It was concluded that since dust *must* be being produced, there must also be some ongoing dust-removal process. Voyager 2's ultraviolet spectrometer provided the answer. Uranus is surrounded by an exosphere of hydrogen which extends all the way from the top of the atmospheric haze out through the ring system and beyond. The hydrogen is escaping from the planet. The presence of this 'planetary wind' flowing out through the ring system exerts a 'drag' on the dust, robbing it of orbital energy and causing it to spiral down to the planet. The broad inner ring that the spacecraft detected is this dust spiralling in. Evidently there is no such 'wind' radiating from Jupiter because, being more massive, it retains its hydrogen and its rings are dusty. Although the Uranian ring system is *structurally* stable because the shepherds keep the larger 'particles' concentrated, the fact that the dust is being continuously lost indicates that the system is being eroded since the dust is derived from wearing down the larger fragments by impacts.[62,63] There was now considerable variety evident in the ring systems of the giant planets, with Jupiter's comprising 50 per cent dust, Saturn's 5 per cent and Uranus's a mere 0.05 per cent. Even if Uranus's ring material could be reaccreted into a single body, this would be just 30 kilometres in diameter, so they have only 1/1,000th the mass of Saturn's ring system.[64]

Voyager 2 discovered eight other new moonlets, all outside the ring system. The largest was spotted on a long-range image when the spacecraft was three weeks inbound, so there was time to update the hectic encounter sequence to snap a single image of the moonlet from a range of half a million kilometres. Orbiting midway between the Epsilon ring and Miranda, the moonlet, now named Puck, is spheroidal, 170 kilometres across and heavily cratered. All of the moonlets are darker than the other moons. Indeed, with an albedo of 0.05 they are as black as soot. One theory is that they are primordial carbonaceous material.[65] Another proposal is that methane clathrate exposed on the surface has been dissociated by protons circulating in Uranus's magnetic field, and the carbon has polymerised to create hydrocarbons.[66] Indeed, when the Giotto spacecraft passed close by the nucleus of Halley's Comet in 1986, it found it to be black with hydrocarbons.[67] It could be that all of the moonlets within Miranda's orbit are the remains of a moon that broke up 3 billion years ago. Perhaps it also provided the raw material for the ring system.

THE TILT

As Voyager 2 flew 81,600 kilometres above Uranus's hazy atmosphere, the gravitational slingshot increased its velocity by 2 kilometres per second and deflected it through an angle of 23 degrees, in so doing placing it on a track that would result in an encounter with Neptune three and a half years later. It had revealed more about the Uranian system in a few days than several centuries of telescopic observation.[68] However, it found no evidence to shed light on why the entire system is tilted over onto its side. Nevertheless, the fact that the orbits of the moons and the rings are in the planet's equatorial plane meant that, whatever the process, it had occurred very early in the Solar System's history. One possibility is that a glancing impact with a large planetesimal during the planet's accretion tipped its axis over and the moons formed from the ejected debris.[69] All of the moons and rings combined constitute only 0.1 per cent of the system's mass, so an insignificant amount of material was required to be retained in orbit. It is not clear, however, why this material formed a series of small moons rather than a single large one. In fact, Uranus is notable for *not* having at least one satellite to rival Saturn's Titan.

THE DISCOVERY OF NEPTUNE

A comprehensive check of old charts by J.E. Bode established that Uranus had been noted as a 'star' as far back as 1690, when observed by John Flamsteed. In all, it had been charted 22 times. P.C. Lemonnier had actually observed it 12 times! The reason that no one had deduced its true nature was that telescopes built prior to Herschel's innovative reflector were incapable of resolving its disk.[70] The computation of Uranus's orbit was enhanced by the fact that there was data spanning a complete circuit of the Sun. However, it was soon realised that the planet was progressively drawing ahead of its predicted position. The simplest strategy was to reject the prediscovery sightings as inaccurate, but their veracity was apparent. Even when this was was done in 1821 by Alexis Bouvard – Simon Laplace's partner in celestial computations – and Uranus's orbit had been computed afresh, the planet promptly diverged again. For a generation of astronomers who considered themselves to have mastered the motions of the planets using Universal Gravity, this was an intolerable situation. After years of progressively accelerating, around 1822 Uranus settled down and then started to fall behind schedule. In 1824, therefore, F.W. Bessel reasoned that there was an unknown planet present whose gravity had been first accelerating Uranus and was now retarding it. An analysis of the 'residuals' in Uranus's motion ought to reveal this perturber's position, but a calculation of this type had never been attempted before.

In 1843, J.C. Adams at Cambridge University in England began to work on the problem, and by October 1845 he had a solution. He communicated this to G.B. Airy, the Astronomer Royal in Greenwich, but Airy firmly believed that the tools available to contemporary mathematics were inadequate to determine the orbit of an unknown object in this fashion.[71] He had received several such propositions over the years. In fact, the first had been in 1834, when he was still the director of the Cambridge Observatory. T.J. Hussey, impressed by the evident success of the Titius–Bode 'law' in predicting the existence of the asteroids, had urged him to apply it to a trans-Uranian planet but Airy had dismissed him. Consequently, Airy was not

326 Planets beyond

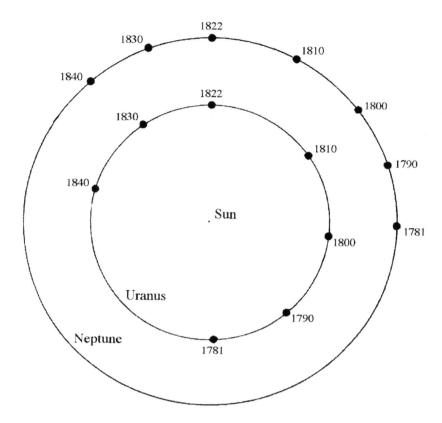

This plot shows the relative position of Neptune in relation to Uranus during the period after Uranus's discovery in 1781 when the planet was being mysteriously accelerated ahead of its calculated position. After this anomalous acceleration diminished in 1822, astronomers were perplexed to find Uranus start to lag behind. A mathematical *tour de force* that calculated the location of the perturbing object led to the discovery of Neptune.

predisposed to believe Adams's thesis.[72] In fact, Adams's calculations had pinned down the mystery planet to within 2 degrees in the sky. If a search had been made during the autumn of 1845, this would almost certainly have identified the planet.

Having informed the French Academy of Sciences on 1 June 1846 that his mathematical analysis had demonstrated there *had* to be a planet perturbing Uranus, U.J.J. Leverrier derived its position. On 31 August, he wrote to J.G. Galle in Berlin informing him of his calculation and, in essence, saying that if a telescope was turned to a particular spot in the sky the planet would become instantly obvious. Galle received the letter on 23 September and that evening he initiated his search. Fortunately, the Berlin Observatory had a brand new star map. Galle, at the telescope, called out the positions of the stars one by one and his assistant, H.L. D'Arrest, checked them off the map. Within half an hour, they had spotted a star that was not on the map. When Galle stepped up the magnification, he resolved a tiny disk.

On 9 July, having heard that Leverrier had predicted the position of the planet, and having noted that this was almost the same as the position that Adams had calculated, Airy had rather belatedly asked J.C. Challis, his successor at the Cambridge Observatory, to make an urgent search. However, Cambridge did not possess a detailed chart of that part of the sky so Challis set out to construct one as a preliminary to his search. On 4 August, and then again on 12 August, Challis charted the planet as a star without recognising its true nature. Only on 29 September, upon switching to a higher magnification, did he finally resolve the disk. His joy was short-lived, however, because the news of Galle's success was published in London on 1 October. "The planet was in my grasp", Challis lamented to Airy.[73] As the many players argued their individual claims for priority of discovery, Leverrier suggested that the planet be named Neptune.

As with Uranus, once the orbit was traced back, it was discovered that Neptune had been marked on old star charts so it was soon possible to refine the orbit.[74] Interestingly, Galileo seems to have seen it while observing Jupiter but his telescope was too crude to have resolved it as a disk.[75] Although he annotated one sketch with a remark that the 'star' appeared to have moved from one night to the next, he did not follow this up.[76] J.J. Lalande also saw Neptune on two nights in May 1795 and, upon noting the inconsistency in the two positions, presumed that his initial sighting had been flawed.

After Neptune had been discovered, Uranus's orbit was recomputed. The old observations by John Flamsteed and P.C. Lemonnier were now a perfect fit and the planet was clearly following its prescribed path. The relative positions of the two planets were such that, prior to 1882, Uranus had been catching Neptune up and so had been accelerated, but after conjunction Uranus had been retarded. It had been this perturbation that had enabled Adams and Leverrier to infer Neptune's existence. If the planets had been on opposite sides of the Sun in the early 19th century, Uranus would not have departed from its predicted path and astronomers would not have been so *driven* to seek yet another planet, in which case it might well have remained undiscovered for a considerable time.[77]

Because it is so far away, Neptune shows only a tiny greenish-blue disk, so surface detail is elusive even to observers with large telescopes. Nevertheless, in 1899 T.J.J. See, using the 26-inch refractor at the US Naval Observatory, reported equatorial belts but E.E. Barnard was unable to confirm their existence using the 40-inch at Yerkes Observatory. T.L. Cragg, using Mount Wilson's 60-inch reflector, reported a bright equatorial zone that set a marked contrast against the darker polar regions, but again this was unconfirmed.

In 1883 Maxwell Hall in Jamaica reported a periodic fluctuation in the planet's brightness with a 7.9-hour period, so for many years this was assumed to be its rotational period.[78] However, J.H. Moore and D.H. Menzel at Lick derived a period 15.7 hours (that is, precisely twice that expected) by measuring the doppler effect.

NEPTUNE'S MOONS

On 10 October 1846, after only a few days of observing Neptune, William Lassell realised that it had a satellite. He named it Triton because in mythology Triton was Neptune's son.[79] The fact that the moon was so immediately obvious implied that it was a large one. Although Lassell reported that the moon was much brighter on one side of its 6-day orbit than on the other, and inferred this to mean that its albedo was patchy and that its rotation was synchronised,[80] such a marked variation was not confirmed by subsequent observers. Nevertheless, C. Wirtz in Strasbourg made a photometric study which identified a slight variation that confirmed that the rotation is indeed synchronous. Triton's retrograde orbit is in a plane inclined at 157 degrees to Neptune's equator, which is inclined at 29.6 degrees to the planet's orbital plane, which is in turn inclined at 1.8 degrees to the ecliptic, so Triton's orbital plane is offset 188.4 degrees from the ecliptic. If the fact that the motion is retrograde is ignored, then Triton's orbit is offset at 23 degrees from the planet's equator. The planet is an oblate spheroid, so perturbations arising from this asymmetric mass distribution cause the plane of moon's orbit to precess over a period of about 665 years.

In 1852, Lassell suspected the presence of another satellite but this proved to be illusory.[81] At Mount Wilson Observatory, W.H.M. Christie undertook a photographic search, but to no avail and so it was concluded that Neptune had only one significant companion. On 1 May 1949, however, G.P. Kuiper using the McDonald Observatory's 82-inch reflector identified a faint moon just below Christie's limiting magnitude. Following the nomenclature, this one was called Nereid after another of the sea-god's helpers. It soon became evident that this small moon was in an extremely elliptical orbit ranging from about 1.5 million kilometres when at periapsis to about 10 million kilometres at apoapsis, with a period of just over a year. If its orbital energy had been a little greater, it would have escaped from the planet long ago. Its orbital plane is inclined both with respect to the planet's equator and to Triton's orbital plane. It cannot be more than a few hundred kilometres across, so while it is undoubtedly a captured comet or asteroid (possibly from a reservoir of such objects in the outer Solar System rather than the asteroid belt between Mars and Jupiter) it is nevertheless fairly sizeable.

Clearly, with everything inclined with respect to everything else, the various components of the Neptunian system are not very tightly coupled. This hints at unusual events early in its history.

NEPTUNE'S RINGS

After reading in the newspaper on 1 October 1846 of Galle's discovery of Neptune, John Herschel urged William Lassell to inspect it using his new 24-inch reflector. Upon turning to the new planet the following evening, Lassell was surprised to see what seemed to be a faint ring close around the planet.[82] J.R. Hind confirmed its existence on 11 December,[83] and J.C. Challis in Cambridge also claimed to see it on

12 January 1847.[84] However, by 1852 Lassell realised that the orientation of the ring changed when he rotated the tube of his telescope, so it was really an artefact of his telescope.[85] The other sightings were evidently cases of seeing what the mind expected to see.[86]

When Voyager 2 was launched, therefore, Neptune was not listed as having rings, but in light of the 1977 discovery of rings around Uranus astronomers decided to check when next Neptune occulted a star.[87] Even if the planet possessed rings, the chance of detecting them by this technique was low, firstly because Neptune is much farther away and so offers a smaller target and, secondly, as the planet's axis is not tilted over so far, a ring system would be projected against the sky as a narrow ellipse rather than as an open bull's-eye. An occultation in 1981 produced ambiguous results: there was a dip in the signal prior to the main event, but no matching dip after the main occultation. The working hypothesis was that (by an incredible chance) an unknown moon had occulted the star.[88] When Neptune occulted a star in 1983 there was no anomalous signal. In 1984, however, two teams saw the same flicker prior to the main event, with nothing afterwards, and because the telescopes were on different mountains, 100 kilometres apart, the cause could not have been a moon; it must have been a diffuse feature. Analysis of a light curve with a time resolution of 1/100th of a second prompted the suggestion of a short 'arc' of material extending no more than about 10 per cent of the way around an orbit located about 42,700 kilometres above the planet's atmosphere.[89] Subsequent analysis suggested that there were at least three distinct 'ring arcs' at different distances from the planet, the first to be noted being both the outermost and the most significant. It was evident that short arcs could not persist for more than a few years because Kepler's laws of orbital motion would soon draw the material into a complete ring. Attempts to explain the ring arcs drew analogies with the Saturnian system. One proposal, inspired by the manner in which Tethys holds Calypso and Telesto in its co-orbital Lagrange points, argued that there was a 200-kilometre moon holding clouds of debris in position.[90] Another theory, inspired by the relationship between Janus and Epimetheus, proposed a shepherd in an *inclined* orbit that was stabilising the ring arcs.[91]

FINAL FLY-BY

To enable Voyager 2 to pursue its 'Grand Tour', the necessary configuration of the outer planets recurred only every 175 years, so it was fortunate that the 'Space Age' did not dawn a decade later than it did, because a spacecraft on a minimum-energy transfer orbit would take 40 years to reach Neptune. By playing 'interplanetary billiards', Voyager 2 was able to reach it in only 12 years. At its launch, the mood at JPL was that the chances of the spacecraft surviving long enough to reach Neptune were low, but it was a mission worth trying, and everything beyond Saturn was a bonus. As a result of being so far from the Sun, Neptune proceeds so slowly on its orbit that as Voyager 2 approached it in August 1989 the planet had yet to complete its first orbit since it had been discovered.

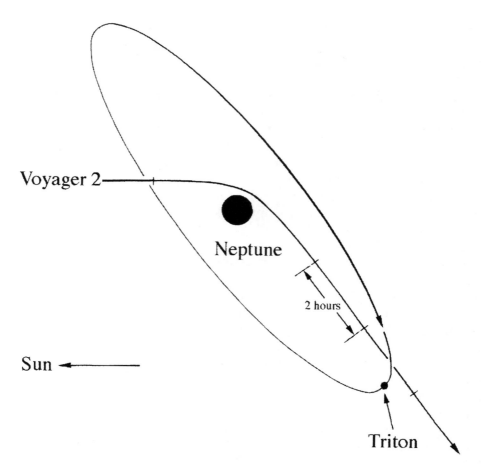

As Triton orbits Neptune in an inclined retrograde orbit, Voyager 2 made a low pass over the planet's pole in order to deflect its trajectory towards the enigmatic satellite for the final encounter of its epic voyage.

The scientific objectives at Uranus (and indeed at Jupiter and Saturn) had been constrained by the need to establish the post-encounter trajectory required to aim for the next planet in the sequence, but Neptune was the *final* objective, and so the planners had rather more freedom of action. Apart from Neptune itself, interest focused upon the primary satellite, Triton. However, this orbited at such an angle to the plane in which Voyager 2 was flying that the spacecraft was obliged to pass very close over the planet's north pole in order to deflect its trajectory south of the ecliptic for a high-resolution imaging fly-by of Triton. Such a trajectory would also enable the spacecraft to study the expected (but as yet unconfirmed) magnetic field, and as the spacecraft passed behind the planet (as viewed from the Earth) it would be possible to use the refraction of the radio signal to study the chemical composition and physical state of the planet's upper atmosphere. On the way to Triton, it should be possible to observe the ring arcs by forward-scattered sunlight. Prior to the discovery of the ring arcs it had been planned for Voyager 2 to perform a 1,300-

kilometre pass over the planet's north pole in order to deflect it to a 10,000-kilometre fly-by of the satellite, but the risk of the spacecraft being disabled by ring material was considered excessive, so the interplanetary cruise was tweeked to open the range to 5,000 kilometres in order to stay clear of the outer ring. If all went well, this would produce a 40,000-kilometre Triton fly-by about five and a half hours later.[92] The refraction of the radio signal as the spacecraft flew behind Triton would provide information on the moon's putative atmosphere. Even with the opened range, this would still be the spacecraft's closest approach to any planet.

Sunlight at Neptune has barely 1/900th of the power it has at the Earth's distance from the Sun. More to the point, the illumination at Neptune is 3 per cent of Jupiter's and 10 per cent of Saturn's. Voyager 2's camera had therefore to be reprogrammed to photograph such faintly illuminated objects, and the spacecraft was instructed to roll during the extended exposures to cancel out the smearing from its velocity. As its speed had been increased by every gravitational slingshot, the spacecraft would fly past Triton at 100,000 kilometres per hour. By the time it reached Neptune, therefore, Voyager 2 had been taught a few tricks that its designers had not originally envisaged.

Because they were alike in size, mass and atmospheric composition, Uranus and Neptune had traditionally been regarded as being near-twins. In fact, as Neptune is slightly more massive and more compact than Uranus its bulk density is significantly greater, and Neptune's interior was therefore expected to be warmer. The first infrared observations in 1972 from a telescope in Hawaii confirmed that Neptune had indeed an excess of energy.

Table 10.1 Giant planet energy budgets

Planet	AU	Diameter (km)	Mass (E=1)	Insolation (% E)	Energy (int/ext)
Jupiter	5.20	142,800	318	3.70	1.7
Saturn	9.54	120,000	95	1.10	1.8
Uranus	19.18	51,200	14.5	0.27	1.1
Neptune	30.06	49,600	17.2	0.11	2.3

Whereas Jupiter and Saturn both radiate almost twice as much energy as they receive from the Sun, and Uranus, being both farther from the Sun and smaller, radiates only slightly more energy than it receives, Neptune was found to radiate proportionately the greatest energy of all of the gas giants. As observed from Neptune, the solar disk is a mere 1 arc-minute across. In fact, if the Sun were suddenly to cease radiating, Neptune would barely react. In contrast, the Earth's weather system is driven by sunlight. Ironically, therefore, Neptune's weather system was expected to be more active than that of Uranus and, indeed, telescopic observers had reported seeing a range of features over the years. Efforts to measure the rotation rate by measuring the doppler on opposing limbs had been frustrated by the fact that the planet's disk is so tiny, and the results had broad uncertainties, but

332 Planets beyond

cyclical photometric variations believed to be due to bright atmospheric features crossing the disk had enabled the rotational period to be defined as 17 hours and 43 minutes. In the early 1980s, large telescopes fitted with CCD cameras detected high-altitude clouds racing across the disk at various latitudes, and thereby confirmed that the atmosphere rotates differentially.

By January 1989, when Voyager 2 started is far-encounter imaging, it was able to resolve considerable atmospheric detail. In March, it resolved a dark elliptical spot located 22 degrees south of the equator. The anticlockwise circulation indicated that this was a storm rising above the general cloud deck. Because it resembled Jupiter's Great Red Spot it was promptly named the 'Great Dark Spot'. A streak of bright white methane ice cirrus cloud was 'stationed' about 100 kilometres above its southern periphery. A smaller spot appropriately dubbed the 'Little Dark Spot', was

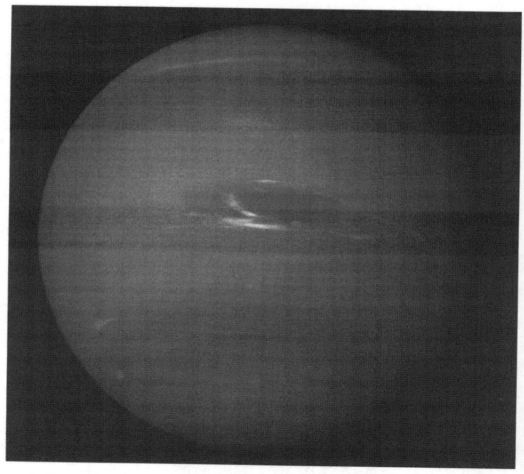

As Voyager 2 viewed Neptune's south pole it snapped this image of the planet showing the 'Great Dark Spot' and its related bright methane streaky clouds, and, further south and near the western limb, the 'Little Dark Spot' and the fast-moving bright 'Scooter'.

later seen at 55 degrees south. It had a patch of high cirrus stationed directly over its centre. As the spacecraft flew over the terminator and peered down on the pole, it saw streaks of bright cirrus casting shadows on the haze layer 100 kilometres below. Whereas the larger spot circled Neptune in 18.3 hours, the small one raced around in just 16.0 hours. This extreme differential rotation explained why telescopic attempts to measure the planet's rotation by photometric variability had produced such a wide variety of results.

As expected, Voyager 2 confirmed that Neptune possesses a magnetic field with a strength that is slightly less than that of Uranus. As in the case of Uranus, the magnetic axis is inclined (47 degrees) to the planet's rotational axis, but by being physically offset by 13,600 kilometres the centre of activity is closer to the planet's surface than its centre. The fact that both Uranus and Neptune have such similar fields banished any remaining suspicion that Uranus's magnetic field had been caught in the process of reversing its polarity. Clearly the fields were generated by the same process, involving flows of ionised water within the superdense atmosphere.[93] In Neptune's case, the divergence between the rotational and magnetic axes was puzzling because the planet has not been tipped right over. The precession of the magnetic axis established that Neptune's interior rotates in 16 hours and 13 minutes.[94]

Combining simultaneous observations of Neptune by the Hubble Space Telescope and by the Infrared Telescope Facility in Hawaii, a team led by L.A. Sromovsky of the University of Wisconsin at Madison blended a series of Hubble images taken in 1996 and in 1998 to make a time-lapse 'movie' of the planet's rotation over three rotational cycles which recorded a variety of discrete cloud features and zonal bands but, to general amazement, no 'Great Dark Spot'; it had disappeared. Further observations showed that large bright clouds developed and faded over a period of weeks.[95] The fact that the Hubble could image a planet as remote as Neptune in such detail opened up the prospect of conducting long-term weather monitoring.[96]

RINGS REVEALED

Voyager 2 first resolved the ring system when it was still three weeks out from the planet. A fortnight later, against expectation, the material was seen to trace a series of *complete* rings rather than short arcs. Once the rings were seen forward-scattering sunlight it became evident that the material is unevenly distributed (there are clumps of material) and it is dustier than that of the Uranian rings. In fact, there are five rings and in order of increasing distance from the planet they have been called Galle, Leverrier, Lassell, Arago and Adams. The Galle and Lassell rings are broad and faint, but the others are thin and relatively bright.[97] A re-analysis of the occultation data revealed that it was the clumpiness of the Adams ring that prompted the belief that the rings were discontinuous.[98] If we discard the possibility that the rings formed recently and are rapidly eroding (and hence that we were lucky to catch them) we are faced with the mystery of how their clumpiness is maintained.

Although Voyager 2 spotted half a dozen moonlets, it did not identify all the shepherds believed to be needed to maintain such fine ring structures. A shepherd about 150 kilometres wide orbits 900 kilometres inside the Arago ring, preventing it from dispersing inwards, and another 180 kilometres wide orbits 700 kilometres inside the Leverrier ring. The outer shepherds for these narrow rings were not detected but they cannot exceed 12 kilometres in diameter, otherwise the spacecraft would have resolved them.

Proteus, the largest of the moonlets identified by Voyager 2, orbits the planet at about twice the radius of the outermost ring. At 400 kilometres across, it is bigger than Nereid. Despite its size it is non-spheroidal. Heavily cratered, it was lucky to survive one impact which excavated a crater spanning almost half of its diameter. Some of the crater rims rise 20 kilometres above their floors, so it is a rugged terrain. Its surface is very dark. If this is due to interactions with charged particles circulating in Neptune's magnetosphere, then the absence of any bright rays indicates that the moon has seen little activity for a considerable time. A smaller moonlet was spotted orbiting just outside the ring system. It, too, is very dark and crater scarred. When its motion was traced backwards, it was realised that this was the satellite (since named Larissa) that had occulted a star in 1981.

As the ring system and all the moonlets orbit in the planet's equatorial plane in a prograde manner, they are probably intimately related to the planet, but Triton and Nereid both appear to be interlopers that were captured. If there were initially any small moons further out, Triton's arrival would have disturbed their orbits and ejected them from the system. Indeed, some of them may have been directed inwards, and been completely disrupted upon straying within the planet's Roche radius, thereby forming the rings.[99] Evidently only the moons much closer in survived Triton's capture. Nereid was not well presented and spanned only 20 pixels in the one image taken. As no surface detail was resolved, its nature remains a mystery. If it turns out not to be an interloper, it could be one of the original moons that narrowly missed being ejected by Triton's appearance in the system.

TRITON

In 1975, infrared observations by Dale Cruikshank of the University of Hawaii identified methane ice on Triton's surface.[100] This implied that the moon would also have an atmosphere, because at the estimated 55 K surface temperature some of the methane ice would turn to gas when in daylight and sublimate again once in shadow. Earlier, G.P. Kuiper had suggested that Triton might possess an atmosphere but he had not been able to secure a reliable indicator of its composition. Continuing infrared studies identified molecular nitrogen, which would also participate in a sublimation cycle. As Voyager 2 made its approach, therefore, Triton was expected to have a nitrogen-rich atmosphere laced with methane, somewhat similar to Titan's, but less dense (a pressure of 0.1 bar was predicted). If a combination of the protons that were expected to be circulating within Neptune's magnetic field and the weak solar ultraviolet were dissociating the methane, there might well be a hydrocarbon haze. It was hoped that by being both thinner and colder this would be sufficiently

A view of Triton's southern hemisphere from the dimpled mid-latitude 'cantaloupe' terrain to the south polar cap. Triton's surface is mostly covered by nitrogen frost mixed with traces of condensed methane, carbon dioxide and carbon monoxide. The southern cap was sublimating (as shown by its irregular and eroded edge) because the moon's orbital and rotational motions cause the Sun to shine directly on the cap for several decades during Neptune's and Triton's lengthy austral summer. Although extremely tenuous, the resulting atmosphere is sufficiently dense to produced eolian streaks of dark hydrocarbon fall-out downwind of the plumes from nitrogen geysers which erupt from the 'Sun baked' cap.

transparent to show the surface – it would have been extremely frustrating to have found Triton (like Titan) 'socked in'. In fact, the low temperature raised the prospect that the hydrocarbons would have condensed and collected in low-lying areas on the surface as a red sludge. Indeed, variations in the infrared light curve in phase with the moon's rotation hinted that Triton had 'highlands' of rock and water ice, with depressions filled with methane ice and hydrocarbons and (if the temperature and the pressure were just right) lakes of nitrogen. Since Triton's orbit is inclined with respect to the plane of Neptune's orbit and the planet takes 165 years to make an orbit of the Sun, at certain times in the planet's orbit one or other of Triton's polar zones is in continuous daylight for 41 years. As the moon's visible disk was progressively dominated by the southern polar region during the 1980s, the infrared surface patterns detected a decade earlier faded and the colour changed from reddish to pink-white.

As the spacecraft emerged from its encounter with Neptune, and headed for the moon, a general surface mottling was gradually resolved into a bewildering variety of distinct features. Actually, it is criss-crossed with fractures up to 35 kilometres in width and 1,000 kilometres in length. There are some relics of large basins several hundred kilometres in diameter, but there is a general paucity of craters. Because they are not modified by impacts, the fractures may be no more than a billion years old. However, they have been modified by an unusual 'dimpled' terrain comprising circular features up to 25 kilometres wide with rims a few hundred metres high. Due to its unique texture, this is also referred to as 'cantaloupe' terrain. It is dominant in the equatorial region. The basins have been filled by a cryogenic fluid which, in some places, has spilled out onto the surrounding plain. Triton's surface relief seems to be limited to a kilometre or so of elevation. Nevertheless, there are 'rough' patches in the centre of some of the basins which may be the result of recent extrusion of viscous fluids. The south polar zone presented a gleaming bright ice cap that was so reflective that its temperature was a mere 38 K. Since nitrogen freezes at 63 K, the cap was primarily nitrogen ice. It was probably no more than a few metres thick and, with the onset of 'summer', was receding to expose a rough dark red material. The low atmospheric pressure meant that the ice sublimated straight to gas, so the anticipated lakes of nitrogen were absent. Once the gas has made its way around to the darkened hemisphere, it will resublimate as frost and build up that polar cap. Frozen nitrogen and methane are both white, but sustained irradiation will dissociate the methane, the hydrogen will escape, and a reddish residue of hydrocarbons will settle and be buried upon the ice's return.[101] It was remarkable that a satellite of this size, located so far from the Sun, could have been so extensively resurfaced, and evidently in the recent geological past. What nobody in their wildest dreams had expected to find was evidence of *ongoing* activity.

L.A. Soderblom of the US Geologic Survey stereoscopically analysed a series of images of the southern polar cap and discovered several dark plumes rising from dark surface spots to an altitude of about 8 kilometres, at which point the material was caught by the prevailing wind and blown horizontally for several hundred kilometres, trailing 'fall out' across the surface. Amazingly, Voyager 2 had caught a cluster of geysers in the act of venting.[102] The dark streaks were indistinguishable

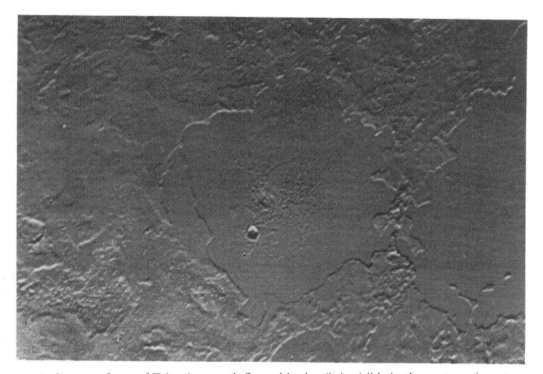

A close up of one of Triton's smooth-floored basins (it is visible in the context view at the top right). It may be an old impact scar that has been extensively modified by flooding, melting, faulting, and collapse. Several episodes of filling and partial removal of material appear to have occurred. The rough area in the middle of the main depression probably marks the most recent eruption of material.

from others seen earlier, near the margin of the receding ice cap. It was significant that the active geysers were clustered at the subsolar point, where the illumination is directly overhead. Furthermore, the earlier sites were further back on the track of the subsolar point's migration.

An analysis of likely processes concluded that the geysers are driven by gaseous nitrogen under pressure.[103] A previous study had revealed that sunlight can pass through several metres of clear ice and heat subsurface materials. If the nitrogen ice cap was laced with hydrocarbons and these were absorbing sunlight, they would heat up and sublimate the immediately adjacent nitrogen, creating pockets of gas which, once the pressure built up, would burst out. Actually, because the moon's atmospheric pressure is only 16 microbars,[104] the venting required only to achieve a few millibars for the plume to rise buoyantly several kilometres. Hydrocarbons are evidently carried in the venting plume, but they condense and form the dark streaks deposited downwind of the vent. If it were not for the fact that the temperature is sufficiently low to freeze methane, Triton's plume activity would not be so obvious. With an annual hydrocarbon cycle at work, the moon's surface is very dynamic. The knowledge that the illuminated pole was so active provided the insight required to interpret telescopic observations. Intriguingly, over the ensuing decade, as Neptune started to draw away from the Sun, Triton continued to warm up.[105]

A variety of estimates of Triton's diameter had been made over the years using a range of assumptions and the 'best guess' was 3,500 kilometres, but when the surface was revealed to be far brighter than expected it was realised that the photometric studies had been misleading. In fact, it is only 2,700 kilometres in diameter, and as such is slightly smaller than the Moon. Given the early uncertainty of Triton's diameter, estimates of its mass had suggested that it had a remarkably high bulk density – indeed, a value of 4 g/cm^3 was considered.[106] Although the actual figure is only 2 g/cm^3, this is nevertheless slightly higher than even Saturn's Titan, and implies that Triton's constitution is about 70 per cent rock. The moon evidently suffered significant gravitational tidal heating, firstly when it was captured by Neptune, and then as its rotation was progressively synchronised. Once its interior warmed up, the rock would have settled towards the centre to form a core.[107] This is probably 2,000 kilometres in diameter, and is englobed by an icy mantle 350 kilometres thick. With the rocky core insulated from the chill of space by the mantle, the heat from the decay of radioactive elements should have kept the interior 'warm'. If the temperature of the lower mantle is 175 K, this would be sufficient to melt any methane ice or ammonium hydrate, either of which would then serve as an effective magma for cryogenic volcanism. However, Triton may be doomed. It is sufficiently massive to raise tides within Neptune's dense atmosphere, and the tidal bulge not only slows the planet's rotation but also slows the retrograde orbital motion.[108] In about 100 million years or so, as it approaches the Roche radius, the gravitational stresses will prompt intense volcanism, and if it continues to spiral in, Triton will eventually break up.[109,110]

MISSION ACCOMPLISHED

As it watched its last planet recede and the Sun fade, Voyager 2 joined its stable mate (and indeed their Pioneer predecessors) in conducting a search for the heliopause, the 'edge' of the Solar System where the Sun's magnetosphere yields to the interstellar magnetic field. It is remarkable that most of our knowledge concerning the outer planets and their retinues of satellites derives from this single spacecraft's epic voyage of exploration.[111,112]

DISTANT PLUTO

Once Neptune had been tracked down, the orbits of the outer planets were monitored for any further irregularities that might indicate the presence of yet another planet.[113] Indeed, even without the benefit of a predicted position, D.P. Todd of the US Naval Observatory started a search in 1877 in the hope of recognising a planet by its disk, but the high magnification that was required to discern a tiny disk restricted the field of view and progress was not only slow, but in the end fruitless.

At the turn of the century, Percival Lowell made a prediction based on the residual motion of Uranus. He concluded that there was a perturber with a mass six times that of the Earth, in an eccentric orbit with a period of 282 years which would reach perihelion in the early 1990s. Although the residuals were tiny, and the uncertainties of his calculations were significant, Lowell launched a search in 1905, but gave it up in 1907 after finding nothing. By 1914, however, he had refined his prediction and C.O. Lampland conducted a new search, but once again to no avail. After Lowell's death in 1916, the effort lapsed. In 1919, Milton Humason conducted a search at the Mount Wilson Observatory, using the prediction by W.H. Pickering based on Neptune's residuals and the orbits of comets that had their aphelia in that region of space, again without result.

In 1929 V.M. Slipher, the new director of Lowell's Observatory, hired Clyde Tombaugh to make a renewed search using a new photographic telescope which had a wide field of view. Two plates taken a week apart in January 1930 revealed a faint object that was slowly moving across the sky. It was six degrees from Lowell's predicted point.[114] At 15th magnitude, it was considerably fainter than expected but it was moving in a suitably distant orbit, so the search was declared successful. Lowell had been correct about the planet having an eccentric orbit with its perihelion later in the century, but he had overestimated its orbital period because, while the planet was beyond Neptune, it was not very *far* beyond, and so it orbited the Sun in 'only' 248 years. The discovery was announced on 13 March, to mark what would have been Lowell's 75th birthday. The new planet was promptly named Pluto, after the god of the stygian underworld.

Ironically, a re-examination of Lowell's original search turned up Pluto on a pair of plates, but because it was much fainter than he had anticipated he had not recognised its significance. Asteroids – cursed as the 'vermin of the skies' –

A view by the Hubble Space Telescope of Pluto and Charon, which orbit a common centre of gravity every 6.4 days and maintain the same hemispheres facing one another. (Courtesy of the Space Telescope Science Institute.)

complicate any search for a faint speck of light moving against the starry background. Humason had simply been unlucky. Two of his plates had included Pluto, but on one the image had been masked by a star and in the other it had been masked by a flaw in the emulsion.[115]

It was soon realised from the 0.25 eccentricity that when Pluto reached perihelion it would be within Neptune's orbit. The 17-degree inclination to the ecliptic meant that their orbits did not actually intersect, so there was no possibility of a collision. It was later calculated that they could *never* come close, because a resonance would force them apart if Pluto approached perihelion while Neptune was in the vicinity. In fact, Pluto's orbit is chaotic, and its inclination, eccentricity and the alignment of the major axis of its orbit vary in a random manner.[116]

As the 100-inch Hooker reflector on Mount Wilson was unable to resolve Pluto's disk, it was not possible to estimate its diameter until a larger telescope had been constructed. When the 200-inch Hale reflector on Mount Palomar became available in 1949, G.P. Kuiper promptly measured Pluto's disk to be a mere 0.2 arc-seconds wide.[117] The 5,800-kilometre diameter that this implied was much smaller than expected. In 1936, A.C.D. Crommelin had suggested that its faintness was due to the fact that its surface was so smooth and reflective that we see only the 'specular' reflection. Photometric observations indicated slight fluctuations in brightness over a 6.4-day period. In 1965, when Pluto failed to occult a star, this was interpreted as evidence that it could be no more than 6,800 kilometres in diameter.[118] Such a small planet could not have caused the perturbations that had led to its discovery: both Pickering and Lowell had predicted a mass several times that of the Earth, but it was much less massive than the Earth. Was Tombaugh's discovery just luck? Was a larger planet in the same area of sky? In fact, he had continued his search for some time after finding Pluto, but found nothing else lurking in the outer Solar System.[119]

For almost half a century, little was learned concerning Pluto's physical characteristics. In 1976, however, infrared photometry indicated the presence of ices, and methane ice was later confirmed.[120]

In 1978 James Christy of the US Naval Observatory was tracking Pluto's motions across the sky in order to refine its orbit. Thanks to the shimmering of the Earth's atmosphere, the planet's image on the photographic plate was a fuzzy blob. Christy noticed that the blobs on several of the photographs were distorted as if an object was orbiting the planet, and realised that Pluto had a satellite,[121] which he named Charon after the ferryman who sailed across the River Styx to Hades, the underworld ruled by Pluto. Centre to centre, the bodies are about 20,000 kilometres apart. Further observations established that Charon orbits Pluto in 6.4 days, which is the same time as the planet itself rotates, so the two bodies are evidently locked, perpetually facing one another. For both bodies to have had their rotations synchronised, Charon's mass had to be a significant fraction of the overall system, so this is effectively a 'double planet'. The synchronisation of both Pluto's and Charon's rotations with Charon's orbital period would have induced considerable tidal stress, so both bodies will most likely turn out to have undergone significant cryovolcanism and they may also be found to be rather similar to Triton. In 1983, a telescope configured for 'speckle interferometry' resolved the two bodies, and

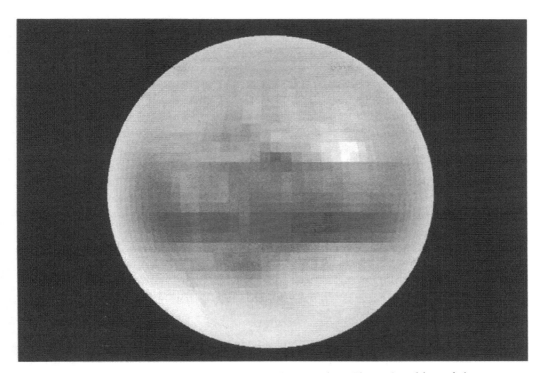

A series of 'mutual events' between 1985 and 1990 when Charon's orbit took it across the front of the planet every few days, as viewed from Earth, enabled the albedo variations across Pluto's Charon-facing hemisphere to be inferred. This is likely to remain the best view of the planet until a spacecraft makes a fly-by. (Courtesy of E.F. Young of the Southwest Research Institute in Boulder, Colorado.)

followed Charon's orbital motion. Charon's orbital plane is inclined 122 degrees to the plane of the ecliptic. Our view changes as the system orbits around the Sun. At its discovery, it was viewed almost pole on. By 1985, however, the system was approaching edge-on and a series of 'mutual events' enabled the sizes of the bodies to be accurately determined. At only 2,300 kilometres across, Pluto is considerably smaller than our Moon, and has only about 1 per cent of the Earth's mass; Charon is 1,190 kilometres across and its mass is 5 to 10 per cent that of its primary. The system's overall density is about 2 g/cm^3, so it must have drawn in a fair amount of rock from the solar nebula. In this respect, it extended the trend that the moons of the outer planets are rather rockier than the predominantly icy satellites of Saturn.

Over time, the transits and eclipses (of the mutual events) enabled the characteristics of the individual surfaces to be inferred. Spectroscopic observations when Charon was visible, and when it was occulted, enabled Pluto's spectrum to be isolated and then 'subtracted' to reveal Charon's spectrum. Pluto's surface contains methane ice and Charon's surface is water ice, but why they have chemically distinct surfaces is not known. An analysis of the entire series of mutual events showed albedo variations that are too small to be directly resolved on the disk. Pluto appears to possess a dark equatorial band and bright polar caps.[122] When the Infrared

Astronomical Satellite had observed the system in 1983, it found that the surface temperatures are somewhat lower than expected on the basis of insolation. In fact, the amount of energy varies around the eccentric orbit, being greatest at perihelion. After crossing within Neptune's orbit in 1979 and making its 30 AU perihelion on 5 September 1989, Pluto reclaimed its status as the outermost planet in 1999. The methane frost evidently sublimated to gas to form an atmosphere, although at only 10 microbars this is tenuous. The light curve during an occultation of a star in 1988 confirmed that a gas was refracting the light, and the timing suggested the presence of layers of haze far above the surface.[123] Pluto's 'summertime' atmosphere could include nitrogen and carbon monoxide. This envelope will have settled onto the surface again long before the 50 AU aphelion in 2113. In fact, whenever one of Pluto's poles is maintained facing away from the Sun for year after year, its 35 K surface is undoubtedly the coldest place in the Solar System.

In the 1930s, R.A. Lyttleton, impressed by the fact that Pluto's orbit is not only eccentric and inclined to the ecliptic, but also strays within Neptune's orbit at perihelion, proposed that it started out as a satellite of Neptune and that as Triton penetrated the system it encountered Pluto and ejected it from the system and settled itself into a retrograde orbit. While this is *possible*, detailed modelling determined that the ejected object would be more likely to escape from the Solar System than settle into an eccentric heliocentric orbit. Of course, the dynamical constraints on this scenario were compounded by the discovery that it has a satellite of its own. The 'double planet' could be the result of 'binary accretion' from the solar nebula (as was once advocated for the Earth–Moon system). Furthermore, it may be the first of a family of protoplanetesimal Kuiper Belt Objects.

Pluto's is the only planetary system not to have been reconnoitred by a robotic explorer. A small probe may be dispatched in the near future, but this project is continually under threat of the budgetary axe, and even if it survives it will be many years in transit.[124]

TEN?

J.D. Anderson of JPL recently re-examined the motions of Uranus and Neptune and drew the conclusion that their residuals vanished around 1910. Many people have suggested likely orbits for the 'missing' planet and predicted where it is, but the searches that have been made have proved fruitless. Noting that some short period comets have orbits which are associated with one or other of the large planets of the outer Solar System, George Forbes of Edinburgh proposed in 1880 that a statistical survey of cometary orbits might indicate the existence of an undiscovered remote planet.[125] As the pairs of Pioneer and Voyager spacecraft depart from the Solar System they are serving a similar role, and their trajectories are being monitored for any sign that they are being perturbed by an unknown gravitational influence.[126] If another planet is finally tracked down, a search of the archives will undoubtedly show that it has been recorded many times without being recognised!

NOTES

1. *William Herschel*, A. Armitage. Doubleday, 1963.
2. *Phil. Trans.*, vol. 77, p. 125.
3. *The scientific papers of Sir William Herschel*, J.L.E. Dreyer (Ed.). Royal Society and Royal Astronomical Society. London, 1912.
4. *Mon. Not. Roy. Astron. Soc.*, vol. 11, p. 248, 1851.
5. 'The fifth satellite of Uranus', G.P. Kipier. *Pub. Astron. Soc. Pacific.*, vol. 61, p. 129, 1949.
6. For a review of the discovery and early observations of Uranus's satellites, including the many false sightings, see *Uranus: the planet, rings and satellites*. E.D. Miner. Wiley–Praxis (second edition), 1998.
7. *Mon. Not.*, vol. 33, p. 164.
8. *Astron. Nach.* No.2526.
9. *Astron. Nach.* No.2545.
10. A. Secchi. *Comptes rendus hebdomaires des Séances de l'Academie des Sciences*, vol. 68, p. 761, 1869.
11. 'Methane in the atmospheres of the great planets', R. Wildt. *Natürwissenschaft*, vol. 20, p. 851, 1932.
12. 'Spectroscopic discovery of the rotational period of Uranus', P. Lowell and V.M. Slipher. *Lowell Obs. Bull.*, vol. 2, p. 17, 1912.
13. 'Diameters and albedoes of the satellites of Uranus', R.H. Brown, D.P. Cruikshank and D. Morrison. *Nature*, vol. 300, p. 423, 1982.
14. 'Uranus: distant giant', D.P. Cruikshank. In *The planets*, B. Preiss (Ed.). Bantam Books, p. 172, 1985.
15. For a review of knowledge about the Uranian system while Voyager 2 was en route, see *Uranus and the outer planets*, G. Hunt (Ed.). Cambridge University Press, 1982.
16. *Rings: discoveries from Galileo to Voyager*, J.L. Elliot and R. Kerr. MIT Press, 1984.
17. 'Uranus: the view from Earth', J.L. Elliot. *Sky & Telescope*, vol. 70, p. 415, November 1985.
18. Starting with the outermost ring and working inwards, Uranus's rings are named epsilon, delta, gamma, eta, beta and alpha, with the innermost three simply being numbered as 4, 5 and 6.
19. *Astronomy*, p. 28, March 1998.
20. 'Voyager 2 and the Uranian rings', C.C. Porco. *The Planetary Report*, vol. 6, p. 11, November/December 1986.
21. *The Space Shuttle: roles, missions and accomplishments*, D.M. Harland. Wiley–Praxis, 1998.
22. 'Clouds and aerosols in the Uranian atmosphere', R.A. West, K.H. Baines and J.B. Pollack. In *Uranus*, J.T. Bergstralh, E.D. Miner and M.S. Matthews (Eds). University of Arizona Press, p. 296, 1991.
23. 'Uranus', A.P. Ingersoll. *Scient. Am.*, vol. 255, p. 38, January 1987.
24. The temperature was within 3 degrees of a mean of 52 K. This was the measured temperature of the 'surface'. The 'black body' effective temperature of the planet's radiation was slightly warmer, at 58 K.
25. 'Voyager infrared observations of Uranus's atmosphere: thermal, structure and dynamics', F.M. Flasar, B.J. Conrath, P.J. Gierasch and J.A. Pirraglia. *J. Geophys. Res.*, vol. 92, p. 15011, 1987.

26 'Baronic instability in the interiors of the giant planets: a cooling history of Uranus', R. Holme and A.P. Ingersoll. *Icarus*, vol. 110, p. 340, 1994.
27 *Uranus: the planet, rings and satellites*, E.D. Miner. Wiley–Praxis, p. 249, 1998.
28 Voyager 2's preliminary particles and fields results were published on 30 December 1987 in a 'special issue' of *J. Geophys. Res.*, vol. 92.
29 'The magnetic field and magnetospheric configuration of Uranus', N.F. Ness, J.E.P. Connerney, R.P. Lepping, M. Schulz and G.H. Voigt. In *Uranus*, J.T. Bergstralh, E.D. Miner and M.S. Matthews (Eds). University of Arizona Press, p. 739, 1991.
30 'Models of Uranus's interior and magnetic field', M. Podolak, W.B. Hubbard and D.J. Stevenson. In *Uranus*, J.T. Bergstralh, E.D. Miner and M.S. Matthews (Eds). University of Arizona Press, p. 29, 1991.
31 'The magnetic field of Uranus', J.E.P. Connerney, M.H. Acuna and N.F. Ness. *J. Geophys. Res.*, vol. 92, p. 15329, 1987.
32 'Structure and evolution of Uranus and Neptune', W.B. Hubbard and J.J. MacFarlane. *J. Geophys. Res.*, vol. 85, p. 225, 1980.
33 *Physics of planetary interiors*, V.N. Zharkov and V.P. Trubitsyn (translated by W.B. Hubbard). Pachart Publishing 1978.
34 The 'ionised water' is actually water which has been dissociated into hydronium and hydroxyl ions.
35 'The magnetic field of Uranus', J.E.P. Connerney, M.H. Acuna and N.F. Ness. *J. Geophys Res.*, vol. 92, p. 15329, 1987.
36 'The rotational period of Uranus', M.D. Desch and J.E.P. Connerney. *Nature*, vol. 322, p. 42, 1986.
37 'Visible and near-IR imaging of giant planets: outer manifestiations of deeper secrets', H.B. Hammel. *Bull. Am. Astron. Soc.*, vol. 28, p. 1056, 1996.
38 'Temporal variations in the Uranus near-IR geometric albedo', C.M. Walter, M.S. Marley and K.H. Baines. *Bull. Am. Astron. Soc.*, vol. 28, p. 1076, 1996.
39 *Astronomy*, p. 28, March 1998.
40 'Spring storms strike Uranus', R. Talcott. *Astronomy*, p. 26, July 1999.
41 'Radio science with Voyager 2 at Uranus: results on masses and densities of the planet and five principal satellites', J.D. Anderson, J.K. Campbell, R.A. Jacobson, D.N. Sweetnam, A.H. Taylor, A.J.R. Prentice and G.L. Tyler. *J. Geophys. Res.*, vol. 92, p. 14877, 1987.
42 'Uranian satellites: densities and composition', T.V. Johnson, R.H. Brown and J.B. Pollock. *J. Geophys. Res.*, vol. 92, p. 14884, 1987.
43 'Voyager 2 in the Uranian system: imaging science results', B.A. Smith *et al. Science*, vol. 233, p. 43, 1986.
44 See *Planets beyond: discovering the outer Solar System*, M Littmann. Wiley, p. 127, 1988.
45 'The moons of Uranus', T.V. Johnson, R.H. Brown and L.A. Soderblom. *Scient. Am.*, vol. 256, p. 48, 1987.
46 It must be noted that this is brighter than the average for the near side of the Moon.
47 'Fluffy layers obtained by ion bombardment of frozen methane', L. Calcogno, G. Foti, L. Torrisi and G. Strazzulla. *Icarus*, vol. 63, p. 31, 1985.
48 'The enigma of the Uranian satellites orbital eccentricities', S.W. Squyres, R.T. Reynolds and J.J. Lissauer. *Icarus*, vol. 81, p. 218, 1985.
49 'Voyager 2 in the Uranian system: imaging science results', B.A. Smith *et al. Science*, vol. 233, p. 43, 1986.
50 'Geology of the Uranian satellites', S.K. Croft and L.A. Soderblom. In *Uranus*, J.T. Bergstralh, E.D. Miner and M.S. Matthews (Eds). University of Arizona Press, p. 561,

1991.
51 Miranda's Verona Rupes cliff is three times deeper than the wall of Valles Marineris on Mars.
52 'Mysteriously muddled Miranda', J. Eberhart. *Science News*, vol. 129, p. 103, 15 February 1986.
53 'The moons of Uranus', T.V. Johnson, R.H. Brown and L.A. Soderblom. *Scient. Am.*, vol. 255, p. 48, April 1987.
54 'Extensional tectonics of Arden Corona, Miranda: evidence for an upwelling origin of coronae', R. Pappalardo, R. Greeley and S.J. Reynolds. *Proc. Lunar Planet. Sci. Conf.*, p. 1047, 1994.
55 'Thermal stress tectonics on the satellites of Saturn and Uranus', J. Hillier and S.W. Squyres. *J. Geophys. Res.*, vol. 96, p. 15665, 1991.
56 'Miranda', R. Greenberg, S.K. Croft, D.M. Janes, J.S. Kargel, L.A. Lebofsky, J.I. Lunine, R.L. Marcialis, H.J. Melosh, G.W. Ojakangas and R.G. Strom. In *Uranus*, J.T. Bergstrahl, E.D. Miner and M.S. Matthews (Eds). University of Arizona Press, p. 963, 1991.
57 Gene Shoemaker.
58 'Towards a theory for the Uranian rings', P. Goldreich and S. Tremaine. *Nature*, vol. 277, p. 97, 1979.
59 'Shepherding the Uranian rings', C.C. Porco and P. Goldreich. *Astron. J.*, vol. 93, p. 724, 1987.
60 Note that recent re-analysis has turned up many more!!
61 'Orbits of the ten small satellites of Uranus', W.M. Owen and S.P. Synnott. *Astron. J.*, vol. 93, p. 1268, 1987.
62 'Origins of the rings of Uranus and Neptune', J.E. Colwell and L.W. Esposito. *J. Geophys. Res.*, vol. 97, p. 10227, 1992.
63 As the Galileo spacecraft later showed to be so in the case of the Jovian rings.
64 'Voyager 2 and the Uranian rings', C.C. Porco. *The Planetary Report*, vol. 6, p. 11, November/December 1986.
65 Robert Hamilton Brown *et al.*
66 Carl Sagan et al.
67 *Giotto to the comets*, N. Calder. Presswork, 1992.
68 The preliminary results of the Uranuan encounter were published in *Science*, vol. 233, pp. 39–109, 1986.
69 'Origin of the Uranian satellites', J.B. Pollack, J.I. Lunine and W.C. Tittemore. In *Uranus*, J.T. Bergstrahl, E.D. Miner and M.S. Matthews (Eds). University of Arizona Press, p. 469, 1991.
70 At best, Neptune shows a disk a mere 2.3 arcseconds in diameter.
71 *Autobiography*, G.B. Airy. Cambridge University Press, 1896.
72 *Mon. Not. Roy. Astron. Soc.*, vol. 16, p. 399.
73 *Mem. Roy. Astron. Soc.*, vol. 16, p. 412.
74 For a review of pre-discovery observations of Neptune, see *The planet Neptune*, P. Moore. Ellis Horwood, p. 33, 1988.
75 *Patrick Moore's history of astronomy*, P. Moore. MacDonald Co. (fifth edition), p. 52, 1977.
76 'Neptune: farthest giant', D.P. Cruikshank. In *The planets*, B. Preiss (Ed.). Bantam Books, p. 204, 1985.
77 Ironically, the Titus–Bode 'law' that had prompted early speculation of a trans-Uranian planet did not serve the search very well. In fact, it was later concluded that it did not

apply in the outer Solar System because Uranus's mean distance from the Sun is almost 60 million kilometres less than the mathematical sequence predicted, and Neptune is 1,280 million kilometres closer to the Sun, which represents an error for the rule of 25 per cent. It is now seen as being no more than an interesting empirical correlation with no physical implications, and thus no real predictive power.

78 *Observatory*, vol. 7, p. 134.
79 *Mon. Not. Roy. Astron. Soc.*, vol. 6, p. 167, 1846.
80 'Observation of Neptune and its satellite', W. Lassell. *Mon. Not. Roy. Astron. Soc.*, vol. 7, p. 30, 1847.
81 *Mon. Not. Roy. Astron. Soc.*, vol. 12, p. 155, 1852.
82 W. Lassell. Letter to *The Times* published on 14 October 1846.
83 J.R. Hind. *Mon. Not. Roy. Astron. Soc.*, vol. 7, p. 169, 1847.
84 'Second report of the proceedings in the Cambridge Observatory relating to the new planet', J.C. Challis. *Astron. Nach*, vol. 25, 1847.
85 W. Lassell. *Mon. Not. Roy. Astron. Soc.*, vol. 13, p. 38, 1852.
86 For a review of 'Lassell's ring', see *The planet Neptune*, P. Moore. Ellis Horwood, p. 39, 1988.
87 'Where are the rings of Neptune?', A.R. Dobrovolskis. *Icarus*, vol. 43, p. 222, 1980.
88 'Occultation by a possible third satellite of Neptune', H.J. Reitsema, W.B. Hubbard, L.A. Lobofsky and D.J. Tholen. *Science*, vol. 215, p. 289, 1982.
89 'Occultation detection of a Neptunian ring-like arc', W.B. Hubbard, A. Brahic, B. Sicardy, L-R. Elicer, R. Roques and F. Vilas. *Nature*, vol. 319, p. 636, 1986.
90 See *Planets beyond: discovering the outer Solar System*, M Littmann. Wiley, p. 153, 1988.
91 'Towards a theory for Neptune's arc rings', P. Goldreich. S. Tremaine and N. Borderies. *Astron. J.*, vol. 92, p. 490, 1986.
92 Triton orbits in a retrograde manner, and so is considered to be 'upside down', and a low pass over Neptune's north pole that deflected the spacecraft southward set up an approach to the moon's north pole.
93 'The magnetic fields of Uranus and Neptune', R. Holme and J. Bloxham. *J. Geophys. Res.*, vol. 101, p. 2177, 1996.
94 'The interior of Neptune', W.B. Hubbard, M. Podolak and D.J. Stevenson. In *Neptune and Triton*, D.P. Cruikshank (Ed.). University of Arizona Press, p. 109, 1995.
95 NASA/STScI press release STScI-PR98-34, 1998.
96 '[Heidi Hammel:] Observer of the gas giants', R. Flanagan. *Astronomy*, p. 50, July 1997.
97 *Neptune: the planet, rings and satellites*, E.D. Miner and R.R. Wessen. Springer-Praxis 2001.
98 'The Neptune system in Voyager's afterglow', R.A. Kerr. *Science*, vol. 245, p. 245, 1989.
99 'Origins of the rings of Uranus and Neptune', J.E. Colwell and L.W. Esposito. *J. Geophys. Res.*, vol. 97, p. 10227, 1992.
100 'Neptune: farthest giant', D.P. Cruikshank. In *The Planets*, B. Preiss (Ed.). Bantam Books, p. 207, 1978.
101 'Physical state of volatiles on the surface of Triton', J.I. Lunine and D.J. Stevenson. *Nature*, vol. 317, p. 238, 1985.
102 *Far encounter: the Neptune system*, E. Burgess. Columbia University Press, p. 113, 1991.
103 *Planets beyond: discovering the outer Solar System*, M. Littmann. Wiley (second edition), p. 247, 1990.
104 Voyager 2 measured Triton's atmosphere to be 99.9 per cent nitrogen, with a trace of methane. Although the composition is similar to Titan's atmosphere, Triton's is much colder and considerably thinner

105 'The warming wisps of Triton', J.L. Elliot. *Sky & Telescope*, p. 42, February 1999.
106 'Satellite masses in the Uranus and Neptune systems', R. Greenberg. In *Uranus and Neptune*, NASA CP-2330, p. 463, 1984.
107 'On the origin of Triton and Pluto', W.B. McKinnon. *Nature*, vol. 311, p. 355, 1984.
108 Triton's situation differs from that of the Moon because Triton orbits its primary in a retrograde manner. The tidal bulge that is raised on the Earth by the Moon is carried 'ahead' by the planet's rotation, and so it serves to accelerate the Moon, causing it to rise. Neptune's rotation carries its tidal bulge 'behind' Triton, decelerating it and causing it to fall.
109 O. von Struve. *Nature*, vol. 49, p. 324, 1894.
110 'Dynamic evolution of the Neptunian system', T.B. McCord. *Astron. J.*, vol. 71, p. 585, 1966.
111 'The triumphant Grand Tour of Voyager 2', M. Littmann. *Astronomy*, vol. 16, p. 34, December 1988.
112 For an excellent review of Voyager 2's tour of the outer Solar System, see *Planets beyond: discovering the outer Solar System*, M. Littmann. Wiley, 1988 (updated in 1990).
113 *Planet X and Pluto*, W.G. Hoyt. Arizona University Press, 1980.
114 C.W. Tombaugh. ASP Leaflet No.209, 1946.
115 *Guide to the planets*, P. Moore. The Scientific Book Club, p. 170, 1954.
116 Gerald Sussman and Jack Wisdom.
117 G.P. Kuiper. *Publ. ASP*, vol. 62, p. 133, 1950.
118 *The planet Pluto*, A.J. Whyte. Pergamon Press, p. 57, 1980.
119 *Out of the darkness: the planet Pluto*, C.W. Tombaugh and P. Moore. Stackpole, 1980.
120 D.P. Cruikshank, C.B. Pilcher and D. Morrison. *Science*, vol. 194, p. 835, 1976.
121 J.W. Christy and R.S. Harrington. *Astron. J.*, vol. 83, p. 1005, 1978.
122 'A two-colour map of Pluto's sub-Charon hemisphere', E.F. Young, R.P. Binzel and K. Crane. *Astron. J.*, vol. 121, p. 552, 2001.
123 'Discovering Pluto's atmosphere', J.K. Beatty and A. Killian. *Sky & Telescope*, p. 624, December 1988.
124 *Pluto and Charon*, A. Stern and J. Mitton. Wiley, 1999.
125 *Proc. Roy. Soc. Edinbugh.*, vol. 10, p. 429.
126 'Dynamic evidence for Planet X', J.D. Anderson and E.M. Standish. In *The galaxy and the Solar System*, R. Smoluchowski, J.N. Bahcall and M.S. Matthews (Eds). University of Arizona Press, p. 286, 1986.

11

Life and death

DIM SUN

An astonishing discovery, announced in early 2001, was that within 100 million years of its formation, the Earth's surface had cooled sufficiently to enable it to support a hydrosphere.[1,2]

Four billion years ago, the Sun emitted energy at only 70 per cent of the rate it does today. As the Earth cooled, why didn't its surface water freeze? James Kasting, a Pennsylvania State University climatologist, argued that the outgassing associated with the rampant volcanism of that time would have so pumped up the atmosphere with carbon dioxide that the 'greenhouse effect' kept the surface sufficiently warm for water to exist in its liquid phase.[3] Furthermore, whereas the ocean is an excellent storehouse for heat, at night continents lose most of the heat they soak up during the day. Four billion years ago, the Earth was dominated by volcanic islands; there were no continents, so the ocean would have retained most of the heat arriving from the dim Sun. In 1993, Greg Jenkins of the National Center for Atmospheric Research in Boulder, Colorado, added a new twist by pointing out that at that time the planet would have been rotating on its axis so rapidly (he estimated a 14-hour day) that the Coriolis effect would have limited the size of the swirling weather systems and confined them to the equatorial zone, thereby allowing more solar radiation to reach the surface at higher latitudes.[4] Given that there was a hydrosphere throughout most of the Hadean, when did the first life forms develop?

BIOGENESIS

The animal heterostegina which lives on the sea floor requires just sea water and sunlight. It is one of several species that hosts small colonies of algae within its semi-transparent coiled shell. As it grows, the heterostegina increases the number of chambers in its shell and small versions with only a few chambers develop within it, each enclosing a fragment of algae. The offspring are released as the parent expires. Its legacy is its old calcareous shell, which sinks and adds to the limestone sediment.

Despite extensive searches, by 1960 half-billion-year-old trilobites were still the oldest known fossils. In fact, evidence of Precambrian life is scarce precisely because evolution had yet to develop the readily fossilised calcareous skeletal structure. This was frustrating for the palaeontologists because although the humble heterostegina is not much to look at from our vantage point on the evolutionary tree, such a complex organism represented an *advanced* state of evolution.[5] Perhaps the origin of life was simply unknowable?

In 1963, while studying the cherts in the Fig Tree Group near the base of the sedimentary phase of the Barberton Mountainland greenstone sequence, John Ramsay discovered ancient microscopic fossils that are at least 3.2 billion years old. In the same strata, shortly thereafter, Elso Barghoorn identified the ancestor of the most primitive type of single-celled plants living today; namely the blue-green algae (cyanobacteria). In the Onverwacht Group which forms the base of the Barberton sequence there were microscopic 'ovoids' 3.4 billion years old that were even more primitive. In the 1980s Maarten de Wit recognised concentric rings in the Barberton's banded iron formation as having been produced by volcanic gases bubbling through mudpools. The rocks around these ancient hot springs contained fossilised bacteria. Evidently, although the Earth teemed with life for several billion years, this comprised only marine microorganisms. Nevertheless, colonies of blue-green algae were able to leave behind macroscopic structures as stromatolites. They are found in the Archean cherts of the South African, Canadian and Australian cratons. About 600 million years ago most of the stromatolites were decimated by grazing marine creatures, but some survived. The biological origin of the rock structures was realised when *living* stromatolites were found on the shore of Shark Bay in western Australia. The microorganism colonies create mats of calcareous material in the shallow water. By growing through the detritus that settles on the mats they develop their characteristic irregular stratified 'pancake' structure as a succession of layers.[6] Some of the ancient stromatolites constructed domes tens of metres across. Living descendents notwithstanding, the discovery of 3.45-billion-year-old stromatolites in the Pilbara[7,8,9] cast doubt on their biogenicity,[10,11,12] because the 'best' evidence for Archean life consisted of the 'ovoids' in the Barberton[13] and microfossils of filamentous bacteria in cherts interlayered with komatiite in the Pilbara.[14] The finding of cone-shaped calcareous stromatolites in the 1990s, however, provided very convincing evidence of their biological origin.[15] Considering the hazards of impacts and volcanic activity of the Archean, it is hardly surprising that stromatolites are only rarely located in the many-kilometre-thick greenstone sequences.

Our ideas concerning life itself have evolved, too. Over recent decades, biologists have repeatedly been perplexed to discover life thriving in habitats that previously they had thought to be incompatible with life. It appears that life is much hardier and more adaptable than we had given it credit for. In 1977 Jack Corliss of Oregon State University was in a submersible studying the rift in the Galapagos Ridge and he discovered a hydrothermal vent emitting a superheated plume of water.[16,17] A vent found on the East Pacific Rise in 1979 was so rich in dissolved minerals that precipitates had constructed a 'chimney' around the vent. Many of these 'black smokers' have now been mapped. Remarkably, they support colonies of hundreds of species of life. While many are related to clams, mussels, shrimps and tube worms, there were others that were new to science. In 1982 it was found that single-celled organisms occupied the base of these isolated food chains. They drew their energy from a reaction between the hydrogen sulphide flowing out of the vent and the carbon dioxide dissolved in the sea water (in the rift, the pressure of sea water is about 300 bars). The DNA of these *thermophyllic* single-celled organisms suggests that they are pre-bacterial, and so they have been named the 'archea' (meaning literally 'the old ones').

The oldest fossils of primitive life were found in association with hydrothermal activity in Archean rocks which look as if they were either the floor of shallow seas or hot springs on the nearby volcanic islands. Might terrestrial life actually have originated in such a niche?[18,19,20,21] It had been believed that complex organic molecules would fragment in such a hot environment. The rampant volcanism of the time would have emitted prodigious amounts of carbon dioxide. As James Kasting pointed out in 1991, carbon dioxide is a 'heavy' gas, the pressure at the Earth's surface would have been ten times greater than it is today, and the ocean would have been rich in dissolved carbon dioxide; in effect, the ocean would have been carbonic acid.[22] Despite the temperatures of hydrothermal vents, at such a high pressure organic molecules would have been stable.[23]

In the mid-1980s, drilling revealed microbes living in the spaces between grains of rock a kilometre or so below the surface of Colorado's Piceance Basin.[24] These subterranean microbes are heterotrophes and lithotrophes. The heterotrophes consumed the remains of plant detritus that was bound up in sedimentary rock. Some resembled microorganisms on the surface, but they could live in a hot anoxic environment. Faced with such a limited supply of energy, they grew very slowly, and tended to remain in their individual niches between rock grains. The lithotrophes (which were more common in the samples) were strikingly similar to archea. They exploited the high temperature, and consumed hydrocarbons that are toxic to 'conventional' life. They would have found the hot and intensely volcanic early Earth to be a veritable paradise. It was once considered axiomatic that life ultimately draws its energy from sunlight, but lithotrophes literally live off the Earth itself. The implications of this discovery were astounding, because it raised the prospect of life developing in *any* hydrothermal environment, no matter how far from the Sun.

However, the development of the ability to utilise sunlight was a great *advance* in the development of life on Earth, because photosynthesis is much more *efficient*. Furthermore, no longer restricted to sites of hydrothermal activity, life was able to colonise the entire ocean.[25] At that time, life would have consisted of algae and primitive plants that photosynthesised carbon dioxide and water into carbohydrates, and liberated oxygen. Upon death, the bodies of the organisms would have 'decayed' by reacting with oxygen and reverted to carbon dioxide and water. In principle this was a fully reversible reaction, and life should have had no net effect on the environment. Over a 200-million-year interval 2.2 billion years ago, however, the

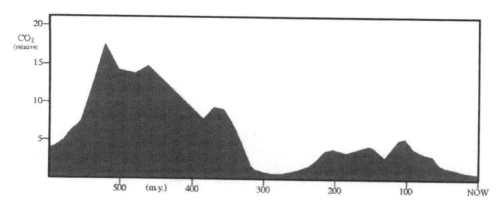

A plot tracing the variation in the amount of carbon dioxide in the Earth's atmosphere (relative to the current ratio of 0.03 percent) throughout the Cambrian. (Adapted from data published by Robert Berner of Yale University.)

proportion of oxygen in the atmosphere rose rapidly from a negligible level to about 15 per cent, and David DesMarais of NASA's Ames Research Center has posited that this came about as a side-effect of the onset of global plate tectonics and the sweeping together of the cratons to make the first large continental masses.[26] The isotopic ratio of the carbon in sedimentary rock indicates whether it was formed from inorganic or organic material. A sudden change in the isotopes showed that a large amount of organic carbon became locked up in rock 2.2 billion years ago. DesMarais has suggested that this occurred because the integration of the cratons would have thrust up fold belts along their sutures, and the mountainous terrain would have prompted vigorous erosion. The chemical weathering would have drawn vast amounts of carbon dioxide out of the atmosphere and locked it up in carbonate. This would have had the effect of increasing the *proportion* of atmospheric oxygen. The organics that were flushed out to sea would have been unable to decay in the anoxic sediment, and this would have further increased the oxygen concentration of the atmosphere.[27]

Banded iron is a chemically precipitated sedimentary rock comprising alternating layers of iron-rich silica and iron-poor sediment such as chert. Massive stacks of banded iron were laid down in shallow water during the late Archean and on through to the mid-Proterozoic, but not thereafter.[28] The banded iron would have consumed the oxygen released by photosynthesising microbes.[29] In 1993, Friedrich Widdel of the Max Planck Institute and Bernhard Schink of the University of Konstanz in Germany discovered that so-called purple non-sulphur bacteria that live in sea sediment could derive energy from a reaction that converts colourless ferrous oxide into red ferric oxide, so it is possible that some of the banded iron formations are the result of direct biological activity.[30] However the banded iron formed, it constitutes some 90 per cent of today's mineable iron.

The next great advance in the development of life was the confining of the genetic material of a cell to a nucleus instead of having it float freely within the cell's membrane – eukaryotic life. This also enabled much larger cells to be formed. It is likely that these cells originated as symbiotic relationships with the membrane enclosing a variety of different organisms. In fact, the relationship between heterostegina and the algae that it contained could be considered to be similar, if on a macroscopic scale. Evolution was severely restricted by the degree to which a single-celled organism could vary its characteristics, but once life developed the ability to form colonies of cells (as multi-cellular organisms) the old single-celled prokaryotes were left behind by the evolutionary process. With greater scope for mutation in multiple-celled organisms, there was an 'explosion' in biological diversity at the start of the Cambrian. At that time, most of the major branches of multi-cellular life which exist today originated. In a sense, a multiple-celled organism is an extreme example of symbiosis in which many cells cooperate. How all these various kinds of cells 'learned' to coordinate their activities is one of the outstanding mysteries of life.[31]

By the start of the Cambrian, life had developed the ability to utilise oxygen constructively and the oxygen concentration was sufficient to promote the development of primitive animals. One of the earliest forms of such life were worms which burrowed in the sediments on the sea floor. Soon, however, the floors of the shallow seas were alive with trilobites! This was remarkably rapid evolution. About 370 million years ago, sea creatures with gills that were capable of drawing oxygen from the air as well as by extracting it from the water moved ashore as amphibians. Some amphibians lost the ability to live in water, remained ashore, and evolved into reptiles. They remained the dominant form of land life for over a hundred million years until receiving intense competition from a type of reptile that had developed mammalian characteristics.[32] How do we know all this?

In 1695, John Woodward in England suggested that peculiarly shaped structures in some rocks were the fossilised remains of creatures of species not included when Noah stocked his ark. Woodward argued that the Flood had been so severe that it had stirred up the surface of the Earth into a suspension, and the forgotten creatures had been trapped in the thick layer of mud that formed when the waters receded. A century later, William Smith, an English canal engineer, realised that fossils of a given type occurred in specific strata. From this he inferred that these strata, even though dispersed geographically, should be part of the same formation. When Smith published a book in 1816 summarising his findings, he argued that if it could be assumed that strata high in a sequence were younger than those below (an assumption that did not allow for the possibility that sequences of strata might have been inverted during folding) then the fossils ought to serve as indicators that would provide geologists with a relative timescale.[33] Furthermore, palaeontologists would be able to chart the manner in which species had changed; or 'evolved', as Charles Darwin would put it a few decades later.[34] The evolution of the multiple-chambered shells of ammonites proved particularly useful in tracing the geological sequence. In 1812 in Paris, George Cuvier noted that some species appeared to have become 'extinct' long before Noah's Flood. In fact, the boundaries between many of the periods that subdivide the Cambrian seem to mark 'extinction events'.

EXTINCTIONS

In 1979, Luis Alvarez, a physicist at the University of California campus at Berkeley, and his geologist son Walter, discovered a thin layer of clay in an exposure marking the boundary between the Cretaceous and Tertiary periods which was rich in iridium, an element that is rare in the Earth's crust but is common in meteorites.[35] A survey confirmed that the iridium layer is present on a global basis.[36] Analysis of the clay revealed that it contained small glassy beads of 'melt' and types of pressure-induced shocked quartz, which served to suggest that the demise of the dinosaurs was prompted by an asteroid striking the Earth.[37,38,39] Critics of the hypothesis countered that this iridium was released during flood basalt eruptions because these draw their magma from the mantle, which is not as depleted in iridium as the crust.[40] Although it is true that Hawaiian-style volcanoes emit iridium, some volcanoes eject glassy melt spherules, and certain high-pressure eruptions produce a kind of shocked quartz, no volcano, either ancient or modern, has produced these in combination, or with the characteristics present in the Cretaceous–Tertiary boundary clay.

Calculations showed that the heat liberated by the impact of an asteroid about 10 kilometres wide would have triggered a global forest fire.[41] The airborne soot would have blocked out the sunlight for months – if not years – which would have induced a global drop in temperature and killed off the plankton that supported the marine food chain by inhibiting photosynthesis. Despite having survived several such catastrophes in the previous 300 million years (some of them even more devastating to the biosphere as a whole) the ammonites expired in this extinction.[42]

An impact on such a scale ought to have left a crater a few hundred kilometres in diameter. The absence of an appropriate site on land suggested that the impact had occurred in the ocean. One survey revealed that the shocked quartz grains are most abundant in Mexico and the American southwest. Studies of fossil pollen and fauna showed a sudden decline in vegetation at that time, especially in the western United States.[43] Flowering plants and some species of tree died out. As primitive ferns occupied this denuded land, they produced so many spores that this feature in the sedimentary sequence has been dubbed the 'fern spike'. However, in the southern hemisphere, where the iridium-enriched clay is thinner and there is less shocked ejecta material, the change in vegetation was less significant. Given this distribution, an impact in the Atlantic seemed unlikely. If it struck off the west coast then the crater no longer exists, because that part of the ocean floor has since been subducted. An analysis of the distribution of shocked material argued against an abyssal strike and in favour of an impact on the continental shelf.[44] Tsunami debris around the Caribbean focused attention on the isthmus linking the Americas. In 1990, after an analysis of all the available evidence, Alan Hildebrand and William Boynton of the University of Arizona concluded that the impact was in the Caribbean.[45,46] As it happened, after noting prominent circular anomalies on the tip of the Yucatan Peninsula in geophysical data in the early 1950s, Petroleos Mexicanos drilled core samples. Cores close to the Mayan village of Chicxulub, near the centre of the feature, recovered silicate rock with an igneous texture that was initially taken to imply volcanic origin, but cores 140 to 210 kilometres from the centre found a

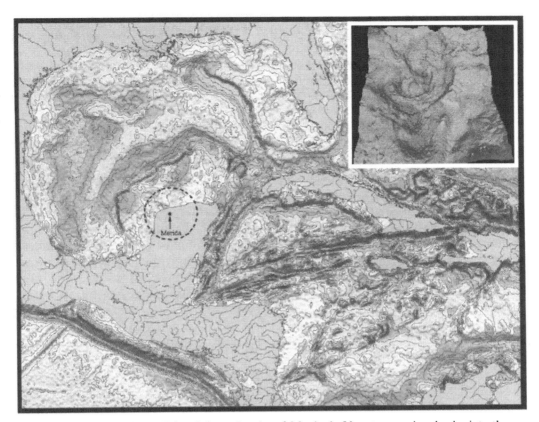

The circle centred on Chicxulub at the tip of Mexico's Yucatan peninsula depicts the outline of the impact crater which was excavated 65 million years ago, and is currently buried by several hundred metres of sediment. The sea floor was mapped by Geosat. (The gravity anomaly map of the crater (insert) is courtesy of V.L. Sharpton at the Lunar Planetary Institute in Houston.)

deposit of brecciated rock several hundred metres thick. In 1990 some of the samples were located in Mexico City, and re-analysed shock metamorphism confirmed the site as an impact crater.[47,48]

Biostratigraphic information indicated that the Chicxulub structure formed at the end of the Cretaceous. Radiometric dating showed that the melt rocks and the associated breccias were of the same age as the tiny spherules of impact glass found within boundary clay deposits in Haiti and the unmelted granitic fragments found in exposures throughout western North America for that time. Furthermore, isotopic ratios demonstrated that the melt rocks and the ejecta spherules originated from the same source rocks. There was therefore clear chemical as well as temporal linkage between the Chicxulub structure and the boundary deposits spanning a broad area. To those favouring the impact theory for the Cretaceous extinction, this constituted the 'smoking gun'.

Although the crater now lies beneath 300 metres of sediment, its presence can be inferred from a pattern of concentric rings in the local gravitational field. By compressing the rock, the impact enhanced the local gravitational field. The local magnetic field displays a similar pattern because the dense basaltic basement is more magnetic than the sediments that have filled in the crater. The presence of multiple rings complicated the task of determining the crater's size, as a preliminary to modelling the energetics of the impact, but it is of the order of 250 kilometres in diameter.

After analysing the gravitational field data, Peter Schultz, a geologist at Brown University on Rhode Island, announced in 1993 that (1) the inner ring is slightly elongated, with its primary axis aligned southeast–northwest; (2) towards the northwest the outer rings are fragmentary; and (3) the cavity comprises a deep pit in the southeast that shallows towards the northwest, effectively forming a ramp. Having simulated high-velocity impacts in the laboratory, Schultz recognised this pattern. He concluded that the projectile had flown in from the southeast on a shallow trajectory of about 30 degrees, struck at an oblique angle, and ejected a fan of debris downrange. Critics have warned that Schultz has over-interpreted the data. V.L. Sharpton of the Lunar and Planetary Institute in Houston (who had secured the gravitational data) insisted that until a higher resolution survey was conducted Schultz's assertions had to be considered speculative.

In 1997 Steven D'Hondt, a palaeontologist at the University of Rhode Island, realised that Schultz's analysis of the Yucatan crater explained why the environmental devastation had been so much worse in North America than in the southern hemisphere, as evidenced by the pollen record. Long before the ash particulates injected into the stratosphere could have produced the 'winter' effects by blocking out sunlight and starving the food chain on a global basis, the fan of superheated gas and vaporised rock from the oblique impact would have seared the western half of the United States.[49] Microscopic spherules from the fan of vaporised rock are found on much of the northern shore of the Caribbean, but the ensuing tsunamis disrupted this deposit. The spherules that fell close offshore in the Atlantic proved to be better preserved, as Richard Olsson and Kenneth Miller of Rutgers University discovered when they drilled into sediment north of Atlantic City, New Jersey, in 1997. Back in the Cretaceous this area had been under 100 metres of water. The end of the Cretaceous is marked by a deposit of spherules 5 centimetres thick. For a site 2,500 kilometres from the impact, this was a rather thicker deposit than had been expected. As hoped, the layer was in a pristine state, and the decrease in biological diversity was striking – 90 per cent of the plankton species that were present below the layer were absent above it.[50]

In 1996 Frank Kyte, a geochemist at the University of California in Los Angeles, found a fossilised remnant of the impactor at the bottom of a 1-metre-thick layer of iridium-rich clay in sediment on the floor of the north Pacific.[51] Although no longer composed of its original minerals, it had retained its original shape and texture. This was the first direct evidence of the type of the impactor. The rock-like imprint confirmed that it was asteroidal, because the materials of a cometary body would have been more porous. The recovery site was consistent with Schultz's conclusion that the impactor had approached Chicxulub from the southeast at a shallow angle and blasted debris downrange.[52]

A series of 'flood basalt' eruptions built up the thick stack of horizontal flows of the Deccan Plateau in India 65 million years ago. (Courtesy of Laszlo Keszthelyi of the Lunar and Planetary Laboratory at the University of Arizona.)

Although the case for an impact 65 million years ago is compelling, there is evidence that some species of dinosaur became extinct *prior to* the deposition of the iridium-enriched clay. Had their extinction occurred gradually over a period of millions of years? Had the impact simply delivered the *coup de grace?* In 1993, the US Geologic Survey's Gary Landis analysed microscopic air bubbles locked in fossilised tree resin and discovered that during the final few million years of the Cretaceous the proportion of oxygen fell from 35 per cent to 28 per cent. Dinosaurs lacked a diaphragm to help to push air in and out of their lungs so Rich Hengst, a physiologist at Purdue University in Indiana, suggested that the larger species of dinosaur had not been able to adapt to this change in the environment. Palaeontologists Keith Rigby of the University of Notre Dame in Indiana and Robert Sloan of the University of Minnesota in Minneapolis agreed, and blamed extreme respiratory failure for reducing the diversity of the North American dinosaurs from 35 genera to 12 genera in the final 10 million years of the Cretaceous.[53] Landis proposed that the change in the relative proportions of carbon dioxide and oxygen in the atmosphere was due to intense volcanic activity.[54] The stack of lava flows which form the Deccan Plateau in western India are 65 million years old.[55] In fact, the iridium layer is within sediment sandwiched between two lava flows, so this volcanism had been underway long before the impact.[56,57] Nevertheless, the largest flow sits just above this layer. French geophysicist Vincent Courtillot discovered that there had been only two reversals in the Earth's magnetic field during the emplacement of the entire stack. This meant that the pulse of volcanic activity had been brief, lasting no more than a few million years.[58] The outgassing from volcanic activity on such a scale would have prompted a sustained 'winter' scenario.

In 1783 a fault 25 kilometres long opened at Laki on southeast Iceland and 12 cubic kilometres of basalt flooded an area of about 500 square kilometres. Although minor in comparison to the eruptions which formed the Columbia River Plateau 16 million years ago, the Deccan Plateau 65 million years ago, or the vast Siberian Traps 250 million years ago, Laki is the worst such eruption witnessed by humans and its environmental effects were devastating. (Courtesy of Laszlo Keszthelyi of the Lunar and Planetary Laboratory at the University of Arizona.)

The largest flood basalt eruption witnessed by humans was Laki in Iceland in 1783 84. As well as lava, it emitted sulphur dioxide, carbon dioxide, hydrogen chloride and other noxious gases. Ash and acid rain destroyed most of the island's crops and poisoned the livestock, and ten thousand people – a quarter of the population – starved to death. The survivors had to temporarily abandon the island.[59] The ash particulates and sulphur aerosols which reached the stratosphere temporarily cut the temperature by 1°C globally. Despite its devastation in human terms, Laki extruded just 12 cubic kilometres of lava. The Deccan eruptions produced several hundred thousand cubic kilometres of lava! The climate must have been in an anomalous state for some time prior to the Chicxulub impact, and the dinosaurs may well have been in decline as a result.[60] Nevertheless, Peter Ward of the University of Washington and Charles Marshall of the University of California at Los Angeles established in 1996 that the ammonites expired simultaneously, so their extinction had a specific cause.

CYCLES?

David Raup and John Sepkoski, palaeobiologists at the University of Chicago, observed in 1984 that there seemed to be a 30-million-year periodicity in the times of the extinctions during the last 250 million years.[61] Walter Alvarez and R.A Muller suggested that there was a similar periodicity in the impact record.[62] A multitude of theories were promptly advanced to account for this. One involved a faint

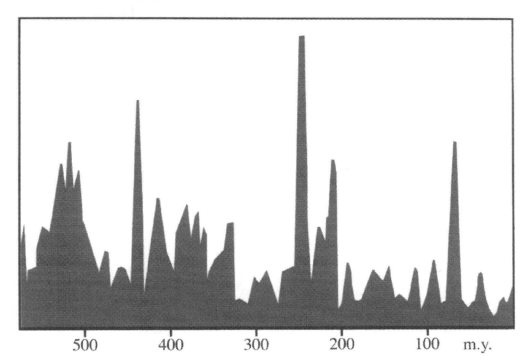

A plot showing the severity in the extinctions (the peaks) during the Cambrian. The greatest 'dying' was at the end of the Permian, 250 million years ago. Notice that many species were short-lived during the early Cambrian, as evolution explored many options. (Based on data by M.R. Rampino of New York University.)

companion for the Sun – appropriately dubbed 'Nemesis' – in an extended orbit that periodically perturbed the comets in the Oort Cloud and sent some of them into the inner Solar System.[63,64,65,66,67] However, the statistical significance of such a periodicity evaporated when the dates of the more recent extinctions were refined.

The most serious of all extinctions was 250 million years ago, at the close of the Permian period, soon after the Pangean supercontinent formed. The amphibians were badly hit, and only one family of the recently evolved mammalian reptiles survived,[68] and this went on to form the root of all mammals. The effect at sea was worse. It claimed more than half of marine families and most of the rest lived on through the persistence of just a few of their individual species. Among those that disappeared were the trilobites. In comparison, 15 per cent of marine life succumbed to the extinction that claimed the dinosaurs.[69] Overall, the Permian extinction wiped out almost 90 per cent of the animal species, and the planet was virtually sterilised.[70] Was an impact involved in this case too? The fossilised sea life in the sediments that were laid down at that time in the Tethys Ocean are now limestone that has been uplifted by the orogeny that dominates the southern margin of Eurasia and runs almost unbroken from Italy to China. The carbon-13 isotope that provides a measure of the abundance of biological activity indicates that there was a decline in sea life at this time; radiometric dating of zircon crystals puts the close of the Permian almost exactly 250 million years ago; and there is a pronounced iridium spike.[71,72]

Michael Rampino of New York University has suggested that the Permian extinction was caused by an impact on Gondwanaland between what has become the southern cape of Africa and the Argentinian coast. Several lines of evidence support

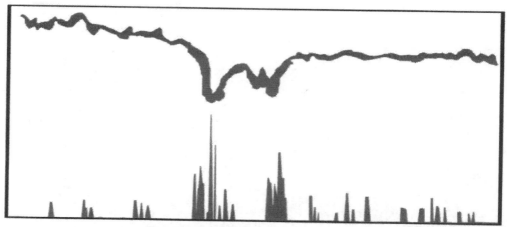

Permian ∎ Triassic

A plot of the concentration of carbon-13 (top) in marine carbonate rock sediments laid down in the late Permian and early Triassic (and thereafter folded into the Alps) and the concentration of iridium (bottom) in those same sediments. There is a striking correlation between the spikes in iridium with the dramatic fall in carbon-13 from reduced marine biological activity. (Based on data by M.R. Rampino of New York University.)

this claim. A survey has identified several circular gravitational anomalies on the continental shelf off South America. Samples of granites retrieved from these sites indicated intense deformation 250 million years ago. Rocks recovered from South Africa, southern Chile and southern Argentina had also suffered severe deformation 250 million years ago. Rampino has calculated that the impact would probably have generated a seismic disturbance measuring at least 12 on the Richter scale and this would have sent crustal undulations with a vertical displacement of about 100 metres radiating outwards. He has suggested that the Cape Fold Belt on the southern tip of Africa (for which there is no obvious tectonic cause) was thrust up by the shock of the impact to the south. A reconstruction of the uneroded mountain range indicated a massive wave-like anticline with its crest facing north.

Rampino has also suggested that if the seismic waves from a massive impact in southern Gondwanaland were focused at the antipode (as is known to have happened opposite major strikes on the Moon and Mercury) then this might have induced intense volcanism in the form of the Siberian Traps north of Lake Baikal. This flood basalt eruption formed a plateau covering 1.5 million square kilometres, in the process drowning the earlier terrain in lava that, in some places, is 3 kilometres thick. Its effect on the environment would have been both pronounced and

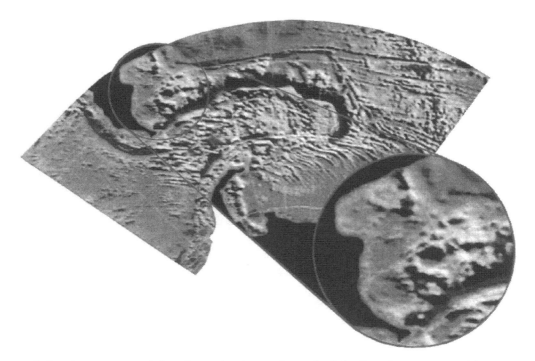

It has been suggested that the extinction at the end of the Permian was caused by an impact on Gondwanaland between what is now the southern cape of Africa and the Argentinian coast. A survey has identified several circular gravitational anomalies on the continental shelf off South America and samples indicate intense deformation 250 million years ago. (Courtesy of M.R. Rampino of New York University.)

sustained.[73,74,75] Although Gerry Czamanske of the US Geologic Survey dated the Siberian Traps at 250 million years old, he preferred the traditional view that flood basalts are due to a mantle plume 'burning' through the continental lithosphere.[76,77,78] The Permian and Cretaceous extinctions, he argued, were caused by the environmental effects of flood basalts which were due to endogenic processes, not impacts. If there is a relationship between terrestrial impacts and antipodal flood basalts, it has not been proved.[79] Rampino has speculated that Gondwanaland's break up may have been triggered by an impact. The traditional view is that continental rifting is caused by mantle plumes,[80] but there is considerable debate as to the role that flood basalts play in this process.[81] Actually, there may be a more direct link between impacts and flood basalts than the possibility of focusing of seismic waves antipodal to an impact. When an impact excavates crustal rock it relieves the pressure on the mantle beneath and allows it to 'decompress', melt, and rise through the faults in the floor of the crater. Once the lava fills the cavity it floods its surroundings, and, in doing so, it masks its source. Might there be a crater *beneath* the Siberian Traps? What if the ocean floor is hit? There are several massive volcanic plateaux in the western Pacific. The Nauru and the nearby and much larger Ontong Java Plateaux were erupted 130 and 122 million years ago, respectively, each over a period of several million years, and they rise some 3 kilometres above the surrounding ocean floor.[82] Seismic studies established that the oceanic lithosphere beneath Ontong Java is about 40 kilometres thick, which is several times the usual thickness. Although traditionally considered to be the result of mantle plumes,[83,84,85] might these submarine volcanic plateaux mark the sites of impacts? Perhaps a longer perspective is needed to understand these rare events.[86]

VIOLENT PLANET

The popular impression of a volcano is a conical mountain with a summit vent from which a succession of lava flows augment the edifice, but not all volcanoes construct mountains. The source of the magma determines its characteristics because the greater the proportion of gas in the magma the more violent is the eruption. As the pressure diminishes as magma rises up the conduit, the included volatiles exsolve and expand in volume by a factor of a thousand. If this occurs sufficiently rapidly it produces an explosion in which the edifice literally 'blow its top'. The most violent of such volcanoes, referred to as Plinian,[87] blast supersonic plumes of ash into the stratosphere. Once the plume loses power and collapses, it sends a hot ash flow rolling down the flank. This hugs the ground just like lava but travels more rapidly, and much farther. The eruption of Mount Pele on the Caribbean island of Martinique in 1902 blasted out thousands of cubic metres of such pyroclastic and smothered the town of Saint Pierre together with its 30,000 inhabitants. The effect is even more devastating when the edifice collapses. In 1912 the flank of Katmai in Alaska failed and vented a *lateral* pyroclastic flow that blanketed 200 square kilometres. The explosive detonations of Tambora in 1815 and Krakatau in 1883 were both powered by steam produced when sea water penetrated their magma

Violent planet 363

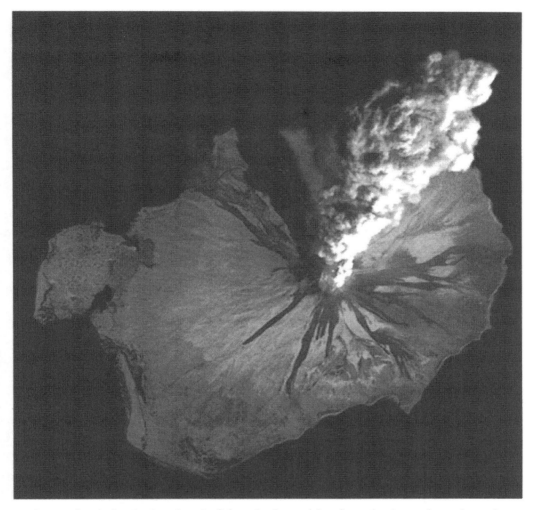

An overhead view by Landsat 5 of the ash plume rising from the Augustine volcano in Alaska in 1986, showing the lava flows on the flanks and the plume's shadow.

chambers, and they literally blew themselves apart. Krakatau was particularly violent; indeed the sound of its detonation was heard 5,000 kilometres away and its tsunami devastated the coasts of Java and Sumatra.[88] In 1967 archaeologists found the ash-smothered remains of a city on the island of Thera in the Aegean. In fact, Thera is just a fragment of a larger volcanic edifice that blew itself apart circa 1450 BC when water penetrated its magma chamber. The tsunami wiped out the Minoan civilisation on Crete 150 kilometres to the south, and may well have given rise to the myth of a 'lost civilisation' of Atlantis.[89] Archaeologist David Keys has recently suggested that a massive eruption in the Sunda Strait between Sumatra and Java in AD 535 significantly impeded the development of human civilisation.[90]

Nevertheless, even peaks that collapse pale into insignificance in comparison to resurgent calderas that blast out not thousands but millions of cubic metres of ash.

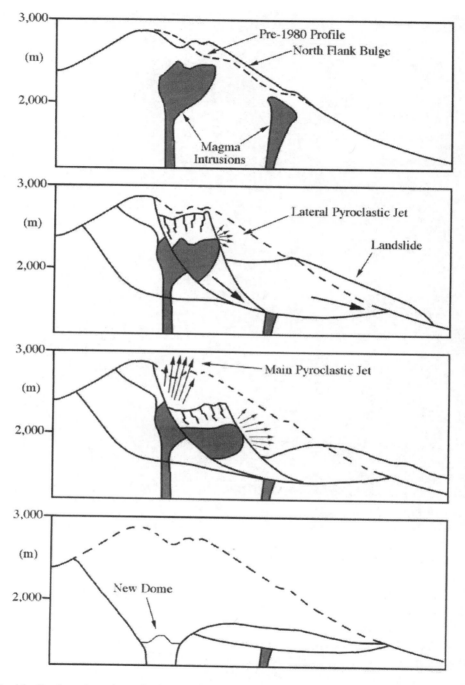

An idealised explanation of why the 1980 eruption of Mount St Helens in the Cascade Range was so violent.

An overhead radar image of the horseshoe-shaped relic of Mount St Helens after the northern flank collapsed during its eruption in 1980.

The sustained partial melting that occurs above a subducting slab generates silicic magma. As this rises towards the surface it pushes up a shallow dome in the crust. As the included volatiles exsolve, the rapid increase in pressure literally causes the magma to explode from the ground. Once the eruption has relieved the pressure, the magma chamber's roof collapses. Ironically, despite being the largest volcanoes, only a dozen were known before satellites started to map the planet, but the ignimbrites of welded tuff that congealed from the superheated ash they expelled were readily distinguished by multi-spectral sensors, and several hundred calderas have now been catalogued. They are all very subtle structures. In many cases ring faults prompted fissure eruptions and parasitic cones. Their ignimbrites have been deeply etched by dendritric drainage channels from runoff. Similar calderas exist on the most ancient terrain on Mars.

Crater Lake in Oregon formed when the magma chamber drained following a catastrophic eruption 7,000 years ago. It blasted out 150 cubic kilometres of rock and made a 10-kilometre-wide caldera that now hosts a lake. And when Toba in Sumatra exploded 75,000 years ago, it ejected 3,000 cubic kilometres of rock and left a caldera 30 kilometres across that now hosts a lake 2 kilometres deep. These eruptions would have produced 'winter' environmental effects, but were clearly insufficient to induce an extinction event. Although rare on the human timescale, an eruption of this type is simply a normal part of the Earth's ongoing surface activity. So when will the next one strike?

In 1872 the Yellowstone National Park was established by the US Congress as America's first national park. It spans northwestern Wyoming, southern Montana and eastern Idaho, and hosts the world's greatest concentration of geothermal features. For a century, it was simply a nature reserve. Only recently was it realised that it hosts a caldera 65 kilometres in diameter – in fact, the world's largest! Although Yellowstone last erupted 650,000 years ago, it is by no means extinct. In 1998, Ken Pierce, a geologist at the US Geologic Survey, realised that every few

366 Life and death

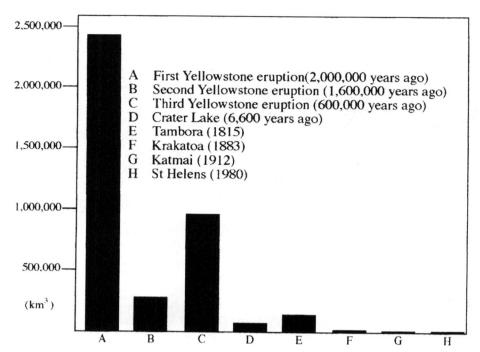

A plot showing the amount of crust expelled by the most powerful volcanoes in the last few million years. Considered in this context, the devastating explosion of Mount St Helens in 1980 was trivial in comparison to the eruptions of the Yellowstone resurgent caldera (which will one day erupt again).

thousand years in the last 10,000 years several hundred square kilometres of the caldera's floor was inflated by about 30 metres and then subsided again. The shore of Yellowstone Lake changed as the land tilted.[91] Pierce has suggested that gases accumulating within the magma feed system raise the floor of the caldera until a vent opens and the land settles once the pressure has been relieved. Such venting would be like releasing the lid of a pressure cooker. It appears that superheated water 'flashes' to steam as it escapes in a hydrothermal explosion. Although the 'Old Faithful' geyser is impressive, it pales into insignificance in comparison to what must occur within the caldera from time to time. In past full-scale eruptions, Yellowstone has blanketed much of the western half of the continent with a deep layer of pyroclastics.

As a result of the Atlantic Ocean's opening, North America is drifting in a southwesterly direction at a rate of a few centimetres per year. However, the mantle plume that produced the Yellowstone caldera complex is static. Unable to penetrate the lithosphere and release primary magma as flood basalts, the plume flattened out against the underside of the plate and induced partial melting that prompted silicic magma to rise through the lithosphere. In fact, just as the Pacific plate's motion over a hot spot beneath Hawaii has made a chain of volcanic islands of progressively greater age, there is a line of relics left by caldera explosions extending to the southwest from Yellowstone across southern Idaho, in the form of the Snake River Volcanic Plain.[92] A batholith has made northern Idaho mountainous so the majority

Violent planet 367

An overhead of Yellowstone National Park (indicated by the thin black line). Only recently was it realised that the Park occupies the floor of a resurgent caldera that was formed by a massive eruption 600,000 years ago.

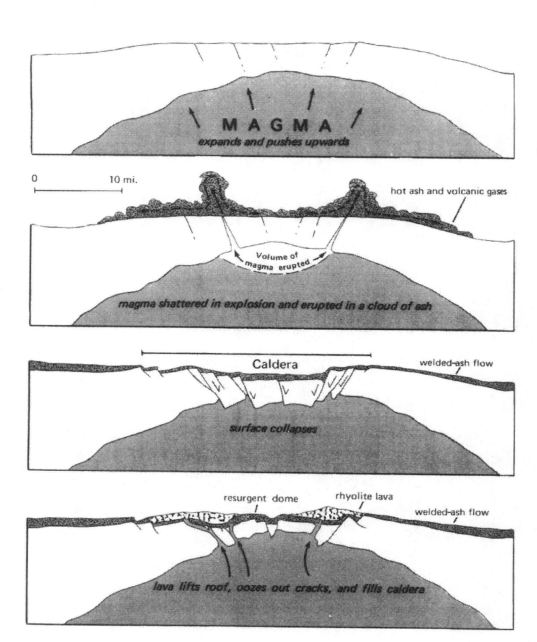

A diagram illustrating the sequence that made the Yellowstone caldera during the third eruptive cycle 600,000 years ago. Resurgent doming occurred early in the last stage. Post-caldera lava eruptions and further uplift occurred slowly over a long time period. Every few thousand years in the last 10,000 years several hundred square kilometres of the caldera's floor have inflated by about 30 metres and then subsided again. (Copyright © W.J. Fritz. *Roadside Geology of the Yellowstone Country*, 1985. Courtesy of the Mountain Press Publishing Company, Missoula, Montana.)

Table 11.1 Volcano Explosivity Index

VEI	Example	Character	Plume (km)	Volume (km^3)
0	Kilauea	Non-explosive	<0.1	0.000001
1	Stromboli	Gentle	0.1–1	0.00001
2	Vulcan	Explosive	1–5	0.001
3	Ruiz (1985)	Severe	3–15	0.01
4	Galuggung (1982)	Cataclysmic	10–25	0.1
5	St Helens (1981)	Paroxysmal	>25	1
6	Krakatau (1883)	Colossal	>25	10
7	Tambora (1815)	Titanic	>25	100
8	Yellowstone (2 mya)	–	>25	1,000

Note: Plumes that ascend more than 25 kilometres are drawn around the world in the stratosphere.

of the residents live on the vast, and more or less flat, crescent-shaped volcanic plain that is the result of millions of years of rhyolitic eruptions. It was later veneered by fissure flows up to 15 metres thick, and peppered with cinder cones built from lava spattered by fire fountains and hundreds of central vent basalt shields up to 100 metres tall whose flanks overlap to form an undulating volcanic terrain. When the hot spot was on the Oregon border in southwestern Idaho 16 million years ago,

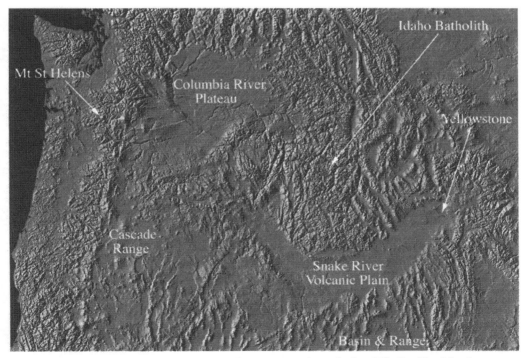

A shaded relief representation of the topography of the US Pacific Northwest. (Courtesy of USGS.)

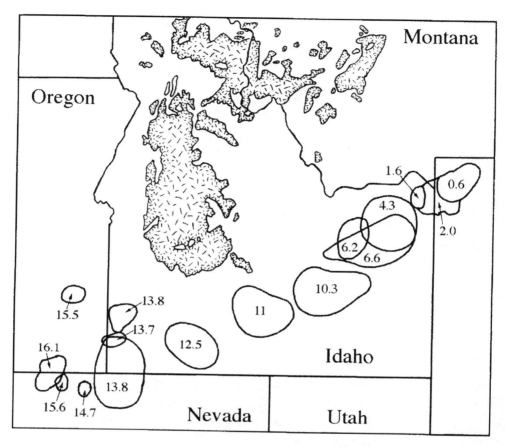

A map showing the locations of volcanic calderas created during the past 16 million years. They trace the motion of the 'hot spot' induced by a rising plume in the mantle. In reality, of course, it was the North American plate that drifted in a westerly direction over the mantle plume. The calderas underly the Snake River Volcanic Plain of southern Idaho. The granite to the north is the Idaho batholith. The hot spot is currently just across the border in northwestern Wyoming, beneath the Yellowstone caldera which has erupted three times in the past 2 million years.

swarms of fissures developed to the north, some of them 100 kilometres in length, and poured forth the flood basalts which, over the next several million years, built up the Columbia River Plateau. This spans some 220,000 square kilometres, and the individual flows average 25 metres thick. It has been proposed that this activity was triggered by an impact and that the flows masked the crater.[93] This, however, is disputed, not least because there are no other indications of an impact in this area at that time. Regardless of the event that triggered it, the construction of the plateau would have had a serious effect on the climate. The passage of the hot spot itself has transformed the northern fringe of the Basin and Range province into a plain.[94] It is a striking example of how the release of heat from the interior controls the form of the Earth's surface.

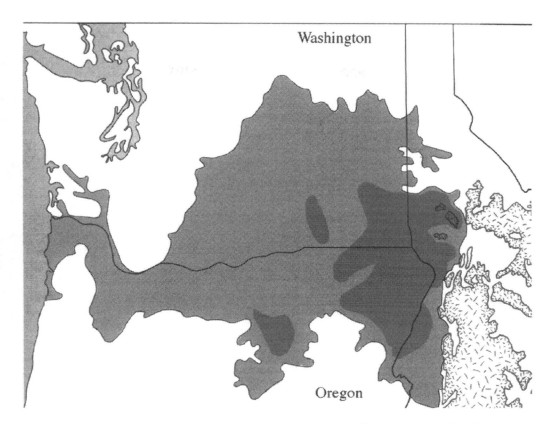

A map showing the sources (darker) and the total extent (lighter) of the 'flood basalts' which built up the Columbia River Plateau in the US Pacific Northwest some 15 million years ago. To the east is the granitic mass of the Idaho batholith.

Clearly, therefore, terrestrial life has had to cope with a variety of endogenic catastrophic events in addition to impacts. For a long time, the uniformitarianist geological community dismissed impacts as catastrophism and denied that they played a significant factor in shaping the Earth, but as the list of impact craters increased Gene Shoemaker retorted that impacts constituted an intermittent but *ongoing* process, and hence were simply part of the Earth's uniformitarianist environment.

DANGEROUS ENVIRONMENT

Most of the debris left over from the accretion of the planets was swept up during the 'late bombardment' which tailed off 3.8 billion years ago – at the close of the appropriately named Hadean epoch. Nevertheless, models of impact rates during the ensuing Archean predicted at least 150 impacts excavating craters exceeding 100 kilometres in diameter, of which two dozen would have been greater than 300 kilometres in diameter.[95,96] Indeed, there is evidence of Archean impacts in the form of layers of glassy spherules in the 3.45-billion-year-old Pilbaran sediment,[97] and in the 3.24-billion-year-old Barberton sediment.[98] Similar layers in the 2.63-, 2.56- and 2.49-billion-year-old sediments in the Hamersley Range (which forms the western part of the Proterozoic fold belt between the Pilbara and the Yilgarn cratons) testify to later impacts.[99] Any lifeforms in the immediate vicinity of an impact would have been vaporised, killed by the thermal flash, or buried by the ejecta. It was a very dangerous environment, even for primitive single-celled organisms.

The surface of the Moon bears witness to the effects of impacts. The rain of cosmic debris may have greatly diminished, but it has not ceased. The 85-kilometre-wide crater that dominates the Southern Highlands and was named after the 16th-century Danish astronomer Tycho Brahe was excavated just 100 million years ago. In the otherwise inert lunar environment it has been preserved in a pristine state. If that intruder had hit the Earth, the crater would long since have been eroded. Nevertheless, impacts on such a scale ought to leave behind tell-tale geological clues to reveal where they once were.

At the start of the 20th century, Daniel Barringer sought the relic of the meteorite which he believed had formed a 1-kilometre-wide crater near Winslow on the Arizona Plateau. In 1903 he was so sure that it marked an impact that he started drilling to locate what he assumed would be a chunk of iron–nickel, which he intended to mine. His rationale was based on the fact that the surrounding plain was littered with small fragments of such alloy. Upon his death in 1929, Barringer had drilled dozens of holes without finding anything. The problem was his assumption that the rock would have buried itself. Only later was it calculated that the energy released by the impact would have vaporised much of the projectile and shattered the rest of it into tiny fragments that would have been ejected with the excavated rock. The inability to find a meteorite prompted the hypothesis that the cavity was a maar from the explosive release of volcanic gases. However, a detailed analysis by Gene Shoemaker in 1960 of the manner in which the layers of rock had been flipped over – just as if by a circular hinge – proved that Meteor Crater had been appropriately named.[100,101] Although fairly small, this 50,000-year-old hole in the desert went on to serve as a 'Rosetta Stone' for the study of impacts on planetary bodies.[102,103]

In an experiment in 1953 in which he subjected quartz to very high pressure, Loring Coes identified a new high-density mineral, which he named coesite. Upon hearing of this, Harvey Nininger suggested that coesite might be created as the shockwave from an impact propagated into the ground. In 1960, L.R. Stieff and John O'Keefe sent a 'glassy' sample taken from the Arizona crater to Ed Chao at the US Geologic Survey to have it analysed for a possible link to 'tektites', which at that time were believed to have come from space. In fact, Chao discovered that the sample contained shocked quartz.[104,105,106]

Soon thereafter Shoemaker visited the Rieskessel in Bavaria, a structure which surrounds the town of Nordlingen. It was generally believed to be some kind of eroded volcanic caldera, but Shoemaker was fairly sure that it was not. He collected a sample of glassy material which (if the structure was an impact crater) ought to be target rock that had been highly shocked and partly melted. Chao confirmed that it contained shocked quartz.[107] A survey found the bedrock of the 26-kilometre-

Meteor Crater near Winslow on the Arizona Plateau is 1.2 kilometres across its rim crest, and 200 metres deep. It was Gene Shoemaker's study of the circular 'hinge' in 1960 which proved that it was excavated by an impact 50,000 years ago. (Courtesy of USGS.)

diameter structure to be seismically disturbed to a depth of 4.5 kilometres, even though the cavity – which has since been partially filled in by erosion – was excavated to no more than 1.5 kilometres.[108]

In 1962 Robert Dietz proposed that an oval basin at Sudbury, Ontario, Canada, measuring some 30 by 60 kilometres, is actually an impact crater.[109] The original circular form has been distorted by tectonic forces. In 1994 Walter Roest and Mark Pilkinton analysed the patterns of deformation in the rocks around the feature and reconstructed its original shape.[110] 'Shatter-cones' confirmed that it resulted from the impact. Evidently, it was excavated by a 5-kilometre-wide rock some 1.85 billion years ago.[111] The fact that 75 per cent of the western world's nickel is mined from this crater suggests that it was probably an iron–nickel body. In 1995 John Spray and Lucy Thompson of the University of New Brunswick in Canada studied the rock exposures surrounding the Sudbury crater and found mineralogical evidence for several outer rings. They concluded that the shock of the impact excavated a hole that was initially *32 kilometres deep*, and induced a set of deep concentric ring faults. In the immediate aftermath, massive blocks of rock had sheared off at these faults and slipped down the wall of the crater, over time creating the alternating ridges and troughs of a multiple-ringed form. As the rock slipped, the friction induced a 'melt' mineral called pseudotachylyte. Although the terrain where the rings stood was eroded long ago, the pseudotachylyte indicated where the faults had been.[112] This was the first direct evidence of a multiple-ringed structure on Earth, and it marked an important discovery because it provided insight into the processes that operate when such a structure is formed.

And what of all the rock that would have been ejected by such a large impact? Much of the lunar surface is really a thick blanket of ejecta. What would eroded terrestrial ejecta look like? After the Apollo missions, some geologists who had studied the Moon's surface in detail took a close look at the Earth with precisely this question in mind. Today's glaciers deposit 'boulder clays' of poorly sorted debris ranging in size from mud to large boulders. On the basis of James Hutton's motto that "the present is the key to the past", similar consolidations called 'tillites' had been assumed to indicate sites of past glaciation. However, lunar impact breccias were similarly poorly sorted deposits and the angular clasts showed similar abrasional markings. When Verne Oberbeck of NASA's Ames Research Center examined deposits at a number of impact sites – including the Ries crater – and concluded that the terrestrial breccias were indistinguishable from tillites, it started to look as if the Earth's surface might have preserved more evidence of ancient impacts than had been believed. Simulations of the deposits likely to have been made by impacts during the last 2 billion years (based on the Moon's cratering rate) left a distribution that was very similar to that of the tillites.[113] Could these two mechanisms be distinguished for any particular deposit? The presence of shocked quartz would rule in favour of an impact. However, the search for shocked quartz has proved frustrating, largely because likely impact sites have yet to be identifed.[114]

The population of the impactors is dominated by small objects, hence, the more lethal the potential impact the less likely it is to occur. Small objects burn up in the atmosphere. A rocky body 50 metres in diameter would explode upon reaching the lower atmosphere. The forest in Tunguska in Siberia that was flattened in 1908 was almost certainly hit by the blast from such an airburst. Objects 100 metres across are estimated to strike every thousand years, and 1,000-metre objects every hundred thousand years. To human civilisation, strikes on this scale would be devastating, but only the rare objects much larger than this are likely to cause extinctions.[115]

A map showing the distribution of impact craters around the globe as of the mid-1980s. (Based on Map 1658A by the Geological Survey of Canada by R.A.F. Grieve and P.B. Robertson.)

Dangerous environment 375

As the STS-100 mission performed a fly-around inspection of the International Space Station in 2001, it snapped a picture of the station set against the circular structure of the Manicouagan impact crater in Canada.

A map produced in 1987 documented 116 confirmed impact craters[116,117]. Of the two dozen craters exceeding 5 kilometres in diameter, only four larger than 50 kilometres were known to have been excavated in the last 250 million years – that is, since Pangea's formation. About 215 million years ago a rock excavated the 65-kilometre crater in the Manicouagan area of Quebec in Canada. Although the crater is still circular, the structure has been vertically deformed by isostatic adjustment and so its form is now evident only by virtue of an annular lake. An impact 74 million years ago excavated the 35-kilometre-diameter crater at Manson in Iowa.[118,119] As the crater had been buried by sediment, it was discovered as a circular geological structure during well-drilling. In 1994, while drilling to find out why the floor of the southern Chesapeake Bay is depressed, C.W. Poag of the US Geologic Survey discovered that it is an eroded 85-kilometre crater. Beneath the sediment, there is a layer of rocky debris. Because it was a coastal site, he initially assumed that this must be tsunami detritus, but the presence of shocked quartz and glassy spherules indicated that it was brecciated ejecta from an impact 35 million years ago.[120] A few million years later, an impact dug the 100-kilometre-wide Popigai crater in the former Soviet Union. Why did these large impacts *not* cause extinctions?

Might it be that 'mass' extinctions occur only if the biosphere is already under great stress for entirely biological reasons and that a change in the environment (whether prompted by an impact or by some endogenic cause) is simply the proverbial straw which breaks the camel's back? At times when the biosphere is more robust, might climatic change simply cause 'mini' extinctions affecting either only a specific area, or type of life? In 1998, while studying a 30-kilometre-long layer of greenish glass and red rocky material in shoreline cliffs in Argentina, Peter Schultz recognised it as a 3.3-million-year-old impact. It had clearly not caused a global extinction, but the regional fall-out was the demise of several types of sloths, hoofed animals and armadillo-like mammals whose fossils are found beneath the glass layer but not above it. Oceanographic studies had previously indicated a sharp drop in temperature 3.3 million years ago. Since smaller impactors are more common than

An artist's impression of an asteroid striking the Earth. Most of the atmosphere is confined to a layer 15 kilometres thick, and this will pose no obstacle to an object 100 kilometres in diameter travelling at a 'cosmic speed' of 30 kilometres per second. (Courtesy of V.R. Oberbeck and K. Zahnle of Ames Research Center in California. Artist: Don Davis.)

larger ones, it is likely that the Earth has suffered many such localised setbacks interspersed with the global extinctions.

That impacts do still happen was strikingly shown in 1994, when the fragmented comet Shoemaker-Levy 9 hit Jupiter. George Wetherill of the Carnegie Institution of Washington has pointed out that if it were not for Jupiter's immense gravitational field attracting stray asteroids and comets and flinging them into the outer Solar System, there would be more debris crossing the orbits of the inner planets, and more impacts. If it were not for Jupiter, therefore, instead of being subjected to a major impact every 100 million years on average, the Earth would be being hit every hundred thousand years or so, in which case life on the planet would be in a continuous state of being winnowed.[121,122] Wetherill ran a computer simulation in which the solar nebula did not produce a planet as massive as Jupiter, and found that even if the largest planet was the size of Uranus this would not have been able to clear out the debris. The implication from this is that even if planets around other stars turn out to be commonplace, the prospects for life might hinge not just upon there being a terrestrial planet in the 'habitability zone', but also upon there being a giant planet to clear out the debris, to produce a benign environment.[123,124]

NOTES

1. 'In the beginning', A.N. Halliday. *Nature*, vol. 409, p. 144, 2001.
2. 'Initiation of clement surface conditions on the earliest Earth', N.H. Sleep, K. Zahnle and P.S. Neuhoff. *Pub. Nat. Acad. Sci.*, vol. 98, p. 3666, 2001.
3. 'Earth's early atmosphere' J. F. Kasting. *Science*, vol. 259, p. 920, 1993.
4. 'A sensitivity study of changes in Earth's rotation rate with an atmospheric general circulation model', G.S. Jenkins. *Global and Planetary Change*, vol. 11, p. 141, 1996.
5. If ever something as advanced as heterostegina is found on another planet, it will be a momentous discovery.
6. *Stromatolites*, M.R. Walter (Ed.). Elsevier, 1976.
7. 'Stromatolites 3400–3500 Myr old from the North Pole area, Western Australia', M.R. Walter, R. Buick and J.S.R. Dunlop. *Nature*, vol. 248, p. 443, 1980.
8. 'Stromatolite recognition in ancient rocks: an appraisal of irregularly laminated structures in an Early Archean chert-barite unit from North Pole, Western Australia', R. Buick, J.S.R. Dunlop and D.I. Groves. *Alcheringa*, vol. 5, p. 161, 1981.
9. 'Stromatolites 3400 Myr-old from the Archean of Western Australia', D.R. Lowe. *Nature*, vol. 284, p. 441, 1980.
10. 'Abiological origin of described stromatolites older than 3.2 Ga', D.R. Lowe. *Geology*, vol. 22, p. 387, 1994.
11. 'An abiotic model for stromatolite morphogenesis', J.P. Grotzinger and D.H. Rothman. *Nature*, vol. 383, p. 423, 1996.
12. 'Abiological origin of described stromatolites older than 3.2 Ga: comment', R. Buick, D.I. Groves and J.S.R. Dunlop. *Geology*, vol. 23, p. 191, 1995.
13. 'Occurrence and potential uses of Archean microfossils and organic matter', M.D. Muir. *Univ. West. Aust. Ext. Serv.*, vol. 2, p. 11, 1978.
14. 'Microfossils of the early Archean Apex chert: new evidence of the antiquity of life', J.W. Schopf. *Science*, vol. 260, p. 640, 1993.
15. 'Origin of 3.45 Ga coniform stromatolites in Warrawoona Group, Western Australia', H.J. Hoffmann, K. Grey, A.H. Hickman and R.I. Thorpe. *Geol. Soc. Am. Bull.*, vol. 111, p. 1256, 1999.
16. 'Submarine thermal springs on the Galapagos rift', J.B. Corliss, J. Dymond, L.I. Gordon, J.M. Edmond, R.P. von Herzen, R.D. Ballard, K. Green, D. Williams, A. Bainbridge, K. Crane and T.H. van Andel. *Science*, vol. 203, p. 1073, 1979.
17. *Fire under the sea: the discovery of the most extraordinary environment on Earth - volcanic hot springs on the ocean floor*, J. Cone. Quill/William Morrow, 1991.
18. 'Life's undersea beginnings', J. Cone. *Earth Mag.*, p. 34, July 1994.
19. 'Submarine hydrothermal systems: a probable site for the origin of life', J.B. Corliss, J. Baross and S. Hoffman. 1980.
20. 'Oasis of life in the cold abyss,' J.B. Corliss and R.D. Ballard. *National Geog. Mag.*, vol. 152, p. 441, 1977.
21. 'The hot dive', W.J. Broad. *Earth Mag.*, p. 26, August 1997.
22. Estimates of the amount of carbon dioxide in the early atmosphere vary from 100 to 1,000 times its present level.
23. 'Earth's early atmosphere' J. F. Kasting. *Science*, vol. 259, p. 920, 1993.
24. 'The biosphere below', D. Grossman and S. Schulman. *Earth Mag.*, p. 34, June 1995.
25. *Earth's earliest biosphere: its origin and evolution*, J.W. Schopf (Ed.). Princeton University Press, 1983.

26 See news item on, p. 13 of *Earth Mag.*, April 1997.
27 Of course, as carbon dioxide was removed from the atmosphere, the proportion of oxygen would have been increased even if the actual amount did not. The atmosphere was initially dominated by carbon dioxide and there was negligible oxygen. The proportion of oxygen in the atmosphere is currently 21 per cent and carbon dioxide, at fraction of a per cent, has been reduced to a minor constituent.
28 'Atmospheric and hydrospheric evolution on the primitive Earth', P.E. Cloud. *Nature*, vol. 160, p. 729, 1968.
29 'Palaeoecological significance of the banded iron formation', P.E. Cloud. *Economic Geol.*, vol. 68, p. 1135, 1973.
30 See news item on, p. 14 of *Earth Mag.*, January 1994.
31 Recently, Paris-based geologist John Saul has argued that if one type of cell whose role was to suppress the reproduction of others began to attack harmless cells the pathology would be similar to that of an auto-immune disease. Similarly, if a cell reproduced out of control it would correspond to a cancer. See news item on, p. 11 of *Earth Mag.*, July 1994.
32 For a lively account of the biological explosion during the Cambrian see *Oasis in space: Earth history from the beginning*, P. Cloud. Norton Co., 1988.
33 The word 'fossil' was coined by Austrian mineralogist Georgius Agricola in a book published upon his death in 1555 which summarised the knowledge of Saxon miners. He meant the term to be anything excavated from the ground, but it soon came to apply only to the remnants of once-living things.
34 *On the origin of species*, C.R. Darwin. London, 1859.
35 'Extraterrestrial cause for the Cretaceous–Teriary extinction', J.W. Alvarez, W. Alvarez, F. Asaro and H.V. Michel. *Science*, vol. 208, p. 1105, 1980.
36 The initial paper cited iridium spikes from only two locales, but iridium enrichment was subsequently found at several hundred sites all over the globe in sediments deposited in shallow and deep seas, in rivers and on land.
37 The propagation of the shock of an impact densifies rock and induces phase changes which alter the minerals. Quartz can be transformed into coesite and stishovite.
38 'Mineralogic evidence for an impact event at the Cretaceous–Tertiary boundary', B.F. Bohor, E.E. Foord, P.J. Modreski and D.M. Triplehorn. *Science*, vol. 224, p. 867, 1984.
39 'Clay mineralogy of the Cretaceous–Tertiary boundary clay', M.R. Rampino and R.C. Reynolds. *Science*, vol. 219, p. 495, 1983.
40 *The great dinosaur extinction controversy*, C. Officer and J. Page. Addison-Wesley, 1996.
41 'Global fire at the Cretaceous–Tertiary boundary', W.S. Wolbach, I. Gilmour, E. Anders, C.J. Orth and R.R. Brooks. *Nature*, vol. 334, p. 665, 1988.
42 *Extinctions*, M.H. Nitecki (Ed.). University of Chicago Press, 1984.
43 'Megafloral change across the Cretaceous–Tertiary boundary, northern Great Plains, USA', K.R. Johnson and L.J. Hickey. In *Global catastrophes in Earth History: an interdisciplinary conference on impact, vulcanism, and mass mortality*, V.L. Sharpton and P.D. Ward (Eds). Geol. Soc. Am. Special Paper no.247, p. 433, 1991.
44 'Shocked quartz in the Cretaceous–Tertiary boundary clays: evidence for a global distribution,' B.F. Bohor, P.J. Modreski and E.E. Foord. *Science*, vol. 236, p. 705, 1987.
45 'Proximal Cretaceous–Tertiary boundary impact deposits in the Caribbean', A.R. Hildebrand and W.V. Boynton. *Science*, vol. 248, p. 843, 1990.
46 'The Cretaceous/Tertiary boundary impact (or the dinosaurs didn't have a chance', A.R. Hildebrand. *J. Roy. Astron. Soc. Canada*, vol. 87, p. 77, 1993.

47 'New links between the Chicxulub impact structure and the Cretaceous–Tertiary boundary', V.L. Sharpton, G.B. Dalrymple, L.E. Marin, G. Ryder, B.C. Schuraytz and J. Urrutia-Fucugauchi. *Nature*, vol. 359, p. 819, 1992.
48 'Chicxulub Multiring Impact Basin: size and other characteristics derived from gravity analysis', V.L. Sharpton *et al. Science*, vol. 261, p. 1564, 1993.
49 See news item on, p. 22 of *Earth Mag.*, April 1997.
50 'Ejecta layer at the K–T Boundary, Bass River, New Jersey (ODP Leg 174AX)', R.K. Olsson, K.G. Miller, J.V. Browning, D. Habib, P.J. Sugarman. *Geology* 1997.
51 'A meteorite from the Cretaceous Tertiary boundary', F.T. Kyte. *Nature*, vol. 396, p. 237, 1998.
52 See news item on, p. 12 of *Earth Mag.*, August 1996.
53 *New Scientist*, vol. 140, 6 November 1993.
54 See news item on, p. 12 of *Earth Mag.*, March 1994.
55 'K–Ar age of the Deccan traps, India', P. Wellman and M.W. McElhinny. *Nature*, vol. 227, p. 595, 1970.
56 'A search for iridium in the Deccan traps and inter-traps', D. Boclet, V. Courtillot and J.J. Jaeger. *Geophy. Res. Lett.*, vol. 15, p. 812, 1988.
57 'Age of the Deccan traps using 187Re-187Os systematics', C.J. Allegre, J.L Birck, F. Capmas, and V. Courtillot. *Israel Earth Planet. Sci. Lett.*, vol. 170, p. 197, 1999.
58 *Evolutionary catastrophes: the science of mass extinction*, V. Courtillot. Cambridge University Press, 1999.
59 *Asimov's new guide to science*, I. Asimov. Penguin, p. 152, 1984.
60 *The influence of continental flood basalts on mass extinctions: where do we stand?*, V. Coutillot, J.J. Jaeger, Z. Yang, G. Feraud and C. Hofman. Geol. Soc. Am. Special Paper no.307, p. 513, 1996.
61 'Periodicity of extinctions in the geologic past', D.M. Raup and J.J. Sepkowski. *Proc. Nat. Acad. Sciences*, vol. 81, p. 801, 1984.
62 'Evidence from crater ages for periodic impacts on the Earth', W. Alvarez and R.A. Muller. *Nature*, vol. 308, p. 718, 1984.
63 *The Nemesis affair: a story of the death of dinosaurs and the ways of science*, D.M. Raup. Norton Co., 1985.
64 *Nemesis: the death star and other theories of mass extinction*, D. Goldsmith. Walker Co., 1985.
65 *Nemesis: the death star*, R. Muller. Weidenfeld & Nicolson, 1988.
66 *Cosmic catastrophes*, C.R. Chapman and D. Morrison. Plenum Press, 1989.
67 'Terrestrial mass extinctions, cometary impacts and the Sun's motion perpendicular to the galactic plane', M.R. Rampino and R.B. Stothers. *Nature*, vol. 308, p. 709, 1984.
68 This was the genus dicynodon.
69 'Extinctions – which way did they go?', S.M. Stanley. *Earth Mag.*, p. 17, January 1991.
70 'Earth's near-death experience', J. Alper. *Earth Mag.*, p. 42, January 1994.
71 'The Permian–Triassic boundary event: A geochemical study of three Chinese sections', L. Zhou and F.T. Kyte. *Earth Planet. Sci. Lett.*, vol. 90, p. 411, 1998.
72 'Search for evidence of impact at the Permian–Triassic boundary in Antarctica and Australia', G.J. Retallack, A. Seyedolali, E.S. Krull, W.T. Holser, C.P. Ambers and F.T. Kyte. *Geology*, vol. 26, p. 979, 1998.
73 'Flood basalt volcanism during the past 250 million years', M.R. Rampino and R.B. Stothers. *Science*, vol. 241, p. 663, 1988.
74 'Volcanic winters', M.R. Rampino, S. Self and R.B. Stothers. *Ann. Rev. Earth Planet. Sci.*, vol. 16, p. 73, 1998.

75 'Flood basalts and extinction events', R.B. Stothers. *Geophys. Res. Lett.*, vol. 20, p. 1399, 1993.
76 'Synchronism of the Siberian traps and the Permian Triassic boundary', I.H. Campbell, G.K. Czamanske, V.A. Fedorenko, R.I. Hill and V. Stepanov. *Science*, vol. 258, p. 1760, 1992.
77 'Mantle and crustal contributions to continental flood basalt volcanism', N.T. Arndt, G.K. Czamanske, J.L. Wooden and V.A. Fedorenko. *Tectonophysics*, vol. 223, p. 39, 1993.
78 'Demise of the Siberian plume: paleogeographic and paleotectonic reconstruction from the prevolcanic and volcanic record, north-central Siberia', G.K. Czamanske, A.B. Gurevitch, V. Fedorenko and O. Simonov. *International Geol. Rev.*, vol. 40, no.2, p. 95, 1998.
79 'Mass extinctions in the last 300 million years: one impact and seven flood basalts', V. Courtillot. *J. Earth Sci.*, vol. 43, p. 255, 1994.
80 'On causal links between flood basalts and continental breakup', V. Courtillot, C. Jaupart, I. Manighetti, P. Tapponnier and J. Besse. *Earth Planet. Sci. Lett.*, vol. 166, p. 177, 1999.
81 'Flood basalts and large igneous provinces from deep mantle plumes: fact, fiction, and fallacy', H.C. Sheth. *Tectonophysics*, vol. 311, p. 1, 1999.
82 For a discussion of the Ontong Java Plateau, see *The oceanic crust, from accretion to mantle recycling*, T. Juteau and R. Maury. Springer–Praxis, p. 297, 1999.
83 'Geochemical features of intraplate oceanic plateau basalts', P.A. Floyd. In *Magmatism in the ocean basins*, Geol. Soc. Special Publication no.42, p. 215, 1989.
84 'Scratching the surface: estimating dimensions of large igneous provinces', M. Coffin and O. Eldholm. *Geology*, vol. 21, p. 515, 1993.
85 'Rapid formation of the Ontong Java Plateau by Aptian mantle plume volcanism', J.A. Tarduno, W.V. Sliter, L. Kroenke, M. Leckie, H. Mayer, J.J. Mahoney, R. Musgrave, M. Storey and E.L. Winterer. *Science*, vol. 254, p. 399, 1991.
86 'Large igneous provinces: a planetary perspective,' J.W. Head and M.F. Coffin. In *Large igneous provinces: continental, oceanic, and planetary flood volcanism*, J.J. Mahoney and M.F. Coffin (Eds). AGU Geophysical Monograph no.100, 1997.
87 Plinian volcanoes are so-called because the historian Tacitus asked a survivor by the name of Pliny to write an account of the eruption of Vesuvius in AD 79. It was during this eruption that pyroclastic flows smothered the towns of Pompei and Herculaneum. Pliny's uncle, the historian Gaius Plinus Secundus (who is referred to as Pliny the Elder in order to distinguish him from his nephew) was killed in the eruption.
88 'Fire and water at Krakatau', S. Carey, H. Sigurdsson and C. Mandeville. *Earth Mag.*, p. 26, March 1992.
89 *The end of Atlantis: new light on an old legend*, J.V. Luce. Book Club Associates and Thames & Hudson, 1969.
90 *Catastrophe: an investigation into the origins of the modern world*, D. Keys. Century, 1999.
91 See news item on, p. 14 of *Earth Mag.*, June 1998.
92 'The Yellowstone hotspot', R.B. Smith. *J. Volcanology and Geothermal Research*, vol. 61, p. 121, 1994.
93 *Northwest exposures: a geologic story of the Northwest*, D. Alt and D.W. Hyndman. Mountain Press, Co., p. 241, 1995.
94 In a few million years, the Yellowstone hot spot will cross the state line from Wyoming into Montana, and threaten Billings.
95 'Mega-impacts and mantle melting episodes: tests of possible correlations', A.Y. Glikson. *AGSO J. Aust. Geol. Geophys.*, vol. 16, p. 587, 1996.

96 'Oceanic mega-impacts and crustal evolution', A.Y. Glikson. *Geology*, vol. 27, p. 387, 1999.
97 'Early Archean silicate spherules of possible impact origin', D.R. Lowe and G.R. Byerly. *Geology*, vol. 14, p. 83, 1986.
98 'Geological and geochemical record of 3400 million year old terrestrial meteorite impacts,' D.R. Lowe, G.R. Byerly, F. Asaro and F.T. Kyte. *Science*, vol. 245, p. 959, 1989.
99 'Revised correlations in the early Precambrian Hamersley Basin based on a horizon of resedimented impact spherules', B.M. Simonson and S.W. Hassler. *Aust. J. Earth Sci.*, vol. 44, p. 37, 1997.
100 'Penetration mechanics of high velocity meteorites, illustrated by Meteor Crater, Arizona', E.M. Shoemaker. International Geological Congress Report, XII Session (Norden) Pt. 18, p. 418, 1960.
101 'Impact mechanics at Meteor Crater', E.M. Shoemaker. A chapter in *The Moon, meteorites and comets*, B.M. Middlehurst and G.P. Kuiper (Eds). University of Chicago Press, 1963.
102 'Arizona's Meteor Crater', R. Burnham. *Earth Mag.*, p. 50, January 1991.
103 'Calamity at Meteor Crater', D.A. King. *Sky & Telescope*, p. 48, November 1999.
104 'Natural coesite: an unexpected geological discovery', *Foote Prints*, vol. 32, no.1, p. 25, 1960.
105 'First natural occurrence of coesite', E.C.T Chao, E.M. Shoemaker and B.M. Madsden. *Science*, vol. 132, p. 220, 1960.
106 For the full story of this critical juncture in the study of impact craters, see *To a rocky Moon: a geologist's history of lunar exploration*, D.E. Williams. University of Arizona Press, p. 44, 1993.
107 'New evidence for the impact origin of the Ries basin, Bavaria, Germany', E.M. Shoemaker and E.C.T. Chao. *J. Geophys. Res.*, vol. 66, p. 3371, 1961.
108 'The Ries impact crater', J. Pohl, D. Stoffler. H. Gall and K. Ernstson. In *Impact and explosion cratering*, D.J Roddy, R.O. Pepin and R.B. Merrill (Eds). Pergamon, p. 343, 1977.
109 'Sudbury structure as an astrobleme', R.S. Dietz. *J. of Geology*, vol. 72, p. 412, 1964.
110 See news item on, p. 16 of *Earth Mag.*, December 1994.
111 'Are we mining an asteroid?', R.S. Dietz. *Earth Mag.*, p. 36, January 1991.
112 'Friction melt distribution in a multi-ring impact basin', J.G. Spray and L.M. Thompson. *Nature*, vol. 373, p. 130, 1995.
113 'Impacts, tillites, and the breakup of Gondwanaland', V.R. Oberbeck, J.R. Marshall and H Aggarwal. *J. of Geology*, vol. 101, p. 1, 1993.
114 As Paul Spudis explained the problem: "Fundamentally, this is a difficult thing to prove. Clastic ejecta tends to be noticably deficient in shocked quartz – the Ries Bunte Breccia (the deposit analogous to tillites of 'ejecta' origin) is almost totally devoid of shock features – all the shocked quartz is found in the melt-bearing suevite, which is volumetrically very minor and on top of the Bunte breccia. So if tillites ARE ejecta, shocked quartz would not necessarily be expected. However, Verne Oberbeck has yet to identify an appropriate source crater, and the burden is on him to do so."
115 'The threat from space', D. Desonie. *Earth Mag.*, p. 24, August 1996.
116 'Terrestrial impact structures', R.A.F. Grieve and P.B. Robertson. Map 1658A, Geol. Survey, Canada, 1987. See also: 'Terrestrial Impact Structures', R.A.F. Grieve. *Ann. Rev. Earth Planet. Sci.*, vol. 15, p. 245, 1987.
117 By the mid-1990s, the number of known impact craters had risen to 150.

118 See news item on, p. 14 of *Earth Mag.*, May 1994.
119 '^{40}Ar–^{39}Ar dating of the Manson impact structure: a Cretaceous–Tertiary boundary candidate', M.J. Kunk, G.A. Izett, R.A. Haugerud and J.F. Sutter. *Science*, vol. 244, p. 1565, 1989.
120 See news items on, p. 11 of February 1995 and, p. 12 of August 1996 of *Earth Mag.*.
121 'Possible consequences of absence of "Jupiters" in planetary systems', G.W. Wetherill. *Astrophys. Space Sci.*, vol. 212, p. 33, 1994.
122 'How special is Jupiter?', G.W. Wetherill. *Nature*, vol. 373, p. 470, 1994.
123 In July 1999, M. Kuerster, M. Endl, S. Els, A.P. Hatzes, W.D. Cochran, S. Doebereiner and K. Dennerl announced the discovery of a planet with a mass several times that of Jupiter in an eccentric orbit of iota Hor, a solar-type star in the southern constellation Horologium, varying between 0.78 and 1.08 AU, with an orbital period of 320 days. It is the first extrasolar planet to be identified within the 'habitability zone' of a solar-type star. While the planet might have a system of satellites comparable in size to the Earth they will be subjected to ongoing bombardment by the debris drawn in by their primary's gravity, so even though the thermal regime is conducive to life, if it has established a foothold on one of the moons its status will be precarious.
124 In January 2000 D. Naef, M. Mayor, F. Pepe, D. Queloz, N.C. Santos, S. Udry and M. Burnet reported the discovery of a planetary companion around the young chromospherically active dwarf star Gliese-Jahreiss 3021 (GJ 3021; also listed as HD 1237, HIP 1292) in the constellation of Hydrus. The inferred minimum planetary mass was 3.32 Jovian masses. The orbital separation between the planet and its parent star ranges from 0.25 to 0.75 AU. This planet resides within the 'habitability zone' most of the time and the equilibrium temperature is estimated to vary from about 260 K when farthest out to 440 K when closest in.

Appendix 1: Planetary data

SUN

Diameter (km)	1,392,000 (equatorial)
(× Earth)	109
Mass (kg)	1.989×10^{30}
(× Earth)	333,000
Volume (cm^3)	1.412×10^{33}
(× Earth)	1,300,000
Mean density (g/cm^3)	1.41
Rotational period	25.4 days (sidereal, at equator)
Escape velocity (km/s)	617.7
Mean surface temperature (K)	5,780
Obliquity of axis (deg)	7.25
Gravity (× Earth)	27.9

The Sun's rotation is differential with latitude, varying between 25 and 34 days, and its axis is tilted 7.25 degrees to the ecliptic.

MERCURY

Diameter (km)	4,878
(× Earth)	0.382
Mass (kg)	3.302×10^{23}
(× Earth)	0.055
Volume (× Earth)	0.058
Mean density (g/cm^3)	5.43
Rotational period	58.6462 days
Escape velocity (km/s)	4.43
Mean surface temperature (K)	452
Maximum surface temperature (K)	700
Minimum surface temperature (K)	100
Semi-major axis (AU)	0.38710
Orbital eccentricity	0.205631
Orbital inclination (deg)	7.0048
Obliquity of axis (deg)	0.1

Gravity (× Earth)	0.376
Orbital period (days)	87.97
(years)	0.2408
Synodic period (days)	116
Albedo	0.11
Atmospheric components	Trace amounts of hydrogen and helium
Surface materials	Basaltic and anorthositic rocks and regolith

VENUS

Diameter (km)	12,104
(× Earth)	0.949
Mass (kg)	4.868×10^{24}
(× Earth)	0.814
Volume (× Earth)	0.86
Mean density (g/cm^3)	5.24
Rotational period	243.0185 days, retrograde
Escape velocity (km/s)	10.36
Mean surface temperature (K)	726
Semi-major axis (AU)	0.72333
Orbital eccentricity	0.006773
Orbital inclination (deg)	3.3947
Obliquity of axis (deg)	177.33 retrograde
Gravity (× Earth)	0.903
Orbital period (days)	224.70
(years)	0.6152
Synodic period (days)	584
Albedo	0.62
Atmospheric components	96% carbon dioxide, 3% nitrogen, 0.1% water vapour
Surface materials	Basaltic rock and altered materials

EARTH

Diameter (km)	12,756
Mass (kg)	5.974×10^{24}
Mean density (g/cm^3)	5.52
Rotational period	23.9345 hours
(days)	1.0
Escape velocity (km/s)	11.19
Mean surface temperature (K)	281
Maximum surface temperature (K)	310
Minimum surface temperature (K)	260
Semi-major axis (AU)	1.0
Orbital eccentricity	0.016710
Orbital inclination (deg)	0.0
Obliquity of axis (deg)	23.45
Gravity (× Earth)	1.0
Orbital period (days)	365.26

Appendix 1: Planetary data

(years)	1.0
Synodic period (days)	–
Albedo	0.38
Atmospheric components	78% nitrogen, 21% oxygen, 1% argon
Surface materials	Basaltic and granitic rock and altered materials

Moon

Orbital period (days)	27.3216
Synodic period (days)	29.53
Rotational period	Synchronous
Escape velocity (km/s)	2.4
Semi-major axis (km)	384,000
Diameter (km)	3,476
Orbital eccentricity	0.0549
Orbital inclination (deg)	18 to 29 (variable)
Obliquity of axis (deg)	6.7
Mass ($\times$ Earth)	0.0123
($\times$ kg)	7.349×10^{22}
Density (g/cm^3)	3.34
Gravity ($\times$ Earth)	0.18
Albedo	0.12

MARS

Diameter (km)	6,792
($\times$ Earth)	0.533
Mass (kg)	6.419×10^{23}
($\times$ Earth)	0.108
Volume ($\times$ Earth)	0.158
Mean density (g/cm^3)	3.93
Rotational period	24.6230 hours
(days)	1.026
Escape velocity (km/s)	5.03
Maximum surface temperature (K)	240
Minimum surface temperature (K)	190
Semi-major axis (AU)	1.52366
Orbital eccentricity	0.093412
Orbital inclination (deg)	1.8506
Obliquity of axis (deg)	25.19
Gravity ($\times$ Earth)	0.380
Orbital period (days)	686.98
(years)	1.8807
Synodic period (days)	780
Albedo	0.15
Atmospheric components	95% carbon dioxide, 3% nitrogen, 1.6% argon
Surface materials	Basaltic rock and altered materials

Appendix 1: Planetary data

Phobos

Discovered	1877, A. Hall
Orbital period (hours)	7.7
(days)	0.319
Rotational period	Synchronous
Semi-major axis (km)	9,400
Diameter (km)	27 × 21 × 19
Orbital eccentricity	0.0151
Orbital inclination (deg)	1.08
Mass (kg)	1.06×10^{16}
Density (g/cm^3)	1.90
Gravity (× Earth)	0.0007
Albedo	0.07

Deimos

Discovered	1877, A. Hall
Orbital period (hours)	30.3
(days)	1.262
Rotational period	Synchronous
Semi-major axis (km)	23,500
Diameter (km)	15 × 12 × 11
Orbital eccentricity	0.0003
Orbital inclination (deg)	1.79 variable
Mass (kg)	2.4×10^{15}
Density (g/cm^3)	2.10
Gravity (× Earth)	0.0004
Albedo	0.07

JUPITER

Diameter (km)	142,984 (equatorial)
(× Earth)	11.21
Mass (kg)	1.898×10^{27}
(× Earth)	317.83
Volume (× Earth)	1,320
Mean density (g/cm^3)	1.32
Rotational period	9.925 hours (core)
(days)	0.40
Escape velocity (km/s)	59.54
Mean surface temperature (K)	120 (cloud tops)
Semi-major axis (AU)	5.20336
Orbital eccentricity	0.048393
Orbital inclination (deg)	1.3053
Obliquity of axis (deg)	3.08
Gravity (× Earth)	2.44
Orbital period (years)	11.856
Synodic period (days)	339

Appendix 1: Planetary data

Albedo   0.48
Atmospheric components   90% hydrogen, 10% helium, 0.07% methane

Rings

1979J1R ('Halo')
Planetocentric distance (km)   100,000–122,800
Radial width (km)   22,800

1979J2R ('Main')
Planetocentric distance (km)   122,800–129,200
Radial width (km)   6,400

1979J3R ('Gossamer')
Planetocentric distance (km)   129,200–214,200
Radial width (km)
85,000

Infrared spectra imply that the rings are tiny dark rock fragments.

Metis (1979J3)

Discovered   1979, Voyager; 1980, S. Synnott
Orbital period (hours)   7.1
  (days)   0.295
Semi-major axis (km)   126,000
Diameter (km)   40
Orbital eccentricity   0.0
Orbital inclination (deg)   0.0
Albedo   0.05

Adrastea (1979J1)

Discovered   1979, Voyager, D.C. Jewitt, E. Danielson, S. Synnott
Orbital period (hours)   7.1
  (days)   0.298
Semi-major axis (km)   128,500
Diameter (km) $25 \times 20 \times 15$
Orbital eccentricity   0.0
Orbital inclination (deg)   0.0
Albedo   0.05

Amalthea

Discovered   1892, E.E. Barnard
Orbital period (hours)   12
  (days)   0.498
Rotational period   synchronous
Semi-major axis (km)   181,300
Diameter (km) $262 \times 146 \times 134$
Orbital eccentricity   0.003
Orbital inclination (deg)   0.45

Appendix 1: Planetary data

Mass (kg)	1.5×10^{19}
Albedo	0.05

Thebe (1979J2)

Discovered	1979, Voyager, S. Synnott
Orbital period (hours)	16.2
(days)	0.674
Semi-major axis (km)	223,000
Diameter (km)	110×90
Orbital eccentricity	0.014
Orbital inclination (deg)	0.8
Albedo	0.05

Io

Discovered	1610, G. Galilei
Orbital period (days)	1.769
Rotational period	Synchronous
Semi-major axis (km)	421,600
Diameter (km)	3,630
Orbital eccentricity	0.004
Orbital inclination (deg)	0.04
Mass (kg)	8.932×10^{22}
($\times$ Earth)	0.0149
($\times$ Moon)	1.21
Density (g/cm^3)	3.55
Gravity ($\times$ Earth)	0.138
Albedo	0.6

Europa

Discovered	1610, G. Galilei
Orbital period (days)	3.552
Rotational period	Synchronous
Semi-major axis (km)	671,000
Diameter (km)	3,130
Orbital eccentricity	0.009
Orbital inclination (deg)	0.47
Mass (kg)	4.8×10^{22}
($\times$ Earth)	0.0081
($\times$ Moon)	0.66
Density (g/cm^3)	3.03
Gravity ($\times$ Earth)	0.138
Albedo	0.6

Ganymede

Discovered	1610, G. Galilei
Orbital period (days)	7.155

Appendix 1: Planetary data

Rotational period	Synchronous
Semi-major axis (km)	1,071,000
Diameter (km)	5,280
Orbital eccentricity	0.002
Orbital inclination (deg)	0.21
Mass (kg)	1.482×10^{23}
($\times$ Earth)	0.0249
($\times$ Moon)	2.03
Density (g/cm^3)	1.93
Gravity ($\times$ Earth)	0.145
Albedo	0.4

Callisto

Discovered	1610, G. Galilei
Orbital period (days)	16.689
Rotational period	Synchronous
Semi-major axis (km)	1,883,000
Diameter (km)	4,840
Orbital eccentricity	0.007
Orbital inclination (deg)	0.4
Mass (kg)	1.076×10^{23}
($\times$ Earth)	0.0179
($\times$ Moon)	1.46
Density (g/cm^3)	1.82
Gravity ($\times$ Earth)	0.124
Albedo	0.2

S/1975J1

Discovered	1975, C. Kowal; 2000, T.B. Spaht *et al*
Orbital period (days)	130
Semi-major axis (km)	7,398,200
Diameter (km)	8
Orbital eccentricity	0.206
Orbital inclination (deg)	45.37

Leda

Discovered	1974, C. Kowal
Orbital period (days)	238.7
Semi-major axis (km)	11,094,000
Diameter (km)	16
Orbital eccentricity	0.148
Orbital inclination (deg)	26.70

Himalia

Discovered	1904, C.D. Perrine
Orbital period (days)	250.6

Appendix 1: Planetary data

Semi-major axis (km)	11,480,000
Diameter (km)	150 × 120
Orbital eccentricity	0.163
Orbital inclination (deg)	27.63
Albedo	0.03

Lysithea

Discovered	1938, S. Nicholson
Orbital period (days)	259.2
Semi-major axis (km)	11,720,000
Diameter (km)	36
Orbital eccentricity	0.107
Orbital inclination (deg)	29.02

Elara

Discovered	1905, C.D. Perrine
Orbital period (days)	260.1
Semi-major axis (km)	11,737,000
Diameter (km)	76
Orbital eccentricity	0.207
Orbital inclination (deg)	25.77
Albedo	0.03

S/2000J11

Discovered	2000, S.S. Sheppard *et al*
Orbital period (days)	289.7
Semi-major axis (km)	12,623,200
Diameter (km)	4
Orbital eccentricity	0.215
Orbital inclination (deg)	28.5

S/2000J10

Discovered	2000, S.S. Sheppard *et al*
Orbital period (days)	590.0, retrograde
Semi-major axis (km)	20,299,700
Diameter (km)	3.8
Orbital eccentricity	0.155
Orbital inclination (deg)	165.6

S/2000J3

Discovered	2000, S.S. Sheppard *et al*
Orbital period (days)	605.9, retrograde
Semi-major axis (km)	20,643,200
Diameter (km)	5.2
Orbital eccentricity	0.269
Orbital inclination (deg)	150

Appendix 1: Planetary data

S/2000J5

Discovered	2000, S.S. Sheppard *et al*
Orbital period (days)	618.1, retrograde
Semi-major axis (km)	20,918,000
Diameter (km)	4.3
Orbital eccentricity	0.2
Orbital inclination (deg)	149.3

S/2000J7

Discovered	2000, S.S. Sheppard *et al*
Orbital period (days)	626.1, retrograde
Semi-major axis (km)	21,098,400
Diameter (km)	6.8
Orbital eccentricity	0.146
Orbital inclination (deg)	146

Ananke

Discovered	1951, S. Nicholson
Orbital period (days)	631, retrograde
Semi-major axis (km)	21,200,000
Diameter (km)	25
Orbital eccentricity	0.169
Orbital inclination (deg)	147 retrograde

S/2000J9

Discovered	2000, S.S. Sheppard *et al*
Orbital period (days)	651.8, retrograde
Semi-major axis (km)	21,672,200
Diameter (km)	5
Orbital eccentricity	0.246
Orbital inclination (deg)	163.5

S/2000J4

Discovered	2000, S.S. Sheppard *et al*
Orbital period (days)	660.6, retrograde
Semi-major axis (km)	21,868,100
Diameter (km)	3.2
Orbital eccentricity	0.346
Orbital inclination (deg)	161

Carme

Discovered	1938, S. Nicholson
Orbital period (days)	692, retrograde
Semi-major axis (km)	21,600,000
Diameter (km)	35

Orbital eccentricity	0.207
Orbital inclination (deg)	163 retrograde

S/2000J6

Discovered	2000, S.S. Sheppard *et al*
Orbital period (days)	703.6, retrograde
Semi-major axis (km)	22,805,000
Diameter (km)	3.8
Orbital eccentricity	0.281
Orbital inclination (deg)	165

S/2000J8

Discovered	2000, S.S. Sheppard *et al*
Orbital period (days)	733.1, retrograde
Semi-major axis (km)	23,439,400
Diameter (km)	5.4
Orbital eccentricity	0.528
Orbital inclination (deg)	151.7

Pasiphae

Discovered	1908, P. Melotte
Orbital period (days)	735, retrograde
Semi-major axis (km)	23,500,000
Diameter (km)	40
Orbital eccentricity	0.378
Orbital inclination (deg)	146

S/1999J1

Discovered	1999, Spacewatch
Orbital period (days)	736, retrograde
Semi-major axis (km)	23,500,000
Diameter (km)	10
Orbital eccentricity	0.206
Orbital inclination (deg)	143.5

Sinope

Discovered	1914, S. Nicholson
Orbital period (days)	758, retrograde
Semi-major axis (km)	23,700,000
Diameter (km)	30
Orbital eccentricity	0.275
Orbital inclination (deg)	153

S/2000J2

Discovered	2000, S.S. Sheppard *et al*
Orbital period (days)	766, retrograde

Semi-major axis (km)	24,136,000
Diameter (km)	5
Orbital eccentricity	0.317
Orbital inclination (deg)	165.8

SATURN

Diameter (km)	120,540 (equatorial)
($\times$ Earth)	9.45
Mass (kg)	5.685×10^{26}
($\times$ Earth)	95.16
Volume ($\times$ Earth)	745
Mean density (g/cm^3)	0.69
Rotational period	10.6562
(days)	0.44
Escape velocity (km/s)	35.49
Mean surface temperature (K)	88 K (1-bar level)
Semi-major axis (AU)	9.53707
Orbital eccentricity	0.054151
Orbital inclination (deg)	2.4845
Obliquity of axis (deg)	26.73
Gravity ($\times$ Earth)	1.12
Orbital period (years)	29.425
Synodic period (days)	378
Albedo	0.47
Atmospheric components	97% hydrogen. 3% helium, 0.05% methane

Rings

D

Planetocentric distance (km)	67,000–74,500
Radial width (km)	7,500

C

Planetocentric distance (km)	74,500–92,000
Radial width (km)	17,500

Maxwell gap

Planetocentric distance (km)	87,500
Radial width (km)	270

B

Planetocentric distance (km)	92,000–117,500
Radial width (km)	25,500

Cassini Division

Planetocentric distance (km)	117,500–122,200
Radial width (km)	4,700

Huygens gap

Planetocentric distance (km)	117,680

| Radial width (km) | 285–440 |

A
| Planetocentric distance (km) | 122,200–136,800 |
| Radial width (km) | 14,600 |

Encke Division
| Planetocentric distance (km) | 133,570 |
| Radial width (km) | 325 |

Keeler gap
| Planetocentric distance (km) | 136,530 |
| Radial width (km) | 35 |

F
| Planetocentric distance (km) | 140,210 |
| Radial width (km) | 30–500 |

G
| Planetocentric distance (km) | 165,800–173,800 |
| Radial width (km) | 8,000 |

E
| Planetocentric distance (km) | 180,000–480,000 |
| Radial width (km) | 300,000 |

The rings are only a few hundred meters thick. The particles are centimetres to decametres in size and are ice (some may be covered with ice), but with traces of silicate and carbon minerals. There are four main ring groups and three more faint, narrow ring groups separated by 'gaps' called divisions.

Pan

Discovered	1980, Voyager; 1990, M. Showalter
Orbital period (hours)	13.8
(days)	0.575
Semi-major axis (km)	133,600
Diameter (km)	9.7
Orbital eccentricity	0.0
Orbital inclination (deg)	0.0

Atlas (1980S15)

Discovered	1980, Voyager, R. Terrile
Orbital period (hours)	14.4
(days)	0.602
Semi-major axis (km)	137,600
Diameter (km)	38 × 34 × 28
Orbital eccentricity	0.002
Orbital inclination (deg)	0.3
Albedo	0.4

Prometheus (1980S14)

Discovered	1980, Voyager, S. Collins, D. Carlson
Orbital period (hours)	14.7
(days)	0.613
Rotational period	Synchronous
Semi-major axis (km)	139,300
Diameter (km)	$145 \times 95 \times 68$
Orbital eccentricity	0.003
Orbital inclination (deg)	0.0
Albedo	0.6

Pandora (1980S13)

Discovered	1980, Voyager, S. Collins, D. Carlson
Orbital period (hours)	15
(days)	0.629
Rotational period	Synchronous
Semi-major axis (km)	141,700
Diameter (km)	$114 \times 82 \times 62$
Orbital eccentricity	0.004
Orbital inclination (deg)	0.06
Albedo	0.5

Epimetheus (1980S10)

Discovered	1979, Pioneer 11, R Walker, J. Fountain, S. Larson
Orbital period (hours)	16.7
(days)	0.695
Rotational period	Synchronous
Semi-major axis (km)	151,422
Diameter (km)	$144 \times 108 \times 98$
Orbital eccentricity	0.009
Orbital inclination (deg)	0.14
Albedo	0.5

Note: Epimetheus is co-orbital with Janus.

Janus (1980S11)

Discovered	1966, A. Dollfus
Orbital period (hours)	16.7
(days)	0.695
Rotational period	synchronous
Semi-major axis (km)	151,472
Diameter (km)	$196 \times 192 \times 150$
Orbital eccentricity	0.007
Orbital inclination (deg)	0.34
Albedo	0.3

Note: Janus is co-orbital with Epimetheus.

Appendix 1: Planetary data

Mimas

Discovered	1789, W. Herschel
Orbital period (hours)	22.6
(days)	0.942
Semi-major axis (km)	185,600
Diameter (km)	396
Orbital eccentricity	0.020
Orbital inclination (deg)	1.53
Mass (kg)	3.7×10^{19}
($\times$ Moon)	5.2×10^{-4}
Density (g/cm^3)	1.15
Albedo	0.8

Enceladus

Discovered	1789, W. Herschel
Orbital period (days)	1.370
Semi-major axis (km)	238,000
Diameter (km)	498
Orbital eccentricity	0.004
Orbital inclination (deg)	0.01
Mass (kg)	7.3×10^{19}
($\times$ Moon)	1.0×10^{-3}
Density (g/cm^3)	1.1
Albedo	0.95

Tethys

Discovered	1684, J.D. Cassini
Orbital period (days)	1.888
Semi-major axis (km)	294,670
Diameter (km)	1,050
Orbital eccentricity	0.0
Orbital inclination (deg)	1.1
Mass (kg)	6.3×10^{20}
($\times$ Moon)	1.0×10^{-2}
Density (g/cm^3)	1.0
Albedo	0.8

Telesto (1981S16)

Discovered	1980, Voyager, B. Smith *et al*
Orbital period (days)	1.888
Semi-major axis (km)	294,670
Diameter (km)	$34 \times 28 \times 26$
Orbital eccentricity	0.0
Orbital inclination (deg)	1.0
Albedo	0.7

Note: Telesto librates about Tethys's trailing (L5) Lagrangian point.

Appendix 1: Planetary data

Calypso (1981S17)

Discovered	1980, D. Pascu *et al*
Orbital period (days)	1.888
Rotational period	Synchronous
Semi-major axis (km)	294,670
Diameter (km)	$34 \times 22 \times 22$
Orbital eccentricity	0.0
Orbital inclination (deg)	1.1
Albedo	0.9

Note: Calypso librates about Tethys's leading (L4) Lagrangian point.

Dione

Discovered	1684, J.D. Cassini
Orbital period (days)	2.737
Semi-major axis (km)	377,400
Diameter (km)	1,120
Orbital eccentricity	0.002
Orbital inclination (deg)	0.02
Mass (kg)	1.1×10^{21}
($\times$ Moon)	1.4×10^{-2}
Density (g/cm^3)	1.45
Albedo	0.6

Helene

Discovered	1980, Voyager, L. Lecacheux, P. Laques
Orbital period (days)	2.737
Rotational period	Synchronous
Semi-major axis (km)	377,400
Diameter (km)	$36 \times 32 \times 30$
Orbital eccentricity	0.005
Orbital inclination (deg)	0.15
Albedo	0.6

Note: Helene librates about Dione's leading (L4) Lagrangian point.

Rhea

Discovered	1672, J.D. Cassini
Orbital period (days)	4.518
Semi-major axis (km)	527,000
Diameter (km)	1,530
Orbital eccentricity	0.001
Orbital inclination (deg)	0.35
Mass (kg)	2.31×10^{21}
($\times$ Moon)	3.4×10^{-2}
Density (g/cm^3)	1.25
Albedo	0.6

Appendix 1: Planetary data

Titan

Discovered	1655, C. Huygens
Orbital period (days)	15.945
Semi-major axis (km)	1,222,000
Diameter (km)	5,140
Orbital eccentricity	0.029
Orbital inclination (deg)	0.33
Mass (kg)	1.346×10^{23}
($\times$ Moon)	1.83
Density (g/cm^3)	1.88
Albedo	0.2

Hyperion

Discovered	1848, W.C. Bond, G.P. Bond, W. Lassell
Orbital period (days)	21.277
Rotational period	Chaotic
Semi-major axis (km)	1,484,000
Diameter (km)	$400 \times 280 \times 220$
Orbital eccentricity	0.104
Orbital inclination (deg)	0.4
Mass (kg)	1.59×10^{21}
Density (g/cm^3)	1.0
Albedo	0.3

Iapetus

Discovered	1671, J.D. Cassini
Orbital period (days)	79.330
Semi-major axis (km)	3,561,000
Diameter (km)	1,460
Orbital eccentricity	0.028
Orbital inclination (deg)	14.72
Mass (kg)	1.88×10^{21}
($\times$ Moon)	2.6×10^{-2}
Density (g/cm^3)	1.2
Albedo	0.08

S/2000S5

Discovered	2000, B.J. Gladman *et al*
Orbital period (days)	448
Semi-major axis (km)	11,270,400
Diameter (km)	16
Orbital eccentricity	0.158
Orbital inclination (deg)	48.5

S/2000S6

Discovered	2000, B.J. Gladman *et al*

Appendix 1: Planetary data   399

Orbital period (days)	452
Semi-major axis (km)	11,356,000
Diameter (km)	13
Orbital eccentricity	0.367
Orbital inclination (deg)	49.3

Phoebe

Discovered	1898, W.H. Pickering
Orbital period (days)	550.45, retrograde
Semi-major axis (km)	12,960,000
Diameter (km)	220
Orbital eccentricity	0.163
Orbital inclination (deg)	150
Mass (kg)	1.0×10^{19}
Density (g/cm^3)	—
Albedo	0.05

S/2000S2

Discovered	2000, B.J. Gladman *et al*
Orbital period (days)	690.25
Semi-major axis (km)	15,063,600
Diameter (km)	24
Orbital eccentricity	0.495
Orbital inclination (deg)	46.2

S/2000S8

Discovered	2000, B.J. Gladman *et al*
Orbital period (days)	730.5, retrograde
Semi-major axis (km)	15,361,100
Diameter (km)	8.5
Orbital eccentricity	0.214
Orbital inclination (deg)	148.6

S/2000S3

Discovered	2000, B.J. Gladman *et al*
Orbital period (days)	791
Semi-major axis (km)	16,496,000
Diameter (km)	48
Orbital eccentricity	0.293
Orbital inclination (deg)	48.6

S/2000S11

Discovered	2000, B.J. Gladman *et al*
Orbital period (days)	881.9
Semi-major axis (km)	17,736,800
Diameter (km)	32

| Orbital eccentricity | 0.387 |
| Orbital inclination (deg) | 34.9 |

S/2000S12

Discovered	2000, B.J. Gladman et al
Orbital period (days)	888, retrograde
Semi-major axis (km)	17,817,900
Diameter (km)	8
Orbital eccentricity	0.0866
Orbital inclination (deg)	174.8

S/2000S4

Discovered	2000, B.J. Gladman et al
Orbital period (days)	889.5
Semi-major axis (km)	17,839,300
Diameter (km)	16
Orbital eccentricity	0.635
Orbital inclination (deg)	35

S/2000S10

Discovered	2000, B.J. Gladman et al
Orbital period (days)	926.8
Semi-major axis (km)	18,334,200
Diameter (km)	10
Orbital eccentricity	0.614
Orbital inclination (deg)	33.2

S/2000S9

Discovered	2000, B.J. Gladman et al
Orbital period (days)	943.3, retrograde
Semi-major axis (km)	18,551,200
Diameter (km)	12
Orbital eccentricity	0.254
Orbital inclination (deg)	169.6

S/2000S7

Discovered	2000, B.J. Gladman et al
Orbital period (days)	1036, retrograde
Semi-major axis (km)	19,751,600
Diameter (km)	8
Orbital eccentricity	0.544
Orbital inclination (deg)	175

S/2000S1

| Discovered | 2000, B.J. Gladman et al |

Appendix 1: Planetary data

Orbital period (days)	1,288, retrograde
Semi-major axis (km)	22,832,400
Diameter (km)	25
Orbital eccentricity	0.367
Orbital inclination (deg)	172.8

URANUS

Discovered	1781, W. Herschel
Diameter (km)	51,118 (equatorial)
($\times$ Earth)	4.01
Mass (kg)	8.683×10^{25}
($\times$ Earth)	14.48
Volume ($\times$ Earth)	67
Mean density (g/cm^3)	1.29
Rotational period	17.240 hours
(days)	0.72
Escape velocity (km/s)	21.33
Mean surface temperature (K)	59
Semi-major axis (AU)	19.1913
Orbital eccentricity	0.047168
Orbital inclination (deg)	0.7699
Obliquity of axis (deg)	97.9 retrograde
Gravity ($\times$ Earth)	1.1
Orbital period (years)	84.01
Synodic period (days)	370
Albedo	0.53
Atmospheric components	83% hydrogen, 15% helium, 2% methane (at depth)

Rings

1986 U2R

Planetocentric distance (km)	38,000
Radial width (km)	~ 2,500

Ring 6

Planetocentric distance (km)	41,840
Radial width (km)	1–3

Ring 5

Planetocentric distance (km)	42,230
Radial width (km)	2–3

Ring 4

Planetocentric distance (km)	42,580
Radial width (km)	2–3

Alpha

Planetocentric distance (km)	44,720
Radial width (km)	7–12

Beta
Planetocentric distance (km)   45,670
Radial width (km)   7–12

Eta
Planetocentric distance (km)   47,190
Radial width (km)   0–2

Gamma
Planetocentric distance (km)   47,630
Radial width (km)   1–4

Delta
Planetocentric distance (km)   48,290
Radial width (km)   3–9

1986U1R
Planetocentric distance (km)   50,020
Radial width (km)   1–2

Epsilon
Planetocentric distance (km)   51,140
Radial width (km)   20–100

Uranus has a system of narrow, faint rings. The ring particles are dark, and could consist of rocky or carbonaceous material.

Cordelia (1986U7)

Discovered	1986, Voyager, R. Terrile
Orbital period (hours)	8
(days)	0.335
Semi-major axis (km)	49,500
Diameter (km)	26
Orbital eccentricity	0.0
Orbital inclination (deg)	0.14
Albedo	0.07

Note: Cordelia is the inner shepherd for the Epsilon ring.

Ophelia (1986U8)

Discovered	1986, Voyager, R. Terrile
Orbital period (hours)	9
(days)	0.376
Semi-major axis (km)	53,760
Diameter (km)	32
Orbital eccentricity	0.010
Orbital inclination (deg)	0.09
Albedo	0.07

Note: Ophelia is the outer shepherd for the Epsilon ring.

Appendix 1: Planetary data

Bianca (1986U9)

Discovered	1986, Voyager
Orbital period (hours)	10.2
(days)	0.435
Semi-major axis (km)	59,160
Diameter (km)	44
Orbital eccentricity	0.001
Orbital inclination (deg)	0.16
Albedo	0.07

Cressida (1986U3)

Discovered	1986, Voyager, S. Synnott
Orbital period (hours)	11.1
(days)	0.464
Semi-major axis (km)	61,780
Diameter (km)	66
Orbital eccentricity	0.0
Orbital inclination (deg)	0.04
Albedo	0.07

Desdemona (1986U6)

Discovered	1986, Voyager, S. Synnott
Orbital period (hours)	11.4
(days)	0.474
Semi-major axis (km)	62,670
Diameter (km)	58
Orbital eccentricity	0.0
Orbital inclination (deg)	0.16
Albedo	0.07

Juliet (1986U2)

Discovered	1986, Voyager
Orbital period (hours)	11.8
(days)	0.493
Semi-major axis (km)	64,360
Diameter (km)	84
Orbital eccentricity	0.001
Orbital inclination (deg)	0.06
Albedo	0.07

Portia (1986U1)

Discovered	1986, Voyager
Orbital period (hours)	12.3
(days)	0.513
Semi-major axis (km)	66,100
Diameter (km)	110

Appendix 1: Planetary data

Orbital eccentricity 0.0
Orbital inclination (deg) 0.09
Albedo 0.07

Rosalind (1986U4)

Discovered 1986, Voyager, S. Synnott
Orbital period (hours) 13.4
(days) 0.558
Semi-major axis (km) 69,930
Diameter (km) 54
Orbital eccentricity 0.0
Orbital inclination (deg) 0.28
Albedo 0.07

Belinda (1986U5)

Discovered 1986, Voyager, S. Synnott
Orbital period (hours) 15
(days) 0.624
Semi-major axis (km) 75,250
Diameter (km) 68
Orbital eccentricity 0.0
Orbital inclination (deg) 0.03
Albedo 0.07

S/1986U10

Discovered 1986, Voyager, S. Synnott; 1999, E. Karkoschka
Orbital period (hours) 15.3
(days) 0.637
Semi-major axis (km) 76,000
Diameter (km) 40
Albedo —

Puck (1985U1)

Discovered 1985, Voyager
Orbital period (hours) 18.3
(days) 0.762
Rotational period Synchronous
Semi-major axis (km) 86,000
Diameter (km) 154
Orbital eccentricity 0.0
Orbital inclination (deg) 0.31
Albedo 0.07

Miranda

Discovered 1948, G.P. Kuiper
Orbital period (days) 1.414

Rotational period	Synchronous
Semi-major axis (km)	129,800
Diameter (km)	470
Orbital eccentricity	0.003
Orbital inclination (deg)	3.40
Mass (kg)	6.1×10^{19}
($\times$ Moon)	1.0×10^{-3}
Density (g/cm^3)	1.15
Gravity ($\times$ Earth)	0.009
Albedo	0.34

Ariel

Discovered	1851, W. Lassell
Orbital period (days)	2.520
Rotational period	Synchronous
Semi-major axis (km)	191,200
Diameter (km)	1,160
Orbital eccentricity	0.003
Orbital inclination (deg)	0.0
Mass (kg)	1.35×10^{21}
($\times$ Moon)	1.8×10^{-2}
Density (g/cm^3)	1.56
Gravity ($\times$ Earth)	0.021
Albedo	0.40

Umbriel

Discovered	1851, W. Lassell
Orbital period (days)	4.145
Rotational period	Synchronous
Semi-major axis (km)	266,000
Diameter (km)	1,170
Orbital eccentricity	0.005
Orbital inclination (deg)	0.0
Mass (kg)	1.17×10^{21}
($\times$ Moon)	1.7×10^{-2}
Density (g/cm^3)	1.52
Gravity ($\times$ Earth)	0.022
Albedo	0.19

Titania

Discovered	1787, W. Herschel
Orbital period (days)	8.705
Rotational period	Synchronous
Semi-major axis (km)	435,800
Diameter (km)	1,580
Orbital eccentricity	0.002
Orbital inclination (deg)	0.0

Appendix 1: Planetary data

Mass (kg)	3.53×10^{21}
($\times$ Moon)	4.7×10^{-2}
Density (g/cm^3)	1.70
Gravity ($\times$ Earth)	0.029
Albedo	0.28

Oberon

Discovered	1787, W. Herschel
Orbital period (days)	13.463
Rotational period	Synchronous
Semi-major axis (km)	582,800
Diameter (km)	1,520
Orbital eccentricity	0.001
Orbital inclination (deg)	0.0
Mass (kg)	3.01×10^{21}
($\times$ Moon)	4.0×10^{-2}
Density (g/cm^3)	1.64
Gravity ($\times$ Earth)	0.028
Albedo	0.24

Caliban (1997U1)

Discovered	1997, B.J. Gladman, P.D. Nicholson, J. Burns and J.J. Kavelaars
Orbital period (days)	579, retrograde
(years)	1.59
Semi-major axis (km)	7,170,000
Diameter (km)	60
Orbital eccentricity	0.082
Orbital inclination (deg)	140

Stephano (1999U2)

Discovered	1999, J.J. Kavelaars *et al*
Orbital period (days)	675, retrograde
(years)	1.85
Semi-major axis (km)	7,940,000
Diameter (km)	30
Orbital eccentricity	0.146
Orbital inclination (deg)	141.5

Sycorax (1997U2)

Discovered	1997, P.D. Nicholson, B.J. Gladman, J. Burns and J.J. Kavelaars
Orbital period (years)	3.5, retrograde
Semi-major axis (km)	12,214,000
Diameter (km)	120

Orbital eccentricity 0.51
Orbital inclination (deg) 153

Prospero (1999U3)

Discovered 1999, J.J. Kavelaars *et al*
Orbital period (years) 5.346, retrograde
Semi-major axis (km) 16,110.000
Diameter (km) 40
Orbital eccentricity 0.327
Orbital inclination (deg) 146.3

Setebos (1999U1)

Discovered 1999, J.J. Kavelaars *et al*
Orbital period (years) 6.42, retrograde
Semi-major axis (km) 18,200,000
Diameter (km) 40
Orbital eccentricity 0.494
Orbital inclination (deg) 148.8

NEPTUNE

Discovered 1846, J.C. Adams, U.J.J. Leverrier and J.G. Galle
Diameter (km) 49,550 (equatorial)
 (× Earth) 3.88
Mass (kg) 1.024×10^{26}
 (× Earth) 17.13
Volume (× Earth) 57
Mean density (g/cm^3) 1.64
Rotational period 16.110 hours
 (days) 0.67
Escape velocity (km/s) 23.7
Mean surface temperature (K) 48
Semi-major axis (AU) 30.058
Orbital eccentricity 0.008586
Orbital inclination (deg) 1.7692
Obliquity of axis (deg) 29.6
Gravity (× Earth) 1.19
Orbital period (years) 164.79
Synodic period (days) 367
Albedo 0.45
Atmospheric components 74% hydrogen, 25% helium, 1% methane (at depth)

Rings

1989N3R, Galle
Planetocentric distance (km) 41,900
Radial width (km) 15

Appendix 1: Planetary data

1989N2R, Leverrier
Planetocentric distance (km)   53,200
Radial width (km)   15

Lassell
Planetocentric distance (km)   55,400
Radial width (km)   –

Arago
Planetocentric distance (km)   57,600
Radial width (km)   –

1989N1R, Adams
Planetocentric distance (km)   62,930
Radial width (km)   < 50

The Adams ring contains concentrations referred to as 'arcs', and these have been individually named: Liberte ('leading'), Egalite ('equidistant'), Fraternite ('following') and Courage.

Naiad (1989N6)

Discovered	1989, Voyager
Orbital period (hours)	7.1
(days)	0.294
Semi-major axis (km)	48,2300
Diameter (km)	58
Orbital eccentricity	0.0
Orbital inclination (deg)	0.0
Albedo	0.06

Thalassa (1989N5)

Discovered	1989, Voyager, R. Terrile
Orbital period (hours)	7.5
(days)	0.311
Semi-major axis (km)	50,080
Diameter (km)	80
Orbital eccentricity	0.0
Orbital inclination (deg)	4.5
Albedo	0.06

Despina (1989N3)

Discovered	1989, Voyager, S. Synnott
Orbital period (hours)	8
(days)	0.335
Semi-major axis (km)	52,530
Diameter (km)	144
Orbital eccentricity	0.0
Orbital inclination (deg)	0.0
Albedo	0.06

Appendix 1: Planetary data 409

Galatea (1989N4)

Discovered	1989, Voyager, S. Synnott
Orbital period (hours)	10.3
(days)	0.429
Semi-major axis (km)	61,950
Diameter (km)	158
Orbital eccentricity	0.0
Orbital inclination (deg)	0.0
Albedo	0.054

Larissa (1989N2)

Discovered	1981, H Reitsema, *et al* 1989, Voyager
Orbital period (hours)	13.3
(days)	0.555
Rotational period	Synchronous
Semi-major axis (km)	73,560
Diameter (km)	208×178
Orbital eccentricity	0.001
Orbital inclination (deg)	0.0
Albedo	0.056

Proteus (1989N1)

Discovered	1989, Voyager, S. Synnott
Orbital period (days)	1.122
Rotational period	Synchronous
Semi-major axis (km)	117,640
Diameter (km)	416
Orbital eccentricity	0.0
Orbital inclination (deg)	0.0
Albedo	0.06

Triton

Discovered	1846, W. Lassell
Orbital period (days)	5.877, retrograde
Rotational period	Synchronous
Semi-major axis (km)	354,800
Diameter (km)	2,706
Orbital eccentricity	0.0
Orbital inclination (deg)	157
Mass (kg)	2.14×10^{22}
($\times$ Moon)	2.93×10^{-1}
Density (g/cm^3)	2.0
Gravity ($\times$ Earth)	0.077
Albedo	0.8

Nereid

Discovered	1949, G.P. Kuiper
Orbital period (days)	360.15
Semi-major axis (km)	5,514,000
Diameter (km)	340
Orbital eccentricity	0.6
Orbital inclination (deg)	29

PLUTO

Discovered	1930, C. Tombaugh
Diameter (km)	2,345
($\times$ Earth)	0.184
Mass (kg)	1.30×10^{22}
($\times$ Earth)	0.0025
Volume ($\times$ Earth)	0.007
Mean density (g/cm^3)	2.0
Rotational period	6.3872 days
Escape velocity (km/s)	1.2
Mean surface temperature (K)	37
Semi-major axis (AU)	39.45
Orbital eccentricity	0.248808
Orbital inclination (deg)	17.1417
Obliquity of axis (deg)	120
Orbital period (years)	247.68
Synodic period (days)	368
Albedo	0.54
Atmospheric components	Perhaps methane and nitrogen
Surface materials	Perhaps methane ice

Charon

Discovered	1978, J.W. Christy
Orbital period (days)	6.387, retrograde
Rotational period	Synchronous
Semi-major axis (km)	19,450
Diameter (km)	1,172
Orbital eccentricity	0.0076
Orbital inclination (deg)	96.16
Mass (kg)	1.6×10^{21}
($\times$ Earth)	5.5×10^{-4}
($\times$ Moon)	3.0×10^{-2}
Density (g/cm^3)	2.03
Albedo	0.34

Note: 1 Astronomical Unit (AU) is 149,597,870 kilometres.

Appendix 2: Space missions

Spacecraft	Launch date	Mission
Sputnik 1	4 Oct 1957	The first artificial satellite in Earth orbit.
Explorer 1	31 Jan 1958	America's first satellite, discovered the Van Allen radiation belts.
Pioneer 1	11 Oct 1958	NASA's first attempt to send a probe on a lunar fly-by, but it fell short.
Pioneer 2	8 Nov 1958	Attempted lunar fly-by frustrated by the failure of the third stage to ignite.
Pioneer 3	6 Dec 1958	Attempted lunar fly-by, but it fell short.
Luna 1	2 Jan 1959	The first Soviet attempt to hit the Moon resulted in a fly-by at a range of 6,000 kilometres.
Pioneer 4	3 Mar 1959	Lunar fly-by at a range of 60,000 kilometres.
Luna 2	12 Sep 1959	First probe to impact on the Moon, striking near the craters Aristillus and Autolycus in Oceanus Procellarum.
Luna 3	4 Oct 1959	The first probe to fly a circumlunar trajectory, to photograph the far side of the Moon.
Pioneer 5	11 Mar 1960	A probe placed into solar orbit to study the solar wind.
–	10 Oct 1960	The first Soviet attempt to dispatch a probe towards Mars was frustrated by a launch failure.
–	14 Oct 1960	A Soviet Mars probe lost during launch.
Sputnik 7	4 Feb 1961	The first Soviet attempt to dipatch a probe towards Venus was frustrated when it became stranded in parking orbit.
Venera 1	12 Feb 1961	Successfully dispatched towards Venus, but fell silent soon thereafter.
Ranger 1	23 Aug 1961	An engineering test flight for a NASA probe designed to send TV imagery of the Moon immediately prior to impact, but it became stranded in parking orbit.
Ranger 2	18 Nov 1961	Stranded in parking orbit.
Ranger 3	26 Jan 1962	An attempt to strike the Moon which turned into a fly-by at a range of 37,000 kilometres.
OSO 1	7 Mar 1962	A satellite placed into Earth orbit by NASA to study the Sun.

Appendix 2: Space missions

Ranger 4	23 Apr 1962	Although this was NASA's first lunar impact, contact with the spacecraft had already been lost.
Mariner 1	22 Jul 1962	NASA's first attempt to send a probe to Venus, but it was lost at launch.
Sputnik 19	25 Aug 1962	An attempt to send a probe to Venus, stranded in parking orbit.
Mariner 2	27 Aug 1962	A successful Venus remote-sensing fly-by at a range of 35,000 kilometres.
Sputnik 20	1 Sep 1962	A Venus probe stranded in parking orbit.
Sputnik 21	12 Sep 1962	A Venus probe stranded in parking orbit.
Ranger 5	18 Oct 1962	An attempt to reach the Moon that resulted in a 724-kilometre fly-by; no data.
Sputnik 22	24 Oct 1962	Mars probe stranded in parking orbit.
Mars 1	1 Nov 1962	Successfully dispatched towards Mars, but fell silent soon thereafter.
Sputnik 24	4 Nov 1962	Mars probe stranded in parking orbit.
Sputnik 25	4 Jan 1963	The first attempt to send a probe to soft-land on the Moon, but stranded in parking orbit.
Luna 4	2 Apr 1963	A lunar soft-lander that missed by 8,500 kilometres.
Cosmos 21	11 Nov 1963	A Venus probe stranded in parking orbit.
Ranger 6	30 Jan 1964	Lunar impact, but the TV system failed.
Cosmos 27	27 Mar 1964	A Venus probe stranded in parking orbit.
Zond 1	2 Apr 1964	A Soviet Venus fly-by, but contact was lost early on.
Ranger 7	28 Jul 1964	NASA's first successful TV imaging immediately prior to lunar impact of the Mare Nubium area.
Mariner 3	5 Nov 1964	NASA's first attempt to dispatch a probe towards Mars, frustrated by the aerodynamic shroud's failure to release.
Mariner 4	28 Nov 1964	The first successful Mars fly-by on 15 July 1965, returning a sequence of 22 images of the planet.
Zond 2	30 Nov 1964	An attempt at a Mars fly-by, but contact was lost early on.
OSO 2	3 Feb 1965	A solar observatory placed in Earth orbit.
Ranger 8	17 Feb 1965	A successful lunar impact mission, returning pictures across the Central Highlands and Mare Tranquillitatis.
Cosmos 60	12 Mar 1965	A lunar soft-lander stranded in parking orbit.
Ranger 9	21 Mar 1965	The last in the series of NASA lunar impact probe, which hit the large crater Alphonsus.
Luna 5	9 May 1965	A lunar soft-lander that crashed.
Luna 6	8 Jun 1965	A lunar soft-lander that missed the Moon by 161,000 kilometres.
Zond 3	18 Jul 1965	An engineering test flight for a Mars probe which was sent on a lunar fly-by to test its camera by photographing the far side of the Moon, but as it flew on into interplanetary space contact was lost.
Luna 7	4 Oct 1965	A lunar soft-lander that crashed.
Venera 2	12 Nov 1965	Contact was lost with this probe as it approached Venus.
Venera 3	16 Nov 1965	This probe fell silent shortly before penetrating Venus's atmosphere.
Cosmos 96	23 Nov 1965	A Venus probe stranded in parking orbit.
Luna 8	3 Dec 1965	A lunar soft-lander that crashed.

Appendix 2: Space missions 413

Pioneer 6	16 Dec 1965	A probe placed into solar orbit to study the solar wind.
Luna 9	31 Jan 1966	The first successful lunar soft-lander, at a site in Oceanus Procellarum.
Cosmos 111	1 Mar 1966	A lunar soft-lander stranded in parking orbit.
Luna 10	31 Mar 1966	The first probe to enter lunar orbit to conduct remote-sensing of the surface and nearby space.
Surveyor 1	30 May 1966	NASA's first lunar soft-lander.
Lunar Orbiter 1	10 Aug 1966	The first of a series of NASA spacecraft placed into lunar orbit to photograph potential Apollo landing sites.
Pioneer 7	17 Aug 1966	A probe placed into solar orbit to study the solar wind.
Luna 11	24 Aug 1966	A lunar orbital remote-sensing mission.
Surveyor 2	20 Sep 1966	A soft-lander that crashed in Sinus Medii.
Luna 12	22 Oct 1966	A lunar orbital remote-sensing mission.
Lunar Orbiter 2	6 Nov 1966	Apollo landing site reconnaissance.
Luna 13	21 Dec 1966	A lunar soft-lander.
Lunar Orbiter 3	4 Feb 1967	Apollo reconnaissance.
Surveyor 3	17 Apr 1967	Successful soft-landing in Oceanus Procellarum.
Lunar Orbiter 4	4 May 1967	Apollo reconnaissance.
Cosmos 159	17 May 1967	A Venus probe stranded in parking orbit.
Venera 4	12 Jun 1967	The first probe to report as it penetrated Venus's atmosphere on 18 October 1967.
Mariner 5	14 Jun 1967	A Venus remote-sensing fly-by at a range of 4,000 kilometres.
Cosmos 167	17 Jun 1967	A Venus probe stranded in parking orbit.
Surveyor 4	14 Jul 1967	A soft-lander that crashed in Sinus Medii.
Lunar Orbiter 5	1 Aug 1967	Apollo reconnaissance.
Surveyor 5	8 Sep 1967	Successful soft-landing in Mare Tranquillitatis.
OSO 4	18 Oct 1967	Solar observatory in Earth orbit.
Surveyor 6	7 Nov 1967	Successful soft-landing in Sinus Medii.
Pioneer 8	13 Dec 1967	In solar orbit.
Surveyor 7	7 Jan 1968	Soft landing on Tycho's ejecta blanket.
Zond 4	2 Mar 1968	Automated test flight of the spacecraft designed for a Soviet manned circumlunar mission, but deliberately directed away from the Moon.
Luna 14	7 Apr 1968	A lunar orbital remote-sensing mission.
Zond 5	14 Sep 1968	An automated test flight of the circumlunar spacecraft.
Pioneer 9	8 Nov 1968	In solar orbit.
Zond 6	10 Dec 1968	An automated test flight of the circumlunar spacecraft.
Apollo 8	21 Dec 1968	The first manned lunar orbital mission.
Venera 5	5 Jan 1969	Penetrated Venus's atmosphere on 16 May 1969.
Venera 6	10 Jan 1969	Penetrated Venus's atmosphere on 17 May 1969.
OSO 5	22 Jan 1969	Solar observatory in Earth orbit.
Mariner 6	24 Feb 1969	Mars fly-by on 31 July 1969.
Mariner 7	27 Mar 1969	Mars fly-by on 5 August 1969.
	27 Mar 1969	Soviet Mars probe lost at launch.
–	14 Apr 1969	Soviet Mars probe lost at launch.
Apollo 10	18 May 1969	Manned lunar orbital mission.
Luna 15	13 Jul 1969	An attempt to soft-land, retrieve a soil sample and return it to Earth, but the spacecraft crashed in Mare Crisium.

Appendix 2: Space missions

Mission	Date	Description
Apollo 11	16 July 1969	The first manned landing on the Moon, on 20 July 1969, on Mare Tranquillitatis.
Zond 7	8 Aug 1969	An automated test flight of the circumlunar spacecraft.
OSO 6	9 Aug 1969	Solar observatory in Earth orbit.
Apollo 12	14 Nov 1969	Manned lunar landing at the Surveyor site in Procellarum.
Apollo 13	11 Apr 1970	Aborted lunar landing mission.
Venera 7	17 Aug 1970	Reached Venus's surface on 15 December 1970.
Cosmos 359	22 Aug 1970	Venus lander stranded in parking orbit.
Luna 16	12 Sep 1970	The first successful Soviet lunar sample return mission.
Zond 8	20 Oct 1970	An automated test flight of the circumlunar spacecraft.
Luna 17	10 Nov 1970	Successful landing of a Lunokhod on the lunar surface, on western Procellarum.
Apollo 14	31 Jan 1971	Manned lunar landing in the Fra Mauro Formation.
Mariner 8	8 May 1971	A Mars fly-by probe lost at launch.
Cosmos 419	10 May 1971	A Mars soft-lander stranded in parking orbit.
Mars 2	19 May 1971	The spacecraft dropped off a direct-entry probe immediately before entering Mars orbit on 27 November 1971. No data was returned from the entry probe. The orbital mission ran for 3 months but little useful data was returned.
Mars 3	28 May 1971	The spacecraft dropped off a direct-entry probe immediately before entering Mars orbit on 2 December 1971. Although the probe survived the parachute descent, it failed after just 20 seconds and no useful data was returned.
Mariner 9	30 May 1971	On 14 November 1971 this became the first spacecraft to enter Mars orbit. In contast to the fly-by missions, it was able to map the entire planet.
Apollo 15	26 Jul 1971	Manned landing at Hadley-Apennine on the eastern rim of the Imbrium basin. A particles and fields subsatellite was released into lunar orbit on 4 August 1971.
Luna 18	2 Sep 1971	A sample-return spacecraft that crashed.
Luna 19	28 Sep 1971	A photographic reconnaissance mission in lunar orbit.
OSO 7	29 Sep 1971	Solar observatory in Earth orbit.
Luna 20	14 Feb 1972	Successful lunar sample-return.
Pioneer 10	3 Mar 1972	After passing through the asteroid belt, this prbe made a fly-by of Jupiter on 3 December 1973.
Venera 8	27 Mar 1972	Landed on Venus on 22 July 1972.
Cosmos 482	31 Mar 1972	A Venus lander stranded in parking orbit.
Apollo 16	16 Apr 1972	Manned landing at Descartes-Cayley in the lunar Central Highlands. A particles and fields subsatellite was released into lunar orbit on 19 April 1972.
Apollo 17	7 Dec 1972	The final Apollo landing in the Taurus-Littrow valley on the southeastern rim of the Serenitatis basin.
Luna 21	8 Jan 1973	Second Lunokhod near Lemonnier on Mare Serenitatis.
Pioneer 11	6 Apr 1973	This spacecraft used a fly-by of Jupiter on 4 December 1974 to make a Saturn fly-by on 1 September 1979.
Explorer 49	10 Jun 1973	Radioastronomy mission stationed in lunar orbit.
Mars 4	21 Jul 1973	A Mars orbiter which failed to make the insertion burn, and made a 2,000-kilometre fly-by on 10 February 1974. As

		it flew past it took some pictures, but they added little to those from Mariner 9.
Mars 5	25 Jul 1973	Entered orbit of Mars on 12 February 1974. It operated for 10 days, during which time it returned 60 pictures.
Mars 6	5 Aug 1973	The bus released a lander on 12 March 1974, but this fell silent just before it reached the surface. The bus made a 16,000-kilometre fly-by.
Mars 7	9 Aug 1973	The bus prematurely released its lander on 9 March 1974, so it missed the planet by 1,300 kilometres.
Mariner 10	3 Nov 1973	This spacecraft used a Venus fly-by on 5 February 1974 to enter a solar orbit that resulted in three Mercury fly-bys.
Luna 22	29 May 1974	A photographic reconnaissance mission from lunar orbit.
Luna 23	28 Oct 1974	An attempted sample return was frustrated when the arm of the coring drill was damaged preventing sample recovery.
Helios 1	10 Dec 1974	A German probe placed in solar orbit to study the solar wind.
Venera 9	8 Jun 1975	Shortly before entering orbit of Venus on 22 October 1975, the bus released a lander, which was successful.
Venera 10	14 Jun 1975	Shortly before entering orbit of Venus on 25 October 1975, the bus released a lander, which was successful.
OSO 8	21 Jun 1975	A solar observatory in Earth orbit.
Viking 1	20 Aug 1975	The main spacecraft entered orbit of Mars on 19 June 1976 and after an orbital survey of possible landing sites released its lander, which successfully touched down on the surface on 20 July 1976. The orbiter operated for 4 years, and the lander for 6 years.
Viking 2	9 Sep 1975	The spacecraft entered orbit of Mars on 7 August 1976 and the lander set down on 3 September 1976. The orbiter was switched off after 2 years and the lander failed after nearly 4 years.
Helios 2	15 Jan 1976	A second German-built interplanetary solar wind monitor.
Luna 24	9 Aug 1976	Lunar surface sample-return.
Voyager 2	20 Aug 1977	A marathon 'Grand Tour' of the outer Solar System which took in a Jupiter fly-by on 9 July 1979, a Saturn fly-by on 26 August 1981, a Uranus fly-by on 24 January 1986 and finally a Neptune fly-by on 25 August 1989.
Voyager 1	5 Sep 1977	A Jupiter fly-by on 5 March 1979 produced a slingshot for a Saturn fly-by on 12 November 1980.
IUE	26 Jan 1978	An ultraviolet astronomy satellite placed in Earth orbit. After many years of operation, it was finally shut down without a replacement.
Pioneer 12	20 May 1978	Pioneer Venus Orbiter entered orbit of Venus on 4 December 1978. It operated until 1992.
Pioneer 13	8 Aug 1978	The bus dispatched one large and three small probes which penetrated Venus's atmosphere, and then itself burned up in the atmosphere, on 9 December 1978.
ISEE 3	12 Aug 1978	After completing its primary mission of monitoring the solar wind in interplanetary space, it was diverted to pass through the tail of Comet Giacobini-Zinner on 11 September 1985.

Appendix 2: Space missions

Venera 11	9 Sep 1978	Landed on Venus on 25 December 1978.
Venera 12	14 Sep 1978	Landed on Venus on 21 December 1978.
SolarMax	14 Feb 1980	A solar observatory placed in Earth orbit. It soon suffered a fault, which was repaired by a Shuttle crew.
Venera 13	30 Oct 1981	Landed on Venus on 27 February 1982.
Venera 14	4 Nov 1981	Landed on Venus on 5 March 1982.
IRAS	26 Jan 1983	An infrared astronomy satellite placed in Earth orbit. It was operational only for a few months (until it ran out of coolant) but it provided astounding results.
Venera 15	2 Jun 1983	Entered orbit of Venus on 10 October 1983 and mapped the northern hemisphere by radar.
Venera 16	7 Jun 1983	Entered orbit of Venus on 14 October 1983 and mapped the northern hemisphere by radar (backing up its mate).
VeGa 1	15 Dec 1984	During a fly-by of Venus the spacecraft released a lander which set down on 11 June 1985 (releasing a package on a balloon during the descent) and then the spacecraft flew on to make a 8,900-kilometre rendezvous with Halley's Comet on 6 March 1986.
VeGa 2	21 Dec 1984	During a fly-by of Venus the spacecraft released a lander which set down on 15 June 1985 (releasing a package on a balloon during the descent) and then the spacecraft flew on to make a 8,000-kilometre rendezvous with Halley's Comet on 9 March 1986.
Sakigake	7 Jan 1985	A Japanese probe which made a distant fly-by of Halley's Comet on 8 March 1986.
Suisei	18 Aug 1985	A Japanese probe which made a 150,000-kilometre fly-by of Halley's Comet on 14 March 1986.
Giotto	2 Jul 1985	An ESA probe that made a close fly-by of Halley's Comet on 14 March 1986, passing withing 600 kilometres of its nucleus.
Phobos 1	7 Jul 1988	A Mars orbiter that was disabled en-route by the uploading of a faulty command sequence in September 1998.
Phobos 2	12 Jul 1988	After entering Mars orbit on 29 January 1989, contact was lost as the spacecraft manoeuvred for a close pass over the moon Phobos on 27 March 1989.
Magellan	5 May 1989	After entering orbit of Venus on 10 August 1990 it mapped virtually the entire planet by radar.
Galileo	18 Oct 1989	After a Venus fly-by on 10 February 1990, Earth fly-bys on 8 December 1990 and 8 December 1992, and fly-bys of the asteroids Gaspra and Ida on 29 October 1991 and 28 August 1993, respectively, the spacecraft entered Jovian orbit on 7 December 1995 and began a mission of exploration which ran into the new century. During its initial approach to that planet, it released a probe which penetrated the atmosphere on 7 December 1995.
Hiten	24 Jan 1990	A Japanese lunar fly-by which delivered the Hagormo probe which entered lunar orbit in March 1990.
Ulysses	6 Oct 1990	It used a Jupiter fly-by on 8 February 1992 to deflect it into a solar orbit steeply inclined to the ecliptic in order to

		perform 'polar passes' over the Sun.
Yokhoh	30 Aug 1991	A Japanese solar observatory in Earth orbit.
Mars Observer	25 Sep 1992	Contact was lost on 25 August 1993 as it prepared to enter Mars orbit.
Clementine	25 Jan 1994	After entering lunar orbit on 19 February 1994 it mapped the surface using multispectral sensors, and then left orbit so as to head for a rendezvous with asteroid Geographos, but was crippled by a computer error soon thereafter that exhausted its propellant supply.
ISO	17 Nov 1995	An infrared space observatory in Earth orbit.
SOHO	2 Dec 1995	A solar observatory stationed at the Earth's inner Lagrange point.
NEAR	17 Feb 1996	After a fly-by of asteroid Mathilde on 27 June 1997, it made a rendezvous with asteroid Eros on 10 January 1999, but an error in the breaking sequence meant that the rendezvous had to be delayed a year. A series of manoeuvres enabed it to settle on the surface of the asteroid.
MGS	7 Nov 1996	After entering initial orbit of Mars using aerobraking on 11 September 1997.
Mars 8	16 Nov 1996	Lost when the escape stage failed.
MPF	2 Dec 1996	Landed on Mars on 4 July 1997 by direct-entry and deployed the Sojourner rover. It fell silent after 84 days.
Cassini	15 Oct 1997	After gravity-assists in the inner Solar System and a Jupiter fly-by on 30 December 2000, it will arrive in the Saturnian system in June 2004. It will later release the Huygens probe to land on Titan.
Prospector	6 Jan 1998	It entered lunar orbit on 11 January 1998 and conducted a remote-sensing survey.
Nozomi	4 Jul 1998	After a malfunction which impaired its departure manoeuvre on 20 December 1998 (denying it the planned trajectory that would have seen it reach Mars in October 1999) it pursued a 'long' route to Mars involving two Earth fly-bys to boost its aphelion for arrival in December 2003. Its objective at Mars is to study the manner in which the solar wind interacts with the planet's upper atmosphere.
DS 1	24 Oct 1998	An ion-drive technology demonstrator which made a fly-by of asteroid Braille and then set off for Comet Borelly.
MCO	11 Dec 1998	Lost during initial aerobraking on 23 September 1999.
MPL	3 Jan 1999	Lost attempting to land on 3 December 1999.
Mars Odyssey	7 Apr 2001	To map the composition of the Martian surface by remote-sensing.
Mars Express	– – 2003	To drop off the Beagle probe for a direct-entry descent just before itself entering Mars orbit.

Glossary

aa – a Hawaiian term for 'a surface on which one cannot walk barefoot' type of basaltic lava that forms a very rough, blocky, jagged, and clinkery surface.

abyssal plain – the essentially flat (slopes of less than 1 degree) part of the ocean floor which formed as sediments deposited by turbidity currents from the shallow continental margins masked the underlying rough rocky crust. The deposits are thickest nearest the continents and thinnest near the mid-ocean ridge system, which are free of sediment.

accretion – the assembly of a body by the incorporation by impact of a large number smaller bodies. This is believed to be how the planets formed out of the Solar Nebula.

acetylene – C_2H_2, present in hydrocarbon hazes.

acid a substance that liberates hydrogen ions in solution. An acid reacts with a base to make a salt and water (only). A chemical with acid properties (opposite to an alkali; a base) is said to be acidic.

adiabatically a hot gas or plasma movement in which there is insignificant loss of heat to the environment.

age, absolute a tricky concept. Depending upon the context, it can mean the interval elapsed since a geological rock unit was formed; the time since a specific rock solidified from a molten state; the time that a rock has been exposed to cosmic rays, etc.

albedo – the reflectivity (expressed as a percentage) of a material, with a dark material having a low value. The darkest Solar System objects are the asteroids that reflect only 1 to 3 percent; their carbonaceous surfaces are as dark as coal.

algae a diverse group of plants ranging from single-celled prokaryotic cyanobacteria to giant kelp that is tens of metres in size.

aggregate a rock that is a consolidation of fragments of other rocks.

alkali – a soluble hydroxide of a metal. A substance with alkali properties (as distinct to those of an acid) is said to be alkaline. The light univalent metals (lithium, sodium and potassium) are called alkali metals. The bivalent metals (beryllium, magnesium and calcium) are referred to as 'alkali earth' metals.

alkali feldspar – sodic plagioclase and orthoclase.

alluvial fan – an outspread gently-sloping mass of detritus deposited by running water where there is an abrupt change in gradient from steep to gentle.

alumina – aluminium dioxide (Al_2O_3).

Amazonian – Mars's most recent geological era, since the Hesperian.

ammonite – an extinct member of the starfish family whose fossilised remains served as the basis for a palaeontological system for dating fossiliferous rock strata.

amorphous having no internal structural order; that is, glassy, not crystalline.

amphibole – a group of ferromagnesian silicate minerals which form an important constituent of metamorphic rocks (amphibolite) but is also found in intermediate and mafic igneous rocks and some felsic plutonic rocks. It is commonly found in granitoids, gneisses and andesites.

amphibolite – a crystalline metamorphic rock containing primarily amphibole and plagioclase feldspar, with little or no quartz.

andesite – a very fine-grained igneous rock of intermediate composition (between felsic and mafic) made largely of plagioclase feldspar, some quartz and pyroxene as the mafic minority constituent. It is the extrusive equivalent of diorite.

angle of repose – the maximum angle at which loose material remains stable on a slope.

anorthosite – a rock comprising more than 90 percent calcic plagioclase feldspar. It is a slow-cooled plutonic rock with a conspicuous crystalline structure (twinning). Pyroxene and olivine are the mafic minority constituents.

ANT – an acronym for Anorthosite, Norite and Troctolite which together form the majority of the endogenic igneous rock types in the ancient lunar crust. Anothosite is by far the predominant member of the group. Anorthosite is extremely pure plagioclase; norite is plagioclase enriched by pyroxene; troctolite is plagioclase enriched by olivine.

antecedent flow – a water course which was established prior to the onset of local uplift, then proceeded to erode a channel at the same rate the land rose, thereby maintaining its course. The Grand Canyon was created by the Colorado River in this manner.

anticline – a downward facing concave fold such that stratigraphically older rocks are nearest the centre of curvature; in contrast to a syncline.

aphanitic a hard rock possessing a homogeneous ground-mass with a texture that is so fine that its crystals are not resolvable to the naked eye; very fine grained.

apoapsis – the farthest point that an object in orbit can be from its primary; the term aphelion is used in the case of the Sun and apogee in the case of the Earth.

aquifer a body of rock that contains filled pore spaces that are sufficiently interconnected to enable water to flow through the rock and collect to form substantial subterranean reservoirs.

archea – literally 'the old ones', thermophyllic single-celled organisms that drew their energy from chemical reactions near hydrothermal vents. An early branch on the 'tree of life'.

Archean – the interval of geological time since the Hadean 3.8 billion years ago to 2.8 billion years ago, after which there was a several-hundred-million-year transition to the Proterozoic.

argillite – a rock derived from siltstone, mudstone or shale that has been consolidated by the heat of 'weak' metamorphism and hence does not break into thin layers in the manner of shale and slate.

artesian flow – when the hydrostatic pressure on an aquifer is sufficient to cause water to rise above the local water table, and perhaps to broach the surface and pool in low-lying areas.

ash – tiny fragments of lava sprayed out of a volcanic vent. Because it has a high surface area to mass ratio, ash cools rapidly, falls as dust, and settles as a fine-grained blanket. On a slope, ash will flow like a fluid.

asthenosphere – a zone of ductile deformation about 200 kilometres below the surface of the Earth in which the temperatures and pressures are sufficient for the dense ultramafic mantle to melt. Its lubricating effect facilitates the motion of the lithospheric plates above.

astrobleme – the surface expression of an eroded impact crater. Even if the cavity and rim are no longer present and the circular form has been tectonically distorted, the impact origin can be inferred from the shocked nature of the rock and the presence of shatter cones.

Astronomical Unit (AU) – defined as the average radius of the Earth's orbit around the Sun; namely 149,599,000 kilometres.

augite – a pyroxene mineral found in mafic igneous rocks such as basalt/gabbro.

aulacogen – the continental arm of a triple junction that did not induce lithospheric spreading, in contrast to the two offshore arms. The East African Rift is such a 'failed' rift.

back-arc – the zone immediately 'behind' a trench where one oceanic plate is being subducted by another. A trench is arcuate because lithospheric plates are parts of a spherical surface. The descending plate stimulates magmatism in the overriding plate.

back-arc basin – an igneously induced back-arc depression (basin) in an oceanic lithospheric plate. The Aegean Sea, for example, marks a back-arc basin behind the Hellenic Trench.

back-arc spreading – the rifting and resultant spreading of the lithosphere in the back-arc zone behind a trench. It is stimulated both by the lithospheric thinning as a hot diapir rises above the subducting plate and by the subduction trench 'pulling' the island arc.

bacteria – single-celled prokaryotic microorganisms which seem to form the base of the 'tree of life'.

banded iron formation – a chemical sedimentary rock, containing a significant fraction of iron of sedimentary origin, and showing marked banding of alternating layers of iron-rich minerals and chert or quartz.

basalt – a dark aphanitic porphyritic mafic igneous rock made mainly of calcic plagioclase and clinopyroxene, and sometimes with olivine. It is the extrusive equivalent of gabbro.

base – a substance that reacts with acid to make a salt. A substance with this property is said to be basic.

'basement' – older, usually crystalline igneous and/or metamorphic rock underlying younger stratified rock formations.

basin and range – a type of regional block faulting where the valleys (basins) and mountains (ranges) are elongate, parallel to one another, and alternating. The basins often are partially or completely filled with sedimentary debris derived from the erosion of the immediately adjacent ranges.

basin, impact – a structure left by the impact of an asteroidal-sized body. Most impact basins formed during the period of bombardment which drew to a close some 3.8 billion years ago.

basin, sedimentary – a geosyncline formed by the lithospheric stretching that occurs when a continental plate overrides an oceanic plate and the trench is sufficiently close offshore to 'tug' upon the continental plate, inducing tensile stress 'behind' the Andean-style coastal range.

batholith – old magma chamber complex that was once within an Andean-style coastal range and has been exposed by extensive erosion. The exposure forms a large area of igneous rock, usually granite.

bedrock – the intact layer of rock immediately below the loosely consolidated surface. It is a regional geological unit in the overall stratigraphic structure.

bench – an inflection on a slope such as a mountain flank or a crater wall forming a relatively flat surface.

Binary Accretion Model – a theory that accepts that the Earth and Moon formed together, by accretion, but claims that they did so independently.

biotite – black to brownish-black mica.

black body – a hypothetical object that perfectly absorbs incident radiation, without reflecting any of it, and radiates it with to a well-defined spectral distribution that can be characterised by a given temperature.

block faulting – a type of normal faulting in which converging faults cause blocks of crust to drop relative to their neighbours. In isolation, it produces a graben. If the faults are nested, the result is a series of grabens within grabens which create a step-like rift valley. If the faults are parallel and shallow dipping, the result is a 'basin and range' topography.

blue-green algae – cyanobacteria.

bolide – an asteroid, meteor or comet that has collided with a planet.

bombardment – the post-accretional period during which most of the remaining large objects were swept up by the newly formed planets. In the inner Solar System, this period concluded about 3.8 billion years ago.

boulder – in terms of field geology, a 'boulder' is defined to be a rock which has at least one dimension in excess of a metre in size.

breccia – a mechanically assembled rock. A fragmental breccia comprises angular fragments of many other shattered rocks. An impact-melt breccia is a fragmental breccia which has been welded by impact melt. A granulitic breccia is a melt that has recrystallised by a high pressure and temperature, and now has a granulitic texture. A regolith breccia is a clod of soil made by the shock of an impact compressing the regolith. A 'one-rock' breccia is a matrix in which the clasts are fragments of homogeneous material. If a breccia is subsequently shattered and then bound into another matrix, the result is 'two-rock'. If this is smashed and bound into another breccia, it is 'three-rock'. In each generation, the consolidating matrix may be different and the clast mix may be varied, so complex rocks can result. On the Moon, most breccias formed by impact processes. On Earth, most breccias are sedimentary in form, but they can also be made by igneous activity (volcanic breccia). A breccia differs from a conglomerate by its fragments having very sharp edges and unworn corners, whereas the components of a conglomerate are rounded.

brine – extremely salty water.

calc – denoting the presence of calcium carbonate.

calc-alkaline rock the basalt andesite–dacite–rhyolite suite of igneous rocks that is typically formed at active continental margins of destructive plate boundaries and island arcs overriding subducting lithospheric plates.

calc-silicate rock – a crystalline metamorphic rock formed mainly of calcium-bearing silicate minerals.

calcareous – any rock that has sufficient carbonate material for it to react with a strong acid to produce bubbles of carbon dioxide. Usually, the carbonate material is calcite or dolomite.

calcareous algae – algae that secrete a calcareous skeleton. Although the blue-green algae of the Precambrian became encrusted with calcium carbonate (forming stomatolites) they did not necessarily secrete a calcareous skeleton. Today, calcareous algae are important members of coral reefs.

calcic – adjective for a mineral that is rich in calcium.

calcite – the major mineral form of calcium carbonate. It is a common rock-forming mineral and is the main constituent of limestone and marble.

calcium carbonate – $CaCO_3$.

calcium magnesium carbonate – $CaMg(CO_3)_2$.

caldera – a volcanic crater formed when the magma chamber drains and its roof collapses. On Earth, calderas were first recognised on the summits of conical volcanoes. It was once thought that many of the large lunar craters were 'open' calderas, but this turned out not to be the case. There are 'open' calderas (called paterae) on Venus, Mars and on Io, however, and it was realised that there are such structures on the Earth, in the form of resurgent calderas.

Cambrian – the most recent 600 million years or so of geological time (prior to which was the Precambrian).

Capture Model – a theory that proposed that the Earth and Moon were formed independently, in different parts in the Solar System, and that the Moon was 'captured' by an encounter with the Earth.

carbonaceous material – primitive icy material that is rich in carbon compounds left over from the solar nebula. If the small grains of carbon are distributed evenly throughout a rock or ice, it dramatically reduces its albedo (indeed, many carbonaceous objects are jet black).

carbonate – any mineral or rock that includes the CO_3 ion.

carbonatite – a magma that contains more than 50 percent carbonate minerals such as calcite or dolomite.
carbonic acid – an acid of carbon dioxide dissolved in water.
central peak – a mountain complex in the centre of a crater which formed when the material at the point of impact rebounded immediately prior to solidifying. As a result, the peak is derived from material drawn from far below the local bedrock.
chemical weathering – the process by which chemical reactions transform rocks and minerals into new chemical combinations. Chemical weathering of feldspar cystals, for example, forms clays.
chert – an extremely fine-grained siliceous sediment formed from interlocking quartz crystals.
chlorite – a dark-green mica mineral, usually formed during 'low grade' metamorphism.
cinder cone – a conical mound of viscous lava or pyroclastic ash that builds up as an annular ring around a volcanic vent.
circular maria – the mare lavas which are confined within impact basins (as opposed to lavas which flooded terrain outwith the basins).
clast – an angular fragment of pre-existing rock caught up in a breccia matrix. Clasts can be rock (lithic) or individual minerals (non-lithic).
clastic – as applied to a sedimentary rock it means composed principally of clasts transported some distance from their parent rocks; in contrast to sediments accumulated up from 'fines' or chemical precipitates.
clathrates – unstable crystalline structures of water ice and a significant proportion of frozen volatiles such as carbon dioxide (dry ice).
clay – an aqueous non-clastic sediment (mud) formed from very small detrital particles whose alumino-silicate composition is determined by the mineral composition of the weathered parent rocks. When the water evaporates, the fines in suspension settle out.
cleavage – the tendency to break along planes of weakness.
clinopyroxene – a form of pyroxene with calcium; as distinct from orthopyroxene.
coesite – an extremely dense type of quartz. It is one of several types of 'shocked quartz' that are formed in impacts.
Collisional Ejection Model – the theory which proposes that the Moon formed from material ejected from the Earth when a Mars-sized body struck the newly-formed Earth; once in orbit, the ejecta rapidly accreted to form the Moon. This theory has the advantage of explaining why the Moon's density is the same as the terrestrial crust.
conglomerate – a coarse-grained clastic sedimentary rock in which relatively large fragments (pebbles, cobbles or boulders) are cemented together by a fine-grained matrix of sand or silt; it is similar to a breccia but with rounded rather than angular fragments.
continental drift – a theory developed to explain the 'fit' between the coastlines of Africa and America which argued that the continents had split and were in the process of drifting apart. It was not until the discovery of sea-floor spreading (and later plate tectonics) that it was realised that this 'crazy idea' held a grain of truth.

continental shelf – the part of the continental margin which runs through a very gentle incline of about 0.1 degree from the shoreline to the start of the continental slope, typically at a depth of several hundred metres. The majority of marine life resides on the shelf. A drop in sea level sufficient to expose the shelf has a catastrophic effect on marine life.

continental shield – see craton.

continental slope the part of the continental margin which descends though a relatively steep 5 degree incline from the continental shelf to the abyssal plain that forms the deep ocean floor.

convection – mass movement of material from one place (usually hotter) to another (usually colder). If cyclic, the flow forms patterns called 'convection cells'.

core the central region of a thermally differentiated planetary body in which the dense iron-rich minerals have sunk to form a metallic nucleus.

cosmic ray – the rain of charged particles pervading space. The solar wind contributes ions of the lighter elements, but there is a flux of relativistic nuclei of heavier elements from beyond the Solar System.

'country rock' – the rock present in any area prior to some other event (such as faulting or an igneous intrusion) affecting it.

crater, impact – a structure made by an impact. A primary is produced by the impact of space debris travelling at 'cosmic speed' (typically 30 kilometres per second). A secondary is caused by the fall of debris ejected from a primary on low energy ballistic trajectories. There is a well-defined relationship between frequencies and sizes running all the way from the smallest pit up to the largest basin. A significant proportion craters up to 25 kilometres across are secondaries from basin formation.

crater curve – a plot of crater size versus frequency.

craton – a part of the Earth's crust which has attained stability and has seen little deformation in recent geological time. It usually refers to the oldest parts of a continent. In effect, craton is synonymous with the term 'continental shield'.

crust – the outer layer of a thermally differentiated planetary body. In the inner Solar System the crust is an alumino-silicate crystalline shell, but ices are present in the frozen outer regions. In the case of the Earth, this outermost layer consists of oceanic crust (8 to 16 kilometres thick and basaltic) and continental crust (80 to 120 kilometres thick and composed of a low-density silicate-rich material).

cryoclastic – a debris flow sustained by the energy of a devolatising ice.

cryovolcanism – volcanic processes involving cryogenic (aqueous) lava.

crystal – as a silicate-rich lava cools, its minerals precipitate out of solution to create a homogeneous rock. The atoms link to form a highly regular lattice. The shape of this pattern depends on the elements involved. Only certain elements can fit into specific shapes, and these are called 'compatibles'; all others are called 'incompatibles'. There can be elements in common, but as soon as a crystal starts to assume a particular form the incompatibles are excluded. Since different crystals form at different temperatures, the magma chemistry evolves; it becomes progressively enriched by a succession of different incompatibles. If gas is released into the solution and becomes trapped, the resulting rock will have cavities. If these are spheres, they are called 'vesicles'; if they are irregularly shaped, they are called 'vugs'. If the rock cools rapidly, these will have smooth walls, but if it cools slowly, the metals in the gas may condense out and form crystals on the wall of the cavity, partially filling it, transforming it into a vug. If the molten magma crystallises too rapidly to permit a lattice to form, the result is a homogeneous mass of glass.

cumulate – an igneous rock formed by concentration of primary minerals, such as may occur by stratification in a slowly cooling magma chamber.

Curie point – the temperature (different for different materials) above which the spontaneous magnetisation transforms from ferromagnetic to paramagnetic; in effect, it loses its magnetism. As the material cools through this temperature it 'locks in' the ambient field upon crystallising.

cyanobacteria – a primitive single-celled prokaryotic microorganism. It is rather like a bacteria that has developed the ability to use chlorophyll for photosynthesis. It predates the 'Cambrian explosion' in biodiversity. In massive colonies, it produced stromatolites. Its descendents still lurk in hot springs, at the edges of saline seas and in other environments that are 'inhospitable' to macroscopic organisms today. Although often called 'blue-green algae', this simply reflects the fact that the first examples to be studied were that colour.

dacite – a fine-grained volcanic rock; the extrusive equivalent of granodiorite.

decompressional melting – the melting of already-hot rock by the release of pressure without an increase in temperature, such as happens when a major impact suddenly excavates a crustal cavity and allows the solid upper mantle immediately beneath to melt and (as a consequence of the resulting expansion) to well up towards the surface.

delta – the nearly flat alluvial tract at the mouth of a river that commonly forms a triangular or fan-shaped plain.

diabase – an intrusive igneous rock in which the interstices between the plagioclase feldspar crystals are filled with grains of pyroxene. Its texture is intermediate between a coarse-grained gabbro and a fine-grained basalt. It is also known as microgabbro or dolerite. As an intrusive basalt-like igneous rock it commonly forms dykes and sills.

diamagnetic – having a magnetic permeability less than unity. In a magnetic field, the induced magnetisation of a diamagnetic material is in a direction opposite to that of the applied field.

diapiric – the rise of a buoyant mass of rock through a denser rock by solid-state flow. In the case of an igneous intrusion, the process is driven by the natural buoyancy of a hot (and hence less dense) rock. Note that diapiric action can also occur in a non-igneous environment, such as when pressure forces a loose material like salt up through a point of weakness in an overlaying strata to form a dome.

differentiation – gravity-driven separation of minerals.

diorite – the intrusive form of andesite.

dip – the amount and direction of tilt of a geologic surface from horizontal, measured in degrees. The geologic surface usually is a layered rock unit or a fault plane.

distal – igneous or sedimentary deposits composed of fine particles deposited far from their sources; contrast with proximal.

dolerite – see diabase.

dolomite – a mineral comprising calcium magnesium carbonate.

dolostone – the rock form of the mineral dolomite, usually including some impurities such as clay. When metamorphosed it forms dolomitic marble.

dome – an uplifted, anticlinal or constructed feature that may be either circular or elliptical in outline. In terms of volcanic activity it is a small mound somewhat similar to a cone, but with an irregular slope.

down-dropped – said of that side of a fault that appears to have moved downward compared with the other side; opposite of 'up-thrusted'.

drift – all material of glacial origin irrespective of whether deposited directly from melting ice (till) or by meltwater streams (outwash).

dunite – an igneous rock composed almost entirely of olivine. It has no extrusive equivalent.

dyke – a relatively thin sheet-like igneous intrusion that transects the structure of the country rocks; in contrast to a sill.

ecliptic – by definition, the plane in which the Earth orbits the Sun.

effusive eruption – the smooth flow of volatile-free lava from a volcanic vent.

ejecta – the debris thrown out during the excavation of an impact crater. Where it piles up in the immediate environ of a crater, it is called the ejecta blanket.

eolian – erosional and depositional processes accomplished by the wind.

epidote – a yellowish-green mineral formed by low-grade metamorphism.

ethane – C_2H_6.

ethylene – C_2H_4.

eukaryote – a single-celled microorganism whose cell contains a distinct nucleus; in contrast to a prokaryote. All multicellular life forms are eukaryotic. Eukaryotic cells are far larger than prokaryotic cells, and it has been speculated that the specialised components of complex cells may have developed from prokaryotic cells which formed the symbiotic relationships required to sustain more complex cells.

evaporite – a non-clastic sedimentary rock composed primarily of minerals produced from a saline solution as a result of extensive or total evaporation.

extrusion – lava which has erupted from a volcanic vent onto the surface.

facia – a distinctive surface texture that can be exploited to delineate the stratigraphic extent of rock unit on a geological map.

fault – a fracture along which there has been differential (horizontal and/or vertical) motion of the opposite sides.

fault scarp – a cliff or escarpment formed by vertical movement along a fault that reaches the surface.

feldspar – 'field crystal' in German. The feldspars are alumino-silicates in combination with a metal such as sodium, potassium or calcium. They comprise the most common type of rock-forming minerals. See orthoclase feldspar and plagioclase feldspar.

feldspathic rock – a coarsely-grained rock with a light-toned granular mass of feldspar laths.
felsic – an igneous rock composed primarily of light-coloured feldspar minerals (orthoclase, plus some sodic plagioclase) usually accompanied by quartz and some biotite.
ferromagnesian minerals – a magnesium-iron-rich mineral such as augite.
fire fountain – the spray of fine droplets of molten rock which are explosively vented when a volatile-rich lava nears the surface and rapidly expands.
fissure – an extensive crack, break, or fracture in rocks.
Fission Model – the model whereby the Earth and the Moon were once a single body. As it condensed from the solar nebula, conservation of angular required the body to rotate faster, this led first to the formation of an equatorial bulge and then, at a critical point, to that bulge splitting off to form a satellite.
flood basalt eruption – a fissure eruption in which extremely low viscosity lava is extruded at a high rate for a sustained time and effectively transforms the pre-existing terrain into a stack of lava flows which build up a plateau. The largest known terrestrial flood basalt is the Siberian Traps. The lunar maria are such extrusions.
fluidised ejecta – ejecta from an impact that behaves like a fluid when it 'splashes' back onto the surface, either because it has been extensively melted or because it is so finely fragmented that it can flow.
fluvial – pertaining to streams and river deposits.
flux – the amount of a quanity (such as energy or the number of meteors) passing through a unit of volume in a unit of time.
formaldehyde – CH_2O.
fossilisation – the process by which inorganic minerals replace organics in once-living things. Although the chemistry is altered the morphology is preserved.
Fra Mauro Formation – a characteristically hummocky terrain peripheral to the lunar Imbrium basin which is a blanket of crustal rock displaced by that impact. Its equivalent for Mercury's Caloris basin is the Van Eyck Formation.
fungi – plants which lack chlorophyll and hence cannot photosynthesise; instead they live on organic matter.

gabbro – plutonic rocks made of plagioclase and clinopyroxene that crystallised sufficiently slowly at depth to create coarse-grained (granular) crystals. It is the intrusive form of basalt.
geological time – a series of time slices which together span the time since the Earth formed, most coarsely divided into the Precambrian and the Cambrian.
geological unit – a stratigraphically distinct body of rock with a single mode of formation and time of deposition.
geosyncline – similar to a synclinal fold but on a far larger scale.
glacier – a body of moving ice, usually at least 30 metres thick so that the ice crystals on the bottom deform to effect solid-state movement.
glacial lobe – a large tongue-like protrusion from the margin of an ice sheet.
glass – if a high-viscosity melt solidifies into a state with no internal order, the result

is glass rather than a crystal. It can also be formed by a pressure-pulse in the immediate vicinity of an impact.

gneiss – a rock formed by regional high grade metamorphism of a granitic mass.

graben – a linear trench formed when a strip of crust down-drops between a pair of parallel normal faults in response to extensional forces.

graded bed – a type of bedding in which each layer displays a gradual change in particle size, usually from coarse at the base to fine at the top.

grain – the individual crystals in a rock, the texture of which ranges from granular through coarse to fine.

granite – a coarse-grained felsic igneous rock composed of quartz and alkali feldspars with some biotite; it is the intrusive form of rhyolite. In effect, it is a silicic gabbro.

granitoid – a term to describe both intermediate and felsic crystalline rocks (that is, non-mafic rocks).

granodiorite – the intrusive form of rhyodacite; similar to granite except that granite contains more alkali feldspar and granodiorite contains a higher proportion of plagioclase feldspar and mafic minerals.

granulite – a granular rock that recrystallised in its solid state (that is, without remelting) as a result of high-grade metamorphism.

gravity-assist – a trajectory to enable a spacecraft to utilise a close fly-by of a planet to either increase or decrease its heliocentric velocity so as to adopt a trajectory that it would otherwise not have been able to pursue; commonly referred to as a 'slingshot'.

greenschist – a medium-grade metamorphic rock. Its green colour is due to the presence of chlorite.

greenstone – an aphanitic igneous rock such as basalt which has been subjected to low-grade metamorphism. Its greenish-black colour is due to the presence of chlorite and epidote.

greywacke – a quartz sandstone that includes significant amounts of mud and/or mica. It is often called a 'dirty sandstone'. Its graded bedding shows that it was deposited in deep water by turbidity currents.

ground water – subsurface water lying below the level of the local water table; often referred to as phreatic water.

group – two or more formations in a stratigraphic column which formed by similar events or processes; see supergroup.

Hadean – the span of geological time since the Earth's formation (some 4.5 billion years ago) to the tail-end of the bombardment 3.8 billion years ago.

headward erosion – the lengthening of a young valley or gully by water erosion (sapping) at the head of its valley.

heliopause – the limit of the Sun's dominance over interstellar space.

hematite – a reddish mineral of iron oxide; Fe_2O_3. Hematite is a common accessory mineral in lava; it is common in hydrothermal veins but rare in plutonic rock. Small amounts of it create red, pink, and reddish-grey hues in rocks, especially sedimentary rocks.

Hesperian age – the Martian age between the Noachian and the Amazonian as derived from a study of cratering record. Unfortunately, different researchers place the transitions at different times.

Hohmann transfer orbit – an elliptical transfer from one orbit to another which requires the minimum of energy.

hydrocarbon haze – a thin cloud layer that forms in the stratospheres of the gas giants and on Titan (Saturn's primary satellite, and the only moon to possess a dense atmosphere). It forms by polymerisation of the carbon that is released when methane is dissociated by the ultraviolet component of insolation.

hydrated minerals – minerals involving hydroxyl (OH) ions.

hydrogen sulphide H_2S.

igneous – relating to material formed within a planetary body from a liquid state. Note that in the inner Solar System, igneous activity involves molten silicate, but on the frozen satellites of the outer planets it includes aqueous solutions.

ignimbrite – a massive blanket of volcanic ash, volcanic breccias, and welded ash-flow tuff which has consolidated and lithified.

immature sediment – a clastic sediment comprising poorly weathered detrital rock with lithic fragments as well as abundant feldspars; as opposed to mature.

impact basin – an impact by an asteroidal-sized body which not only excavated a large cavity but also produced a multiple-ringed structure by shock effects. Such impacts were common in the later phase of accretion, and during the bombardment that tailed off about 3.8 billion years ago.

impact melt – that part of the target rock that is melted as the shockwave of the impact passes through it. It is the zone between the rock which is vaporised and the underlying rock which is merely fractured. Although the melt coats the cavity of the crater, in the case of a large impact the melt can pool and leave a surface similar to a lava flow. It was once thought that the lunar maria were vast sheets of impact melt, but this is not the case.

inclination, orbital – the angle at which a planet's orbital plane is offset from the ecliptic.

insolation – a measure of the energy in the light from the Sun (Sol).

intermediate – an igneous rock with a composition is transitional between felsic and mafic.

intermediate-silica rock – a quartz fraction of 50 to 60 percent corresponds to diorite/andesite.

intrusion – subterranean magma that penetrates already-existing rocks to form plutons, dykes or sills. Intrusive activity can also occur in the absence of melting, if solid-state flow prompts diapiric mobilisation.

iridium – a rare earth element that is common in primitive meteorites.

island arc – an arcuate chain of volcanic islands rising from the ocean floor due to magmatism prompted by the subduction of an oceanic lithospheric plate.

isostasy – the forces which govern the level at which materials of different densities settle in the crust of a planetary body.

isostatic rebound – the adjustment of the lithosphere to maintain equilibrium among

Glossary 431

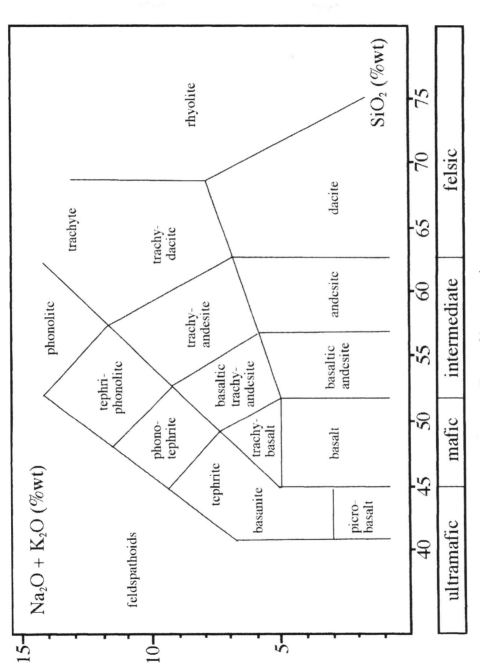

Types of igneous rock.

units of varying mass and density. Excess mass above is balanced by a deficit of density below, and vice versa. Mass added onto the surface depresses the lithosphere slowly, and mass removed permits the lithosphere to rise slowly (although often not to the same elevation it had before it was depressed). Scandanavia, for example, is slowly rising as it adjusts to the withdrawal of the ice sheet at the end of the last Ice Age.

karst – a characteristic landscape that is produced by the subsurface dissolution of the soluble fraction of a permeable rock, usually limestone. Valleys in karst often end at a sink hole where the eroding flow of water disappeared underground.

komatiite – a very-high-temperature orthopyroxene-rich alkali-poor ultramafic lava named after the Komati River in South Africa's Barberton Mountainland, where it was first recognised. It is the extruded form of peridotite. Massive komatiite extrusions formed the lower levels of the greenstone sequences in the early Archean, but it has not been erupted since. However, such a high-temperature lava is erupted from Io's calderas.

Kuiper Belt – a disk of planetesimals beyond Neptune's orbit, generally coincident with the ecliptic. Some of these objects probably dip into the inner Solar System as comets. The Pluto–Charon system may well be the primary member of this population.

Lagrange points – in 1772 J.L. Lagrange undertook a mathematical study of orbital dynamics and discovered that in multiple-body systems there are 'stable' points where the gravitational fields of the various bodies balance, and an object that is stationary at such a point will not be perturbed. For a planet orbiting the Sun, there are two co-orbital 'Lagrange points'. One is 60 degrees ahead of the planet and the other is 60 degrees behind it. A few years later, asteroids were discovered orbiting between Mars and Jupiter, and in 1904 asteroidal bodies were found co-orbiting with Jupiter. Following the naming convention that was adopted for these bodies, they are collectively known as Trojans. Therefore the co-orbital Lagrange points are sometimes referred to as 'Trojan points'. There are several cases of small bodies co-orbiting with some of the moons of the giant planets.

lahar – a water-based landslide or mudflow on a volcano's flank consisting of cooled chunks of pyroclastic material.

laminations – layers, often very thin.

lath – a mineral with a lenticular shape.

lava – magma that is extruded from a volcanic vent or fissure. In the inner Solar System lava is usually silicate rock (sulphur and carbonatite flows are also possible, although rare) but on the frozen moons of the outer planets lava can be aqueous.

lava channel – the landform in which low-viscosity lava flows downhill from a volcanic vent. On Earth, the flow rate is modest, the flow is laminar, and the lava exploits existing terrain. However, on the Moon, the rate was greater and the flow was turbulent, so the lava excavated the terrain to make a sinuous rille far larger than its terrestrial equivalent (Hadley Rille, which the Apollo 15 crew visited in 1971, is one of the largest lunar examples).

lava tube – a lava channel where congealed lava on the walls progressively creates a roof; the lava then flows down an enclosed tube.
lichen – a hardy form of vegetation comprising a symbiotic relationship between fungi and algae.
limestone – a non-clastic sedimentary rock consisting mainly of the mineral calcite, with clay, chert or dolomite as common impurities.
line of dichotomy – the morphological boundary which marks the transition on the Martian surface between the generally elevated southern highlands and the low-lying northern plains.
lithosphere – the layer of rigid rock above the asthenosphere. It comprises both the crust and the solid upper mantle, which has an ultramafic composition, so it is a structural classification rather than a boundary in terms of chemical differentiation. It is fragmented into a dozen or so plates which move with respect to one another in accordance with the theory of plate tectonics. It is important to understand that because the lithosphere includes the outer mantle, the term is *not* synonymous with crust.
lobate scarp – the small lobe-shaped (scalloped) cliff that forms at the end of a lava or debris flow.

maar – a cavity left by a phreatic eruption.
mafic – a mineral such as olivine and pyroxene which is enriched in iron and magnesium, at the expense of silicon and aluminium.
maghemite – an iron oxide mineral.
magma ocean – in the case of some planetary bodies the entire surface was molten to a depth of several hundred kilometres for a short time immediately following accretion. In the case of the Moon, this permitted buoyant plagioclase minerals to rise to crystallise as the anorthositic crust.
magmatism – the formation of magma and its transport through the lithosphere.
magnesian-suite – plutonic rocks having a high fraction of magnesium-rich silicates.
magnetite – a strongly magnetic iron oxide mineral; $(Fe,Mg)Fe_2O_4$.
magnetosphere – if a planetary body possesses a global magnetic field, the magnetosphere is the volume of space that this field dominates.
mantle – the layer of a differentiated planetary body that is sandwiched between the core and the crust. In the case of the Earth, it is predominantly magnesium-rich silicate rock, and is the source of primary magma for igneous activity.
mantle plume – a large-scale upwelling of hot mantle which impinges upon the underside of the lithosphere.
marble – metamorphosed limestone; dolomitic marble is metamorphosed dolostone.
mare – a dark extrusion of mafic lava.
marginal basin – a regional depression on the margin of a continental plate formed by rifting in the back-arc zone of an island arc behind a trench some distance offshore. The Sea of Japan is a marginal basin.
massif – an isolated coherent block within a group of mountains. In terrestrial terms, it means that a mountain is a segment of the crystalline basement, or an intruded pluton which survived the severe erosional processes which wore down the rock

that originally surrounded it. On the Moon, however, a massif is a mountain which formed by the sudden uplift of a block of crust during a major impact event.

matrix – in a breccia it is the 'ground-mass' of fine material that consolidates the clasts.

mature sediment – a clastic sediment that has evolved from its parent rock and is composed of stable minerals which are well rounded and sorted; as opposed to immature.

mechanical weathering – the process of weathering by which physical processes break down a rock into fragments, involving no chemical change.

megaregolith – on planetary bodies which have not undergone extensive resurfacing, it is the outer few kilometres of crust that was pulverised by the impacts of the bombardment early in the Solar System's history.

melt percentage – temperature determines the fraction of the rock that melts; a low percentage (say 15 percent) will yield a magma that is rich in volatiles; a high percentage will yield a more mafic magma; a very high percentage (75 percent) will yield an ultramafic magma approaching peridotite.

metamorphic complex – the metamorphic rocks constituting a whole group closely related on a regional and/or stratigraphic basis.

metamorphic rock – a rock which is formed from reprocessing existing rock by application of heat sufficient to result in mineralogical, structural and chemical change. In the Earth's crust, a silicate rock which is recrystallised under high temperature and pressure takes on a granulitic texture (as limestone is turned into marble for example). There is great variety because igneous, sedimentary, or older metamorphic rocks can all be reprocessed, and to different degrees (or 'grades'). On planetary bodies without active crustal processes, metamorphism can take place when rock is heated and shocked by impacts.

metasediment – a sediment or sedimentary rock that has been subjected to metamorphism.

metasomatism – selected addition and removal of chemical elements by hot briny fluids which soak rocks during metamorphism. Sometimes, the 'parent' rock is altered to resemble granite, even though it never was igneous.

metavolcanic – a volcanic rock that has been subjected to metamorphism.

methane – CH_4.

mica – minerals characterised by low hardness and breakage in one direction along thin layers. This occurs when the intraplane bonds are strong the interplane bonds are weak.

mineral – the naturally occurring substances possessing crystalline structures which form the raw material of rocks. A mineral containing water in its structure is said to be hydrated. These are common in the Earth's crust, but conspicuously absent in the lunar crust.

Mohorovicic Discontinuity (Moho) – the transition in density between the crust and the upper mantle.

monocline – a local increase in an otherwise uniform gentle dip of layers of rock; a 'draping' of rocks over something, often a buried fault.

moraine – a mound or ridge of unstratified and unsorted glacial till deposited by direct action (melting) of glacial ice.
mudstone – a fine-grained sedimentary rock consisting mainly of clay mineral particles.

Noachian – Mars's earliest geological era, prior to the Hesperian.
norite – an intrusive coarsely-grained plutonic rock akin to gabbro but with orthopyroxene as the dominant mafic mineral.
normal fault – a fault along which downward vertical motion occurs on an inclined plane due to crustal extension; also known as a 'gravity fault'.
nuée ardent – a 'hot avalanche'; a turbulent pyroclastic cloud which rapidly descends down a volcano's flank. The cities of Herculaneum and Pompeii were engulfed by such flows from an eruption of Vesuvius in AD 79.

obliquity, axial – the angle at which a planet's rotational axis is offset from the plane of its orbit around the Sun.
ocean ridge system – a ridge with a median rift in the lithosphere that winds its way for some 60,000 kilometres around the Earth; the site of sea-floor spreading.
olivine – an ultramafic magnesium–iron silicate; $(Mg,Fe)_2SiO_4$.
Oort Cloud – a spherical shell of dormant comets hypothesised to exist far out from the Sun, such that when another star passes nearby, its gravity perturbs the comets, ejecting some into interstellar space and sending swarms of others into the inner Solar System.
ophiolite – a sequence of mafic and ultramafic igneous rocks (namely of basaltic pillow lavas, diorite dykes, gabbros and peridotites) which represent a vertical section through a piece of an oceanic lithospheric plate which became detached and failed to subduct. The Troodos Massif of Cyprus is an overturned piece of the floor of the ancient Tethys Ocean. When found embedded in continental marginal terranes, ophiolite complexes reveal the locations of extinct subduction trenches.
organic molecule – a molecule involving chains of carbon atoms. Although life as we know it is based on organics, their presence does not necessarily imply the existence of life.
orogeny – an episode of 'mountain building' stimulated by plate tectonics, and usually lasting several millions to tens of millions of years.
orthoclase feldspar – a member of the feldspar group with the formula $(K,Na)AlSi_3O_8$.
orthopyroxene – a form of pyroxene without calcium; as distinct from clinopyroxene.
outcrop – a naturally occurring exposure of bedrock.
outwash – glacial debris re-eroded and re-deposited as sediment by meltwater streams.

palimpsest – a craterform that has been either partially buried, completely buried and partially exhumed, or isostatically adjusted to such an extent that it has lost its profile; a 'ghost crater'.
pahoehoe – a Hawaiian term for 'a surface on which one can walk barefoot' type of

basaltic lava that is slowly extruded and assumes a smooth 'ropey' texture.

paleosol – (literally 'soil of the past'); a buried soil. It indicates that a land surface was above sea level for a period sufficiently long for weathering to form a mature soil, which was buried by subsequent rock formation.

paramagnetic – having a magnetic permeability greater than unity. In a magnetic field, the induced magnetisation of a paramagnetic material is directly proportional to that of the applied field.

pegmatite – very coarse-grained igneous rock. It represents the last and most water-enriched portion of a magma to crystallise and so contains high concentrations of minerals present only in trace amounts in typical granites.

periapsis – the closest point that an object in orbit can be from its primary; the term perihelion is used in the case of the Sun and perigee in the case of the Earth.

peridotite – a very high-temperature orthopyroxene-rich ultramafic magma; the intruded form of komatiite. It crystallises deep within an oceanic spreading ridge, and so forms the basement of an oceanic lithospheric plate.

'period' – a geological time subdivision during which the rocks of a 'system' were formed.

permafrost – subsurface crust that is permanently frozen.

phenocryst – a conspicuous mineral crystal in a porphyritic rock.

phonolite – a fine-grained alkaline volcanic rock that is primarily composed of alkali feldspar and feldspathoid minerals.

phreatic eruption – if an igneous intrusion encroaches upon a substantial reservoir of phreatic water, this will flash to steam and explosively erupt from the ground, forming a surface cavity (maar) which can be distinguished from an impact crater by the lack of shock effects. Such an eruption blasts out primarily shattered country rock, rather than magma.

phreatic water – another term for a 'ground water' reservoir.

phraeto-magmatic eruption – an explosive volcanic eruption resulting from the interaction of magma with a phreatic water.

pillow lava – a general term for lavas formed under water that display a pillow-like structure.

plagioclase feldspar – an alumino-silicate mineral; $(Ca,Na)(Al,Si)_4O_8$. It forms 60 percent of the outermost 15 kilometres of the Earth's crust. Much terrestrial plagioclase is sodic, but that on the Moon is calcic because the Moon is deficient in sodium.

planet – from 'planetes asteres' meaning 'wandering stars'

planetesimal – a primitive condensate body, left over from the solar nebula, which comprises ices and grains of rock.

plate tectonics – a theory devised in the late 1960s to explain the gross structure of the Earth's surface in terms of lithospheric plates which jostle one another. By considering the Earth as an integrated system, it accounted for mid-ocean ridge, oceanic trenches, earthquakes, volcanoes, island arcs, rift valleys, mountain ranges and peculiarities of paleo-ecology.

plasma – a hot ionised gas, sometimes referred to as the fourth state of matter (after the solid, liquid and gaseous phases).

playa – the shallow basin of a desert plain containing a layer of silt, clay and salty evaporite deposits.
pluton – any body of rock (magma, or a buoyant mass which moves by solid-state flow) that rises through country rock but whose motion ceases prior to it broaching the surface.
plutonic rock – a coarsely-grained igneous rock which crystallises below the surface (strictly, at great depth, because rock that crystallises at shallow depth is called hypabyssal).
porphyry – an extremely finely-grained basaltic matrix hosting phenocrysts. The presence of phenocrysts implies two stages (speeds) of cooling of the original magma.
Precambrian – the span of geological time prior to the Cambrian, comprising (in sequential order) the Hadean, Archean and Proterozoic.
prokaryote – a single-celled microorganism whose cell does not contain a distinct nucleus; in contrast to a eukaryote. The earliest life forms were prokaryotic.
proplyd – a protoplanetary disk around a newly formed star.
Proterozoic – the span of geological time from the transition at the end of the Archean (some 2.8 to 2.6 billion years ago) to the Cambrian.
protoplanet – an accreted body that might either accrete smaller such bodies and grow into a planet, or be consumed during such a collision.
province – a collection of geographically distributed but closely related geological units.
proximal – igneous or sedimentary deposits composed of coarser particles deposited near to their source areas; contrast with distal.
pyroclastic – literally 'fire-shattered fragment'; ash and clastic rock fragments immersed in an incandescent gaseous cloud which is explosively expelled from a volcanic vent.
pyroxene – an ultramafic magnesium iron silicate; $(Ca,Fe,Mg)_2Si_2O_6$. Its principal forms are clinopyroxene (with calcium) and orthopyroxene (without calcium).
pyroxenite – an intrusive rock chiefly composed of pyroxene, but with olivine as a minority constituent.

quartz – a silica mineral; SiO_2. It is metamorphosed chert. Although common in the Earth's crust, it is extremely scarce on the Moon.
quartzite – metamorphosed sandstone.
quartzoze – a quartz-rich sediment.

radiogenic heating – the heating within a planetary body (possibly to melting point) resulting from the decay of radioactive elements.
radiometric dating – age determination based on nuclear decay of various naturally occurring radioactive isotopes which are impurities in mineral grains. Given assumptions of primordial abundances, the ratios of radioactive decay products can measure the age of a rock. Elements with half-lives of billions of years are accurate to about 100 million years.
rare earth elements – a group of elements that are rare in the Earth's crust because

they sank into the interior during the process of differeniation which formed the crust; that is: lanthanum, cerium, neodymium, praseodymium, promethium, samarium, europium, gadolinium, terbium, erbium, thulium, dysprosium, holmium, ytterbium and lutetium.

ray   light-toned material deposited in a streaky radial pattern around a newly formed impact crater. It is comparatively bright because it is finely pulverised rock and when crystalline rock is shattered it exposes fresh cleavage planes which are highly reflective. Over time, a layer of dust covers the exposed crystals and the rays fade.

refractories – compounds with high melting points; as opposed to volatiles.

regolith – a seriate distribution of impact debris with individual sizes ranging down to a few microns. As an unconsolidated blanket formed by 'gardening' of the bedrock, it is the proper name for the lunar 'soil'.

remanent magnetism – the magnetic field frozen into crystallising rock when its iron minerals cool below the Currie point and adopt the alignment of the prevailing magnetic field.

resurgent caldera – as a very large body of magma approaches the surface and a large amount of volatiles boil out of solution the resulting explosion lays down a vast blanket of pyroclastics and opens a huge caldera which repeatedly revives and erupts. These eruptions are sometimes called supervolcanoes.

reverse fault – a fault along which upward vertical motion occurs on an inclined plane due to crustal shortening.

rhyodacite – the extrusive form of granodiorite; similar to granite except that granite contains more alkali feldspar.

rhyolite – a felsic volcanic rock; the extuded form of granite.

rift – a deep fracture in the Earth's lithosphere due to separation of the two sides by tensional forces.

Roche radius – the radius within which an object making a close approach to a planet will be disrupted by the tidal stresses of the primary's gravity field. It depends upon the density of the intruder, but for rocky bodies such as large asteroids the range is about 3 radii measured from the centre of a planet such as the Earth.

rock – an aggregate of minerals.

formation – any rock unit, which is lithologically distinct and capable of being mapped.

rootless volcano – a small cone that forms when a lava flow crosses water-saturated ground and prompts it to flash to steam, with the lava flow being disrupted by the resulting explosion. The term is used to signify that the eruption is not driven by its own feed pipe. Sometimes referred to as a pseudo-volcano.

salt – the result (together with water) of a reaction between a metal compound and an acid in which the metal atom replaces the hydrogen.

saltation – the movement of individual particles as variable leaps or jumps powered by wind or water; a 'roll-and-bounce' motion; how a dust storm gets off the ground.

sandstone – a clastic sedimentary rock composed of fragments of sand (grains of quartz) set in a fine-grained matrix of silt or clay.

sapping – the process by which subterranean gas or water undermines the surface and causes it to collapse.

schist a medium-to-coarse grained crystalline rock that is strongly foliated due to shearing pressures during high-grade metamorphism.

scoria – a vesicular cindery lava which texturally resembles pumice but is darker and denser; such a lava is said to be scoraceous.

sculpture – a term coined to describe all manner of ruts and grooves radial to an impact basin formed by ejecta on low-angle trajectories.

sea-floor spreading the process by which upwelling magma forms a succession of dykes in the rift of the mid-ocean ridge system, literally filling the gap as two oceanic lithospheric plates diverge.

sedimentary rock – a rock formed from the consolidation of sediments transported by water, wind or ice, or indeed deposited by organisms.

serpentinite – an ultramafic rock such as pyroxene, olivine or amphibole that was chemically modifed by hydrothermal alteration during low-grade metamorphism and it became hydrated; $(Mg,Fe)_3Si_2O_5(OH)$. It is commonly found in ophiolite complexes. Its name derives from its snake-like soft scaly texture.

shale – a very fine-grained thinly layered sedimentary rock formed by the consolidation of clay and/or mud.

shatter cone – a striated pattern imposed on a finely-grained rock by the propagation of an intense shock wave. A funnel-shaped fracture forms with the tip aimed towards the source of the shock. Often several nested cones form. They were first observed in rock which had been shattered by underground nuclear explosions, and then found in lunar rock shattered by impact; only later were they it seen in terrestrial landforms identified as impact craters.

shield, continental – see craton.

shield, volcanic – an edifice constructed by a succession of basaltic lava flows. The terrestrial exemplar is Hawaii's Big Island. Although it can grow very tall (and measured from the ocean floor Hawaii is taller than Mt Everest) such an edifice has a very broad base, giving it a profile that resembles a Viking warrior's shield.

shock wave – a transient increase in pressure characterised as 'megabars for microseconds'.

siderophile elements literally 'iron loving'; elements (such as nickel, cobalt and iridium) that readily combine with iron to make compounds, and tend to 'follow' iron in igneous processes.

silica – crystalline silicon dioxide; SiO_2; common sand. It has a high melting point and forms a variety of silicates in rock.

silicate – a group of minerals in which the silicon tetrahedron forms the basis of the structure. It forms the most common minerals in the Earth's crust because silica readily combines with calcium, aluminium, iron, magnesium, sodium and potassium to make a variety of minerals.

silicate melt – molten alumino-silicate.

silicic – rich in silica (most often present in the form of quartz).

sill – an igneous intrusion which parallels the layering in the country rock that it intrudes; in contrast to a dyke.

siltstone – a layered sedimentary rock similar in composition to shale or mudstone, but rather grittier.

slate – a compact fine-grained rock formed from shale that is intensely layered because it was subjected to shearing pressures during low-grade metamorphism.

slump – the downward sliding of a mass of rock or unconsolidated material moving as a unit along a spoon-shaped surface, usually with backward rotation.

solar nebula – the cloud of interstellar gas, dust and rocky grains from which the Sun and its planets condensed.

Solar System – the volume of space (and the objects within it) that is dominated by the Sun (Sol).

solar wind – the plasma that flows from the Sun. It is mostly ionised hydrogen (free protons and electrons), but also contains heavier nuclei.

solid-state flow movement within a solid by means of the rearrangement of the component particles; as opposed to the flow of a melt.

sorted – separated by grain size.

spall – the sudden ejection of material from a surface as a result of focusing a seismic shock wave. This can occur on any scale all the way from a scab from a rock to the formation of the 'chaotic terrain' antipodal to a major impact basin.

spatter cone – a small volcanic cone build up around a 'fire fountain' from globs of molten magma which spatter across the surface and solidify.

stishovite – a form of shocked quartz.

stratigraphic column – an idealised illustration of a sequence of rock units ordered so that the oldest occurs on the bottom of the stack. As the sequence may be nowhere present in its entirety, isolated sections are pieced together like a jigsaw to reconstruct it.

stratigraphic unit – a body of adjacent rock strata recognised as a unit in the classification of a rock sequence with respect to any of the many attributes that rocks may possess.

stratigraphy – the study of a planetary surface by the layering of the rock units exposed at the surface.

stratosphere – the non-convecting outer part of an atmosphere within which, as a result, the thermal profile is essentially static.

stratovolcano – a central-vent volcano which erects a tall conical edifice by erupting ash and viscous silicic lava in successive phases of activity (and hence is sometimes called a composite volcano).

strike – the trend of a linear feature such as the surface expression of a fault plane on a map.

strike-slip fault – a fault in which the primary displacement is parallel to the strike of the fault plane.

stromatolite – a layered calcareous structure resembling a wrinkled stack of pancakes created by colonies of lime-secreting cyanobacteria.

subduction – the process of a dense lithospheric plate is deflected under a more buoyant plate and plunges back into the mantle. The angle at which the diving plate descends can be shallow (30 degrees) or steep (80 degrees).

sulci – a terrain that is etched by parallel ridges and grooves.

supercontinent – an amalgamation of more 'typically-sized' continents into a single landmass as a result of a series of continental collisions.

supergroup – an assemblage of related rock groups which had significant lithological features in common.

superposition, law of – an empirical law of stratigraphy which states that newer geological units overly older ones. In practice, intense crustal folding can negate this relationship.

supracrustals – rocks laid down atop the crust.

suture – the collision feature created by the convergence of continental plates. The rocks and sediments caught up in the suturing event are intensely folded, thrust faulted, and (in the core of the resulting orogen) metamorphosed and selectively melted to form batholiths.

syenite – an intrusive rock containing alkali feldspar, with plagioclase and mafic minerals as minority constituents but little quartz. It is similar to granite but less quartz-rich.

syncline – an upward-facing concave fold such that stratigraphically older rocks are farthest from the centre of curvature; in contrast to an anticline.

T Tauri stars – a class of newly born star in which the radiation pressure is driving a strong 'stellar wind' that rapidly blows away the remnant of the nebula from which the star (and its planets) condensed.

tectonics – the process of deformation of a planetary crust to create surface relief.

tektites – literally 'glassy rocks'. Although initially suspected of having been blasted off the Moon by impacts and heated upon penetrating the Earth's atmosphere, they are now known to have been ejected on high ballistic arcs by major terrestrial impacts and then heated upon re-entry.

temperature – a measure of the kinetic energy of the molecules. It can be measured in several ways. The *effective* temperature that a black body would need to radiate the same total thermal flux as that observed.

terminator – the line of longitude corresponding to sunrise or sunset on a planetary surface.

terrace – a relatively level bench or step-like surface breaking the continuity of a slope.

terrane – a distinctive group of related rocks and to the area in which they outcrop. Adjacent terranes do not have rocks formed in the same geographical location, or by similar geological processes. For example, the rocks either side of the Yellowknife fault on the Canadian shield represent two distinct terranes.

tessera – a mosaicked pattern of irregularly intersecting ridges and grooves.

thermal history – the rate at which heat is produced and lost by a planetary body plotted over geological time.

tholeiitic basalt – an alkali-poor form of basalt characterised by orthopyroxene, in addition to calcic plagioclase and clinopyroxene. It is found in the mid-ocean ridge, and some continental flood basalts.

thrust fault – a reverse fault with a shallowly dipping plane which results in one block riding up and encroaching upon the other.

till – debris deposited directly by melting ice underneath a glacier.

tillite – a rock in which a jumble of debris is consolidated by a fine-grained matrix. Although generally interpreted as cemented glacial till, it has been proposed (as yet only as a hypothesis) that the debris is impact ejecta.

tonalite – an intrusive igneous rock like granodiorite primarily composed of quartz and alkali feldspar, but with few mafic minerals.

trachybasalt – an extrusive igneous rock that is intermediate in composition between trachyte and basalt.

trachyte – a fine-grained, porphyritic volcanic rock, consisting mainly of alkali feldspar and minor mafic minerals; the extrusive equivalent of syenite.

transform fault – a strike-slip fault that forms an integral part of the plate tectonics system of spreading ridges and subduction trenches.

trap eruption – a flood basalt eruption.

trench, subduction – an elongate depression of the floor of the ocean down which marks the surface expression of the line where a lithospheric plate is being subducted. Athough they are the deepest points in the ocean, they are fairly broad and so their profiles are rather shallower than depicted in topographic maps.

triple junction – the point where three lithospheric rifts meet, typically at roughly 120 degree angles. They form during the initial stages of continental rifting as the lithosphere is domed by the pressure of the upwelling mantle plume. However, while two 'arms' progress sufficiently during the ongoing rifting process to start to form a new ocean basin, the third invariably 'stalls' and leaves a broad nested graben cutting across a continent as a 'failed' rift (an aulocogen).

troctolite – a coarsely-grained plutonic rock in which the plagioclase feldspar combines with olivine as the mafic component.

tuff – a general term for all consolidated pyroclastic rocks.

turbidite – debris deposited from an underwater landslide.

turbidity current – a landslide on the ocean floor in which the suspended debris (the turbidite) is transported for a long distance despite extremely shallow slopes and deposited as a graded bed. Such flows are particularly common on the relatively steep continental slope and fan out across the abyssal plain below.

ultramafic – a magnesium-rich igneous rock such as dunite, peridotite or pyroxenite that has less than 45 percent silica. It is the typical composition of the Earth's mantle.

unconformity – a stratigraphic gap in the geological record due to a period of erosion which removed a section of the record before subsequent deposition buried that surface.

vent – the opening (either a centralised pipe or a fissure) in the crust through which volcanic materials are extruded onto the surface.

vesicle – as lava approaches the surface its pressure rapidly decreases and the volatiles 'boil' out of solution. If the gas is trapped in the rock as it solidifies, it forms cavities.

viscosity – a measure of the resistance of a fluid to flow. A runny fluid has a low value and a sticky fluid has a high value.

vitrophyre – a porphyritic igneous rock whose matrix is amorphous (glassy, and also known for this reason as a glass porphyry).

volatiles – compounds with low melting (or boiling) points; as opposed to refractories.

volcanic rock – magma which solidifies after being extruded onto the Earth's surface as lava.

volcano – any vent through which lava is extruded onto the surface.

volcanogenic   of volcanic origin.

water table – the subterranean level which marks the transition between the zone of saturation (all rock pores permanently filled with water) and the zone of aeration (openings having water only temporarily, or not at all).

weathering – the mechanical and/or chemical erosion of solid rock on the surface, producing loose detritus.

welded ash-flow tuff   recrystallised ash.

wrinkle ridge – the crumpling of a mare lava plain due to compressional forces induced either when the lava cools, densifies and sinks to reflect the pre-existing topography, or because the ground on which it resides is subjected to tectonic deformation.

xenolith – literally 'strange rock'. A fragment of rock from deep within the lithosphere which is torn off the wall of a magma feed pipe and carried to the surface by the flow. Such magma inclusions are very rare and much prized for the insight they yield into otherwise inaccessible depths.

zircon – a silicate with zircon as the metal which forms in felsic igneous rocks and their metamorphic derivatives; it is virtually immune to erosion; zircon is a common accessory in felsic rock. It is not dissolved by erosional processes. It is considered to be a semi-precious mineral.

Index

Page numbers in **bold** are illustrations.

Adams, John Couch, 325, 326, 327, 333
Adams, Walter S., 72, 147
Adel, Arthur, 147
Agricola, Georgius, 378
Airy, George Biddell, 325
Albarede, Francis, 227
Albritton, Claude, 46
Aldrin, E.E. (Buzz), 47
Alfven, Hannes, 5
aliens, 60
Allegheny Observatory, 286
Alvarez, Luis, 354
Alverez, Walter, 354, 359
Amalthea (satellite of Jupiter), 237, 238, 245
Ames Research Center, 154, 242, 352, 374, 376
Anderson, J.D., 342
Andromeda Nebula (M31), 2, 4, 9
Anglo Australian Observatory, 8
Antoniadi, Eugene Michael, 73, 78, 145, 184, 188, 197, 238
Applied Physics Laboratory, 201
Arago, Francois Jean Dominique, 333
Arecibo Observatory, 148, 158, 184
Ariel (satellite of Uranus), **318**
  bulk density, 311
  craters, 319
  cryovolcanism, 319
  discovery, 310
  grabens, 319
  resonances, 319
  resurfacing, 319
  tidal heating, 319
Armstrong, Neil A., 47
Arrhenius, Svante, 71, 146

asteroids, 241, 302, 319, 328, 354, 356, 373, 376
  5 Astraea, 241
  belt, 185, 242
  basalt, 106
  'celestial police', 240, 241
  1 Ceres, 241
  diameters, 241, 242
  8 Flora, 241
  6 Hebe, 241
  10 Hygeia, 241
  7 Iris, 241
  3 Juno, 241
  9 Metis, 241
  'minor planets', 242
  2 Pallas, 241
  'planetoids', 241
  shattered planet, 241
  total mass, 242
  'vermin of the skies', 241, 339
  4 Vesta, 106, 241
  'zones of avoidance', 242, 287
Astronomical Unit (AU), 144, 421
Athens Observatory, 44
Australian National University, 226

Bacon, Francis, 14
Baldwin, Ralph B., 46, 47, 74
Ball, William, 285
Barghoorn, Elso, 350
Barker, R., 145
Barnard, Edward Emerson, 72, 145, 237, 238, 241, 242, 284, 286, 327
Barringer, Daniel M., 372
Bartlett, J.C., 144, 145

446 Index

Baum, R.M., 145, 286
Baumgardner, John, 227
Bean, Al, 48
Beatty, Kelly, 168
Becquerel, Antoine Henri, 13
Beer, Wilhelm, 44, 70
Benioff, Hugo, 26
Berner, Robert, 352
Bessel, Friedrich Wilhelm, 325
Bethe, Hans Albrecht, 4, 13
Bianchini, Francesco, 144
Bible, 11, 14
Birr Castle, Ireland, 4
Birt, William R., 45
Blackwell, Alex, 130
Blaney, Diana, 252
Blichert-Toft, Janne, 227
Bode, Johann Elert, 240, 309, 325
Bond, George P., 236, 286, 287
Bond, William Cranch, 286, 287
Boon, John, 46
Bouvard, Alexis, 325
Boyer, C., 146, 148
Boynton, William, 354
Brahe, Tycho, 1, 2, 372
Briden J.C., 37
British Astronomical Association, 45, 145
Brown University, 119, 153, 356
Brunhes, Bernard, 24
Buffham, J., 310
Buick, Roger, 217
Bullard, Edward C., 15
Bulmer, Mark, 259
Bunge, Hans-Peter, 227
Burke, Bernard F., 237
Butler, R. Paul, 8

California Institute of Technology, 30, 245
Calvert, Andrew, 228
Callisto (satellite of Jupiter), xix, 275–278, 291, 316
  age of surface, 277
  albedo variations, 237, 238, 275
  Asgard basin, 276
  atmosphere, 239
  basins, 275, 276, 277
  bulk density, 240, 275
  craters, 275, 276
  diameter, 238
  discovery, 235
  Doh Crater, 276
  erosional processes, 276
  Galileo (spacecraft), 276, 277, 278
  impacts, 277
  internal structure, 278
  isostasy, 276, 277
  magnetic field, 278
  moment of inertia, 277
  Pioneer image, **243**
  radiogenic heating, 278
  resonances, 276, 278
  rotation, 238
  salty fluid layer, 278
  size, 275
  surface properties, 238
  tectonism, 275
  tidal heating, 278
  Valhalla basin, 275, 276
  Voyager (spacecraft), 275, 276, 277
  water layer, xix
Calypso (satellite of Saturn), 304, 329
Cameron A.G.W., 61
Campani, 286
Capen, Charles F., 78
Carey, Warren, 15
Carnegie Institution, 376
Carr M.H., 94, 118, 126, 268
Cassen, Patrick, 247, 261
Cassini, Giovanni Domenico (Jean Dominique), 69, 144, 145, 285, 287, 300
Cassini, J.J., 144
Cave, T.R., 286
'celestial sphere', 1, 2, 143
Cernan, Gene, 52
Chaikin, Andrew, 168
Challenger, Glomar, 19
Challenger, HMS, 18
Challis, J.C., 327, 328
Chamberlain, Thomas Chrowder, 4
Chao, Edward Ching-T, 372
Charon (satellite of Pluto), **339**, 340
Christie, W.H.M., 310, 328
Christy, James W., 340
Clarke, Arthur C., 283, 300
Coes, Loring, 372
Cold War, 21
Columbo, Guiseppe, 184
comparative planetology, xxi
Conrad, Charles (Pete), 48
comets, 2, 148, 158, 185, 241, 268, 277, 301, 309, 319, 328, 342, 356, 376
Cook, Captain James, 143
Copernicus, Nicolaus, 1, 3, 143, 175, 235, 279
Coriolis, Gaspard, 160
Corliss, J.B. (Jack), 350
Courtillot, Vincent, 357
Cox, Allan V., 24

Crabtree, William, 143
Cragg, T.L., 327
Crocco, Gaetano Arturo, 245
Crommelin, A.C.D., 340
Cruikshank, Dale, 334
Crumpler, Larry S., 168
Curie, Marie, 13
Curie, Pierre, 18
Curie point, 18, 38, 424
Curtin University, 226
Cuvier, Dagobert (George), 353
Czamanske, Gerry, 362

d'Arrest, Heinrich L., 290, 326
d'Hondt, Steven, 356
Daguerre, L.J.M., 64
Daly R.A., 61
Dana, J.D., 223
Dante, xviii
Darwin, Charles, 59, 353
Darwin, George, 59
Davis, Don, 376
Dawes, William Rutter, 238, 286
'Death Star', 293
de Beaumont, Elie, 14
de Buffon, Compte Georges Louis Leclerc, 1, 11, 14
de Gasparis, 241
de Heen, P., 145
de la Rue, Warren, 45
DeMarcus, W., 237
de Medici, Cosimo, 235
Derham, 144
DesMarais, David, 352
de Vaucouleurs, Gerard, 71, 72, 73
de Vico, F., 144
de Wit, Maarten J., 207
Dietz, Robert S., 23, 46, 373
Dione (satellite of Saturn), **296**
 albedo, 295
 Amata, 296
 bulk density, 291, 296
 cratered plains, 295
 cryovolcanism, 296
 discovery, 285, 305
 highland terrain, 295
 lineaments, 296
 pyroclastics, 296
 radiogenic heating, 296
 resonances, 290, 297, 304
 smooth plains, 295
 tidal heating, 297
 whispy white streaks, 296
Dollfus, Audouin, 73, 145, 146, 184, 238, 304

Draper, J.W., 45
Duke, Charlie, 52
Duluc, J.A., 14
Dunham, Theodore H., 72, 147
du Toit, Alexander L., 15, 16
Dyce, Rolf B., 184

Earth, 11–41, 240
 abyssal plains, 16, 21, 117, 163, 204, 419
 Acasta, 226
 Aegean Sea, 363
 Africa, 14, 15, 16, 17, 21, 31, 32, 33, 34, **35**, 36, 113, 167, 207, 213, 224, 228, 350, 360, 361
 age, 11, 12, 13, 23
 Akilia supracrustals, 212
 Alaska, 362, 363
 albite, 40
 alkaline lavas, 107
 Alps, 14, 33, 34, 73, 204, 360
 American Southwest, 104, 105, 354
 Ameralik fjord, 212, 214
 Ameralik dykes, 212, 213
 Amitsoq gneisses, 212
 amphibole, **210**, 420
 amphibolite, 207, 213, 214, **222**, 420
 Andes, xvii, 222, 223, 224
 andesite, 209, **210**, 420
 Antarctica, 16, 17, 21, 25, 137, 226
 anticline, 361
 Appalachian Range, 33, 204, 223, 224, **225**
 Arabian Sea, 21, **35**, 113
 Archean, 58, 60, 62, 66, 205–, 223, 226–228, 256, 283, 350, 352, 371, 420
 Argentina, 360, 361, 375
 argon, 24
 Arizona, 46, 74, 372
 Ascension Island, 19
 ash deposits, 88, 212, 356, 358, 359, 362, 363, 365, 443
 ash shields, 86
 Asia, 34
 asthenosphere, 30, 31, 32, 33, 421
 astroblemes, 46, 421
 Atlantic Ocean, 14, 16, 18, 21, 27, 33, 34, 211, 213, 224, 360, 366
 'Atlantis' (lost continent), 14, 363
 atmosphere
  'acid rain', 358
  carbon dioxide, 160, 349, 351, **352**, 357, 358, 377, 378
  circulation, 160
  early, 349, 357
  oxygen, 352, 353, 357, 378

Earth, atmosphere, *cont.*
  stratosphere, 359, 362, 363, 369, 440
  sulphur dioxide, 358
atolls, 23
Augustine, **363**
aurorae, 251
Australia, 14, 16, 21, 118, 163, 214, 226, 298, 350
Azores, 15, 19
back-arc processes, 30, 32, 219, 223, 224, 421
Baffin Island, 204
Balkans, 19
Baltic Sea, 204
banded iron, 208, 212, 213, 217, 350, 352, 421
Barberton, 207
Barberton Mountainland, 207, **208**, 209, 212, 214, 229, 256, 350, 372
basalt, 12, 17, 19, 23, 24, 28, 31, 33, 40, 47, 107, 207, **210**, 214, 227, 356, 369, 421
Basement Complex, 207
'basement', 203, 204, 422
Basin and Range province, 167, 370, 422
batholiths, 219, 222, 223, 366, 422
Belt Basin, 224
Belt Mountains, 224
Berlin, 44, 240, 290, 326
bimodal surface distribution, 204
'black smokers', 351
blueschist, **222**
Brazil, 15, 16, 33, 70
Bremen, 44, 241
Britain, 33
British Columbia, 224
bulk density, 188
Bushveld Complex, 207
calc-alkaline minerals, 209, **210**, 220, 423
calcic plagioclase feldspar, 31, 40, **210**
Caledonian belt, 33, 34, **36**
California, 24, 27, 269
Cambrian, 203, 351, 353, 359, 378, 423
Canada, 15, 18, 28, 46, 204, 205, 206, 224, 350, 373, 375
Canadian shield, 204, **205**, 206, 207, 213, 226, 228
Cape Fold Belt, 361
carbonate, 160, 352, 360, 423
Carboniferous, 146, 224
carbonatite flows, 174, 424
Caribbean Sea, 354, 356, 362
Carlsberg Ridge, 35
Cascade Range, 113, 364, **369**
catastrophism, 11, 14, 219, 371

Ceylon, 13
Chassigny, France, 106
chemical alteration, 12, 31, 40, 203, 204, 212, 217, 227
chlorite, 40
chert, 16, 208, 212, 217, 352, 424
Chesapeake Bay, 375
Chile, 221, 223, 238, 361
China, 227, 360
Chicxulub, 354, **355**, 356, 359
Churchill Province, 205, **206**
clasts in sediments, 203, 208, 210, 212, 217, 219, 226, 374
clay, 203, 354, 357, 424
climate, 15, 17
coal, 17, 224
coastlines, 14, 15, 21, 38, 224
coesite, 372, 378
Colorado River, 136
Columbia River Plateau, 94, 103, 358, **369**, 370, **371**
conglomerate, 208, 212, 220, 424
conservative plate margins, 27
constructive plate margins, 27
continental collisions, 33, 204
continental drift, 14, 16, 17, 18, 19, 26, 31, 59, 366, 424
continental margins, 15, 21, 224
continental platforms, 16, 163, 204, 210, 212, 224
continental rifts, 27, 213, 224, 362, 366
continental shelves, 15, 16, 18, 21, 354, 361, 425
continental shields, 39, 204, 228, 425
continental slopes, 16, 18, 117, 425
Coonterunah Group, 217
Copenhagen, 227
core, 24, 61, 227
Coriolis effect, 160, 349
Crater Lake, 365, 366
cratons, 205, 207, 208, 212, 213, 217, **218**, 223, 227, 228, 350, 352
Cretaceous, 354
Cretaceous-Tertiary boundary, 354, 355, 356, 357, 362
Crete, 363
crust, 19, 61, 138, 425
  original, 204, 210, 214, 217, 219, 226, 227, 228
crystallisation processes, 12, 203, 227
cumulates, 32, 426
Cyprus, 32
dacite, 217, 426
Danzig, 43

Earth, *cont.*
  Davis Strait, 211
  Dead Sea, 35, 269
  Death Valley, 269
  Deccan Plateau, 357, 358, 359
  destructive plate margins, 27
  diorite, **210**, 426
  distance from Sun, 143, 144
  dolerite, 16, 427
  dolomite, 16, 427
  Dominion Reef, 207
  drainage systems, 82
  Dresden, 44
  dunite, 2**10**, 427
  dykes, 23, 31, 32, 212, 427
  earthquakes, 19, 26
  East African Rift, 35, 111, 113, 169, 174
  East Pacific Rise, 21, 23, **25**, 26, 27, 28, **30**, 33, 34, 350
  Ecoglite, **222**
  Edinburgh, Scotland, 11, 342
  Ellice Islands, 40
  Emperor Seamounts, 28, **30**, 97
  Enderby Land, 226
  epidote, 40
  erosional processes, 11, 12, 14, 23, 33, 103, 105, 203, 208, 209, 212, 217, 219, 220, 221, 222, 223, 224, 227, 352, 434
  erratics, 15, 16
  Eurasia, 14, 32, 360
  Europe, 32, 33, 34, 36, 224
  expanding planet, 14
  feldspars, 203, 427
  feldspathisation, 204
  felsic minerals, 208, 209, **210**, 212, 214, 219, 227
  Fig Tree Group, 350
  flood basalts, 47, 89, 94, 217, 220, 354, 357, 358, 361, 362, 366, 370, 371, 428
  formation, 11
  fossils, 203, 204, 350, 351, 353, 354, 360, 378, 428
  fracture zones, 21, 24, **25**, 26, 27, 28
  gabbro, 32, **210**, 227, 428
  Galapagos Ridge, 350
  geological timescale, 11, 33, 61, 203, 204, 207, 219, 226, 353, 428
  Geosat (satellite), 22, 355
  geosyncline, 223, 224, 428
  geothermal gradient, 222, 234
  Gilbert Islands, 40
  glaciation, 15, 16, 374, 428
  Glen Tilt, Scotland, 12
  'global ocean', 12, 203, 353

gneiss, 12, 204, 207, 209, 210, 212, 213, 217, 222, 226, 429
Godthaab (Nuk), 212
Gondwanaland, 14, 16, 17, 34, 38, 360, 361, 362
Grand Canyon, 136
granite, 12, 17, 19, 31, 203, 204, 205, 207, 209, 2**10**, 212-, 214, 217, 219, 226–228, 355, 361, 370, 371, 429
Granite Falls, Minnesota, 213
granite-gneiss-greenstone complexes, 207, 208, 209, 214, 220, 221, 223
granodiorite, 212, 226, 429
granulite, 207, **222**, 429
Great Plains, 204, 213, 224, 2**25**
Great Slave Lake, 206
'greenhouse effect', 349
Greenland, 15, 33, 211, 212, 213, 226
greenschist, 207, **222**, 429
greenstone, 207, 209, 211, 213, 217, 219, 226, 229, 256, 350, 429
greenstone belts, 214, 216, 219, 220, 222, 223, 228
greywacke, 203, 204, 208, 217, 226, 429
Gulf of Aqaba, 35
Gulf of California, 21, 27
Gulf of Finland, 204
Gulf of Mexico, 224
Hadean, 226, 227, 349, 371, 429
Haiti, 355
Hamersley Range, 214, 372
Hanover, 240, 309
Harz Mountains, 203
Hawaii, xviii, 19, 28, 61, 82, 134, 174, 250, 257, 354, 366
Hawaiian Ridge, 28, **30**, 97
heat flow, 21, 23, 39, 163, 219, 227, 253, 261, 366, 370
hematite, 38, 429
Hercynian belt, 33, 34, **36**
Himalayan Range, 163
Horn of Africa, 113
'hot spots', 228, 366, 370
Hudson Bay, 204, 205
hydrosphere, 61, 227, 349
hydrothermal activity, 105, 212, 227, 350, 351, 365, 366
ice ages, 15, 103
Iceland, 25, 26, 27, 358
Idaho, 91, 103, 117, 224, 365
Idaho Batholith, 224, 366, **369**, 370, 371
igneous processes, 12, 430, **431**
ignimbrites, 365, 430
impact melt, 354, 355, 372, 373, 375

Index 449

Earth, *cont.*
  impacts, 45, 46, 74, 219, 226, 354–357, 359–362, 370–373, **374**, 375, **376**
India, 14, 17, 18, 38, 106, 163, 357
Indian Ocean, 17, 21, 34, 35, 113
Iowa, 375
Ireland, 19, 33
iridium, 354, 356, 357, 360, 378, 430
iron-bearing minerals, 18, 23, 24, 38, 107, 203
island arcs, 26, 28, **30**, 430
isostasy, 19, 23, 210, 219, 220, 223, 430
Isua, Greenland, 212
Isuan supracrustals, 212, 214, 226
Italy, 43, 360
Jack Hills metasediments, 226, 227
Jamaica, 327
Japan, 30, 222
Java, 363
Johannesburg, 78, 207
Kaapvaal craton, 207, **208**, 210, 228
karst, 82, 432
Katmai, 362, 366
Keewatin Group, 205
Kimberley Plateau, 214
Kola Peninsula, 204, 205
Komati River, 207
komatiite, 107, 207, 208, **210**, 219, 227, 256, 283, 350, 432
Kona, Hawaii, 61
Krakatau, 362, 363, 366
Labrador, 211, 212, 213
Lake Baikal, 361
'Lake Bonneville', 117
'Lake Missoula', 103
Lake Superior, 204, 205, 213
Lake Winnipeg, 205
Laki, **358**, 359
large-ion lithophiles, 137
Laurasia, 34, **36**
lead, 13
Liberia, 213
life
  algae, 349, 350, 351, 353, 419
  ammonites, 354, 359, 420
  amphibians, 353, 360
  'archea', 351, 420
  bacteria, 350, 421
  'blue-green' algae (see cyanobacteria)
  'Cambrian explosion', 353, 378
  cyanobacteria, 350, 426
  dinosaurs, 354, 357, 359, 360
  eukaryotes, 352, 353, 427
  evolution, 353, 378
  extinction events, 353–355, 357, 359, 360, 362, 365, 374, 375, 376
  'fern spike', 354
  fungi, 428
  heterostegina, 349, 353, 377
  heterotrophes, 351
  lithotrophes, 351
  mammals, 353, 360
  origin, 204, 349, 350, 351
  'ovoids', 350
  plankton, 354, 356
  photosynthesis, 351, 354
  pollen, 354, 356
  prokaryotes, 353, 372, 437
  reptiles, 353, 360
  stromatolites, 66, 217, 440
  thermophyllic species, 351
  'tree of life', 351
  trilobites, 350, 360
Lilienthal, 44, 240
limestone, 12, 204, 212, 223, 350, 360, 433
Limpopo Fold Belt, 207, **208**, 210, 228
lithosphere, 26, 27, 28, 30, 31, 107, 111, 138, 219, 222, 224, 227, 228, 433
Los Angeles, 27, 29, 45, 72
'low velocity zone', 30
maar, 46, 372, 433
Madagascar, 17
mafic minerals, 23, 24, 30, 107, 207, **210**, 212, 213, 227, 228, 433
magma ocean, 57, 226, 227
magnesium-bearing minerals, 23, 107, 203, 207
'magnetic anomalies', 24, **25**
magnetic field, 14, 18, 24, 25, 196, 251, 315, 356
magnetic polarity reversals, 24, 25, 357
magnetite, 38
Malene supracrustals, 212
Manicouagan, 375
Manson, 375
mantle, 19, 21, 24, 25, 26, 28, 30–34, 59, 61, 97, 107, 138, 207, 219, 220, 226, 227, 228, 354, 362, 433
mantle plumes, 26, 28, 30, 31, 33, 97, 111, 224, 228, 362, 366, 370, 433
marginal basin, 32, 219, 221, 223, 224, 433
McDonald Lake, Canada, **206**
Mediterranean Sea, 34
Mendocino fracture zone, **25**, 27, 28, **30**
metamorphic belts, 222
metamorphic processes, 12, 32, 204, 207, 209, 210, 212–214, 219, 220, 222, 226, 434

Earth, *cont.*
  metasediments, 207, 226, 434
  metavolcanics, 204, 434
  Meteor Crater, 46, 74, 372, **373**
  Mexico, 28, 354
  Mexico City, 355
  mica, 203, **210**, 434
  Michigan, 213
  microfauna, 25
  Midcontinental Rift, 213
  mid-Atlantic ridge, 18, 19, 21, 24, 25, 26, 27, 31
  mid-ocean ridge system, **20**, 21, **22**, 23, 28, 33, 113, 224, 435
  mid-ocean rift, 21, 23, 26, 27, 31, 32, 156, 350
  Minnesota, 213
  Minnesota River Valley, 213
  'moho', 32
  Mohorovicic Discontinuity, 19, 32, 52, 138, 228, 434
  Montana, 103, 205, 224, 365
  Morton, Minnesota, 213
  Mount Etna, 70, 97, 305
  Mount Everest, xix, 19
  Mount Narryer, 226
  Mount Pele, 362
  Mount Scones, 226
  Mount St Helens, **364**, **365**, 366, **369**
  Murray fracture zone, 24, **25**, 26, 27, 28, **30**
  Nakhla, Egypt, 106
  Naples, 69
  Nauru Plateau, 362
  neptunism, 12
  Nevada, 117
  Newfoundland, 19, 33
  New York, 45, 223
  New Zealand, 21
  Nordlingen, 372
  North America, 14, 18, 21, 33, 34, 36, 224, 356, 366, 370
  north pole, 14, 18
  North Sea, 18
  North Star Basalt, 214
  Norway, 15, 33, 211
  Nuk gneisses, 212
  ocean floor, xxi, 14–16, 18, 19, **20**, 21, **22**, 23, 24, 26, 30–34, 39, 204, 210, 217, 222, 226, 228, 353, 362
  ocean trenches, 16, 19, 21, 23, 26, 28, 32, 34, 163, 204, 219, 222, 224, 228, 442
  'Old Granite', 207
  Oldoinyo Lengai, 174
  olivine, 30, 31, 32, 40, 107, **210**, 435

Ontario, 46
Ontong Java Plateau, 362, 380
Onverwacht Group, 207, 208, 209, 350
ophiolites, 32, 223, 228, 435
Oregon, 28, 103, 365
orogeny, 15, 16, 18, 19, 21, 33, **34**, 39, 163, 204, 208, 219, 223, 224, 228, 259, 352, 360, 361, 435
orthopyroxene, 31, 207, 256, 261, 435
Pacific Ocean, 14, 19, 23, **25**, 26, 28, 30, 34, 59, 97, 143, 356, 362
pack ice, 265
Padua, 43, 235
palaeomagnetic studies, 18, 24, 37
pali, 134
Pangea, 15, 33, 34, **36**, 37, 203, 204, 224, 360, 375
partial melting, 31, 32, 107, 210, 219, 222, 226, 227, 362, 365, 366
Paris, 143, 183, 309, 353
Permian, 359
Permian-Triassic boundary, **360**, 361, 262
peridotite, 32, 40, **210**, 436
Perth, Australia, 227
Piceance Basin, Colorado, 351
Pilbara craton, 214, **215**, 216, 219, 256, 350, 372
Pilbara Supergroup, 217
pillow lavas, 31, 32, 207, 208, 210, 212, 217, 436
plate motions on a sphere, 27
plate tectonics, xvii, 26, 28, 30, 33, 34, 163, 174, 204, 224, 227, 228, 261, 268, 269, 352, 361
plutonism, 12
polar wandering, 18
Popigai, 375
potassic plagioclase feldspar, 31, 107, 137, **210**
potassium, 24
Prairies, 205
Precambrian, 107, 203–234, 350, 437
precession, 110
Pretoria, 207
Proterozoic, 205, 207-, 209, 212-, 215, 217, 219, 223, 226, 228, 352, 437
proto-continents (from cratons), 217, 219, 222, 223, 228, 352
pyroclastics, 119, 120, 208, 217, 362, 363, 364, 366, 369, 437
pyroxene, 30, 32, 40, 107, **210**, 437
quartz, 203, **210**, 378, 437 (see also shocked quartz)
quartzite, 16, 226, 437

Earth, *cont.*
   radiometric dating, 13, 15, 24, 38, 203, 212, 213, 355, 360, 437
   rare earth elements, 354, 437
   Red Sea, **35**, 111, 113
   remanent magnetism, 18, 24, 438
   resurgent calderas, 87, 253, 363, 365, 366, 367, 370, 438
Reykjanes Ridge, 25
Rhodesia, 207
rhyolite, **210**, 217, 438
Rieskessel, 372, 374
River Amazon, 134, 136
River Niger, 15, 16
Rochas Verdes Complex, Patagonia, 221, 223
Roche radius, 59, 60, 438
Rocky Mountains (Rockies), xvii, 204, 205, 224, **225**
roots (of mountain ranges), 19, 33
rotational axis, 18
rotational rate, 57, 58, 59, 66, 349
Russia, 144
Saglek, Labrador, 212
Saint Pierre, Martinique, 362
salt flats, 105
San Andreas fault, 27, **28**, **29**, 268, 269
San Francisco, 27
Sand River, 207, 228
sandstone, 11, 204, 208, 223, 226, 439
   immature sandstone (see greywacke)
Saxony, 12, 203
Scablands, 103
schist, 12, 204, 222, 439
Scotland, 15, 33, 211, 217
sea floor spreading, 23, 24, 25, 26, 27, 28, 31, 39, 113, 223, 224, 227, 228, 439
Sea of Galilee, 35
seamounts, 19, 23
sedimentary basins, 17, 223, 224, 226, 422
sedimentary processes, 11, 12, 14, 19, 21, 25, 32, 33, 203, 204, 208, 212, 217, 219, 223, 224, 227, 350, 352, 355, 360, 374, 430, 434
seismic propagation, 19, 21, 24, 26, 28, 30, 32, 219, 228, 361, 362, 373
serpentinite, 23, 32, 40, 439
shale, 11, 12, 204, 208, 212, 223, 439
Shamvaian-Bulawaysan-Sebakwian Formation, 207
Shark Bay, 350
shatter cones, 46, 373, 439
Shaw Batholith, 214
Shergotty, India, 106

shocked quartz, 354, 355, 372, 374, 375, 378, 381
shrinking planet, 14
Siberia, 358, 374
Siberian Traps, 358, 361, 362
Siccar Point, Scotland, 11
Sicily, 70, 240, 241, 305
Sierra Leone, 213
silicic magma, 365, 366
siltstone, 11, 440
Slave Province, 205, 226
Snake River Volcanic Plain, 91, 248, 254, 366, **369**, **370**
sodic plagioclase feldspar, 31, 107, 137, **210**
South Africa, 78
South America, 14, 15, 16, 17, 33, 213, 224, 361
south pole
Spain, 43
Spitsbergen, 33
St Petersburg, 309
'stable platform' (see continental platform)
stishovite, 378, 440
Strasbourg, 328
stratovolcanoes, 91, 260, 440
strike-slip faults, 26, 27, 28, **29**, **35**, 205, **206**, 268, 269, 440
subduction, 23, 26, 28, 31, 32, 34, 174, 219, 221–223, 227, 228, 440, 442
'subterranean fires', 12
Sudbury structure, 46, 373
Suevite, 381
sulphur flows, 174
Sumatra, 363, 365
Sundra Strait, 363
supercontinents, 14, 33, 34, 203, 224, 360, 441
Superior Province, 213, 214, 219, 228
'sutures', 33, 163, **206**, 219, 228, 352, 441
Swaziland, 207, **208**, 229
Swaziland Supergroup, 229
Sweden, 71, 146
syncline, 219, 441
Tahiti, 143
Tambora, 362, 366
Tasmania, 16
tektites, 372, 441
Tertiary, 354
Tethys Ocean, 32, 34, 360
Thera (Santorini), 363
tholeiite, 31, 32, 156, 441
Tibetan Plateau, 163
tidal braking, 319, 347
tides, 57, 58, 59, 66

Index   453

Earth, *cont.*
  tillites, 16, 374, 381, 442
  Toba, 365
  tonalite, 210, 212, 217, 222, 223, 226, 442
  transform faults, 27, **29**, **35**, 442
  Transvaal, 207
  triple junctions, 35, 113, 442
  Troodos massif, 32
  tsunamis, 354, 356, 363, 375
  tuff, 212, 442
  Tunguska, 374
  Uivak gneisses, 212
  ultramafic minerals, 30, 32, 40, 107, 207, 208, 209, **210**, 212, 213, 219, 220, 228, 256, 442
  unconformities, 11, 217, 219, 224, 442
  uniformitarianism, 11, 219, 371
  United States, 33, 34, 204, 205, 213, **225**, 354, 356
  Urals, 34
  uranium, 13
  Utah, 117
  Van Allen Belts, 75, 237
  Venezuela, 213
  vertical displacements, 14, 26, 217
  Vesuvius, 380
  volcanic shields, 19, 28, 97, 228, 369, 439
  volcanism, 12, 19, 21, 23, 24, 26, 27, 28, 30, 207, 208, 209, 212, 214, 217, 219, 228, 349, 351, 354, 357, 358, 362, 365, 369, 370, 372, 443
  volcanogenic sediments, 208, 209, 212, 213, 214, 217, 219
  Walvis Ridge, 15
  Warrawoona Group, 217
  Washington State, 28, 103, 113, 224
  Watersmeet, Michigan, 213
  'wet' melting, 32, 174, 222
  Winslow, 45, 74, 372
  'winter' effects, 356, 357, 365
  Wittenberg, 240
  Wyoming, 88, 205, 253, 365, 370
  Yellowknife Province, 205, **206**
  Yellowstone, 88, 253, 365, 366, **367**, **368**, 369, 370, 380
  Yilgarn craton, 214, **215**, 226, 227, 372
  Yucatan Peninsula, 354, **355**, 356
  zeolite, **222**
  Zimbabwe, 207
  Zimbabwean craton, **208**, 210, 228
  zircons, 213, 214, 226, 227, 360, 443
Earth sciences, xxi
ecliptic plane, 143, 240, 285, 310, 312, **313**, 328, 340, 341, 427

Ecole Normale Superieure of Lyon, 227
Edgett, K.S., 115, 126
Einstein, Albert, 184, 198
Elizabeth (Queen), 18
Enceladus (satellite of Saturn), **294**
  albedo, 291, 293
  bulk density, 291
  cratered plains, 293
  cryovolcanism, 293, 294
  discovery, 287, 305
  grabens, 293
  isostasy, 293
  radiogenic heating, 293
  relationship to 'E' ring, 293, 306
  resonances, 290, 293
  ridged plains, 293
  smooth plains, 293
  tectonism, 293
  tidal heating, 293
Encke, Johann Franz, 143, 286
Encyclopaedia Britannica, 14
Epimetheus (satellite of Saturn), 302, 304, 329
Euler, Leonard, 27, 40
Europa (satellite of Jupiter), xix, 262–271, 311
  age of surface, 264, 268, 271, 277
  albedo variations, 237, 238, **262**
  anti-Jovian region, **263**, 265
  Astypalaea Linea, 268, 269, 270
  atmosphere, 239, 262
  bright plains, 264
  bulk density, 240
  'chaos zones', 265, 268, 270
  Conamara, 265, **266**, **267**, 268, 270
  core, 265
  craters, 262, 264, 265, 268
  cycloids, 269
  dark plains, 264
  diameter, 238
  discovery, 235
  elliptical orbit, 269
  fractured plains, 264, 265
  Galileo (spacecraft), 264, 265, **266**, **267**, 268, 269, 270, 271
  hydrothermal activity, 265
  icebergs, 265, **266**, **267**, 268
  icy surface, 239, 262
  infrared studies, 239
  isostasy, 264
  Libya Linea, 270
  life, xix, 265, 283
  lineaments, 262, 265, 268, 270
  lithosphere, 265, 268, 271

Europa, *cont.*
  maculae, 264, 269
  magnetic field, 265, 271
  mantle, 265
  moment of inertia, 265
  mottled terrain, 262, 264, 265
  ocean, xix, 264, 265, 268, 269, 270, 271
  palimpsests, 264
  Pioneer image, **243**
  Pwyll Crater, 265, 268
  radiogenic heating, 265
  resonances, 237, 247, 265, 271
  rotation, 238
  silicate, 275, 271
  south polar region, 268, 269
  spreading, 265
  strike-slip fault, 268
  sub-Jovian region, 269
  surface properties, 238, 239
  Thera, **270**
  thickness of ice, 264, 265, 268, 269
  Thrace, **270**
  tidal bulge, 269, 270
  tidal heading, 265
  undifferentiated plains, 264
  volcanism, 265, 268
  Voyager (spacecraft), 262, 263, 265, 269, 270
  wedges, 264, 265

field geology
  on Io, 260
  on Mars, 126
  on the Moon, xxi, 36, 47, **50**, 63
Firsoff, V.A., 147, 149
Fisher, Osmond, 14, 59
Flammarion, Camille, 70, 128
Flamsteed, John, 325, 327
Flandro, Gary, 245, 280
Fontana, Francesco, 69, 144, 197, 236
Forbes, Edward, 18
Forbes, George, 342
Fournier, G., 286, 305
Franklin, Kenneth L., 237
French Academy of Sciences, 326

Galilei, Galileo, 43, 69, 143, 175, 183, 197, 235, 245, 279, 284, 285, 304, 327
Galle, J.G., 286, 326, 327, 328, 333
Ganymede (satellite of Jupiter), 271–275, 290, 298
  age of surface, 273, 277
  albedo variations, 237, 238, **238**, 271, 273
  atmosphere, 239
  Barnard Regio, 274
  basins, 273
  bulk density, 240, 271
  core, 275
  craters, 268, 274
  cryovolcanic activity, 274
  dark cratered terrain, 271, 277, 284
  diameter, 238
  discovery, 235
  Galileo Regio, 271, **272**, 273, 274, 275
  Galileo (spacecraft), 274
  Gilgamesh basin, 273, 274, 277
  grooved terrain (see sulci)
  impacts, 268, 273, 274
  infrared studies, 239
  isostasy, 273
  magnetic field, 274
  mantle, 275
  Marius Regio, **272**, 273, 274
  Nicholson Regio, **272**, 274
  Osiris Crater, **272**, 274
  palimpsests, 273
  Perrine Regio, 274
  Pioneer image, **243**
  'polar spot', 238, 239, 243, 273
  radiogenic heating, 275
  resonances, 237, 247, 275
  rotation, 238, 275
  salty fluid layer, 275
  silicate, 275
  size, 271
  spreading, 272, 273
  strike-slip fault, 273
  sulci, 271, 272, 273, 274, 441
  surface properties, 238, 239
  tectonism, 274
  temperature, 271
  tidal heating, 275
  Tiamat Sulcus, **272**, 273
  Uruk Sulcus, **272**, 273, 274
  Voyager (spacecraft), 271, 274, 275
Gass, Ian G., 32
Gassendi, Pierre, 143, 183
Gauss, Carl Friedrich, 241
geology, 14
George III (King), 309
Gerstenkorn H., 60
Gilbert, Grove Karl, 33, 45, 60
Gilbert, William, 18
Giotto (spacecraft), 324
Glatzmaier, Gary, 24
Goldstone antenna, 79, 148
'Grand Tour'
  final objective, 330, 338

Index 455

'Grand Tour', *cont.*
  'Mariner Mk 2', 245
  multiple-planet missions', 245
  slingshots, 245, 278, 279, 280, 284, **288**, 291, 304, 312, **313**, 320, 324, 329
  Voyager 2 (spacecraft), **244**
Gregory, James, 143
Green, Charles, 143
Green, Nathaniel, 70
Greenwich Observatory, 309, 325
Grimaldi, Francesco Maria, 44
Gruithuisen, Franz von Paula, 144, 145
Gutenberg, Beno, 30
Guthnick, P., 237, 238, 239, 290

Hadley, George, 160
Hadley, John, 286
Hale, George Ellery, 5, 72, 239, 287, 340
Hall, A, 207
Hall, Asaph, 289
Hall, James, 223
Hall, Maxwell, 327
Halley, Edmund, 158
Halley's Comet, 158, 324
Hanlon, Michael, xx
Harding, Karl Ludwig, 144, 241
Harding, K.L., 241
Harper, Charles, 226
Hartmann, William K., 61, 85, 86, 126
Harvard College Observatory, 45, 287
Harvard University, 226
Haystack antenna, 79
Head, James W., 118, 153, 168
Heezen, Bruce Charles, 21
Helene (satellite of Saturn), 304
Helmholtz, Hermann Ludwig Ferdinand von, 4, 13
Hencke, K.L, 241
Hengst, Rich, 357
Herschel, Frederick (Sir) William, 1, 4, 44, 69, 70, 144, 183, 237, 238, 241, 287, 289, 300, 309, 312, 326
Herschel, (Sir) John Frederick William, 305, 328
Hess, Harry Hammond, 23, 24, 32
Hevelius, Johannes, 43, 197
Hildebrand, Alan, 354
Hind, John Russell, 241, 328
Hoffman, Nick, 118, 126
Hohmann transfer orbit, 73, 278, 329, 430
Hohmann, W., 73
Holmes, Arthur, 13, 19, 31
Hooke, Robert, 44
Hooker, John D., 9, 45, 72, 145, 340

Hopkins, Evan, 14
Horrocks, Jeremiah, 143, 175
Housden, C.E., 146
Howard, W.E., 184
Hoyle, (Sir) Fred, 5, 147
Hubble, Edwin Powell, 9
Hubble Space Telescope, 8, 141, 197, 254, 256, 298, 316, 333, 339
Huggins, (Sir) William, 70
human space exploration, xx, xxii, 62, **63**
Humason, Milton, 339, 340
Humbolt, Alexander von, 14
Hurley, Patrick M., 15
Hussey, Thomas John, 325
Hutton, James, 11, 12, 217, 374
Huygens, Christiaan, 4, 69, 285
Hyperion (satellite of Saturn)
  bulk density, 291
  craters, 299
  discovery, 287, 305
  irregular shape, 299
  progenitor, 299
  resonances, 290, 300
  rotation, 299, 302

Iapetus (satellite of Saturn), **300**
  albedo, 301
  brightness variations, 300
  bulk density, 291, 301
  carbonaceous material, 301
  Cassini (spacecraft), 301
  Cassini Regio, 300, 301, 307
  craters, 301
  cryovolcanism, 301
  discovery, 285, 305
  inclination, 302
  radiogenic heating, 301
  resonances, 290
  rotation, 300, 307
  tar, 301
Imperial College London, 25
'independent thinkers', 38, 60, 85, 148
Infrared Astronomical Satellite, 342
Infrared Telescope Facility, 250, 252, 253, 333
Institute of Geophysics and Planetology, Hawaii, 261
International Astronomical Union, 79, 178, 184, 188, 237, 245, 269, 280
International Ultraviolet Explorer (satellite), 315
Io (satellite of Jupiter), xvii, 246–261
  albedo variations, 237, 238, 246
  analogy for early Earth, 234, 256, 261

Io, *cont.*
  anti-Jovian region, **255**
  atmosphere, 239, 240, 247, 254
  Babbar Patera, 256
  basalt, 249
  Boosaule Mons, 281
  bulk density, 240
  calderas, 249, 250, 252, 253, 254, **255**, 256, 257, 260
  Carancho Patera, **248**, 254
  Cassini (spacecraft), 259
  Colchis Regio, **255**
  core, 252
  craters, 246, 250, 261
  desiccated, 247, 249
  diameter, 238
  discovery, 235
  elliptical orbit, 247
  Emakong Patera, 257
  erosional processes, 250, 260
  Euboea Mons, **248**, 251, 259, 260
  fire fountains, xix, **258**, 259
  fissure eruptions, 252, 258
  'flux tubes', 251, 252
  Galileo (spacecraft), 252, 253, 254, 255, 256, 257, 258, 259, 261
  gravitational field, 252, 260
  Haemus Mons, 249
  heat flow, 247, 251, 253, 260, 261
  'hot spots', 247, 251, 252, 256, 261
  inter-vent plains, 250, 253, **255**
  'Io-quakes', 260
  ionosphere, 243, 251
  'Jupitershine', 246
  komatiite, 256, 261
  lava lakes, xix, 250, 252, 253, 254
  layered plains, 250
  lithosphere, 247, 249, 250, 261
  lithospheric recycling, 174, 260, 261
  Loki, 178, **248**, 250, 252, 253, 256, 259, 280
  magma ocean, 247, 249, 261
  magnetic field, 252
  mantle, 251, 252, 260, 261
  moment of inertia, 252
  mountains, xix, 248, 249, 259, 260, 261, 281
  Pioneer image, **243**
  plains material, 250
  plumes, 246, 247, 250, 252, 253, 254, **255**, 256, 280
    as caldera emissions, 254
    as frost sublimation, 257
    as geysers, 249, 251, 257
  Ra Patera, 133, **248**, 249, 254
  Pele, 250, 253, 254, 256, 259, 260, 280
  Pillan Patera, 256
  plasma torus, 251, 252
  plate tectonics, 251, 261
  Prometheus, **255**, 256, **257**
  pyroclastics, 249, 254, 256, 259, 261
  radiogenic heating, 247
  reddish hue, 240, 249
  resonances, 237, 247
  resurfacing, 246, 250, 260
  rotation, 238, 247
  silicate magma, 247, 249, 250, 253, 254, 256, 257, 259, 261
  south pole, 245
  sub-Jovian region, **248**, 249, 254
  sulphur chemistry, 249, 250
  sulphur dioxide, 247, 249, 250, 252, 253, 257
  sulphurous magma, 247, 249, 254, 257, 261
  surface properties, 238
  tectonism, 249, 259
  'tidal bulge', 247, 249
  tidal heating, 247
  Tvashtar Catena, 257, **258**
  ultramafic minerals, xix, 107, 256, 260, 261
  vent-related materials, 250
  volatiles, 247, 249
  volcanic shields, 251, 254
  volcanism, xviii, 246, 247, 248 254, 256-, 161
  Voyager (spacecraft), 245-, 248, 252, 254, 259, 261
Irwin, James B., xxi, 48, 51

Jacobson, Stein, 226
James (King), 11
Janssen, Jules, 70, 73
Janus (satellite of Saturn), 304, 329
Jeans, James Hopwood, 4
Jeffreys, Harold, 4, 5, 19, 59, 236, 287, 290
Jenkins, Greg, 349
Jet Propulsion Laboratory (JPL), 159, 245, 276, 279, 291, 293, 313, 329, 342
Johns Hopkins University, 201
Joly, John, 12
Jones, (Sir) Harold Spencer, 148
Jupiter, xvii, 4, 145, 183, 235–286, 302, 376
  atmosphere, 236
    ammonia, 236
    helium, 236
    hydrocarbon haze, 314, 430
    hydrogen, 236
    latitudinal circulation, 236, **239**, 242, 245, **246**
    methane, 314
    spots, 236

Jupiter, *cont.*
  aurorae, 251, 252
  bulk density, 236
  core, 236, 237
  discovery of satellites, 235
  energy budget, 236, 331
  'failed star', 236
  Galileo (spacecraft), xvii, xix, xxi, 251, 252
  'gas giant', 236
  Great Red Spot, **246**, 289
  impactor rate in Jovian space, 268, 277
  internal structure, 236, 305
  magnetic field, 237
  magnetosphere, 237, 242, 243, 251, 252, 264, 274
  obliquity, 236
  physical properties, 236
  Pioneer (spacecraft)
    Pioneer 10, 242, 251
    Pioneer 11, 242
    slingshot, 243
  radio emissions, 237
  rings, 324, 345
  rotation, 236, 237
  satellites (see Amalthea, Callisto, Europa, Ganymede, Io)
  Voyager (spacecraft)
    Voyager 1, 245, 246
    Voyager 2, 245

Kaiser, Frederick, 70
Kant, Immanuel, 2, 4, 57
Kasting, James, 349, 351
Keck Telescope, 259
Keeler, J.E., 286
Kelvin (Lord), 13
Kepler, Johann, 1, 143, 175, 183, 235, 236, 240, 286, 290, 302, 329
Keszthelyi, Laszlo, 259, 261, 357, 358
Keys, David, 363
Kieffer, Susan, 251
Kirch, 144
Kirkwood, Daniel, 242, 287
Klavetter, James, 307
Kozyrev, Nikolai A., 146
Kubrick, Stanley, 300, 307
Kuiper Airborne Observatory, 312
Kuiper Belt, 328, 342, 432
Kuiper, Gerard Peter, 47, 60, 73, 146, 198, 239, 297, 310, 328, 334, 340
Kyte, Frank T., 356

Lagrange, Joseph Louis, 304, 329, 432
Lalande, Joseph Jerome le Francais, 309, 327

Lampland, C.O., 339
Lamont-Doherty Observatory, 21, 24, 26
Landis, Gary, 357
Langrenus, Michael Florentius, 43
Laplace, Pierre Simon, 2, 4, 286, 309, 325
Larissa (satellite of Neptune), 329, 334
Lassell, William, 286, 287, 310, 328, 329, 333, 345
LaTrobe University, 118
Lemonnier, P.C., 325, 327
Leverrier, Urbain Jean Joseph, 242, 326, 327, 333
Lexell, Anders Johan, 309
Liais, Emmanuel, 70
Lick Observatory, 145, 146, 237, 327
light speed, 236, 279, 312
Lilienthal, Theodore Christopher, 14
Lippershey, Hans, 43
Lohrmann, Wilhelm Gotthelf, 44
Lomonosov, Mikhail Vasilievitch, 144
Longhi J., 56
Los Alamos National Laboratory, 24, 227
Lowell Observatory, 78, 147, 310, 339
Lowell, Percival, 72, 74, 76, 136, 145, 146, 183, 238, 262, 286, 338, 341
Lucas, George, 293
Lunar and Planetary Institute, 51, 55, 259, 355, 356
Lunar and Planetary Laboratory, 198, 259, 298, 357, 358
Lunar Science Conference, 61
Lyell, Charles, 12
Lyman, C.S., 144
Lyot, Bernard Ferdinand, 146, 147, 238, 262, 280, 297
Lyttleton, R.A., 342

Macgregor, A.M., 207
Madler, Johann Heinrich, 44, 70, 144, 237
Malin, Michael, 108, 115, 118, 123
Malin Space Science Systems, 115
Maraldi, Giacomo, 69
Marcy, Geoffrey W., 6, 7, 8
Marius, Simon, 2, 235, 237, 279, 284
Mars, xviii, 69–141, 240, 242, 262
  Acidalia Planitia, **122**, 141
  Alba Patera, 82, 91, **92**, 93, 94, 133, 135
  albedo variations, 69, 70, 74, 79
  Albor Tholus, 98, **99**
  aluminous silicates, 95
  Amazonis Planitia, 74, 90, 91, **92**, 98, **121**, 126
  Amphitrites Patera, 87
  'ancient Mars', 72, 146

Mars, *cont.*
  andesite, 108
  Apollinaris Patera, 88
  Apollonius, 91
  'Arabia shoreline', 117
  Arabia Terra, 81
  Arcadia Planitia, 91
  Ares Vallis, 101, **102**, 107, **109**, 138
  Argyre basin, 84, 85, 104, 131
  Arsia Mons, 79, **92**, 94, 98, 134
  artesian flow, 101, 117, 123, 126
  Ascraeus Lacus, 78
  Ascraeus Mons, 79, **92**, 95, 98
  ash deposits, 71, 84, 86, 87, 92, 96, 98
  Atlantis, 74
  atmosphere
    carbon dioxide, 73, 105, 106
    composition, 70, 73, 75, 106
    nitrogen, 75, 106, 127
    oxygen, 73, 75, 106
    pressure, 70, 73, 75, 76, 92, 100
    water vapour, 70, 71, 73, 75, 106
  aureole, 96, 97, **121**
  Aurorae Sinus, 76
  'Barnacle Bill' (rock), 108, **109**
  basalt, 95, 108, 123
  base surges, 92, 96
  basins, 80, 84, 85
  Biblis Patera, 95
  Big Crater, 108
  bombardment, 86
  'Borealis basin', 80, 129
  breccia, 108
  brine, 131
  calcic clinopyroxene, 107
  calcite, 105
  calcium carbonate, 105, 118, 123
  calderas, 79, 84, 86, 89, 92, 94, 95, 98, 365
  'canali', 72
  canals, 72, 74, 136
  Candor Chasma, 100, **116**
  canyons, 74, 93, 100
  Ceraunius Tholus, 89
  chaotic terrain, 77, 101, 120
  Chartitum Montes, 84
  Chryse basin, 80, 84, 101
  Chryse Planitia, 84, 85, **92**, 100, 101, 104, 107, 118, 141
  Claritas Fossae, 93, 113
  clathrates, 126
  clays, 123, 141
  climate
    'cold and dry', 82, 100, 103, 118, 119, 123
    'warm and wet', 82, 88, 98, 123
  clouds, 70
    white, 77, 84
    yellow, 77
  cones, 90, 95
  conglomerates, 108
  Coprates Chasma, 100, 101, **125**, 136
  core, 80, 110
  cratered terrain, 77, 79, 194
  craters, 74, 76, 81, 83, 85, 101, 126, 184, 199
  crustal thickness, 81
  cryoclastic flows, 103, 108, 119, 126, 425
  Cydonia Mensae, **122**
  Daedalia Planum, 84, **92**
  Dao Vallis, 123
  Deucalionis Regio, 76
  'Deuteronilus shoreline', 118
  domes, 91
  drainage systems, 82
  'dry' melting, 108
  dust storms, 77, 78, 79, 104, 108, 117, 128
  dynamic instabilities, 82
  Echus Chasma, 101
  ejecta, 83, 108
  Electris, 75, 104
  elliptical orbit, 69, 74, 78, 128
  Elysium Fossae, 98, **99**
  Elysium Mons, 98, **99**, 101
  Elysium Planitia, **88**, 98, **99**, 100, 118, 126
  Elysium province, 86, 91, 98, 104, 111, 129
  Eos Chasma, 100, 101
  erosional processes, 75, 80, 82, 84, 85, 88, 91, 94, 98, 100, 102, 103, 105, 118, 123, 126
  evaporites, 105, 118, 123, 141, 427
  'featureless' terrain, 77, 84
  ferromagnesian clays, 105
  flood plains, 107, 108
  flow fronts, 89
  fluidised ejecta, 83, 86, 130
  garnet, 107
  glaciers, 100, 103, 108
  Gordii Dorsum, 113
  grabens, 93, 94, 96, 98, 100
  gravity, 76
  gullies, 100, 123, **124**, 126
  Gusev Crater, 82
  Hadriaca Patera, 87
  heat flow, 80, 81, 87, 98, 108, 126
  Hecates Tholus, 98, **99**
  Hellas basin, 76, 77, 78, 84, 85, 86, 87, 89, 101, 104, 130
  Hellespontica Depressio, 76

Mars, *cont.*
  Hellespontus, 76, 78
  Hellespontus Montes, 84
  hematite (coarse grained), 123
  hematite (fine grained), 123
  hydrated minerals, 105, 123
  hydrosphere, 75, 82, 83, 86, 87
  hygroscopic salts, 71
  hydrothermal activity, 118, 123, 126
  igneous rocks, 108
  ignimbrites, 86, 96
  impacts, 74, 77, 80
  intercrater plains, 83, 86
  ionosphere, 110
  iron oxide, 73
  iron-bearing minerals, 95, 105, 107
  Isidis basin, 80, 84, **88**, 89
  isostasy, 80, 93, 95
  Juventae Chasma, 101
  Kaiser Sea, 126
  Kasci Vallis, 101, 136
  lakes, 101
  launch windows, 73, 74
  lava fllow rates, 94, 134
  lava flows, 84, 89, 92, 93, 94, 95, 101, 126
  lava lakes, 95
  Libya Montes, 84, **88**
  leveed channels, 89, 94
  life, xviii, 70, 71, 72, 73, 74, 77, 85, 104, 283
  line of dichotomy, 79, 80, 81, 82, 84, 86, 88, 93, 98, 101, 102, **112**, 113, 117
  lithosphere, 80, 85, 86, 93, 97, 105, 108, 111, 126
  Lunae Planum, 89, 91, **92**, 101, 136
  Lycus Sulci, **121**
  Ma'adim Vallis, 82
  mafic minerals, 95, 105
  maghemite, 110
  magma chambers, 95, 96
  magnetic field, 75, 110, 200
  magnetic polarity reversals, 111
  Malea Planum, 87
  magnesium-bearing minerals, 95, 105
  Maja Vallis, 101
  mantle, 80, 81, 107
  mantle plumes, 80, 93, 97
  Mare Acidalium, 75
  Mare Chronium, 75
  Mare Hadriaticum, 76
  Mare Sirenum, 74
  maria (see ridged plains)
  Mariner (spacecraft)
    Mariner 3, 127

  Mariner 4, 74, 75, 77, 127, 184, 185
  Mariner 6, 75, 76, 77, 78
  Mariner 7, 75, 76, 77, 78, 84
  Mariner 8, 77
  Mariner 9, 77, 78, 79, 81, 84, 85, 101, 136
  Mariner Crater, 74, 128
  Mars (spacecraft)
    Mars 1, 127
    Mars 2, 78, 103, 104
    Mars 3, 78, 104
    Mars 4, 104
    Mars 5, 104
    Mars 6, 104
    Mars 7, 104
  Mars Global Surveyor (spacecraft), xvii, 81, 90, 92, 110, 112–117, 119, 121–126, 128, 131, 135, 141
  Mars Pathfinder (spacecraft), 106, 107, 108, 110, 138, 141
  marsquake, 106
  megaregolith, 83, 86, 434
  Melas Chasma, 100
  Memnonia, 74
  Meridiani Terra, 79, 123, 128
  Meroe Patera, **88**
  'Middle Spot', 78, 129
  moment of inertia, 110
  mountains, 72, 79
  Mutch Memorial Station (see Viking 1)
  Nanedi Valles, **114**
  Nereidum Montes, 84
  Nili Fossae, **88**
  Nili Patera, **88**
  Nirgal Vallis, 123
  Nix Olympica, 76, 78, 79
  Noctis Labyrinthus, 93, 94, 100, 135
  Nodus Gordii, 78
  nomenclature, 70, 71, 79
  'North Spot', 78, 129
  northern plains, 79, 80, 81, 84, 89, 91, 101, 115, 117, 119, 123, 140
  Novissima Thyle, 127
  oases, 74
  ochre tracts, 70
  obliquity, 69
  'Oceanus Borealis', 117, 118, **119**, **120**, 121–123
  olivine, 107, 123
  Olympus Mons, xviii, 79, **92**, 95, 96, 97, 98, 121, 134, 135
  Ophir Chasma, 100
  oppositions, 73, 78
  orthopyroxene, 107

Mars, *cont.*
  outflow channels, 80, 91, 98, 100, 101, **102**, 103, 104, 107, 113, 115, 118, 120, 123, 136
  parasitic vents, 95, 134
  partial melting, 105, 108
  Pavonis Lacus, 78
  Pavonis Mons, 79, **92**, 95, 98
  Peneus Patera, 87
  permafrost, 82, 83, 87, 90, 98, 101, 117, 126, 131
  Phaetontis, 74, 104
  Phlegra Montes, 80, 113
  phraeto-magmatic eruptions, 86, 87, 91, 111, 436
  phreatic eruptions, 90, 91, 436
  planetary datum, 76, 128
  plate tectonics, 97, 111, 113
  polar caps, 69, 70, 71, 78, 126
    carbon dioxide frost, 73, 75, 76
    temperature, 75, 76
    water ice, 69, 73, 75, 77, 140
  potassium-bearing minerals, 95, 105
  precesion, 110
  pseudocraters, 90
  pyroclastic flows, 86, 92, 96, 119
  pyroxene, 123
  Pyrrhae Regio, 76
  radar observations, 78, 79
  regolith, 105, 108
  remanent magnetism, 110
  ridged plains, 80, 86, 89, 91, 100, 107, 111
  Rima Angusta, 127
  Rima Australis, 127
  rivers, 82
  rootless cones, **90**, 92, 438
  rotation, 69
  runoff channels, 82
  Sabaeus Terra, 79
  Sagan Memorial Station (see Mars Pathfinder)
  sapping channels, 82, 87, 98, 439
  'sculpture', 80, 84, 439
  'seas', 70, 71, 73, 82
  seasons, 69, 70, 72, 73, 77, 128
  sedimentary processes, 79, 80, 84, 86, 91, 102, 103,, 108, 113, 115–123, 126
  seepage, 124
  Shalbatana Vallis, 101
  shorelines, 117–120, **121**, **122**, 123
  silica, 105, 108, 113
  Simud Vallis, 101, **102**
  Sinus Meridiani, 76, 128
  SNC meteorites, 80, 106, 108, 110, 113
  Sojourner (rover), 106, 107, **109**
  'South Spot', 78, 129
  southern highlands, 79, 81, 82, 83, 84, 86, 89, 93, 115, 123
  stratification (large scale), **125**, 126
  subduction trench, 113
  sulphate, 105, 123
  sulphur, 105, 108, 123
  Syria Planum, 91, 93, 100
  Syrtis Major Planum, 84, **88**, 113, 126
  telescopic view, 72
  Tempe Terra, 80, 89, 91, **92**, 93, 94
  Tharsis Montes, 93, 98
  Tharsis province, 79, 80, 84, 86, 89, 91, **92**, 94, 95, 96, 100, 101, 111, 113, 118
  Tharsis Tholus, 95
  Thymiata, 76
  titanium-bearing minerals, 105
  Tithonium Chasma, 126
  Tithonius Lacus, 72
  Tiu Vallis, 101, **102**, 138
  topographic map, **112**
  Trivium Charontis, 74
  Twin Peaks, **109**
  Tyrrhena Patera, 86, 87, 89
  Ulysses Patera, 95
  Utopia Planitia, 84, 104, 107
  Valles Marineris, 91, 93, 98, 100, 101, 111, 113, 116, 123, 126, 136, 345
  Vastitas Borealis, 80, 91
  vegetation, 71, 73, 74, 77
  Viking (spacecraft), **xxiii**, 79, 81, 84, 87, 88, 95, 97, 99, 105, 108, 110, 114, 117, 121, 123
    Viking 1, 104, 106, 107
    Viking 2, 104, 106
  volcanic shields, 79, 86, 87, 88, 93, 95, 96
  volcanism, 71, 72, 79, 81, 82, 84, 86, 87, 88, 89, 93, 111
  water (liquid), 75, 76, 82, 83, 86, 101, 113–126, 140
  water flow rates, 108, 136
  'wave of darkening', 71
  weather system, 77
  'White Mars' hypothesis, 119, 123
  wrinkle ridges, 85, 89, 94, 95, 96, 118, 443
  Xanthe Terra, 101
  'Yogi' (rock), 108, **109**
  Zephyria, 74
  Zond (spacecraft)
    Zond 1, 127
    Zond 2, 127
Marshall, Charles, 359
Maskelyne, Nevil, 143, 309

Mason, Ronald, 25
Massachusetts Institute of Technology, 15, 58
Masson-Smith, David, 32
Masursky, Harold, 156
Matthews, Drummond H., 24
Mayer, Tobias, 44
Max Planck Institute, 352
Maxwell, James Clerk, 4, 178, 286
McDonald Observatory, 146, 239, 297, 310, 328
McEwen, Alfred, 249, 261
McEwen, Henry, 145
McGregor, Vic R., 212
McKenzie, Dan P., 26, 32
McLaughlin, Dean B., 71
Menard, H.W. (Bill), 21
meteorites, 44, 354, 372
   ALH77005, 137
   ALH84001, xviii
   basalt, 107
   carbonaceous ('C') type, 301
   Chassignites, 107
   dunite, 107
   EET79001, 137
   Nakhlites, 106, 118
   nickel-iron, 46, 372, 373
   olivine, 107
   ordinary chondrites, 13
   pyroxene, 107
   radiometric dating, 13, 106
   Shergottites, 106, 137
   SNC meteorites, 80, 106, 108, 110, 113
Menzel, Donald H., 147, 327
Mercury, xvii, 13, 145, 151, 183–201, 240, 271, 298
   albedo variations, 183, 188, 199
   atmosphere, 184, 197, 239
   base surge, 189
   basins, 84, 85, 193, 194
   bulk density, 185, 188
   Caloris basin, 85, **187**, 189, **191**, **192**, 193, 194
   Caloris Montes, 189, **191**, **192**
   Caloris Planitia, **191**, **192**
   core, 62, 185, 188, 194, 196
   craters, 83, 184, 188, 193, 194, 195, 198
   ejecta, 83, 188, 189, 193
   elliptical orbit, 183
   elongation, 184, 197
   escape velocity, 184, 239
   fluidised ejecta, 193
   geological timescale, 193, 194
   global expansion, 196
   global shrinkage, 195, 196
   grabens, 193
   grooves, 193
   'hot poles', 184, 189, 198
   heat flow, 196
   heavily cratered terrain, 188, 194
   impacts, 184, 188
   intercrater plains, 189, 194
   iron-bearing minerals, 188
   isostasy, 193, 195
   lineations, 183, 195
   lithosphere, 196
   mafic magma, 194
   magnetic field, 196
   mantle, 188, 194
   Mariner (spacecraft)
      Mariner 10, 151, 184, 185, 188, 189, 193, 194, 197
      'slinghot', 151, 176, 185
   mass, 198
   meridian, 184
   Messenger (spacecraft), 201
   microwave emissions, 184
   nomenclature, 188
   Odin Planitia, **191**, **192**
   original crust, 189, 194
   phases, 183, 197
   radar studies, 184, 193, 197
   regolith, 184, 438
   ridges, 193, 195
   rotation, 183, 184, 195, 196
   scarps, 195
   sculpture, 189, 193, 439
   seasons, 183
   seismic propagation, 194, 361
   smooth plains, 190, 191, 192, 193, 195
   spin-orbit coupling, 184, 195
   Suisei Planitia, **190**
   surface gravity, 199
   temperature, 184, 239
   Tir Planitia, **192**
   transits of Sun, 183
   Van Eyke Crater, **191**
   Van Eyke Formation, 193
   volatiles, 183, 188, 197
   volcanism, 194, 195, 196
   Vostok Rupes, 195
   'weird' terrain, 194
Messier, Charles, 2
Meudon Observatory, 145, 184
Mignard, Francois, 307
Miller, Kenneth, 356
Mimas (satellite of Saturn), **292**, 304, 320
   bulk density, 291

Mimas, *cont.*
  craters, 292
  discovery, 287
  Herschel Crater, 292, 293, 294
  resonances, 290, 303
Miranda (satellite of Uranus), xix, **320**
  Arden Corona, **320**, 321, 322
  craters, 319, 320, 321
  'chevron', 321
  coronae, 321, 322
  discovery, 310
  elliptical orbit, 322
  Elsinor Corona, **320**, 321
  Inverness Corona, **320**, 321
  'ovoids', 320, 321, 322
  resonances, 322
  rolling plains, 320, 322
  scarps, 321
  tidal heating, 322
  Verona Rupes, 321, **322**, 345
  Voyager 2 (spacecraft), 312, 320
Mitchell, Edgar D., 48
Mohorovicic, Andrija, 19, 21
Montaigne (of Limoges), 145
Montreal Ecole Polytechnique, 228
Moon, xxi, 43–67, 185, 290
  Aitkin basin, 57, **58**
  albedo variations, 194, 199
  aluminous silicates, 47, 51, 52, 53, 57, 194
  angular momentum, 59, 61, 319
  anhydrous minerals, 47
  anorthosite, 51, 52, 53, 55, 56, 57, 61, 62, 194, 420
  ANT, 55, 420
  Apennine Bench Formation, 57
  Apennine Range, xxi, 36, 48, 51, 57, 84, 189
  Apollo missions, xxi, 47, 60, 62, 137, 374
    Apollo 11, 48
    Apollo 12, 48
    Apollo 14, 48
    Apollo 15, 48, **49**, 51, 55, 57
    Apollo 16, 52, **63**
    Apollo 17, 52
  Archimedes Crater, 57
  Aristarchus Crater, 45
  basalt, 47, 48, 52, 53, 56
  basins, 46, 53, 54, 56, 57, 84, 85
  breccias, 48, 52, 422
  bulk density, 47, 59, 61
  calderas, 46
  calcic plagioclase feldspar, 47, 53, 61
  capture theory, 59, 60
  Cayley Formation, 52, **63**, 193
  Central Highlands, 52
  'circular maria', 46, 47, 424
  clasts in breccias, 48
  Clementine (spacecraft), 58, 62, 64, 67
  'cold' Moon hypothesis, 47
  Cone Crater, 48
  Cordillera Range, **54**
  core, 52, 53
  'cratering curve', 46, 47, 189
  craters, 43, 44, 74, 83
  crystallisation process, 57
  cumulates, 56
  Descartes Formation, 52
  domes, 47, 52
  double planet hypothesis, 60
  'Earthshine', 145
  ejecta, 45, 52, 83, 189
  feldspathic minerals, 48, 428
  fire fountains, 52
  fission hypothesis, 59
  Flamsteed Crater, 52
  flow fronts, 47
  fluidised ejecta, 52, 428
  formation theories, 14, 59
  Fra Mauro Formation, 48, 52, 193, 428
  'gardening', 36
  geological timescale, 48, 56, 185
  giant impact theory, 61, 62
  Gruithuisen hills, 52
  Hadley-Apennine landing site, 36, **49**, 51
  Hadley Rille, 48, **49**, **50**, 52, 432
  'heat engine', 52
  highlands, 43, 55
  igneous processes, 44
  ilminite, 48
  Imbrium basin, xxi, 47, 48, 49, 51, 53, 84, 85, 131, 189, 193
  impact melt, 47, 57
  impacts, 44, 45, 46, 52, 54, 74, 381
    primary, 45
    secondary, 45
  iron-bearing minerals, 52
  KREEP, 57, 65
  lava flows, 47
  lava tubes, 47
  low-alkali basalt, 47
  Luna 9 (spacecraft), 47
  Lunar Orbiter (spacecraft), 53
    Lunar Orbiter 4, 54
  Lunar Prospector (spacecraft), 53, 57, 62
  Lunik, 3 (spacecraft), 52
  Lunokhod (spacecraft), xxi
  mafic minerals, 52, 55, 56
  magma ocean, 55, 56, 57, 61, 247, 433

Moon, *cont.*
  'magnesian suite' minerals, 55, 56, 57, 433
  magnesium-bearing minerals, 52, 55
  magnetic field, 53, 196
  mantle, 55, 56, 61
  Mare Imbrium, xxi, 46
  Mare Marginis, 53
  Mare Tranquillitatis, 47, 48, 53
  maria, 43, 46, 47, 52, 57, 58, 80, 89, 91, 193, 194, 219
  Marius hills, 52
  mass, 57, 198
  megaregolith, 52, 56, 373
  moonquakes, 53
  Mount Hadley, xxi
  Mount Hadley Delta, 49, 51
  'Mountain of Eternal Light', 64
  mountains, 43, 46, 47
  non-mare basalt, 56, 57
  norite, 55, 56, 435
  North Complex, 49
  Oceanus Procellarum, 47, 48, 52, 53
  olivine, 48, 55
  'orange soil', 52
  Orientale basin, 53, 54, 55, 58, 276
  poles, 62
  potassium, 56
  pristine state, 47
  pyroclastics, 52
  pyroxene, 48, 55
  rare earth elements, 59
  recession, 319, 347
  regolith, 36, 48, 53, 62, 438
  Reiner Crater, 53
  remanent magnetism, 53
  rilles, 44, 45, 48, 49, 50, 51
  rhyolite, 52
  'rockbergs', 56
  Rook Range, 54
  'sculpture', 46, 439
  'seas', 43
  seismic propagation, 52, 53, 195, 361
  Serenitatis basin, 51, 52, 53
  Shorty Crater, 52
  silicic lava, 52, 53
  sodic plagioclase feldspar, 47, 59
  South Cluster, 49
  Southern Highlands, 53, 372
  Spur Crater, 52
  St George Crater, 49
  surface gravity, 199
  Surveyor (spacecraft), 47, 48, 137
    Surveyor 1, 52
    Surveyor 7, 53
  'swirls', 53
  Taurus Mountains, 52
  thorium, 56
  troctolite, 55, 56, 442
  Tycho Crater, 53, 372
  titanium-bearing minerals, 47, 48, 59
  uranium, 56, 59
  Van de Graaff Crater, 53
  volatiles, 47, 59, 61
  volcanism, 44, 45, 46, 47, 48, 52, 53, 74, 193
  water ice, 62
  wrinkle ridges, 47, 193, 443
Moorbath, Stephen, 212
Moore, J.H., 327
Moore, (Sir) Patrick, 73
Morabito, Linda, 246, 250
Morgan, W. Jason, 26, 33
Moulton, Forest Ray, 4
Mount Palomar Observatory, 72, 239, 287, 340
Mount Wilson Observatory, 5, 8, 45, 72, 73, 75, 145, 146, 147, 197, 310, 327, 328, 339, 340
Muller, R.A., 359
Munich Observatory, 144
Mutch, Tim, 106

NASA, 35, 47, 53, 60, 75, 77, 104, 107, 127, 149, 154, 155, 161, 201, 242, 243, 245, 250, 312, 352, 374
National Academy of Sciences, 32
National Air and Space Museum, 259
National Center for Atmospheric Research, 349
National Ocean and Atmospheric Administration, 22
Naugle, John, 243
Neison-Neville, Edmund, 45
Neivert, Peter, 119
Neptune, xix, 59, 320, 324, 325, 338, 339, 342
  atmosphere, 332
    Great Dark Spot, 332, 333
    hydrocarbon haze, 333, 430
    Little Dark Spot, 332
    methane, 332
    transient features, 327
  discovery, 325, 326
  energy budget, 331
  internal structure, 333
  magnetic field, 330, 333, 334
  rings, 328, 329, 330, 333, 334, 346
  Roche radius, 334, 338
  rotation, 327, 333

Neptune, *cont.*
  satellites (see Larissa, Nereid, Proteus, Triton)
  Voyager (spacecraft), 291, 305, 331
Nereid (satellite of Neptune)
  captured object, 328
  discovery, 328
  elliptical orbit, 328
New York University, 359, 360
Newton, (Sir) Isaac, 1, 236, 291
Nicholson, Seth B., 145, 284
Niepce, J.N., 64
Nininger, Harvey Harlow, 372
Noah, 11, 14, 353

Oberbeck, Verne, 374, 376, 381
Oberon (satellite of Uranus), **318**
  basins, 316
  bulk density, 311
  craters, 316
  discovery, 309
  Hamlet Crater, 316
Occam's Razor, 184
O'Keefe, John, 372
Olbers, Heinrich Willhelm Matthias, 241
Olsson, Richard, 356
Oort Cloud, 360, 435
Oort, Jan H., 360
Öpik, Ernst J., 74
Oregon State University, 350
Orion Nebula (M42), 4, 8
Owen, Richard, 14
Oxford University, 212

Paul III (Pope), 1
Pandora (satellite of Saturn), 302
Parker, Robert L., 26
Parker, Timothy, 117
Parsons, William (Earl of Rosse), 4
Patterson, Claire C., 13
Peale, Stanton, 247, 261, 307
Peirce, B., 286
Pennsylvania State University, 349
Perrine, Charles, 284
Pettengill, Gordon J., 184
Pettit, Edison, 145
Phobos (satellite of Mars), 306
Phoebe (satellite of Saturn)
  albedo, 301
  captured body, 302
  carbonaceous material, 301
  discovery, 287, 305
  inclination, 302
  retrograde orbit, 287, 301

Piazzi, Giuseppe, 240, 241
Picard, Jean, 286
Pic du Midi Observatory, 145, 146, 152, 238, 262, 297, 304
Pickering, William H., 45, 145, 146, 287, 339, 341
Pickering, Edward Charles, 290
Pierce, Ken, 365, 366
Piggot, C.S., 19
Pilkinton, Mark, 373
Placet, Francois, 14
planetesimals, 4, 5, 436
planetology, xxi
Playfair, John, 12
Pluto, xix, 338–342, **339**
  bulk density, 341
  cryovolcanism, 340
  discovery, 338, 339
  infrared studies, 340
  methane ice, 340
  orbit, 340
  origin, 342
  perihelion passage, 339, 342
  satellite (see Charon)
  seasons, 342
  surface detail, **341**
  temperature, 342
  size, 339, 340, 341
Poag, C.W., 375
Porco, Carolyn C. v
Princeton University, 23, 73, 147, 310, 314
Proctor, Richard Anthony, 45, 69, 236
Project Stratoscope II, 314
Prometheus (satellite of Saturn), 302
Proteus (satellite of Neptune), 334
Ptolemaeus, Claudius (Ptolemy), 1, 143, 235
Puck (satellite of Uranus), **318**, 324
Purdue University, Indiana, 357
Puerto Rico, 184
Pythagoras, 43

radioactivity, 13
Raff, Arthur, 24
Rampino, Michael R., 359, 360, 361, 362
Ramsay, John, 350
Ramsey, W.R., 237
Ranyard, A. Cowper, 73
Raup, David M., 359
Rayleigh (Lord), 13
Reese, E.J., 238
Reynolds, Ray, 247, 261
Rhea (satellite of Saturn)
  albedo, 297
  basins, 297

Rhea, *cont.*
  cryovolcanism, 297
  bulk density, 291
  discovery, 285, 305
  isostasy, 297
  resonances, 290
Riccioli, Giovanni Battista, 44, 144, 285
Richards, Mark, 227
Richards, Paul, 24
Richardson, R.S., 146
Rigby, Keith, 357
Rio de Janeiro Observatory, 70
Ristenpart, F., 238
Rittenhouse, R., 144
Roberts, Gregory, 78
Roberts, Paul, 24
robotic exploration, xvii, xxi, 106, 158, 240, 291, 324, 342
Roche Eduard, 59, 286, 287
Roemer, Olaus, 235
Roentgen, Wilhelm Konrad, 13
Roest, Walter, 373
Ross, F.E., 146
Ross, M., 261
Roth, Gunter, 145
Royal Society, 11, 13
Runcorn, Keith, 18
Russell, Henry Norris, 144
Rutgers University, 356
Rutherford, Ernest, 13

Sagan, Carl E v, 107
Saheki, Tsuneo, 72
Sandwell D.T., 20, 22
Santiago Observatory, 238
Saturn, xix, 240, 285–307, 309, 341
  atmosphere, **287, 289**
    hydrocarbon haze, 289, 314, 430
    methane, 314
    transient features, 289
  bulk density, 287
  Cassini/Huygens (spacecraft), xix, 292
  energy budget, 289, 331
  internal structure, 289
  magnetosphere, 296
  Pioneer 11 (spacecraft), 243, 291, 302, 304
  'Saturnus triformis', 285
  Voyager (spacecraft), 279
  rings, 4, 285, **289**, 290, 302, **303**
    'A' ring, 285, 286, 303, 304
    'B' ring, 285, 286, 303
    'C' (Crepe) ring, 286, 302
    'D' ring, 302
    'E' ring, 293
    'F' ring, 302
    'G' ring, 292, 302
    'braided' appearance, 304
    Cassini's Division, 285, 287
    Encke's Division, 286
    nature of rings, 286, 287, 290, 324
    resonances, 287, 303
    shepherds, 302, 303, 304
  Roche radius, 286, 287, 292, 302, 303
  rotation, 289
  satellites (see Calypso, Dione, Enceladus, Epimetheus, Helene, Hyperion, Iapetus, Janus, Mimas, Pandora, Phoebe, Prometheus, Rhea, Tethys, Telesto, Titan)
  Voyager (spacecraft), 288, 289, 291, 292
Saul, John, 378
Schaber, G.G., 169
Schaer, M.E., 286, 305
Schafarik, 145
Schenk, Paul, 259
Schiaparelli, Giovanni Virginio, 69, 70, 71, 72, 73, 76, 127, 145, 183, 184, 262, 310
Schink, Bernhard, 352
Schmidt, Julius, 44
Schmitt, Harrison H. (Jack), 52
Schreiner, C., 285
Schroter, Johann H., 44, 45, 144, 183, 237, 240, 241
Schubert, Gerald, 261
Schultz, Peter, 356, 375
Scott, David R., xxi, 36, 48, 51
Scripps Institute of Oceanography, 22, 23, 26, 46
Secchi, (Father) Angelo, 310
See, T.J.J., 184, 327
Seelinger, H. von, 286
Selenographic Society, 45
Sepkoski, John J., 359
Sharpton, V.L., 355, 356
Shepard, Alan B., 48
Shoemaker, Eugene Merle (Gene), 46, 48, 53, 268, 277, 317, 371, 372, 373
Shoemaker-Levy, 9 (comet), 376
Short, James, 145
Slipher, Vesto M., 310, 339
Sloan, Robert, 357
Smith A.G., 37
Smith, Peter, 298
Smith, William, 353
Smith W.H.F., 20, 22
Snider, Antonio, 14
Soderblom, Laurence, 249, 336

466 Index

Solar System, xvii, 1–9, 440
  accretionary process, 60, 62, 186, 188, 371
  age, 13
  angular momentum, 2, 5
  collision hypothesis, 4, 8, 13
  companion (Nemesis), 360
  formation, 1
  'habitability zone', 376, 382
  heliopause, 243, 338, 429
  interstellar medium, 8, 243, 338
  nebular hypothesis, 2, 5, 13
  nucleation and collapse hypothesis, 5
  protoplanets, 60, 185, 188, 437
  protostar, 5
  solar nebula, 5, 59, 60, 162, 185, 188, 197, 271, 292, 299, 301, 313, 341, 376, 440
  solar parallax, 143
  solar wind, 53, 75, 196, 242, 243, 251, 440
Song, Xiao-dong, 24
Southwest Research Institute, 268, 341
Soviet Academy of Sciences, 153
Soviet Union, xxi, 47, 52, 78, 103, 127, 146, 149, 150, 158, 290, 375
Space Shuttle, 159, 245, 313
Space Telescope Science Institute, 298, 339
Spitzer, Lyman, 5
Spray, John, 373
Spudis, Paul D., 55, 64, 66, 381
Sromovsky, L.A., 333
'Star Wars', 293
stars
  79 Ceti, 9
  double stars, 309
  extrasolar planets, 6, 7, 9, 376
  globular clusters, 4
  HD16141, 9
  HD46375, 9
  iota Hor, 382
  Milky Way, 1
  nova, 4
  proper motions, 1
  protoplanetary disks, 8
  supernova, 5, 143
  T Tauri type, 8, 441
  Upsilon Andromedae, 8, 9
Steavenson, W.H., 145
Stieff, Lorin R., 372
Stone, Edward C., v
Stoney, G. Johnstone, 73, 146
Strutt, R.J. (see Lord Rayleigh)
Struve, Friedrich George Wilhelm von, 285, 290, 291
Struve, Herman von, 290, 291
Suess, Edward, 14, 16, 17

Suess, H, 147
Sun, 1, 59
  age, 13
  early output, 162, 349, 351
  magnetic field, 5, 243, 338
  'red giant', 299
  source of energy, 4, 13, 38
  sunspot cycle, 5
Sykes, Lynn R., 26

'2001: A Space Odyssey', 300
Taylor, Jeffrey, 261
telescope, 64
Telesto (satellite of Saturn), 304, 329
Tethys (satellite of Saturn), 287, **295**
  bulk density, 291
  cryovolcanism, 295
  discovery, 285, 305
  hilly cratered terrain, 295
  isostasy, 295
  Ithaca Chasma, 295
  Odysseus Crater, 294, 295
  resonances, 290, 304, 329
  tectonism, 294
Tharp, Marie, 21
Thomson, William (see Lord Kelvin)
Thompson, Charles W., 18
Thompson, Lucy, 373
Titan (satellite of Saturn), xvii, 287, 290, 302
  albedo, 291, 297, 298
  atmosphere, xix, 346
    acetylene, 298
    ammonia, 299
    carbon, 299
    clouds, 297, 299
    ethane, 298, 299
    ethylene, 298
    hydrocarbon rain, 299
    methane, 297, 299
    nitrogen, 297, 298
    orangey hue, 297
    photochemical hazes, 297
    pressure, 298
    temperature, 299
  bulk density, 291, 298
  Cassini (spacecraft), 299
  discovery, 285, 305
  erosional processes, 299
  Huygens (spacecraft), xix, 299
  life, 283, 299
  near-infrared studies, 298
  oceans, 299
  radiogenic heating, 299
  resonances, 290, 300

Titan, *cont.*
  size, 290
  surface, xix, **298**
  Voyager (spacecraft), 288, 291, 304
Titania (satellite of Uranus), **318**
  bulk density, 311
  craters, 317
  discovery, 309
  erosion, 317
  grabens, 317, 319
  isostasy, 317
  palimpsests, 317
  resurfacing, 317
  temperature, 317
Titius, Johann Daniel, 240, 280
Titius-Bode 'law', 240, 325, 345
Todd, David Peck, 338
Tombaugh, Clyde W., 74, 339, 340
Toronto University, 26
Triton (satellite of Neptune), xix
  atmosphere
    nitrogen, 334, 346
    methane, 334, 346
    pressure, 334
  bulk density, 338
  basins, 336
  discovery, 328
  'cantaloupe' terrain, **335**, 336, **337**
  cryovolcanism, 336, 340
  geysers, xix, **335**, 336, **337**
  hydrocarbon cycle, 337
  lakes of nitrogen, 336
  methane ice, 334
  orbit, 327, 330, 336, 346, 347
  origin, 334, 338, 342
  polar cap, 334, **335**, 336, **337**
  precession, 328
  radiogenic heating, 338
  size, 328, 338
  temperature, 334, 336
  tidal heating, 338
Trouvelot, Etienne L., 144
Tscherednitschenko, V.I., 290
Turkevich, Anthony L., 48

Umbriel (satellite of Uranus), **318**
  albedo, 318
  bulk density, 311
  craters, 318
  discovery, 310
  Wunda Crater, 318
University of Arizona, 198, 259, 354, 357, 358
University of California
  Berkeley, 6, 7, 8, 228, 354
  Los Angeles, 24, 356, 359
  San Diego, 20
University of Cambridge, 13, 15, 24, 26, 32, 37, 51, 325, 328
University of Chicago, 359
University of Hawaii, 130, 334
University of Konstanz, 352
University of Michigan, 71, 184
University of Minnesota, 357
University of New Brunswick, 373
University of Notre Dame, Indiana, 357
University of Rhode Island, 356
University of Western Australia, 217
University of Washington, 359
University of Wisconsin, 333
Uranus, xix, 145, 309–325, **311**, **314**, 376
  'anomalous motion', 326, 327, 339, 342
  atmosphere
    aurorae, 315
    'dayglow', 315
    helium, 313
    hydrocarbon haze, 314, 430
    hydrogen, 310, 313, 324
    hydrogen coma, 315, 324
    methane, 310, 314
    seasons, 310
    temperature, 315, 343
    transient features, 310, 314, 316
    winds, 315
  bulk density, 316
  discovery, 309
  energy budget, 315, 331
  'Georgium Sidus', 309
  'Herschel's Planet', 309
  impactor rate in Uranian space, 319, 320
  interior structure, 315, 316
  magnetic field, 315, 316, 318
  obliquity, 310
  occultation of star, 312
  'planetary wind', 324
  rings, 312, 316, 318, 323, 343
    dust-removal, 324
    Epsilon ring, 323, 324, 343
    shepherds, 323
  rotation, 310, 314, 315, 316
  satellites (see Ariel, Miranda, Oberon, Puck, Titania, Umbriel)
  ultraviolet excess, 315
  Voyager 2 (spacecraft), 291, 305, 311, 312, 315, 316, 323, 324
Urey, Harold C., 47, 57, 60, 61
US Air Force, 242
US Geologic Survey, 24, 46, 155, 156, 225, 336, 357, 362, 365, 369, 372, 375

US Naval Observatory, 289, 327, 338, 340
US Navy, 21, 24, 25
Ussher, Archbishop James, 11, 12

Venus, xviii, 13, 104, 143–182, 228, 240, 291
   Akna Montes, **164**, 171
   albedo variations, 144, 145
   aluminous silicates, 159
   Aphrodite Terra, 159, 167, 168, 169
   Artemis Chasma, 169
   'ashen light', 144
   asthenosphere, 175
   Atalanta Planitia, **166**, 167
   Atla Regio, 167, 172
   atmosphere, 144, 160, **161**, 162, 239
      aerosols, 147, 160
      carbon dioxide, 147, 150, 151, 160
      carbon monoxide, 160
      carbonic acid, 147
      circulation, 160
      clouds, 144, 146, 149, 154, 158, 160
      composition, 146, 147
      deuterium, 154, 161, 162
      formaldehyde, 147
      nitrogen, 150, 151
      opacity, 153, 154
      oxygen, 147, 151
      pressure, xviii, 149, 150, 151, 160
      sulphates, 160
      sulphur, 160
      sulphur dioxide, 160, 161
      sulphuric acid, 160, 161, 177
      water vapour, 146, 151, 154, 158, 161
      wind speeds, 149, 151, 153, 154, 158
   auroral activity, 145
   Baltis Vallis, 172
   basalt, 153, 156
   Beta-Atla-Themis Regiones, 167, 172
   Beta Regio, 153, 167, 168, **170**
   canali, 172
   carbonate, 160, 162
   carbonatite flows, 174
   Cleopatra Crater, 165
   climate
      'hot and arid', 147, 148
      'hot and wet', 146, 148, 149
   Colette Patera, **164**, 178
   core, 163
   coronae, 171, 172, 228
   craters, 158, 165, 174
   cratons, 171
   'cusps', 144
   'delamination', 175
   Devana Chasma, 168, **170**
   'dry' melting, xviii, 174
   Earth's near-twin, xviii, 160
   Eistla Regio, 168
   elongation, 143
   erosional processes, xviii, 147, 153, 156, 157, 172, 173
   Freyja Montes, **164**
   granite, 151, 156, 159, 169
   Hadley cells, 160
   heat flow, 163, 168, 174
   highlands, 163, 164, 165
   'hoods', 144, 145
   hydrocarbons (oil), 147
   impacts, 149, 158, 165
   inferior conjunction, 143
   infrared, 146, 147
   Ishstar Terra, 163, **164**, 165, **166**, 167, 169, 171, 174
   isostasy, 174
   Lakshmi Planum, 163, **164**, 165
   landing sites, **157**, 159
   launch windows, 150
   lava channels, 172, **173**
   lava flows, 153, 156, 167, 172
   leveed channels, 172
   life, 145, 146, 148
   lithospheric stresses, 167, 168, 169, 171
   lithospheric recycling, 174, 175
   lithospheric thickness, 174, 181
   lowland plains, 163, 167, 171, 172
   mafic lavas, 172
   Magellan (spacecraft), 159, 166, 170
   magnetic field, 147, 149, 163, 196, 200
   mantle, 167, 168, 169, 171, 174, 175
   maria, 167
   Mariner (spacecraft)
      Mariner 1, 127, 149
      Mariner 2, 127, 149, 150
      Mariner 5, 150
      Mariner 10, 151, 152
   Maxwell Montes, 163, **164**, 165, 171, 178
   microwave emissions, 147, 149
   mountains, 144
   North Basin, **164**, 166
   north pole, **166**, 167
   oceans, 145, 146, 147, 149, 162
   obliquity, 145, 146
   phases, 143, 175, 183
   Phoebe Regio, 156
   Pioneer (spacecraft), 154, 155, 160, 161, 163, 168, 177
   plate tectonics, xviii, 159, 163, 167, 174
   polar caps, 144, 146
   potassium-rich minerals, 151, 156

Venus, *cont.*
  pyroclastics, 162
  radar observations, xviii, **148**, 149, 158
  radio altimetry, 154, **155**, 163
  resurfacing, xviii, 158, 161, 174
  Rhea Mons, 167, **170**
  ridge belts, **166**, 167, 169
  ridged plains, 167
  rilles, 172
  rolling upland plains, 155, 156, 163, 167, 172
  rotational state, 62, 144, 145, 146, 148, 149, 152, 163, 196
  'runaway greenhouse', 162
  Rusalka Planitia, 172
  Sacajawea Patera, 164, **165**, 178
  Sapas Mons, 167
  satallite, 145
  spreading ridges, 167, **168**
  'stagnant lid', 174
  subduction trenches, 169, 171, 172, 174
  sulphur flows, 174
  superior conjunction, 143
  tectonism, 167, 168, 169, 171
  temperature
    cloud tops, 152, 160
    surface, xviii, 147, 149, 150, 151, 157, 161, 171
    thermal gradient, 157, 160, 161
  tesserae, 165, 441
  Theia Mons, 167, 168, **170**
  tholeiite, 156
  transits of Sun, 143
  troughs, 167, **168**, 169, 172
  ultramafic lavas, 172
  ultraviolet appearance, 146, 149, 151, **152**, 160
valley networks, 172
VeGa (spacecraft)
  VeGa 1, 158
  VeGa 2, 158, 159
Venera (spacecraft)
  Venera 1, 149
  Venera 2, 150
  Venera 4, 150
  Venera 4, 150
  Venera 5, 150
  Venera 6, 151
  Venera 7, 151
  Venera 8, 151, 152
  Venera 9, **153**, 158
  Venera 10, **153**

  Venera 11, 154
  Venera 12, 154, 160
  Venera 13, **156**
  Venera 14, 156, **157**, 160
  Venera 15, 158
  Venera, 16, 158
Vesta Rupes, 163, **164**
volcanic rises, 167, 168, 169, 171
volcanic shields, 167, 169
volcanism, xviii, 153, 156, 158, 161, 162, 164, 165, 167, 168, 169, 171, 172, 174
Volkova, 167
'young Venus', 146
Zond (spacecraft)
  Zond 1, 150
Vine, Fred J., 24, 25, 26
Vogel, Hermann Carl, 70, 184

Walcott, Charles, 12
Wallich, George C., 18
Ward, Peter, 359
Wegener, Alfred Lothar, 14, 16, 18, 31, 33
Weizsacker, Carl F. von, 5
Werner, Abraham Gottlob, 12, 203
Wetherill, George, 376
Whipple, Fred L., 147
Whipple, J.A., 45
Widdel, Friedrich, 352
Wilde, Simon, 226
Wildt, Rupert, 73, 147, 236, 310
Wilson, J. Tuzo, 26, 33
Wirtz, C., 328
Wisdom, Jack, 307
Wolff, Christian, 280
Woodward, John, 353
Wright, W.H., 146
Wyoming Infrared Observatory, 253

Yale University, 352
Yerkes Observatory, 286, 327
Young, Charles Augustus, 310
Young, E.F., 341
Young, John W., 52, **63**

Zach, Franz Xaver von, 240, 241
Zahnle, K., 376
Zeeman, Pieter, 5
Zenger, 144, 145
Zollner, Johann K.F., 184
Zuber, Maria T., 58
Zupus, Giovanni, 183

Made in the USA
San Bernardino, CA
24 January 2018